PREHARVEST FOOD SAFETY

Contributors

Achyut Adhikari
School of Nutrition and Food Sciences
Louisiana State University
Baton Rouge, Louisiana

Walid Q. Alali
College of Public Health
Hamad bin Khalifa University
Doha, Qatar

Lis Alban
Danish Agriculture and Food Council
Copenhagen, Denmark

Elaine D. Berry
USDA, Agricultural Research Service
U.S. Meat Animal Research Center
Clay Center, Nebraska

Jagpinder Brar
Department of Food Science
Purdue University
West Lafayette, Indiana

Pardeepinder K. Brar
Department of Food Science and Human Nutrition
Citrus Research and Education Center
Institute of Food and Agriculture Sciences
University of Florida
Lake Alfred, Florida

Robert L. Buchanan
Center for Food Safety and Security Systems
Department of Nutrition and Food Science
University of Maryland
College Park, Maryland

Benjamin Chapman
Department of Agricultural and Human Sciences
NC State University
Raleigh, North Carolina

Shi Chen
Department of Population Health and Pathobiology
College of Veterinary Medicine
North Carolina State University
Raleigh, North Carolina

Sailaja Chintagari
Department of Food Science and Technology
University of Georgia
Griffin, Georgia

Anna Colavecchio
Department of Food Science and Agricultural Chemistry
Food Safety and Quality Program
McGill University
Ste Anne de Bellevue, Quebec
Canada

Michelle D. Danyluk
Department of Food Science and Human Nutrition
Citrus Research and Education Center
Institute of Food and Agriculture Sciences
University of Florida
Lake Alfred, Florida

Jitender P. Dubey
U.S. Department of Agriculture
Agricultural Research Service, Northeast Area
Animal Parasitic Diseases Laboratory
Beltsville Agricultural Research Center-East
Beltsville, Maryland

Genevieve Edwards
School of Nutrition and Food Sciences
Louisiana State University
Baton Rouge, Louisiana

Arpeeta Ganguly
Department of Animal and Food Sciences
University of Delaware
Newark, Delaware

Lawrence D. Goodridge
Department of Food Science and Agricultural Chemistry
Food Safety and Quality Program
McGill University

Ste Anne de Bellevue, Quebec
Canada

Chris Gunter
Department of Horticulture Sciences
NC State University
Raleigh, North Carolina

Eduardo Gutierrez-Rodriguez
Department of Food, Bioprocessing, and Nutrition Sciences
North Carolina State University
Raleigh, North Carolina

Nicole Hazard
School of Nutrition and Food Sciences
Louisiana State University
Baton Rouge, Louisiana

Dolores E. Hill
U.S. Department of Agriculture
Agricultural Research Service, Northeast Area
Animal Parasitic Diseases Laboratory
Beltsville Agricultural Research Center-East
Beltsville, Maryland

Charles L. Hofacre
Department of Population Health
Poultry Diagnostic and Research Center
University of Georgia
Athens, Georgia

David Ingram
Food and Drug Administration
Center for Food Safety and Applied Nutrition
College Park, Maryland

Ravi Jadeja
Department of Animal Science
Oklahoma State University
Stillwater, Oklahoma

Marlene Janes
Department of Animal Science
Oklahoma State University
Stillwater, Oklahoma

Rolf D. Joerger
Department of Animal and Food Sciences
University of Delaware
Newark, Delaware

David H. Kingsley
U.S. Department of Agriculture

Agricultural Research Service
Food Safety and Interventions Research Unit
Delaware State University
Dover, Delaware

Kalmia E. Kniel
Department of Animal and Food Sciences
University of Delaware
Newark, Delaware

Deepak Kumar
Department of Veterinary Public Health & Epidemiology
College of Veterinary and Animal Sciences
Govind Ballabh Pant University of Agriculture & Technology
Pantnagar, Uttarakhand
India

Cristina Lanzas
Department of Population Health and Pathobiology
College of Veterinary Medicine
North Carolina State University
Raleigh, North Carolina

Sarah M. Markland
Department of Animal and Food Sciences
University of Delaware
Newark, Delaware

Shirley A. Micallef
Department of Plant Science and Landscape Architecture
Center for Food Safety and Security Systems
University of Maryland
College Park, Maryland

Annette M. O'Connor
Department of Veterinary Diagnostic and Production Animal Medicine
Iowa State University College of Veterinary Medicine
Ames, Iowa

Russell Reynnells
University of Maryland Eastern Shore
Department of Agriculture, Food, and Resource Science
Princess Anne, Maryland

Jan M. Sargeant
Center for Public Health and Zoonoses and Department of Population Medicine
Ontario Veterinary College
University of Guelph
Guelph, Ontario
Canada

Manan Sharma
U.S. Department of Agriculture
Agricultural Research Service
Beltsville Area Research Center
Environmental Microbial and Food Safety Laboratory
Beltsville, Maryland

Manpreet Singh
Department of Food Science
Purdue University
West Lafayette, Indiana

David R. Smith
Mississippi State University
College of Veterinary Medicine
Mississippi State, Mississippi

Patrick Spanninger
Department of Animal and Food Sciences
College of Agriculture and Natural Resources
University of Delaware
Newark, Delaware

Siddhartha Thakur
Department of Population Health and Pathobiology
College of Veterinary Medicine
North Carolina State University
Raleigh, North Carolina

Mary E. Torrence
Office of Applied Research and Safety Assessment
Center for Food Safety and Applied Nutrition
U.S. Food and Drug Administration
Laurel, Maryland

James E. Wells
USDA, Agricultural Research Service
U.S. Meat Animal Research Center
Clay Center, Nebraska

Preface

Assuring the safety of the world's food supply continues to be a priority, a necessity, and a challenge. The challenges are created by increasing globalization, reductions in trade barriers, and an ever-growing human population that has developed a craving for fresh, diverse foods. Foodborne pathogens may enter the food supply at the preharvest or harvest phases of the farm-to-fork continuum. However, classic control measures and regulations are instituted primarily at the postharvest phase. High-profile outbreaks still occur, from *E. coli* O157:H7 in ground beef and spinach and *Listeria* in cantaloupes to *Salmonella* in tomatoes, affecting morbidity and mortality. The economic impact of foodborne outbreaks is vast, with significant impact on humans, industry, and our society as a whole.

We turned our attention to the work being done to prevent food contamination in the preharvest phase, which has been increasingly recognized as an important step in the food continuum. Preharvest food safety encompasses the measures that are taken to ensure that food products are produced in a safe and wholesome manner on the farm, thereby ensuring an optimally safe commodity all the way to slaughter, packing, and/or processing. Because many enteric pathogens enter the food chain during production, and some can even proliferate during this phase, it is crucial to understand their preharvest ecology and epidemiology in order to identify and evaluate appropriate intervention strategies. Preharvest control measures have expanded and improved, so it is timely to summarize the recent developments and consider the needs and opportunities for the future in the preharvest food realm.

Our goal in creating this book is to provide the scientific community and stakeholders in the food industry with a knowledgeable resource that discusses the developments and challenges of preharvest food safety, focusing on a variety of microbiological hazards in a variety of foods. The chapters in this book address the current state of knowledge and practice, emerging issues, and emerging solutions, with a focus on both research and control measures. A key aspect of the book is the inclusion of multiple food commodities e.g., food animals, produce, grains, and seafood, and the relevant pathogens for each commodity. In so doing, this book will serve as a comprehensive volume that can be used by food safety scientists in general, as well as those focused on a particular commodity. An important objective of this work is to facilitate understanding of the importance of complex microbial ecology dynamics to preharvest food safety. In the past, little effort was made to undertake an ecological or systems-based approach to understand the broad issues relating to food safety with specific examples, and this book aims to remedy that. Indeed, it is obvious that many of the emerging food safety pathogens or the manifested outbreaks are directly (or indirectly) related to changes in the agri-food systems themselves over the past several decades.

The first section of this book examines the issues associated with preharvest food safety in broad agriculture sectors. We identify major foodborne pathogens of concern for specific products; what is known about the source, prevalence, and transmission of these pathogens in the candidate products; discuss multiple control measures; and identify critical data gaps for the development of future targeted controls. There is information about a vast array of pathogens, including bacteria, viruses, fungi, and protozoa, that are relevant to each of the major commodity types.

In the middle section, we address the emerging issues that impact the preharvest food safety area. We discuss the use of antimicrobials as growth promotants in the food animal industry and its implications. The critical role of the environment, including the potential impact of global climate anomalies on the emergence and transmission of foodborne pathogens, is also highlighted. Finally, we address the growth and challenges of the organic production system. State-of-the-art information on risk assessment, risk management, and emerging preharvest food safety issues in both developed and developing countries is included.

The third and final section aims to provide information on emerging solutions and novel intervention methods that can be employed at the preharvest level to reduce the burden of foodborne pathogens and other potential hazards. We focus not only on the challenges in preharvest food safety, but also on intervention and pathogen reduction strategies, e.g., Good Agricultural Systems, testing, HACCP. We take a holistic approach that treats production agriculture as a system with complex interactions between the environment, the microbe, and the food that are largely driven by ecological considerations and the actions of humans and animals.

This collection will be of use to scientists whose research includes foodborne pathogens, those working in the food industry seeking the latest verified research on food safety, and those interested in our food supply and its environmental impact.

This work required a deep well of knowledge, so sought out and we recruited the very best experts in preharvest food safety to contribute. We are deeply indebted to our many colleagues who worked diligently to write their reviews in spite of their busy schedules. This book would not have been possible without their invaluable input and participation. We also thank ASM Press for being patient with us during this process.

Siddhartha "Sid" Thakur
Kalmia "Kali" Kniel

About the Editors

Siddhartha "Sid" Thakur is a Professor in the College of Veterinary Medicine at North Carolina State University. He is also the Associate Director at the Comparative Medicine Institute where he leads the Emerging and Infectious Diseases Research program. He received his Bachelor of Veterinary Science and Animal Husbandry degree from Govind Ballabh Pant University of Agriculture and Technology (Udham Singh Nagar, India) and his Master's of Veterinary Science in Veterinary Public Health at the Indian Veterinary Research Institute (Izatnagar, India). He earned his Ph.D. in Population Medicine at the College of Veterinary Medicine, NC State University. Prior to joining the faculty at NC State University, Dr. Thakur was an Oak Ridge Research Fellow at Center for Veterinary Medicine, FDA, Maryland. He espouses the concepts of "One Health" and seeks to understand how antimicrobial resistance develops in "superbugs" that affect animal and human health. He has won numerous awards including the Larry Beuchat Young Researcher Award by the International Association for Food and the International Global Engagement award by NC State. He is currently a NC State Chancellor faculty scholar. Dr. Thakur has authored or co-authored 45 peer-reviewed publications and runs a well-funded extramural research program.

Kalmia "Kali" E. Kniel is Professor of Microbial Food Safety in the Department of Animal and Food Sciences at the University of Delaware. She received her Ph.D. from Virginia Tech in Food Science and Technology with a focus on Food Microbiology. Her doctoral work focused on protozoan parasites. Prior to joining the faculty at the University of Delaware, she was a postdoctoral microbiologist at the USDA Agricultural Research Service's Animal Parasitic Diseases Laboratory where she worked in the area of food safety and animal health. She is now nationally recognized as a leading expert in transmission of viruses, protozoa, and bacteria in the preharvest environment. Dr. Kniel has been active in researching the mechanisms behind the survival and inactivation of norovirus, hepatitis A virus, and other enteric viruses prevalent in our water and foods. She is an active advocate for teaching food safety at all levels and has been involved with elementary and secondary education. At the University of Delaware, she teaches courses on foodborne outbreak investigations and the basics of food science and food safety to hundreds of students each year.

OVERVIEW OF THE PREHARVEST FOOD SAFETY PROBLEM

Introduction to Preharvest Food Safety 1

MARY E. TORRENCE[1]

INTRODUCTION

Food safety remains a global public health and agricultural priority. Since 1998, major funding programs and national initiatives have been developed and implemented within the United States and elsewhere. The emphasis of these programs has often been variable depending on the emerging new pathogen or outbreak, the changes in technology and method development, or funding mechanisms. Even terminology has been debated, i.e., farm-to-fork or plate-to-table, preharvest to postharvest. What remains constant is that accomplishments can be cited, yet the solutions for food safety issues remain elusive and changing. The reasons remain multifactorial, both from a scientific and from a strategic perspective. Scientifically, food safety is a complex issue that involves multiple and diverse food production practices, wide food distribution patterns, various and distinct consumers with assorted behaviors, and evolving foodborne pathogens and contaminants. Foodborne illnesses and outbreaks can be caused by not only microbial pathogens, but also by viruses, parasites, chemical agents, and toxins. Foodborne pathogens do not recognize specific barriers as humans, animals, and environments interact and as more food products are distributed globally. In addition to international travel, there

[1]Office of Applied Research and Safety Assessment, Center for Food Safety and Applied Nutrition, U.S. Food and Drug Administration, Laurel, MD 20708.

Preharvest Food Safety
Edited by Siddhartha Thakur and Kalmia E. Kniel

doi:10.1128/microbiolspec.PFS-0009-2015

has been an increase in world trade, and the United States continues to increase the importation of food products. In 2009, imports of various products such as grains and grain products, fruits and vegetables, nuts, and fish and shellfish were 17%, or 358 pounds, per capita (http://ers.usda.gov/topics/international-markets-trade/us-agricultural-trade/import-share-of-consumption.aspx). Importation can increase the access to new and familiar foods year-round but also can increase the possibility of introductions of additional food contaminants and pathogens.

Strategically, food safety is a complex issue that involves multiple federal agencies with distinct regulatory or policy mandates and a diversity of committed food safety stakeholders such as industry, consumers, state and local governments, international organizations, and public health and agricultural researchers and decision makers. This increases the convoluted path for change and for finding consensus on the most important directions for research or policy. For example, the first Presidential Food Safety Working Group was formed by President Clinton in 1998 along with the Food Safety and Produce Initiatives (1). The goals for the interagency senior management working group were to provide specific goals and actions that would address food safety and reduce the risk of foodborne illness. In 2011, President Obama re-established the President's Food Safety Working Group to re-evaluate the food safety regulations and regulatory enforcement and to coordinate food safety efforts in the government (http://www.foodsafetyworkinggroup.gov). Both working groups recommended similar principles involving prevention, effective food safety inspections, surveillance and enforcement, and the need for risk assessment. Numerous comprehensive studies, books, and reports have been published over the years from well-respected scientific panels and organizations, e.g., the American Academy of Microbiology, the Institute of Medicine, and national working groups such as the Presidential Food Safety Working Group (2–4). These reports have reviewed food safety from a scientific, policy, and regulatory perspective, and most provide important future directions and goals for food safety and preharvest food safety alike. It is surprising that despite our scientific accomplishments, new initiatives, and regulatory guidelines, many research directions and goals remain the same. With our accomplishments in food safety over the past 15 years, it is time to evaluate the gaps in knowledge or in action and provide innovative approaches and strategic directions for solutions. The signing into law of the Food Safety Modernization Act (FSMA) in 2011 may provide one opportunity for food safety research and policy to move forward (5).

GENERAL PREHARVEST ISSUES

Preharvest, postharvest, and retail are natural categories for scientists and policy makers to target to reduce foodborne illnesses. Each phase provides unique research areas and challenges in understanding the source and flow of foodborne pathogens and contaminants, risk factors, and management strategies. Postharvest issues have always been at the forefront of new regulations and for established standards (e.g., Hazard Analysis and Critical Control Point [HACCP]) and proposed interventions. This is in part because of the belief that the postharvest phase is the closest to the finished product and consumer risk. In addition, postharvest production seems more controlled and linear during food processing, unlike the open and complex environment of preharvest food production. Yet preharvest is the beginning of the food production system and, as such, demands our renewed efforts to understand foodborne pathogens and to develop and implement prevention, control, and intervention strategies. This is especially true as new or re-emerging zoonotic pathogens are identified. The majority of foodborne pathogens are zoonotic, which makes preharvest research a critical phase for determining the source of the microbial pathogen, the transmission along the food chain, the disease mechanisms in the animal or plant, and how it affects humans. Epidemiologic and ecologic studies remain important in providing a basis for developing, implementing, and evaluating intervention, control, and prevention strategies in addition to potential policy or guidelines. Conversely, it is critical from a policy and research perspective to determine whether the pathogen is foodborne or more likely to be occupational, environmental, or accidental. This often fuels lively debates. One example is methicillin-resistant *Staphylococcus aureus* (MRSA) since it has evolved from a human hospital-acquired illness to now causing community-acquired illnesses (6). Numerous prevalence studies in animals have found different strains of MRSA in animals, especially swine, both in the United States and internationally (7–10). More directed research will continue, especially with the new sequencing methods and specific study designs. To date, MRSA is not considered a foodborne pathogen but rather a potential occupational or environmental risk.

Preharvest food safety research and activities have advanced over time with the recognition of the importance and the complicated nature of the preharvest phase of food production. Research and policies have expanded as knowledge has increased about the pathogenesis, virulence, and transmission of foodborne pathogens and contaminants. With newer methods and technologies, more efforts have been directed toward developing rapid, sensitive, and specific methods for detection or screening of foodborne pathogens at preharvest. In addition, epidemiologic research gained momentum with funded research programs (11) that provided the opportunity to conduct longitudinal cohort, large case-control, or ecological studies. Results from these studies continue to be published and have provided important data for understanding preharvest food safety and for implementing potential intervention, control, or prevention strategies (12, 13). The USDA's Food Safety and Inspection Service (FSIS) has moved toward establishing more guidelines at the preharvest level (*Escherichia coli* and *Salmonella*). Internationally, the European Union has set performance standards at the preharvest level, especially in poultry and targeting *Salmonella*. Both the World Health Organization and Agriculture Organization (WHO/FAO) of the United Nations and Codex have conducted expert consultations, reports, and now standards programs for preharvest action on specific foodborne pathogens in food animals.

These multiple reports, findings, and recommendations often emphasize common themes, that is, potential management strategies, intervention and control measures, and the recognition that preharvest phases are important. At the same time, research and other cited data for interventions have been slow to be implemented, either from industry or from a government perspective. To name just a few, the Institute of Medicine published reports in 1998 and in 2003, and the American Academy of Microbiology published a 2005 report on preharvest food safety and security (2–4). The reports provided recommendations about the divergence of surveillance, research, and collaborations among the different food production phases, the fragmentation of the regulatory systems, and the use of HACCP and Good Agriculture Practices. The reports also provided research recommendations on enhancing our understanding of foodborne pathogens at the preharvest level, bridging our understanding of pathogenesis and transmission at the preharvest level to production and retail, and developing effective intervention, control, and prevention strategies. The most critical steps still remain: the ability to define a "criteria" or performance objective at the preharvest level and the ability to demonstrate what interventions and controls or preventive strategies are successful. On an ambitious and positive note, these reports also provide relevant directions for us in the future, both from a research and strategic approach in "traditional" areas of focus, such as meat and poultry, and more expanded areas of focus such as produce. To accomplish this, it will take collaborative strategic planning and funding. Most importantly, we must identify what research has been successful and complete, what priority research is needed, and how to recognize outcomes and potential impacts to food safety.

TRADITIONAL AREAS OF FOCUS IN PREHARVEST

There are numerous review books, summary reports, and workshop recommendations on preharvest food safety that have been published over the last 15 years (2–4, 14–16). These references provide a comprehensive range of topics in preharvest food safety that cover a multitude of food animals, foodborne pathogens, and foodborne contaminants. Topics covered include the epidemiology, ecology, and mechanisms of disease (e.g., pathogenesis, virulence) as well as the detection of the pathogen or diagnosis of the disease. Also, data are included on the molecular and genetic aspects, laboratory methods, and prevention, intervention, mitigation, and control strategies for food safety. This introductory chapter does not cover all of the current preharvest safety issues that are already extensively published. However, there are significant scientific advancements and areas of concentration that deserve highlighting. These advancements are a culmination of collaborative efforts (both public health and agriculture) and significant research results. Most importantly, these advancements provide the foundation for exploring future preharvest areas and for improving and focusing on more specific intervention, control, and prevention strategies.

E. coli and Cattle

The 1998 Food Safety Initiative and other related initiatives are largely a result of the highly publicized and serious foodborne outbreak involving *E. coli* O157:H7, ground beef, and Jack in the Box fast-food restaurant (17). This outbreak forced agriculture, industry, and the government to establish research programs and initiate potential intervention, prevention, or control programs. The impact of industry efforts, government regulations, and continued research has significantly reduced foodborne outbreaks associated with ground beef. This preharvest collaborative response among industry, government, and academia remains a collaborative model for future initiatives. In a USDA FSIS inspection program started in 1994, and in 1996, *E. coli* O157:H7 was listed as an adulterant (18). The USDA FSIS program has evolved to cover other beef parts, and in 2012, six new specific serotypes of non-O157 Shiga toxin–producing *E. coli* (O26, O45, O103, O111, O121, O145) were added as adulterants (http://www.fsis.usda.gov/OPPDE/rdad/FRPubs/2010-0023.pdf; http://www.usda.gov/wps/portal/usda/usdahome?contentid=2012/05/0171.xml).

In 2012, another USDA agency, the National Institute of Food and Agriculture, awarded a $25 million grant to the University of Nebraska to lead collaborative and broad research, education, and extension efforts among academia, industry, and government (http://www.nifa.usda.gov/newsroom/news/2012news/01231_nebraska_food_safety.html).

Research continues on genomic sequencing and serotyping of *E. coli* to better understand the role of non-O157 *E. coli* in foodborne outbreaks, the identification of risk factors across the food chain, and whether the six serotypes present the most critical public health risk. A 2011 foodborne outbreak in the European Union involving O104:H4 in produce may indicate more research is needed in understanding the role of different serotypes, as well as the role of fruits and vegetables as risk factors for potential increased public health risk (19). Research has expanded to the preharvest areas of produce and how interactions among the environment, humans, and animals may increase *E. coli* prevalence or transmission. It will take innovative research and guidance based on emerging scientific data to keep reducing *E. coli*–related outbreaks. This food safety area may be an ideal model for developing microbiological criteria, performance standards, and food safety objectives at the preharvest level that can be used in parallel with the whole food production system. For example, a specific microbial criterion set for cattle arriving at the slaughter facility and/or slaughter door would be innovative.

Salmonella, *Campylobacter*, and Poultry

The poultry industry has long been challenged with *Salmonella*- and *Campylobacter*-attributed foodborne outbreaks. *Salmonella enterica* serotype Enteritidis (*S*. Enteritidis) in eggs was addressed by multiple approaches. The FDA implemented the Egg Rule in 2009 (20). The Egg Rule focused on different phases of egg production, including preharvest food safety. At the same time, industry and academia developed and implemented new production practices and a vaccine for *S*. Enteritidis. Although *S*. Enteritidis in eggs has decreased dramatically over time, new serotypes of *Salmonella* have emerged in foodborne outbreaks both in the United States and internationally (21) (http://www.eurosurveillance.org/ViewArticle.aspx?ArticleId=20804). The European Union implemented one of the most dramatic preharvest strategies for *Salmonella* by establishing a Zoonosis Directive in 1992 and revising it in directive 2160/2003 with established performance standards for *Salmonella* spp. in poultry and swine (http://www.thaipoultry.org/Portals/5/eudirective/2003_2160.pdf). The directive set preventive controls at the preharvest level and throughout the food chain and instituted regulations on the hygiene of foodstuffs for food of animal origin. Although the action was aggressive, the complexities of the food chain and food production make results variable. The USDA's FSIS has set performance standards, but at the processing level for *Salmonella* and *Campylobacter*. Interestingly, the FDA's Center for Veterinary Medicine is now looking to control *Salmonella* at the preharvest level through the regulation of animal feed (http://www.fda.gov/AnimalVeterinary/NewsEvents/CVMUpdates/ucm360834.htm).

Efforts in controlling *Salmonella*-associated foodborne outbreaks demonstrate, again, that singular or even multiple attempts at prevention and control can fall short of ideal goals. Although *S*. Enteritidis has decreased, other *Salmonella* serotypes have increased. Perhaps this is because our previous interventions or controls have created a new microbial niche or is simply because when one *Salmonella* serotype decreases, another evolves. *Campylobacter* spp. are another foodborne pathogen that has increased, often in parallel with *Salmonella* reductions. Yet there is no scientific data that shows *Campylobacter* has replaced *Salmonella* in a specific food animal microbial niche. Preharvest research continues in several promising areas, e.g., the risk factors for *Salmonella* introduction and movement in cattle (22) and the effect of strategies of poultry production and slaughtering practices on finished product (12).

Interventions, Prevention, and Control Programs

A primary and central theme for preharvest food safety is focused on interventions, controls, prevention, and mitigation strategies. Logically, this critical preharvest goal will decrease foodborne pathogens and decrease foodborne contamination farther down the food production chain and ultimately decrease public health risk. Over the past 15 years, preharvest research has hypothesized and studied multiple mitigations, interventions, control, and prevention of the major pathogens in food animals. These strategies are based on the basic tenet that the preharvest production system involves the food animal and its microbial ecology as well as the interaction of the food animal and its surrounding environment and microbial ecology. Depending on the food animal, both the production practices and the environment (water, soil, wildlife, birds) influence the introduction of potential foodborne pathogens or contaminants to the food animal, the environment, and the potential movement to the postharvest and food products. Unlike the processing environment, which is more contained, preharvest food production involves the environment along with the movement of people, animals, insects and birds, and equipment, which can aid the transmission and movement of foodborne pathogens and contaminants.

Interventions within the animal depend on the food animal as well as the specific microbial pathogen. There has been an increase in research over the years to improve the animal's immunity, alter the gut flora, or even decrease the shedding of a specific pathogen. Strategies have been studied to enhance an animal's immune status through vaccinations, dietary supplements, or genetic breeding while also trying to understand the role of stress on the animal's immunity and the microbial gut flora. Vaccinations have a long history of being used both for animal health and for food safety, e.g., *S.* Enteritidis in poultry and more recently, for *E. coli* O157:H7 in cattle (23, 24). The cost-effectiveness and efficacy, at least for the *E. coli* vaccine, is still unclear. Vaccines are also being developed for *Campylobacter* spp. in poultry.

A popular focus of research for many years has been changing diets. One landmark study by Diez-Gonzalez in 1998 demonstrated that changing cattle's diet from grain to grass in relation to slaughter could decrease shedding of *E. coli* O157:H7 (25). This study prompted numerous new studies, debates, and recommendations. Other research over the years has involved feeding various products both for cheap sources of protein and for the product's role in the commensal and pathogenic flora. Such products include pre- and probiotics and competitive exclusion products. However, all of these products are limited if the cost–benefit ratio is too costly for the industry. The use of distiller grains as a cost-effective source of protein has led to research that reports an increase in *E. coli* shedding and concerns about antimicrobial resistance genes. Research needs to continue (26, 27). Despite all the research on various diet-related production practices at the preharvest phase, the impact on foodborne illness or public health risk is indeterminate.

Interventions concentrated on a specific foodborne pathogen often incorporate bacteriocins, bacteriophages, vaccines, and even antibiotics. However, these treatments affect the entire microbial flora and do not target specific pathogens; they have also had conflicting results. The limiting factor for bacteriocins, other than determining their true effectiveness, is the inability to produce them in large quantities. Bacteriophages have shown promise because they can be produced to target a specific bacteria common in the microbial flora. For example, there is a bacteriophage spray used against *Salmonella* in poultry which has shown some effectiveness (28). As recent recommendations have shown, combinations of certain strategies (e.g., bacteriophage spray, vaccination) may be one effective approach for *Salmonella* spp. in various food animals.

Developing and implementing control, prevention, and mitigation strategies for the environment is more of a challenge. Pathogens are introduced or transmitted through a wide range of possibilities including water, soil, manure, air, vehicles, workers, wildlife, birds, insects, and rodents. Numerous studies have demonstrated the role of insects (flies, darkling beetles), birds (starlings), and wildlife in the persistence and transmission of *E. coli*, *Salmonella*, and other pathogens. Naturally, a feces-contaminated environment is a source of infection for animals. Research over the years has included changing the shape of the feedlot water trough, exchanging or treating poultry litter, and treatments of the soil (13, 29). No matter the specific strategy, biosecurity and control strategies are critical to animal health management as well as to preharvest food safety.

These highlighted preharvest efforts illustrate that food safety research and decision-making cannot be singular in focus or solvable in isolation. The strategies of control, prevention, and mitigation must be considered in a holistic perspective. One flawed concept is that a vaccination or single intervention will be the answer. To produce results with a public health impact, multiple interventions or controls, in series or in parallel, are needed. Finally, well-designed studies focused on evaluating effectiveness and impacts are critical. These efforts will require researchers and policy-makers to provide collaborative directions and efforts. Although this article does not include imported control programs, the USDA and FDA maintain requirements for the safety of food imported into the United States, and both agencies rely on the verification of international food systems and "equivalence" to the United States.

"NONTRADITIONAL" PREHARVEST AREAS

Organic and Free-Range Production

Biosecurity measures and management practices have remained the standard for reducing pathogen introduction and movement. But the latest trend for "natural" or organic foods creates new challenges. Consumers have expressed the desire for food free of pesticides and antibiotics, and animal welfare advocates are demanding more humane animal production. This trend has increased the need to focus on and compare organically raised or free-range production practices with conventional management in terms of their effect on foodborne contamination and even public risk. Current research is being conducted in multiple food species, such as egg production, poultry, cattle, and swine (30–33). Topics of emphasis are numerous and still have applicability to "traditional" food safety. For example, what is the effect of differing management or production practices on foodborne pathogen load, prevalence, or on public health risk? Does free-range production increase the possibility of wildlife or environmental introduction of pathogens? Are organically raised animals free of antibiotic-resistant pathogens or genes? It is critical to continue the research in this area as the demand for organic products and for safer and healthier foods increases. Some current studies on swine raised outdoors have indicated that animal pathogens and parasites that have previously been controlled, e.g., toxoplasmosis, will increase (34).

Water, Manure, and Produce

The emphasis in recent years on consuming fresh fruits and vegetables has created opportunities for foodborne outbreaks related to fruits and fruit juices and leafy greens (35–37). The first large foodborne outbreak in produce (38) shifted the focus of preharvest food safety from meat and poultry to other food commodities such as produce. More epidemiologic and ecologic studies were focused on the environment, the role of wildlife and livestock production in close proximity, and water and manure or fertilizer. The signing of the FSMA in 2011 has presented even more incentive for broader, innovative research directions, specifically in areas of manure, water, and interactions between wildlife, food animal production, and the environment (http://www.fda.gov/Food/GuidanceRegulation/FSMA/default.htm). FSMA gives researchers the opportunity to provide scientific data for impending guidelines and to develop strategies to better measure the effectiveness and the impact of these new guidelines, such as the Produce Rule.

The two most critical areas emphasized in the Produce Rule are water and manure. Both topics are relevant preharvest issues since the major source of foodborne pathogens in fresh produce is often due to animal or human fecal matter. Pathogen reduction in the animal and the production environment is critical to the reduction of pathogens farther down the food production chain. The use of untreated (or inadequately composted) manure as fertilizer, or contaminated irrigation water, can lead to higher microbial loads for the animal, the environment, and most importantly, plants and plant crops (e.g., produce). Both the timing of, and method for, irrigation and fertilization are critical (39–42). Two papers were published that provide a framework for study in these areas as well as potential research directions (43, 44). Recent European Food Safety Authority scientific opinions on *Salmonella* contamination of melons and tomatoes also offer critical research directions for preharvest. Examples of research areas include risk factors such as environmental and climatic conditions (e.g., proximity to animal rearing operations), contact with domestic or wildlife animal reservoirs, use of inadequately or untreated manure or fertilizers, use of contaminated water for irrigation, and cross-contamination with farming equipment or humans (http://www.efsa.europa.eu/en/press/news/141002.htm). More importantly, the European Food Safety Authority suggests setting food safety criteria (or a microbial criterion) at the preharvest stage to meet the process hygiene and food safety criteria for *E. coli* in precut melons and *Salmonella* in retail environments (45, 46).

Genomics, Sequencing, and Bioinformatics

The explosion of new molecular techniques, especially whole genomic sequencing, has influenced food safety research from pre- to postharvest, as well as globally. Expenditures, resources, and research have expanded beyond even the ability to analyze the data. The Smithsonian launched the Global Genomic Initiative in 2011 (https://ggi.si.edu/) to provide a repository for genomic information on the diversity of life and to encourage collaborations. The FDA has developed several national and international public-private partnerships for genomic sequencing in food safety research (http://www.fda.gov/Food/FoodScienceResearch/WholeGenomeSequencingProgramWGS/).

The majority of the programs have multiple partners and take a holistic approach to foodborne outbreaks. For example, Genome Trakr Network is a network of state and federal public health laboratories that are collecting genomic and geographical data on clinical, environmental, and food samples which are stored in the Biotechnology Information (NIH). Researchers and public health officials use these data sequences for a “real-time” comparison and analysis to facilitate outbreak investigations and trace-backs. The Global Microbial Identifier (GMI) initiative involves more than 30 countries. The goal is to provide global collections of genomic data for epidemiologic study, surveillance, and sharing of outbreak strains worldwide. This will provide the opportunity for academia, industry, and government to more accurately characterize and trace microbial pathogens. There are several smaller and more targeted collaborative research projects (FDA, CDC, state laboratories) that involve sequencing clinical, food, and environmental isolates for certain foodborne pathogens, e.g., *Listeria* and *Salmonella*. These are just a few examples of the many sequencing projects occurring in the United States and globally.

Sequencing is also occurring collaboratively at the preharvest level (http://www.stecbeefsafety.org). Current research involves sequencing on-farm environmental and animal samples to better describe serotype (or strain) differences among *E. coli* 0157:H7 and non–Shiga toxin–producing *E. coli* and to understand disease pathogenesis and virulence (47). These analytical and computational tools can be used to better characterize the movement of genes, foodborne pathogens, or even antimicrobial resistance genes as they move through different production stages (e.g., poultry from farm to retail) (48). Genomic studies will contribute to our understanding of virulence, transmission, and survival of pathogens. Obviously, the matching of sequences found in large repositories of outbreak strains will allow public health officials to identify the potential source of the outbreak and intervene to reduce future exposures and illnesses. However, as more resources are devoted to this technology, several cautionary questions need to be answered. There are numerous debates on replacing all culture and other traditional methods with sequencing and then storing only the sequences. Although sequencing technology seems to be a "goldmine" of information, are we going to lose important sources of data and original samples for future hypotheses? Most importantly, the field of bioinformatics has expanded, but expertise and infrastructure still lag behind. Computational analytics need more experts and supercomputers and the ability to transmit large datasets among researchers. Finally, there is a possibility that researchers can incorrectly or over-interpret the data from several populations that may not be related and attempt to make causal links where there are none. Linking large population datasets with epidemiologic data requires expertise in epidemiology, bioinformatics, and multiple other disciplines. It also requires collaborative efforts among public health and agriculture and industry, academia, and government. The impact of this technology on policy, research, and public health risk is still undetermined.

Antimicrobial Resistance: Old and New

Antimicrobial resistance remains a highly visible but polarizing focus for agriculture and public health, and by extension for food safety and the preharvest area. New initiatives and research are critical to providing true evidence-based solutions and to measuring the impact of new intervention, control, or prevention strategies. Success depends on learning from past efforts and building on current networks and projects. The National Antimicrobial Resistance Monitoring System has been a successful collaboration among the FDA, CDC, and USDA since 1996 and has provided historical and temporal data and culture collections for foodborne bacteria in humans, retail foods, and animals (http://www.fda.gov/AnimalVeterinary/SafetyHealth/AntimicrobialResistance/NationalAntimicrobialResistanceMonitoringSystem/default.htm), but the need for better information about the prevalence of resistance in food animals and the role of preharvest in the prevalence of antimicrobial resistance through the food chain required changes in the program. In 2011, the USDA, at the request of the FDA, revised the animal sampling program and piloted a new on-farm sampling scheme. This was possible with a newly formed "consortium" which consisted of university-based ongoing research programs and specific Agricultural Research Service (ARS) research centers. The goal was to bring together epidemiologists involved in long-term on-farm research projects with microbiologists. Like any large surveillance program or research consortium, there were several challenges, including coordinating a large project, developing common goals and laboratory methods, developing a sampling protocol that would meet the objectives, and developing a system to ensure confidentiality. This consortium has provided data on *E. coli*, *Salmonella*, and *Campylobacter* for poultry, feedlots, and dairies to the FDA's Center for Veterinary Medicine for antimicrobial resistance testing. The infrastructure created by the consortium enabled the design of longitudinal cohort and cross-over studies and the opportunity to evaluate several hypotheses. For example, researchers could follow dairy cattle from the farm through a sales barn to slaughter to evaluate the change in foodborne pathogen prevalence and antimicrobial resistance patterns. The advantages of the consortium are the ability to build collaborations among industry, academia, and government; the ability to use an existing framework for specific emerging hypotheses; the ability to have a surveillance and research network; and the ability to leverage resources, but, as previously mentioned, the maintenance of a research consortium is challenging. If this consortium continues, it could provide the opportunity for genomic sequencing projects, the ability to assess the prevalence and change of antimicrobial resistance and foodborne pathogens from on-farm to slaughter, and the opportunity to measure the effect of interventions or prevention strategies.

Other areas require research efforts, e.g., the impact of new FDA guidelines (http://www.fda.gov/downloads/AnimalVeterinary/GuidanceComplianceEnforcement/GuidanceforIndustry/UCM299624.pdf; http://www.fda.gov/downloads/AnimalVeterinary/GuidanceComplianceEnforcement/GuidanceforIndustry/UCM216936.pdf), specifically for the removal of growth promoter use. Multiple papers have documented the efforts in Denmark (49–50). This is the perfect opportunity to measure the impact on animal health, antibiotic usage, and antimicrobial resistance in the United States. This would require a baseline as well as a longitudinal study to measure the change in antimicrobial usage, animal morbidity (or even mortality), antimicrobial resistance, and foodborne pathogens. The time to plan and implement this is before the growth promoter ban is implemented. Finally, it is critical to evaluate the effect of food safety interventions in preharvest and postharvest on resistance. For example, the effectiveness of vaccinations and bacteriophage sprays or processing rinses is currently measured by the reduction of foodborne pathogens (e.g., *E. coli* and *Salmonella*). These interventions could also be evaluated for changes in resistant organisms or even resistance genes.

Evidence-Based Directions and Performance Metrics

Policy-makers traditionally use risk assessments to evaluate regulatory actions and even research directions. Although risk assessments will continue to be accepted practice, food safety researchers in agriculture and veterinary medicine are expanding their use of evidence-based medicine tools. Evidence-based medicine has been used in human clinical medicine for many years, but recently, agriculture and veterinary medical scientists have adopted and revised evidence-based methods to evaluate current published research, identify existing data gaps, and standardize study designs for preharvest food safety (51–54). "Filling data gaps" is a common phrase among researchers and policy-makers, yet the process is often a combination of expert opinions and scientific meetings. Ensuring that there is not duplication of research efforts is less standardized or clear. Systematic reviews (and other evidence-based strategies) are growing in popularity among food safety researchers and should be considered by policy-makers and those setting research and strategic directions (55–58).

Performance metrics and public health impact are a growing priority for federal agencies, although though these topics were highlighted in Institute of Medicine (IOM) reports over 10 years ago (3). Measuring the impact of food safety research and food safety initiatives on public health risk is an ideal goal but is realistically difficult. Assigning causality to singular actions that occur in a complex network of food production and food distribution seems impossible. However, establishing specific criteria, objectives, and intermediary endpoints or outcomes will facilitate objective measurements. For example, the implementation of the HACCP program at the postharvest level in 1997 has reduced the foodborne pathogen levels for finished products (http://www.fda.gov/Food/GuidanceRegulation/HACCP/ucm2006801.htm). Establishing industry objectives and critical control points creates a process for verification, measurement, and even interventions for the reduction of pathogen contamination. At preharvest, there are several established best management or good agricultural practices but not any specific measures for foodborne pathogen reduction or critical control points that can be evaluated. For example, the National Poultry Improvement Plan (http://poultryimprovement.org/default.cfm) is a collaborative long-standing program that has succeeded in reducing poultry pathogens. For preharvest food safety, there is still no clear recognition or definition of prevalence versus pathogen load on-farm or of what impact these microbiological criteria or critical control measures may have farther down the food production continuum. Defining this preharvest critical control level would provide a framework for additional research and potential standards or guidelines.

A critical yet challenging issue is regaining momentum as discussions continue on food safety priorities being aligned strategically with public health outcomes and the need to clarify the relationship of preharvest food safety and public health risk. This issue was raised in the well-distributed reports by the Institute of Medicine (2, 3) and the American Academy of Microbiology 2004 report (4). The 2003 report recommended setting a progressive series of criterion, performance standards, and food safety objectives to evaluate the public health impact. This report articulated the importance of both the initial Good Agricultural Practices for produce and the HACCP principles that were implemented but also acknowledged the complexity of food production and the difficulty of applying criteria and standards across the food production continuum. Thus, it is incumbent for preharvest systems to find a "level," a microbiological criteria, or pathogen load at the preharvest level that will translate into a critical control standard level whereby the postharvest processing can decrease the pathogen or contaminant even more, thus lowering the public health risk.

Odds and Ends

There are several new areas of study for preharvest researchers to ponder in the future. They include niche food markets where consumers are demanding food products such as bison, goat, lamb, and ostrich. Raw milk and raw milk cheeses are an emerging consumer demand and have already been linked to foodborne illnesses. However, this is another polarizing topic between public health officials that insist on pasteurization or consistent farm standards or inspections and consumers and producers that want the freedom to choose a product they consider natural and healthy. Climate change continues to be an area that should receive more attention and funding. Climate change can influence food safety and security in areas such as water use, protein sources, and mycotoxins. An increase in aflatoxins in feeds would have a direct impact on humans and food animals. Finally, there are the evolving areas of genetically modified animals, clones, and nanotechnology.

CONCLUSION AND THE FUTURE

There is no shortage of food safety issues to be addressed or research to be conducted—only a limitation of funding and expertise. This heightens the need to strategically plan for future research, strengthen collaborations, and develop approaches to measure the scientific and regulatory impact of research and policy. This book is comprehensive, and it would be impossible for any of the authors to predict the next emerging pathogen or the magical solution for a persistent foodborne pathogen. There are several common themes that industry, government, and academia could continue to address in preharvest food safety.

It is critical to coordinate and leverage research efforts in parallel with strategic goals and directions for the future. This effort would incorporate an evaluation of what research has been accomplished, the outcome, and the potential impact. The appraisal could be a retrospective of funding programs, intramural research programs, or a broader evidence-based study (e.g., systematic review). The result of the review could impact our future decisions on funding and mechanisms for conducting research. No amount of funding, established research program, or numerous strategic plans will be successful unless we define and measure the effect, outcome, and impact of what we have accomplished in food safety.

It is important to take advantage of current or impending policies or initiatives that would enable researchers to proactively measure the impact of implementation of these regulatory policies or guidelines. For example, the implementation of the Produce Rule within FSMA or the Center of Veterinary Medicine's (FDA) implementation of the elimination of growth promoters is timely. Designing and implementing a broad research effort would allow researchers and policy-makers to measure the impact of such policies and provide a foundation for future prevention or control measures.

Researchers and decision-makers must continue to approach preharvest food safety systematically, both at the micro level (microbial flora, genes, genomic sequences) and at the macro level (animals, environments, plants). Each component of the micro or macro level persists, moves, and interacts, not in a vacuum but within and among different environments. Thus, interventions, prevention and control programs, or even guidelines focused at the preharvest level also have an effect on the entire food production chain. In the end, research and policy decisions on preharvest activities must envision and analyze how that food or potential foodborne pathogen moves from preharvest to postharvest and to the final food product.

Despite the recognized role of epidemiology in food outbreak and trace-back investigations, the scientific approach and expertise is still underutilized in the areas of food safety and preharvest food safety. Epidemiology not only provides crucial approaches for surveillance and monitoring, but can provide the foundation for setting research sampling programs for specific hypotheses or setting industry and government sampling standards for the environment or production process. There continues to be uncertainty around the best approach for sampling large produce fields, large animal production facilities (environment versus animal), and even shipping containers or imported lots. Epidemiology provides the scientific analytical approach to conducting and understanding risk assessments or for implementing evidence-based methods to critically evaluating research directions and methods. Most importantly, epidemiology can provide the critical study approaches for implementing field or population-type studies and for more crucial longitudinal studies.

Food safety research, whether at the pre- or postharvest level, will continue to be a fascinating complex web of foodborne pathogens, risk factors, and scientific and policy interactions. Food safety priorities and research must continue to evolve with emerging global issues, emerging technologies, and methods but remain grounded in a multidisciplinary, collaborative, and systematic approach.

ACKNOWLEDGMENTS

This chapter reflects the opinions of the author and is not an official policy or opinion of the U.S. Department of Health and Human Services, the U.S. Food and Drug Administration, or the Center for Food Safety and Applied Nutrition.

CITATION

Torrence ME. 2016. Introduction to preharvest food safety. Microbiol Spectrum 4(5): PFS-0009-2015.

REFERENCES

1. **U.S. Food and Drug Administration, Department of Agriculture, Environmental Protection Agency, and Centers for Disease Control and Prevention.** 1997. Food safety from farm to table: a national food safety initiative. A report to the president.
2. **Institute of Medicine, National Research Council.** 1998. p 20055. *In Ensuring Safe Food: From Production to Consumption.* National Academy Press, Washington, DC.
3. **Institute of Medicine, National Research Council.** 2003. *Scientific Criteria to Ensure Safe Food.* National Academies Press, Washington, DC.

4. **Isaacson RE, Torrence M, Buckley MR.** 2004. Pre-harvest food safety and security. AAM, Washington DC. http://asmscience.org/content/report/colloquia/colloquia.34.
5. **U.S. Food and Drug Administration.** 2014. *Food Safety Modernization Act.* http://www.fda.gov/Food/GuidanceRegulation/FSMA/default.htm.
6. **Boucher HW, Corey GR.** 2008. Epidemiology of methicillin-resistant *Staphylococcus aureus. Clin Infect Dis* **46**(Suppl 5):S344–S349.
7. **Cox LA Jr, Popken DA.** 2014. Quantitative assessment of human MRSA risks from swine. *Risk Anal* **34:**1639–1650.
8. **Osadebe LU, Hanson B, Smith TC, Heimer R.** 2013. Prevalence and characteristics of *Staphylococcus aureus* in Connecticut swine and swine farmers. *Zoonoses Public Health* **60:** 234–243.
9. **Weese JS, Zwambag A, Rosendal T, Reid-Smith R, Friendship R.** 2011. Longitudinal investigation of methiciliin-resistant *Staphylococcus aureus* in piglets. *Zoonoses Public Health* **58:**238–243.
10. **Verhegghe M, Pletinckx LJ, Crombé F, Vandersmissen T, Haesebrouck F, Butaye P, Heyndrickx M, and Rasschaert G.** 2013. Methicillin-resistant *Staphylococcus aureus* (MRSA) ST398 in pig farms and multiple species farms. *Zoonoses Public Health* **60:**366–374.
11. **Torrence ME.** 2003. U.S. federal activities, initiatives and research in food safety, p 3–10. *In* Torrence ME, Isaacson RE (ed), *Microbial Food Safety in Animal Agriculture.* Iowa State Press, Ames, IA.
12. **Berghaus RD, Thayer SG, Law BF, Mild RM, Hofacre CL, Singer RS.** 2013. Enumeration of *Salmonella* and *Campylobacter* spp. in environmental farm samples and processing plant carcass rinses from commercial broiler chicken flocks. *Appl Environ Microbiol* **79:** 4106–4114.
13. **LeJeune JT, Besser TE, Hancock DD.** 2001. Cattle water troughs as reservoirs of *Escherichia coli* O157:H7. *Appl Environ Microbiol* **67:**3053–3057.
14. **Callaway TR, Edrington TS.** 2012. *On-Farm Strategies to Control Foodborne Pathogens.* Nova Science Publishers, New York, NY.
15. **Bier RC, Pillai SD, Phillips TD, Ziprin RL (ed).** 2004. *Pre-Harvest and Postharvest Food Safety: Contemporary Issues and Future Directions.* Blackwell Publishing, Ames, IA.
16. **Torrence ME, Isaacson RE.** 2003. *Microbial Food Safety in Animal Agriculture: Current Topics.* Iowa State Press, Ames, IA.
17. **Bell BP, Goldoft M, Griffin PM, Davis MA, Gordon DC, Tarr PI, Bartleson CA, Lewis JH, Barrett TJ, Wells JG, Baron R, Kobayashi J.** 1994. A multistate outbreak of *Escherichia coli* O157:H7-associated bloody diarrhea and hemolytic uremic syndrome from hamburgers. The Washington experience. *JAMA* **272:**1349–1353.
18. **U.S. Department of Agriculture, Food Safety and Inspection Service.** 1996. Pathogen reduction: hazard analysis and critical control point (HACCP) systems; final rule. Part II, 9 CFR Part 304 et al. *Fed Regist* **61:**38805–38956.
19. **Rasko DA, Webster DR, Sahl JW, Bashir A, Boisen N, Scheutz F, Paxinos EE, Sebra R, Chin CS, Iliopoulos D, Klammer A, Peluso P, Lee L, Kislyuk AO, Bullard J, Kasarskis A, Wang S, Eid J, Rank D, Redman JC, Steyert SR, Frimodt-Møller J, Struve C, Petersen AM, Krogfelt KA, Nataro JP, Schadt EE, Waldor MK.** 2011. Origins of the *E. coli* strain causing an outbreak of hemolytic-uremic syndrome in Germany. *N Engl J Med* **365:**709–717.
20. **U.S. Food and Drug Administration.** 2009. Prevention of *Salmonella enteritidis* in shell eggs during production, storage, and transportation. *Fed Regist* **74:**33029–33101.
21. **Jackson BR, Griffin PM, Cole D, Walsh KA, Chai SJ.** 2013. Outbreak-associated *Salmonella enterica* serotypes and food commodities, United States, 1998-2008. *Emerg Infect Dis* **19:**1239–1244.
22. **Gragg SE, Loneragan GH, Brashears MM, Arthur TM, Bosilevac JM, Kalchayanand N, Wang R, Schmidt JW, Brooks JC, Shackelford SD, Wheeler TL, Brown TR, Edrington TS, Brichta-Harhay DM.** 2013. Cross-sectional study examining *Salmonella enterica* carriage in subiliac lymph nodes of cull and feedlot cattle at harvest. *Foodborne Pathog Dis* **10:**368–374.
23. **Fox JT, Thomson DU, Drouillard JS, Thornton AB, Burkhardt DT, Emery DA, Nagaraja TG.** 2009. Efficacy of *Escherichia coli* O157:H7 siderophore receptor/porin proteins-based vaccine in feedlot cattle naturally shedding *E. coli* O157. *Foodborne Pathog Dis* **6:**893–899.
24. **Vogstad AR, Moxley RA, Erickson GE, Klopfenstein TJ, Smith DR.** 2013. Assessment of heterogeneity of efficacy of a three-dose regimen of a type III secreted protein vaccine for reducing STEC O157 in feces of feedlot cattle. *Foodborne Pathog Dis* **10:**678–683.
25. **Diez-Gonzalez F, Callaway TR, Kizoulis MG, Russell JB.** 1998. Grain feeding and the dissemination of acid-resistant *Escherichia coli* from cattle. *Science* **281:**1666–1668.
26. **Jacob ME, Paddock ZD, Renter DG, Lechtenberg KF, Nagaraja TG.** 2010. Inclusion

of dried or wet distillers' grains at different levels in diets of feedlot cattle affects fecal shedding of *Escherichia coli* O157:H7. *Appl Environ Microbiol* **76:**7238–7242.

27. **Jacob ME, Parsons GL, Shelor MK, Fox JT, Drouillard JS, Thomson DU, Renter DG, Nagaraja TG.** 2008. Feeding supplemental dried distiller's grains increases faecal shedding of *Escherichia coli* O157 in experimentally inoculated calves. *Zoonoses Public Health* **55:**125–132.
28. **Borie C, Sánchez ML, Navarro C, Ramírez S, Morales MA, Retamales J, Robeson J.** 2009. Aerosol spray treatment with bacteriophages and competitive exclusion reduces *Salmonella enteritidis* infection in chickens. *Avian Dis* **53:**250–254.
29. **Davies RH, Wray C.** 1996. Studies of contamination of three broiler breeder houses with *Salmonella enteritidis* before and after cleansing and disinfection. *Avian Dis* **40:**626–633.
30. **Fox JT, Reinstein S, Jacob ME, Nagaraja TG.** 2008. Niche marketing production practices for beef cattle in the United States and prevalence of foodborne pathogens. *Foodborne Pathog Dis* **5:**559–569.
31. **Gast RK, Guraya R, Jones DR, Anderson KE.** 2014. Contamination of eggs by *Salmonella* Enteritidis in experimentally infected laying hens housed in conventional or enriched cages. *Poult Sci* **93:**728–733.
32. **Jones DR, Anderson KE, Guard JY.** 2012. Prevalence of coliforms, *Salmonella*, *Listeria*, and *Campylobacter* associated with eggs and the environment of conventional cage and free-range egg production. *Poult Sci* **91:**1195–1202.
33. **Tadesse DA, Bahnson PB, Funk JA, Thakur S, Morrow WE, Wittum T, DeGraves F, Rajala-Schultz P, Gebreyes WA.** 2011. Prevalence and antimicrobial resistance profile of *Campylobacter* spp. isolated from conventional and antimicrobial-free swine production systems from different U.S. regions. *Foodborne Pathog Dis* **8:**367–374.
34. **Dubey JP, Hill DE, Rozeboom DW, Rajendran C, Choudhary S, Ferreira LR, Kwok OC, Su C.** 2012. High prevalence and genotypes of *Toxoplasma gondii* isolated from organic pigs in northern USA. *Vet Parasitol* **188:**14–18.
35. **Vojdani JD, Beuchat LR, Tauxe RV.** 2008. Juice-associated outbreaks of human illness in the United States, 1995 through 2005. *J Food Prot* **71:**356–364.
36. **Kendall ME, Mody RK, Mahon BE, Doyle MP, Herman KM, Tauxe RV.** 2013. Emergence of salsa and guacamole as frequent vehicles of foodborne disease outbreaks in the United States, 1973-2008. *Foodborne Pathog Dis* **10:**316–322.
37. **Kozak GK, MacDonald D, Landry L, Farber JM.** 2013. Foodborne outbreaks in Canada linked to produce: 2001 through 2009. *J Food Prot* **76:**173–183.
38. **Jay MT, Cooley M, Carychao D, Wiscomb GW, Sweitzer RA, Crawford-Miksza L, Farrar JA, Lau DK, O'Connell J, Millington A, Asmundson RV, Atwill ER, Mandrell RE.** 2007. *Escherichia coli* O157:H7 in feral swine near spinach fields and cattle, central California coast. *Emerg Infect Dis* **13:**1908–1911.
39. **Berry ED, Woodbury BL, Nienaber JA, Eigenberg RA, Thurston JA, Wells JE.** 2007. Incidence and persistence of zoonotic bacterial and protozoan pathogens in a beef cattle feedlot runoff control vegetative treatment system. *J Environ Qual* **36:**1873–1882.
40. **Hutchison ML, Walters LD, Avery SM, Munro F, Moore A.** 2005. Analyses of livestock production, waste storage, and pathogen levels and prevalences in farm manures. *Appl Environ Microbiol* **71:**1231–1236.
41. **Islam M, Doyle MP, Phatak SC, Millner P, Jiang X.** 2004. Persistence of enterohemorrhagic *Escherichia coli* O157:H7 in soil and on leaf lettuce and parsley grown in fields treated with contaminated manure composts or irrigation water. *J Food Prot* **67:**1365–1370.
42. **Gerba CP, Smith JE Jr.** 2005. Sources of pathogenic microorganisms and their fate during land application of wastes. *J Environ Qual* **34:**42–48.
43. **Harris LJ, Bender J, Bihn EA, Blessington T, Danyluk MD, Delaquis P, Goodridge L, Ibekwe AM, Ilic S, Kniel K, Lejeune JT, Schaffner DW, Stoeckel D, Suslow TV.** 2012. A framework for developing research protocols for evaluation of microbial hazards and controls during production that pertain to the quality of agricultural water contacting fresh produce that may be consumed raw. *J Food Prot* **75:**2251–2273.
44. **Harris LJ, Berry ED, Blessington T, Erickson M, Jay-Russell M, Jiang X, Killinger K, Michel FC, Millner P, Schneider K, Sharma M, Suslow TV, Wang L, Worobo RW.** 2013. A framework for developing research protocols for evaluation of microbial hazards and controls during production that pertain to the application of untreated soil amendments of animal origin on land used to grow produce that may be consumed raw. *J Food Prot* **76:**1062–1084.
45. **EFSA Panel on Biological Hazards.** 2014. Scientific opinion on the risk posed by path-

ogens in food of animal origin. Part 2. *Salmonella* in melons. *EFSA J* **12:**3831.

46. **EFSA Panel on Biological Hazards.** 2014. Scientific opinion on the risk posed by pathogens in food of non-animal origin. Part 2. *Salmonella* and Norovirus in tomatoes. *EFSA J* **12:**3832.
47. **Jung WK, Bono JL, Clawson ML, Leopold SR, Shringi S, Besser TE.** 2013. Lineage and genogroup-defining single nucleotide polymorphisms of *Escherichia coli* O157:H7. *Appl Environ Microbiol* **79:**7036–7041.
48. **Hoffmann M, Zhao S, Pettengill J, Luo Y, Monday SR, Abbott J, Ayers SL, Cinar HN, Muruvanda T, Li C, Allard MW, Whichard J, Meng J, Brown EW, and McDermott PF.** 2014. Comparative genomic analysis and virulence differences in closely related salmonella enterica serotype heidelberg isolates from humans, retail meats, and animals. *Genome Biol Evol* **6:**1046–1068.
49. **Grave K, Jensen VF, Odensvik K, Wierup M, Bangen M.** 2006. Usage of veterinary therapeutic antimicrobials in Denmark, Norway and Sweden following termination of antimicrobial growth promoter use. *Prev Vet Med* **17:**123–132.
50. **Wierup M.** 2001. The Swedish experience of the 1986 year ban of antimicrobial growth promoters, with special reference to animal health, disease prevention, productivity, and usage of antimicrobials. *Microb Drug Resist* **7:**183–190.
51. **Sargeant JM, Torrence ME, Rajić A, O'Connor AM, Williams J.** 2006. Methodological quality assessment of review articles evaluating interventions to improve microbial food safety. *Foodborne Pathog Dis* **3:**447–456.
52. **Torrence ME.** 2012. Pre-harvest food safety: the past and the future, p 5–15. *In* Callaway TR, Edrington TS (ed), *On-Farm Strategies to Control Foodborne Pathogens*. Nova Science Publishers, New York, NY.
53. **Sargeant JM, O'Connor AM, Gardner IA, Dickson JS, Torrence ME, Consensus Meeting Participants.** 2010. The REFLECT statement: reporting guidelines for randomized controlled trials in livestock and food safety: explanation and elaboration. *Zoonoses Public Health* **57:**105–136.
54. **O'Connor AM, Sargeant JM, Gardner IA, Dickson JS, Torrence ME; Consensus Meeting Participants, Dewey CE, Dohoo IR, Evans RB, Gray JT, Greiner M, Keefe G, Lefebvre SL, Morley PS, Ramirez A, Sischo W, Smith DR, Snedeker K, Sofos J, Ward MP, Wills R.** 2010. The REFLECT statement: methods and processes of creating reporting guidelines for randomized controlled trials for livestock and food safety by modifying the CONSORT statement. *Zoonoses Public Health* **57:**95–104.
55. **Sargeant JM, Amezcua MR, Rajic A, Waddell L.** 2007. Pre-harvest interventions to reduce the shedding of *E. coli* O157 in the faeces of weaned domestic ruminants: a systematic review. *Zoonoses Public Health* **54:**260–277.
56. **Sargeant JM, O'Connor AM.** 2014. 'One-stop shopping' for information on conducting systematic reviews and meta-analysis in animal agriculture and veterinary medicine. *Zoonoses Public Health* **61**(Suppl 1):2.
57. **Wisener LV, Sargeant JM, O'Connor AM, Faires MC, Glass-Kaastra SK.** 2014. The use of direct-fed microbials to reduce shedding of *Escherichia coli* O157 in beef cattle: a systematic review and meta-analysis. *Zoonoses Public Health* **62:**75–89.
58. **Young I, Rajić A, Wilhelm BJ, Waddell L, Parker S, McEwen SA.** 2009. Comparison of the prevalence of bacterial enteropathogens, potentially zoonotic bacteria and bacterial resistance to antimicrobials in organic and conventional poultry, swine and beef production: a systematic review and meta-analysis. *Epidemiol Infect* **137:**1217–1232.

Preharvest Farming Practices Impacting Fresh Produce Safety

2

EDUARDO GUTIERREZ-RODRIGUEZ[1] and ACHYUT ADHIKARI[2]

OVERVIEW OF PREHARVEST FOOD SAFETY

Foodborne illness associated with fresh fruits and vegetables and growing and harvesting practices that impact the microbial safety of produce have been under the scrutiny of federal and state regulators for over 40 years. In 1998 the Food and Drug Administration published the Guide to Minimize Microbial Food Safety Hazards for Fruits and Vegetables (GMMFSH) in response to the U.S. President's Food Safety Initiative, which launched, among other programs, the "Fight Bac" campaign. This GMMFSH guideline was created based on outbreak investigations that identified fresh produce as "an area of concern" (1) and summarized the then-current body of research and understanding of the risk factors associated with agricultural practices and how they could impact contamination of fresh produce with human pathogens. At the time, the GMMFSH guidelines did not address practices for risk elimination, supply chain, or environmental contaminants; however, they clearly identified the foundation for future research and new federal policies (the Food Safety Modernization Act [FSMA] and the Produce Safety Rule signed into law in 2011, which are currently impacting the farm to fork continuum). Twenty years after the development of these guidelines it is

[1]Department of Food, Bioprocessing, and Nutrition Sciences, North Carolina State University, Raleigh, NC 27695; [2]School of Nutrition and Food Sciences, Louisiana State University, Baton Rouge, LA 70803.

Preharvest Food Safety
Edited by Siddhartha Thakur and Kalmia E. Kniel

doi:10.1128/microbiolspec.PFS-0022-2018

evident that previously identified sources of contamination linked to agronomic practices continue to be implicated in many of the more than 140 produce-related outbreaks investigated by federal and state agencies (2).

During these 2 decades, there has been an increase in produce contamination events that could be attributed to any of the following factors: (i) improved detection platforms, (ii) increased consumption of fresh produce and fresh-cut products, (iii) transport and distribution of fresh produce from many regions around the world, and (iv) production and potentially harvest and packing practices that either did not present any past problems or that continue to present issues due to continuing lack of understanding of the factors that impact contamination despite industry-driven practices that have improved food safety systems (good agricultural practices, hazard analysis of critical control points, and other third-party food safety schemes).

Better surveillance and detection systems have significantly impacted our ability to identify different microbial contaminants and impacted the increase in the number of pathogen contamination events detected each year. A good example that highlights our increased ability to detect different pathogens is reflected in the number of *Cyclospora* outbreaks detected between 2013 and 2018 and *Listeria monocytogenes* between 2011 and 2018 (3) because of the use of modern technologies to generate genome sequences, identify pathogenic mechanisms, and provide evidence-based risk assessments. This increase in detection capacity has not translated into an increased ability to identify sources of contamination or risk factors impacting pathogen outbreaks or into reducing trends in foodborne diseases. Table 1 summarizes areas of concern within soil, plant, water, and packinghouse environments that impact survival, internalization, and interactions of human pathogens in different anthropogenic environments.

This review is not intended to provide a detailed description of grower production practices and all the potential interactions that impact fresh produce safety. It should not be used as a guide for safely growing fruits and vegetables. Instead, this review largely focuses on those farming practices and research components needed to elucidate pathogen interactions at the farm level, summarizes some of the most important past and current research topics, and provides additional information based on observation and experience of the sources of contamination associated with different farming practices. We advise the reader to also review other publications (4–13) to gain a broader perspective of the issues addressed here.

Fruits and Vegetables in the United States Impacted by Microbial Contamination

According to the U.S. 2012 Census of Agriculture, fruit, tree nut, and berry sales together amounted to $25.9 billion, while vegetable sales were estimated at $16.9 billion. Ten states account for 94% of all fruit, tree nut, and berry sales, with California, Washington, and Florida controlling 87% of these sales. Oranges and grapefruits are by far the largest citrus crops. Among non-citrus fruits, berries, grapes, apples, peaches, sweet cherries, blueberries, avocados, and strawberries are the largest commodities. Almonds, pecans, walnuts, and pistachios account for almost all the tree nut acreage in the United States. The top 10 vegetable-producing states accounted for 77% of sales in 2012, with California alone accounting for 38% of total sales, followed by Florida and Washington. The major vegetable crops are potatoes, sweet corn, tomatoes, lettuce, and watermelon (14).

Fresh produce represents one of the major causes of foodborne illnesses (46%) in the United States (15). The U.S. Centers for Disease Control and Prevention (CDC) estimates that each year 48 million Americans get foodborne diseases, resulting in 128,000 hospitalizations and 3,000 deaths. According to the U.S. Food and Drug Administration (FDA),

TABLE 1 Challenges within the fresh produce continuum

Environment or Condition	Challenge areas
Soil	Soil properties, biological soil amendments, compost, raw manure, herbal teas, and persistence of human pathogens vs. crop contamination
Farm history	Farms formerly used for the rearing of animals or those in close proximity to small, medium, and large CAFOs
	Wild and domestic animal field intrusion and field contamination
Crop	Growth and survival of human pathogens on crop inputs, including organic and chemical fertilizers and pesticides
	Growing practices and whether organic, conventional, biodynamic, sustainable, or other agronomic practices impact pathogen contamination, survival, persistence, and distribution within the farm
Water	Microbial quality based on source, region, activity, and practice and the equipment used to deliver water to the crop (pumps, drip tape, pipeline, emitters, sand and plate filters, and injectors of sanitizers
Worker	Health and hygiene, especially in areas where field and harvest crews move from state to state or between countries while receiving multiple food safety trainings in different languages that in many cases cause fatigue and confusion between growing and harvesting practices
Packinghouse	Harvest equipment and containers
	Packinghouse equipment, sanitation, and pest control
	Proper use of sanitizers and understanding the mechanism and best practices to reduce cross-contamination within sanitation and washing systems
Climate change and globalization	Population growth and demographics, especially in countries with aging populations and with immune-compromised conditions
	Globalization, which allows continuing movement of individuals carrying multiple microorganisms across the world
	Changing eating habits (consumption of raw or lightly cooked food)
	Climate change, water loss, and continuing loss of wildlife habitat converted into farmland that increases the impact of pest control able to support large-scale farming practices

from 1996 to 2010, produce was related to approximately 131 reported outbreaks, causing 1,382 hospitalizations and 34 deaths. In 2015, approximately 37 outbreaks resulting in 1,800 illnesses were associated with fresh produce. The most outbreak-associated illness was from seeded vegetables (e.g., cucumbers and tomatoes; 1,121 illnesses) and vegetable row crops (e.g., leafy vegetables; 383 illnesses) (3). Leafy vegetables were responsible for 22% of all foodborne illness in the United States and were the second most frequent cause of hospitalizations (14%) between 1998 and 2008 (15). In general, the fresh produce most frequently associated with foodborne illnesses included leafy greens, cantaloupes, tomatoes, green onions, and herbs, accounting for more than 80% of all produce-associated outbreaks (16, 17). Despite the significant number of illnesses detected each year, a larger and significant number of foodborne diseases go unreported because of the process of reporting and identifying the disease, where health care providers need to test stool samples for the right pathogen and then report the infection to local health departments. This process accounts for about 20,000 cases specific to food poisoning that are properly diagnosed, while the majority of cases remain unreported or undiagnosed (2, 3, 15, 18).

Foodborne pathogens commonly implicated in the majority of the outbreaks include Shiga toxin-producing *Escherichia coli*, *Salmonella*, and *L. monocytogenes*. Preharvest contamination of produce with foodborne pathogens has been a major food safety issue. Potential sources for foodborne pathogens are fecal materials and soil (19–21), irrigation water (20, 22, 23), improperly composted manure (20–22), air, wild and domestic animals, birds, rodents and flies, equipment, and human handling (24, 25). Several factors such as seasonal and environmental conditions, temperature, and type of produce may influence the prevalence of pathogens at preharvest stages (26, 27). Since fresh produce is consumed raw, understanding the potential sources of pathogens during growing, harvesting, and packing is vital to develop risk management strategies.

FSMA was developed following the scientific evidence available as of 2011, when it was signed into law. Within these regulations, guidelines specific to the microbial quality of water; sampling size and testing methods; the prevalence and persistence of pathogens in soil; management practices and applications of raw, aged, and composted manure; soil remediation; and sanitation practices may not provide sufficient hurdles to prevent contamination. Consequently, we continue to need the development of funding modules for researchers that can perform multidisciplinary research on industry practices to provide pathogen reduction solutions at the farm level.

PREHARVEST FOOD SAFETY: POTENTIAL ROUTES OF MICROBIAL CONTAMINATION

The diversity and scale of farming operations at local, regional, and international locations coupled with various environmental factors and plant, irrigation water, and soil characteristics provide a significant obstacle to listing, describing, and evaluating the complexity of interactions and their impacts on enteric human pathogen risk assessment and management in fresh produce. Such compounding conditions also impact the implementation, development, and adequate assessment of government regulations that look to standardize food safety guidelines across the globe.

It is well recognized that fresh produce-related foodborne illnesses in the United States are associated with microbial pathogens of animal origin, while imported produce tends to be associated with human contamination (28). These potential differences present two questions. The first is, What is the impact of growing practices on actual contamination and transmission of human pathogens to produce on imported food? The second is, Irrespective of whether produce is imported or produced domestically, if both have similar growing and packaging processes; can we infer that pathogen contamination is mainly associated with growing conditions and length in transport instead of human intervention? In both systems, contamination could be occurring from growing or packaging practices and only becoming significant to consumers once extended transport and length of storage, mixed with consumer manipulation, provide the necessary conditions to cause human disease. Based on these potential outcomes, what we could be missing is a clear understanding of how human pathogens, once in produce (field contamination), interact with packing environments, temperature, and phyllosphere or rhizosphere communities and which of these factors masks and/or prepares enteric pathogens to persist during storage and transport.

The presence of a variety of nutrient-rich exudates in the phyllosphere, primarily composed of sugars and organic and amino acids, are known to impact well-established microbial communities and potentially human pathogens. It has been suggested that microbial populations in plants could be modified by directly changing the nutrient availability on the leaf surface. For epiphytic and plant pathogenic bacteria, it is recognized that changes in leaf nutritional composition, particularly the abundance of carbon sources (sugars), coupled with alterations in leaf morphology, significantly influence the capacity of bacteria to colonize leaves and determine their population size and spatial distribution. Fertility management, particularly nitrogen levels and forms, influences the abundance of these sugars and the size and shape of intercellular spacing, water congestion, and subsequent bacterial colonization, including human pathogens (73). For epiphytes and human pathogens to gain access to these nutrient-rich islands after harvesting, handling, and packing, the proximity and time of exposure to these nutrient oases, in combination with moisture content, impact the balance toward increased persistence. Current scientific evidence has identified different sources of contamination; however, the specific influence and interaction among production environments and crop management practices continues to be misunderstood and subject to scientific debate.

One of the least-studied issues concerning fresh produce safety is related to stress conditions (sanitizers, desiccation and low temperatures, frost protection, high UV index, and long-distance transport), which injures cells and triggers the viable but nonculturable (VBNC) state, and human pathogen fitness. Over 35 human pathogens are known to enter the VBNC state under stress conditions in produce (29), but the impact on human pathogen outbreaks is unknown. Dinu and Bach (30) identified different cells of *E. coli* O157:H7 induced into the VBNC state by exposing them to low temperature on the surface of lettuce and spinach plants. Masmoudi et al. (31), Dinu and Bach (30), Gião and Keevil (32), and Liu et al. (33) found similar conditions that would trigger the VBNC state in *L. monocytogenes* and *Staphylococcus aureus*, especially under low temperatures, suggesting that similar environmental conditions, including frost protection events, dry and wet conditions, transport, and storage, could also trigger VBNC under open field environments.

There is no clear evidence to prove that resuscitation of foodborne pathogens is directly associated with human diseases, but many pathogens retain virulence in the VBNC state (34, 35), and some may have been implicated in foodborne illnesses, as suggested by Makino et al. (36) with *E. coli* O157:H7 and *Salmonella* and with *E. coli* O104:H4. Infectious doses for pathogens such as *E. coli* O157:H7, *Salmonella*, and *L. monocytogenes* range in the 10 to 1,000 cell threshold; however, it is not known if there is a similar threshold for VBNC cells from these pathogens and whether they can directly cause foodborne outbreaks. Further research is needed to understand the pathogenicity of VBNC cells, their infectious dose, and their interaction with farming practices, including their survival in water, soil, biological soil amendments, and packinghouse environments.

Water

Food security and safety are intertwined with water, one of our most valuable natural resources. Our ability to provide reliable access to sufficient, affordable, safe, and good-quality food is determined mainly by land and water availability (37). Water, either from rain or commercial methods for sourcing, using, and applying this natural resource, is essential for food security at a global scale. Water withdrawal and consumption is dominated by agriculture, which controls between 70 and 90% of its use (38). Key components of our ability to use water in agriculture include source, volume, and microbial quality to prevent the contamination of food with human pathogens. In the United States, the current standards for agricultural water established by the FSMA Produce Safety Rule identify generic *E. coli* as the indicator of choice to predict water contamination with any human pathogen. Current standards for generic *E. coli* establish the following numerical criterion: geometric mean of ≤126 CFU/100 ml of water and a statistical threshold value of ≤410 CFU/100 ml of water. Both of these values need to be met irrespective of the water source used for irrigation if this water is to be applied to the harvestable portion of a crop (39). These standards were established following the Environmental Protection Agency's recreational water criteria, which take into consideration epidemiological studies and current and past scientific evidence describing how different individuals become ill by swallowing recreational water that is contaminated with feces and by direct human body contact with this water (18, 39). Despite these important efforts to standardize the microbial quality of water, it is clear that these standards were created based on recreational water parameters, with little information on how this microbiological criterion affects pathogen concentrations, survival, and persistence on produce and what impact it may have on consumer safety. National and international farmers will need to comply with these standards in the near future, but if this is not immediately possible, the FDA established three alternative provisions that growers can follow while they address the nature of contamination and develop corrective actions as soon as is practicable, but no later than the following year (39). Farmers using water that does not meet these microbiological criteria can (i) allow a maximum 4-day interval between the last irrigation and the harvest of the crop to allow potentially dangerous microorganisms to die off in the field, (ii) allow time between harvest and the end of storage for potentially dangerous microorganisms to die off or to be removed during commercial washing activities within appropriate limits, or (iii) treat the water. Most of these alternative provisions have not been tested by the scientific community, and it is unknown whether they are safe practices for growers and consumers.

When comparing the FSMA microbial water quality standards to WHO or Codex Alimentarius guidelines, there are major differences in the approach to defining "low-risk irrigation water." However, they all seem to agree that the best practice is to evaluate the risk of contamination before using irrigation water that will be in direct contact with fresh fruits and vegetables. Since 1989, the WHO approach has used fecal coliforms and intestinal nematodes as the indicators of choice, with numerical standards following a geometric mean of ≤1,000 CFU/100 ml of water and ≤1 CFU/1,000 ml of water, respectively. WHO revised these standards in 2015, and their current understanding and guidelines no longer define a specific microbiological criterion. Instead, they point to the use of a risk analysis approach specific to each country to determine the actual risk of contamination of water used for irrigation in primary production (37, 40). In many regions of the world, this approach is unknown, and the microbiological criterion varies by country and in some instances are unknown.

The Codex Alimentarius has addressed the issue of microbial water quality in primary production in a different manner, providing only general guidelines to risk assessment instead of providing a specific numerical value. In these standards, the only defined numerical value refers to the use of potable/clean water when it will be in direct contact with the edible portion of fruits and vegetables (41, 42), regardless of source and method of application. Irrespective of the numerical criteria used to assess the microbial quality of irrigation water, correlations between the presence of generic *E. coli* at those numerical standards and the presence of other human pathogens in the water column or in water sediments coming from wells, rivers, or ponds used for irrigating produce is not significant (43–46) and fluctuates based on time of day, depth, location, rain, water flow, die-off rates of *E. coli*, and the overall diurnal distribution of this microorganism in the water column and sediments (47).

Die-off rates of generic *E. coli* and other enteric microorganisms have been shown to have a biphasic nature. The exponential nature of the reduction of these microorganisms under different environmental conditions suggests that relying on microbial die-off rates alone as a mitigation practice to reduce pathogen contamination could be problematic. Further, sample size, test method, location within the water reservoir impacting nutrient content in the sample (48), and growing season compound even more the potential determination and description of inactivation of generic *E. coli* and other enteric pathogens (12). These factors are known to impact how different microorganisms enter the VBNC state on different matrixes and whether their survival and virulence are enhanced or maintained under this cellular condition (29, 32).

The microbial quality of water sources (wells, rivers, and ponds) has also been shown to vary significantly, mainly from environmental inputs and intrusions, especially in rivers and ponds due to the proximity to wildlife, other animal inputs, runoff, and adjacent farming activities (49, 50). It is well established that surface waters (39, 51, 52) pose the biggest challenge for fruit and vegetable production because of numerous routes of contamination that producers cannot control. Rainfall events tend to increase or reduce the concentration of indicator and pathogenic microorganisms in the water column due to water runoff, sediment disturbance, dilution, and other inputs such as wastewater influents (9, 53, 54). Which factors increase or decrease concentrations seems to vary based on region and water source, and no clear associations can be used to describe overall patterns across regions and irrigation practices. In the case of well water, the main issues associated with enteric pathogen contamination are attributed to soil type, height of the water column, proximity to animal operations, and the design and construction of the well. Deeper wells tend to be less contaminated than shallow wells due to greater microbial infiltration through the soil profile (9).

Frost protection

In the southeastern United States and other global growing regions, the use of frost protection to reduce crop losses during freezing temperatures is a common practice (55). This treatment uses massive quantities of water, in some instances, for days or weeks depending on the weather pattern, with water demands in the rate of 60 gallons/acre/minute or more, depending on the size of the area that needs to be protected. To maintain such levels of water requirements, growers depend on surface water sources to protect their crops. Frost events can occur during blossom, fruit development, or close to harvest, and this method is typically used for strawberries, blueberries, apples, cherries, and peaches, among other crops (56). To date, there seems to be little to no scientific literature describing the impacts of these events on the overall safety of these crops. For example, it is not known whether enteric pathogens are present in those water sources during these cold weather events, and, if present, whether they are able to persist during single or multiple frost events. If present, do these enteric pathogens remain viable until harvest, and if they do, could they transfer to mechanical harvesting equipment, human hands (harvest crew), harvesting and picking baskets, or the packinghouse? These questions are of special importance with certain fruits such as fresh strawberries and blueberries, where further downstream, no commercial washing practices are available that could reduce the presence of enteric pathogens.

It is known that most of the human pathogen outbreaks associated with berries from 1983 to 2013 were linked to frozen rather than fresh berries. The majority of outbreaks have an unknown source of contamination; however, hands and field and processing practices have been suggested as the most common routes of contamination. Typically, consumers of fresh fruits and vegetables, including berries, gain access in supermarkets and other retail establishments to U.S. no. 1 produce, while any other grades tend to be used for processing and frozen product manufacturing. This does not mean that the safety and quality of the product is compromised, but further handling, selection, and in some instances, fruit washing (mainly for blueberries and strawberries) is performed before freezing. It is not known whether these outbreaks related to berries can be associated with field or processing contamination and whether frost protection was used or impacted the safety of the frozen product.

Fresh fruits, including peaches, apples and cherries, also receive further washing during postharvest handling and storage before they are sent to market. Under such conditions, deficiencies in packinghouse sanitation and equipment conditions could provide reservoirs and sources of enteric pathogen contamination, as seen in the 2014 caramel apple or stone fruit Listeriosis outbreaks (57, 58). To date, the origin of the initial contamination event in those two outbreaks is unknown, and further risk assessment and epidemiological studies are needed to identify those sources of contamination.

Overhead cooling and crop protection sprays

Water sources used for field overhead cooling and crop protection sprays mainly determine the potential route and source of contamination of fresh fruits and vegetables with enteric pathogens. In general, overhead cooling is used in orchards to reduce sunburn of exposed fruit and/or to enhance color development on certain fruits (59, 60), while crop sprays encompass applications of pesticide, growth regulators, bioactive natural compounds, or other chemicals that are used to modulate crop diseases and growth. Regardless of the final goal of the application method, the microbial quality of water and the developmental stage of the crop impact enteric pathogen contamination. As discussed previously, well and surface water sources pose different risk levels based on multiple environmental and anthropogenic inputs. Current harmonized good agricultural practices and the Produce Safety Rule have different microbial water quality standards when water will be in direct contact with the edible portion of the crop. In the case of harmonized good agricultural practices, this water, if used for cooling or crop sprays close to harvest, must be "microbial safe water," which for growers and auditors tends to be represented by the use of potable water in these applications. However, the Produce Safety Rule provides a different microbiological water standard for water that will be used under the same conditions. This discrepancy is important since it has been shown that fungicides and other pesticides have little to no negative impact on the survival and persistence of human pathogens in these solutions (61, 62). Further, there is little information on the persistence of enteric pathogens in fruits or vegetables if present in and applied through pesticide solutions. Recent work by Lopez-Velasco et al. (63) showed that *Salmonella* applied to field-grown tomatoes through pesticide application was able to survive up to 15 days postinoculation and that 80% of the samples remained contaminated with this human pathogen even after postharvest washing with a sodium hypochlorite solution. These findings are remarkable when looking at the very low levels of *Salmonella* (log 2 CFU per fruit) inoculated onto the surface of tomatoes and the low levels of generic *E. coli* present in the system, suggesting that this enteric pathogen may be using these pesticides as nutrient sources, allowing it to potentially remain viable even under UV stress and desiccation. These results highlight two very important aspects: (i) our lack of understanding of and the predictability between the presence of generic *E. coli* and *Salmonella* and (ii) the different risk patterns that could be associated with water source, time of application, and physiological stage of the crop. It is unknown whether these two conditions impact the survival of enteric pathogens at different physiological stages of multiple crops and whether the VBNC state in crop protection sprays, pesticide applications, or under field conditions is impacting enteric pathogen survival and persistence.

Soils, Manure, and Biological Soil Amendments

When describing the presence, survival, persistence, and proliferation of human pathogens in soils, manure, biological soil amendments, and a combination of these, it is important to distinguish the potential contribution of each source to the overall level of contamination within each matrix. In general, soil is not considered to be a significant source of enteric pathogens for fresh produce (64), and contamination is generally associated with the presence of manure, treated or untreated biological soil amendments of animal and potentially plant origin, contaminated water, wastewater, land application of biosolids, and other anthropogenic activities (65).

Within soil, physicochemical properties including soil texture, pH, organic matter content, cation exchange capacity, porosity, and organic and inorganic nutrient sources impact the microbial ecology of all soilborne bacteria, fungi, and enteric pathogens (66, 67). In general, it has been described that soil types and lab or field experimental scales impact soil microbial population dynamics, with contradicting results between the influence of sandy soil, loam, and clay loam on pathogen survival (13, 68–72, 178). Soil moisture, pH, and organic matter content have been suggested to be more predictive of microbial community composition and enteric pathogen survival in soil (71, 73–76).

Soil microbial community interactions with enteric pathogens could be a meaningful factor in understanding the survival, persistence, and growth of these pathogens. There have been contradicting studies that either support or refute the idea that greater soil microbial diversity will negatively impact enteric pathogen survival and persistence, with little to no information on the impact on growth of enteric pathogens in soil (13, 69, 70). Among these studies, the most complete evaluation of the impact of microbial communities on the survival and persistence of *E. coli* O157:H7 was performed by van Elsas et al. (13). In this study, an inverse relationship between soil microbial diversity and survival of *E. coli* O157:H7 was established in fabricated microcosms. This was indicative of resistance to invasion due to greater soil species diversity and decreased ability to compete for resources by *E. coli* O157:H7. These results are not supported by reports from Ibekwe et al. (69) and Gagliardi and Karns (70), who used other generic and pathogenic strains of *E. coli* in their studies. Results from these experiments suggest that other microbial species influence these interactions, and further studies are needed to elucidate if nutrient acquisition, predation, or inhibition impacts enteric pathogen survival and persistence. Interactions of this nature are especially important with other human pathogens such as *L. monocytogenes*, which tends to behave like a true soil microorganism. Recently, Delgado-Baquerizo et al. (7) published the first comprehensive genome list of the 500 most common true soil microorganisms across multiple soil types and physicochemical characteristics. This information would help in the development of microcosm, mesocosm, and potentially, field studies in which the interactions of these 500 soil phylotypes with enteric pathogens could be compared to determine the impact on nutrient acquisition, predation, or inhibition and the survival, persistence, and growth of enteric pathogens in soils.

The contributions of soil and air to the spread and dissemination of foodborne infectious diseases may be considered low to minimal (77). Key to this statement is our ability to determine which human pathogens or foodborne pathogens are true soil microorganisms and which are periodic, transient, or incidental soil colonizers. True soil microorganisms capable of infecting humans through food consumption are microbes capable of completing their entire life cycle in soil. Periodic, transient, or incidental soil colonizers are capable of surviving in soil for prolonged periods of time without completing their entire life cycle in this matrix (6, 77). *L. monocytogenes* is one of those microorganisms that is considered to have an unusually broad ecological niche and host range (78) and that has been reported to exist as a "soil resident" of decomposing organic matter (68) and has been found on multiple animal hosts (67, 79). Human outbreak investigations have demonstrated the ability of *L. monocytogenes* to survive and persist under quite a remarkable range of conditions (57) due to the presence of multiple gene products that facilitate the utilization of multiple carbon and other nutrient resources (79, 80). This environmental saprophyte is capable of remaining in soil and other environments for prolonged periods of time and even of growing in cold and other harsh environments (67). Its infectious or intercellular life cycle is typically completed inside a host cell; however, based on its known ability to grow in an array of environmental conditions, including soil, can we consider *L. monocytogenes* to be a true soil microorganism? It is crucial to understand how *L. monocytogenes* and other enteric pathogens interact with native soil microbial communities. This information must be gathered through comprehensive studies with multiple soils in microcosm, mesocosm, and open field environments to develop a broad and detailed characterization of how soil impacts the short- and long-term survival of enteric pathogens.

Environmental and management factors that typically introduce enteric pathogens to soil include the use of raw manure, wastewater, human biosolids, compost, and wild and domestic animal intrusion. Raw manure has been used for thousands of years as a soil amendment to increase or maintain the soil nitrogen and carbon balance and to increase the aggregate stability of agricultural soils. The presence and persistence of enteric pathogens in raw and composted manure has been studied extensively, and we currently have a good understanding of what types of pathogens can be found on different animal excreta (81–83). A significant number of studies also looked at the survival of these pathogens if manure is applied to different soil types and at different rates and correlated it with their survival and persistence, which seems to fluctuate between 30 to over 350 days depending on the soil type and source of manure (84–86). Based on these results, it is important to question three important aspects of the presence of enteric pathogens in soil, manure, compost, and a combination of these. First, once manure is applied to soil, results suggest that enteric pathogen survival and persistence are enhanced (86), probably because of the addition of nutrients that are important for pathogen survival. Second, it is not clear if the addition of manure to soil overrides the resistance to invasion of soil microbial communities by enteric pathogens irrespective of greater soil species diversity and decreased competitiveness for resources as suggested by van Elsas et al. 2012 (13). Finally, properly composted manure has been shown to significantly reduce the presence of enteric pathogens (85, 87–89). It is not clear if this process also impacts the transfer of antimicrobial resistance (AMR) from thermotolerant coliforms and other heat-resistant microorganisms such enterococci to human pathogens present in compost or soil. These pressing questions will significantly impact the viability and continuity of the national organic program 90-120 rule for the use of raw or aged manure incorporated into the soil, since numerous results indicate that this approach may not provide the necessary level of protection to reduce contamination of the edible portion of the crop (84–86). Currently, several national studies are looking at the impact of this rule on public health and how proper composting practices impact the continuation of this standard or the development of new guidelines for the application of raw, aged, and composted manure at specific crop intervals that will significantly reduce produce contamination. Lastly, we are currently facing significant challenges at all scales of production and within our health system, because of the wide spread of AMR microorganisms. With this in mind, we lack understanding on how different approved composting methods impact inactivation and/or transfer of AMR between organisms and to food if compost is used as a source of nutrients for plant growth. Further studies looking at this conundrum are needed to determine how antibiotic use in animals and composting processes are impacting fruit and vegetable contamination with microorganisms presenting broad resistance to different antimicrobials.

Wildlife and Domestic Animals

Contamination of fresh produce with enteric pathogens has been associated with fecal contamination coming from wild and domestic animals (4, 27, 90). Prevention of fecal contamination from wildlife has been suggested to be the most effective approach to reducing contamination of fresh produce; however, controlling wildlife intrusion on farms in some instances is not feasible, especially when dealing with birds, small mammals, and insects. The prevalence of enteric pathogens in the environment and wildlife populations plays a significant role in the levels of contamination in fresh produce, especially when buffer zones and wildlife corridors are removed from agricultural farmland (11). Gorski et al. (10) investigated the prevalence of enteric pathogens in produce environments and wildlife around the Salinas Valley, and found different levels of *Salmonella* contamination in produce, water, and wild animals during the cropping cycle, suggesting that these pathogens were persistent in the environment. Subsequently, losses in wildlife habitat coupled with the natural prevalence of enteric pathogens in wildlife, will significantly impact the level of contamination in fresh produce. The prevalence of many enteric pathogens on farmland or wildlife across the United States is unknown, and this information is needed for proper risk assessment of produce contamination and to determine if remediation practices are available and viable to reduce pathogen contamination.

Proximity to small or large confined-animal operations has also been suggested to be a potential source of enteric pathogen contamination due to aerosol contamination and water runoff when produce fields are in close proximity to these operations. Berry et al. (91) and Jahne et al. (92), discovered aerosol distribution of enteric microorganisms from large animal operations and manure/dust to fresh produce up to distances of over 150 meters, suggesting that the extent of produce contamination is greater in locations where large-scale animal production is adjacent to produce fields. No conclusive evidence of foodborne pathogen transfer or human illness is available that links the proximity of animal operations, size of animal operations, aerosols dispersal, and pathogen contamination with a known foodborne outbreak; however, results from these experiments and recent work by Kumar et al. (93) in tomatoes, suggest that aerosols are a plausible source of contamination. One key component missing from these evaluations is the scale of the animal operation and whether it impacts pathogen transmission. Many organic and diversified farming systems utilize work animals or source manure from small herds within the farm to fertilize their fields with raw, aged, or composted manure. It is unknown what the level of enteric pathogen contamination is in these small-scale animal systems and whether aerosols and proximity to these animal activities represents a significant risk of produce contamination.

Packinghouse, Equipment, and Tools

Outbreak investigations of multiple crops, including melons, watermelons, apples, stone fruit, and other fruit and vegetables (4, 57, 58, 90, 94, 95) have been linked to contaminated packinghouses or equipment used to handle and transport produce. Contamination of equipment or areas within the packinghouse has also been suggested to originate from field contamination, poor sanitation practices, traffic flow, worker health and hygiene, or the overall design of the facility (4, 5, 95). Irrespective of the source of contamination, numerous questions remain unanswered about the fitness of enteric pathogens in those environments, under stress conditions, low populations, and the VBNC stage and how sanitizers impact survival, persistence, and in some cases, resistance to these chemical substances either through biofilm formation or pathogen adaptation (96, 97). The 2014 caramel apple outbreak associated with listeriosis left a significant imprint in our current knowledge on the adaptability of *L. monocytogenes* to a wide range of environments. It became clear that this pathogen could persist in dry packinghouse environments for long periods of time under conditions previously not known to support the survival and persistence of this pathogen (57). During the investigation of this outbreak, over 110 samples were collected from multiple locations, and seven samples were positive for *L. monocytogenes*, six of them coming from food contact surfaces including polishing brushes, drying brushes, a conveyor, and inside a wooden bin. In general, all these locations are difficult to clean and sanitize because of poor design of facilities and equipment, highlighting gaps in our understanding of microbiological risks associated with tree fruit production, especially activities related to maintaining, cleaning, and sanitizing of food contact surfaces and how biofilms may persist for prolonged periods of time under suboptimal conditions (dry and low nutrient availability). Similar observations have been identified for stone fruit (58) and melon packinghouses (95) and have triggered significant changes to industry practices.

Traffic flow within packinghouses also plays a significant role in spreading filth and enteric pathogens in different areas across the facility (5). Typical vehicles for distribution of these contaminants include pallet jacks, pallets, and worker and visitor traffic. One peculiar practice in many facilities is to move visitors from dirty to clean areas when describing the different activities that take place in the facility. This simple and yet influential practice could transfer contamination from receiving areas to finished product (clean) areas if no further steps in the process are available to reduce new contamination events. In a recent study performed in multiple food facilities in Europe by Muhterem-Uyar et al. (98), researchers found three major sources/areas of contamination. These were reception of raw materials, processing areas, and widespread contamination of different microorganisms across the processing facilities. Although these studies were performed on processing facilities instead of fresh produce packinghouses, results also point out traffic flow as one of the major factors impacting the distribution of contamination.

Similar studies on fresh produce packinghouses or fresh-cut conditions are limited. Erickson (8) followed outbreak investigations of cabbage, carrots, celery, onions, and deli salads and found that at these facilities and irrespective of geographical region, transfer of contamination could also be associated with traffic flow. Consequently, further studies looking at the transfer of contaminants across packinghouse environments due to traffic flow are necessary, especially where linear flows within the facility are not possible and where movement of pallet jacks, pallets, and other equipment is not controlled. It is also unknown how these conditions impact the survival and persistence of enteric pathogens in low concentrations (<100 cells) and whether different food-grade lubricants facilitate pathogen persistence on equipment and tools.

Sanitation of packinghouse equipment and tools continues to be a major source of misunderstanding and confusion within different scales of fruit and vegetable production. At the core of the problem lie four major factors: (i) facility and equipment design, (ii) lack of understanding of the proper use of chemical sanitizers and of the development of sound sanitation practices, (iii) new-buyer guidelines that in some instances push for environmental monitoring plans that are not required under the FSMA Produce Safety Rule, and (iv) a lack of technical background in the development, implementation, and assessment of environmental monitoring plans. A good example that highlights some of these deficiencies was observed in a multistate outbreak associated with *Salmonella enterica* serotype Newport linked to mangoes (94). During the investigation of this outbreak, it was determined that a lack of sanitation practices, coupled with the use of hot water treatments without sanitizers and water coming from canals from a river 26 km away from the packinghouse were the major factors linked to the source of contamination. Since this outbreak, the U.S. mango industry has worked on updating and developing new cleaning and sanitation practices and updating equipment and other tools that could reduce or eliminate similar problems (99).

The proper use of sanitizers, including adjustments of pH, temperature, and turbidity when appropriate; monitoring tools; and intervals that can guarantee the effectiveness of the system are scarce in the production of many fruits and vegetables, except for fresh tomato flume systems (19, 63, 100). Numerous outbreaks associated with enteric pathogens (mainly *Salmonella*) in tomatoes have generated the need to establish mitigation practices that could reduce pathogen contamination (4, 101). One of these mitigation practices targets packinghouse sanitation and the correct use of a "fruit disinfection treatments", since postharvest washing was once considered a potential decontamination step in the process instead of an activity to reduce cross contamination between fruit loads and is now clearly identified as a high-risk practice when performed incorrectly. In general, tomato flume systems maintain adequate chlorine levels in the dump tank to prevent pathogen survival, transfer, and internalization of the fruit. However, continuous fluctuations in chlorine concentrations are observed due to rapid accumulation of organic matter that reacts with free chlorine, causing a significant decline in sanitizer concentrations and leaving washing systems vulnerable to pathogen survival and persistence. For these reasons, maintaining chlorine concentrations and its effectiveness relies on maintaining the water and tomato pulp temperatures within 5.5°C and a contact time of no longer than 2 minutes (100). These parameters depend on numerous factors, including the load of debris on the fruit, the quantity of fruit dumped in the tanks, the speed with which the fruit is moved inside the tanks, and the ability of the system and operators to react to constant changes (100). Similar factors can also determine the effectiveness of apple, cherry, pear, melon, watermelon, mango, papaya, and other dump tank systems and highlights the inherit difficulty that many growers are currently facing with their washing systems. To date, several laboratory methods and parameters have been reported that could be used for the sanitation of packinghouse and crop disinfection treatment systems, but further research is needed to fully understand how chlorine or other sanitizers effectively reduce human pathogen contamination, especially when penetration of protective sites on fruit and vegetable surfaces is limited by water tension properties of the sanitizer solution.

PRODUCTION PRACTICES IMPACTING FRESH PRODUCE SAFETY

The wide diversity and size of farm management systems looking for ways to simplify certain farm practices to increase yields and lower costs through fertilizer and pesticide applications, combined with the use and design of equipment, environmental conditions, and federal, state, buyer, and consumer demands within regional and local markets impacts the safety of fresh produce. Farm management systems vary with the farming practices (organic, conventional sustainable, diversified) employed during the growing season, and the microbial risks associated with each system and its crops are linked to grower, packer, shipper, and handler practices in different geographical locations and personal business philosophies. Nonetheless, fruit and vegetable farming systems and their impact on fresh produce safety can be separated based on farming practices.

Conventional versus Organic Farming Systems

At the core of these two systems lies the fundamental difference in the use of synthetic fertilizers and pesticides to promote plant growth and control the incidence of pests. This simple and yet fundamental difference between farming activities mainly impacts farming equipment, seed sources, resistance to specific pests, crop diversity, postharvest disinfection, and some farm-related activities including the timing, method, and frequency of the application of different materials and substances used for pest control and crop fertilization. Conventional farming systems tend to be monocultures since the use of fertilizers and chemical pesticides allows for proper nutrient and pest management. However, organic systems cannot rely on these inputs and follow diversified farming principles that allow them to better control pests and potentially manage nitrogen if nitrogen-fixing cover crops are used within the distribution of crops. Aside from these variances, any other inputs, including water sources, harvest crews, transportation, handling, and storage conditions, of conventional and organic produce tend to be almost identical. To date, very few studies have compared the two production systems in terms of their microbial characteristics and the safety of fresh produce (102, 103).

Hoogenboom et al. (104) evaluated the level of contaminants and microorganisms in Dutch organic and conventional food products and found that they had similar scores with regard to food safety; however, the levels of antibiotic-resistant microorganisms were lower in organic systems. Human pathogen outbreaks have been linked to more conventional than organic produce, but these differences could be explained by the sheer volume of conventional produce sold each year compared to organic produce. Further, results from Williams and Hammit (105), Magkos et al. (106), and Hoogenboom et al. (104) support the idea that there seem to be no significant differences in the incidence of enteric pathogens between organic and conventional systems. Looking at the main differences between farming systems, two practices can significantly influence the presence, survival, persistence, and growth of enteric pathogens and impact the risk of contamination of fresh produce in organic systems: (i) the use of raw, aged, and composted manure and the time and quantity of these amendments applied either to the soil or through the irrigation systems (as in the case of compost teas, fish emulsion, and other substances) and (ii) the use of acetic acid (organic origin), alcohol (organic origin), chlorine, hydrogen peroxide, and peroxyacetic acid in organic packinghouse environments, which have been registered for organic packinghouse surface sanitation, while only ozone, chlorine, and peroxyacetic acid have been registered for fruit and vegetable disinfection (107). How these sanitizers impact the survival and persistence of enteric pathogens and the microbial communities within these systems and whether these parameters are different in conventional packinghouse systems is unknown. Essential oils from thyme, lemongrass, eucalyptus, rosemary, oregano, and sweet basil have also been used to sanitize food contact surfaces and for fresh produce disinfection. However, their use is limited due to transfer of flavor to produce or food contact surfaces, since effective antimicrobial doses exceed acceptable organoleptic levels in produce (108).

Sustainable Farming Systems

Sustainable farming, at its core, strives to combine and utilize the best practices from organic and conventional systems that will allow the rational use of farming inputs while maintaining yields and reducing chemical inputs into the environment. This type of farming system could also utilize raw, aged, and composted manure and farm animals to reduce carbon footprints within the system. To date, there is no clear understanding of whether these farming systems pose equal, lower, or higher risk factors for enteric pathogenic contamination and how fresh produce safety is impacted. Similar to organic systems, the main risk associated with some sustainable farming practices is linked to the proximity of animals and the use of raw or aged manure to fertilize crops. Concerns similar to those for organic practices exist specific to the presence, survival, and growth of enteric pathogens and the impact of small animal herds on the transfer of these pathogens through aerosols or runoff. Leff and Fierer's (102) bacterial community analysis of organic and conventional produce yielded significant differences in the relative abundance of the *Enterobacteriaceae* taxa in organic and conventional produce, with the latter presenting a higher presence of this taxa than organic produce. These differences could be attributed to produce source since most foods were purchased in retail establishments, where further handling and processing occurs. Despite this potentially important difference for food safety, overall, based on their results, consumers are exposed to a substantially different bacterial community when consuming organic produce. Similar observations were made by Maffei et al. (103) when looking at published data describing microbial composition and contamination of organic and conventional produce from different countries. Whether these differences are important for risk assessment of foodborne illnesses and transfer of AMR has yet to be defined, and further efforts are needed to understand these potential similarities.

Greenhouses

Fruit and vegetable greenhouse production exceeded $3 billion in sales in 2013, and it is projected to continue growing and surpass $4 billion in the next 3 to 5 years. The biggest incremental growth in production (2-fold) has been associated with leafy greens and culinary herbs, while fruit production has almost tripled since 2009 (14). Leafy green, culinary herb, and fruit greenhouse production systems utilize different water sources and hydroponic systems; however, in most of them, nutrient solution recirculation is a common practice that minimizes water use, improves nitrogen use, and in some instances reduces pesticide applications (109). Recirculation of the nutrient solution could potentially become a food safety hazard, depending on the microbial quality of the water used for irrigation, human contamination, contaminated substrates used for seed germination, and plant growth (110). Few greenhouse operations currently monitor the microbial quality of water associated with the contamination of human pathogens, and most efforts center on controlling plant pathogens. To date, only one clear outbreak associated with hydroponically grown cucumbers has been investigated by the CDC, and the investigation indicated poor sanitation and agricultural practices as potential sources of contamination. The report also looked into whether cross-contamination could have occurred during handling, distribution, and transport of the fruit; however, this association was not established (2).

Numerous studies have also looked at the potential of fruit or leaf internalization of enteric pathogens through the root system in different crop physiological stages. Although some evidence has been reported, most of the studies (except for tomato [111]) support that little to no known internalization and translocation of enteric pathogens from the root system to leaves or other organs has been established (73, 112–115). Consequently, in hydroponic systems, special attention to reduce or eliminate enteric pathogen contamination should focus on preventing these pathogens from entering greenhouse operations through sick workers, poor sanitation practices, contaminated substrates, and the use of surface or rain water sources that could carry enteric pathogens into the system.

SUSTAINABLE PRACTICES TO MITIGATE FOOD SAFETY RISKS ON PRODUCE FARMS

To date, there is a clear absence and need for the development of sustainable mitigation practices that can reduce or eliminate enteric pathogen contamination from farmland within a relatively short timeframe for the long-term sustainability and stability of the produce industry. Mitigation techniques that follow sound and sustainable farming practices are potential alternatives to chemical-based remediation methods that could eventually reduce soil health and crop production (116, 117). Short-term cover cropping, compost application, and solarization have been proposed as potential alternatives to reduce pathogen and weed contamination from soils while maintaining soil health and crop productivity (118–121). The key to the use and implementation of these techniques is cost, effectiveness, and the ability to implement them across a wide range of cropping systems.

Efforts toward the development of practices that reduce contamination from multiple sources have focused on reducing runoff from confined-animal feedlot operations (CAFOs) and on preventing intrusion from wildlife and domestic animals into production areas (5, 122, 123). Runoff water from feedlots and pastures can be a major source of microbial contamination of surface water and produce farms (5). Potential environmental risk factors associated with the 2007 spinach outbreak were traced back to several factors, including previous land history, the presence of wild pigs, and irrigation wells near surface waterways exposed to feces from cattle and wildlife. The outbreak strain *E. coli* O157:H7 was identified in river water, cattle feces, and wild pig feces on the ranch and within a mile of the implicated field (90).

Comanagement of food safety practices that will reduce enteric pathogen contamination of produce with environmental policies, the economy, and consumer expectations remains a significant challenge across the farm to fork continuum. One of the biggest challenges in this process relates to the limited number of studies looking at this topic at different levels in the food system and the difficulty of extrapolating specific or local results to multiple regions and cropping systems. Here, we present a number of sustainable practices that could be implemented within comanagement of farm food safety practices and that have been suggested to reduce the microbial food safety hazard at the farm level.

Buffer Zones

There are a number of different definitions of buffer zones, but in general terms and based on National Organic Program regulations, a buffer zone is "an area located between a certified production operation or portion of a production operation and an adjacent land area that is not maintained under organic management" (124). It also refers to a strip of land adjacent to production areas or bodies of water. In general, its purpose is to protect farmland from prohibited substances in the case of organic farming systems or to protect waterbodies from pollutants, excess sediments, and microbial contaminants that may negatively impact water quality. Buffer zones also serve as a source of food, nesting cover, and shelter for many wildlife species and can serve as wildlife corridors that could minimize wildlife intrusion into production areas (125, 126). Buffer zones may be natural land where existing vegetation is intact, or they can be developed by altering the slope of the land and adding vegetation and trees. Vegetable production areas that are in close proximity to CAFOs are at higher risk of microbial contamination directly or indirectly from animals, runoff water, or other vectors associated with these types of operations. Dillaha et al. (127), suggested that prevention of produce field contamination from runoff from CAFOs could be achieved by constructing diversion ditches and establishing vegetation barriers such as riparian buffers, filter strips, and grassed waterways.

Several studies have looked at the impact of buffer zones on reducing or augmenting the transmission of human pathogens from CAFOs, animal grazing lands, and large bodies of water (128–131). These studies suggest that buffer zones tend to have a protective effect against the transmission of human pathogens to produce fields. Recently, Karp et al. (11) looked in the Salinas Valley at the impact of removing buffer zones from adjacent produce farmland on the incidence of human pathogens in the finished product. The study found that pathogen prevalence increased the most on farms where noncrop vegetation was removed. These results suggest the importance of potentially keeping or expanding these zones, especially in growing areas in which intense wildlife pressure or proximity to large-scale animal operations poses a significant risk of contamination. It is worth noting that these buffer zones, when located between CAFOs and farmland, tend to accumulate dust and other particles coming from the animal units, and it is not known whether under those conditions buffer zones could become sinks and sources of contamination that could spread enteric pathogens to adjacent farmland.

Vegetative Filter Strips (Buffer Strips)

Vegetative filter strips (VFSs) are another type of buffer zone and refers to lands with a variety of grasses and forage species located downslope from the potential source of contaminants and upslope from the area being protected (132–134). The main purpose of installing VFSs is to protect surface water quality by reducing the level of physical, chemical, and microbial pollutants in agricultural runoff (135). Several studies reported that VFSs can effectively reduce the level of coliforms, *E. coli* (136, 137), nitrogen, and phosphorus (138) in waterways that comes from contaminated runoff from the source of pollutants (181).

VFSs are installed by planting a band of vegetation of a single species or a mixture of grasses, legumes, and/or other forbs, also known as herbaceous flowering plants not including graminoids (127, 135). Once a band of vegetation is installed, it works naturally. The vegetation interrupts overland flow and reduces surface runoff velocity. The reduced flow allows the surface runoff to sit on the soil for a longer period of time, increasing the rate of deposition of sediments, nutrients, and human pathogens. It also increases the rate of infiltration into the soil and adsorption of sediments and some nutrients on leaves and stems. The plant roots and soil microorganisms degrade nutrients and chemical pollutants (139). All these mechanisms collectively help reduce the level of surface water pollution through contaminated surface run-off.

Water flowing through the VFS decreases in velocity, and this consequently decreases the sediment-carrying capacity of the runoff as particles settle (140). The nutrients and microorganisms which are attached to sediment particles are retained in the VFS (127). Conflicting results have been reported on the effectiveness of VFSs in minimizing microbial runoff. Some studies indicated that VFSs are effective in trapping nutrients/solid particles and reducing microbial runoff (91% for fecal coliform bacteria and 74% for fecal streptococci) from pasture or feedlot areas (136, 141–143), while a few others have found limited beneficial effects in reducing fecal contamination in swine wastewater moving through vegetative areas regardless of vegetation type or season (137, 140, 144–146).

Variable results have been reported on the effectiveness of VFSs in removing the pollutants in feedlot runoff manure-applied pasture and cropland runoff (140). Factors impacting the efficiency of these systems include the amount and type of incoming pollutant, slope, runoff volume, type of flow (concentrated or diffusive), and type of vegetation (147, 148). Table 2 describes some of the major pollutants and important parameters within VFSs that improve filtration. There should be an adequate level of plant density for the efficient retention of pollu-

TABLE 2 Examples of VFSs evaluated for eliminating the pollutants

Vegetation	Major pollutants	Source	VFS parameters	References
Kikuyu grass (*Pennisetum clandestinum* Chiov) and Napier grass (*Pennisetum purpureum* Schumach)	*E. coli*	Cattle manure (cowpat)	VFS slope: 15% VFS length: 44 m VFS width: 4 m	175
Fescue (*Festuca*)	Fecal coliforms, *Streptococci*, *E. coli*, nitrogen, phosphorus	Feedlot	VFS slope: 2% VFS length: 30 m VFS width: 15 m Soil type: Newtonia silt loam soil	176
Mixed grass buffer strip	Phosphorus, ammonium, nitrogen, potassium	Feedlot	VFS slope: 2% VFS length: 12 m	140
Blue fescue (*Festuca ovina* L. 'Glauca') and white clover (*Trifolium repens* L.) mixture	Fecal coliforms	Manure	VFS slope: 20% VFS length: 6 m VFS width: 2 m Soil type: sandy loam or clay loam	177

tants. Infiltration is one of the main mechanisms for removing soluble pollutants. However, an increase in the duration of flow may result in decreases in the rate of removal of pollutants and could explain why in some evaluations, the effectiveness of reducing pathogen contamination from waterways was not observed. Overall, these VFSs have shown significant potential for pathogen removal. Questions remain as to their effectiveness over time, especially with sediment accumulation due to weather events or continued water movement from different farm areas. It has also been suggested that over time, these areas could become sinks and sources of contamination for different pollutants, but their impact on human pathogen accumulation has not been well described.

Soil Solarization

Solarization is a convenient nonchemical, nonfumigant sustainable farming technique that utilizes daily solar heating cycles to manage weeds, nematodes, diseases, and insects in soil. Depending on the region, this process is also known as polyethylene mulching, soil trapping, solar pasteurization, and solar soil heating. Soil solarization reduces the survival and persistence of mesophilic microorganisms and plant pathogens within the first 4 to 5 cm of the soil profile by raising the soil temperature to over 40°C (120, 180). Critical to achieving these temperatures is the application of water, keeping soil moisture contents above field capacity but below saturation, and applying this technique during the summer months when longer photoperiods (>11 h), elevated temperatures (>30°C), and higher UV indexes increase the efficiency of this process (149–154).

Survival of enteric pathogens in soil for an extended period of time has been observed under multiple experimental conditions (22, 86, 155). However, survival may be affected when the organisms are exposed to adverse environmental conditions such as limited nutrients and exposure to UV light. Every organism has an optimum temperature range for growth. At temperatures above 50°C the growth of most enteric pathogens is affected, and die-off is expedited. Soil solarization practices are capable of reaching these temperatures and could potentially reduce the population of enteric pathogens and other transient soil colonizers. To achieve significant inactivation of pathogens, weeds, and other diseases, several farm practices should be implemented, including soil ripping and tilling, incorporation of organic matter including compost, which will increase aeration, conductivity and heat transfer via mechanisms of exothermic microbial activity, and thermal conductivity within the top 20 to 40 cm of the soil profile (152, 154). An effective solarization process, in addition to reducing plant pathogens, weeds, and other disease, is conducive to increased nutrient availability, plant growth, and beneficial changes in the population dynamics of soil bacteria and fungi (121, 154, 156, 157), which makes this an excellent sustainable practice to maintain soil health that potentially could reduce the presence of enteric pathogens. To that end, soil solarization has been reported to be effective in reducing generic *E. coli* in soil (119) and in feedlot pen surface material (158) after a 6-week treatment, suggesting that this practice could potentially be used to control other human pathogens, including *Salmonella* and *L. monocytogenes*.

Short-Term Cover Cropping

Enteric pathogen (*Salmonella* spp., *E. coli* O157:H7, *Shigella* spp., and *Campylobacter jejuni*) survival and persistence in soil has been reported to depend on temperature (159, 160), soil moisture content (73, 159), soil physicochemical properties (161, 162), soil C:N ratio (75), microbial community composition (13, 163), and UV light exposure (164). Prolonged survival of enteric pathogens in soils threatens the sustainability of farming because there are currently no remediation practices to remove them from contaminated soil, and the current industry best practice is to abandon these fields and move to other growing regions (117). Consequently, developing and implementing soil remediation practices that reduce or eliminate human pathogen contamination is crucial for the long-term sustainability of the produce industry.

Soil physicochemical properties, nutrient composition, and the microbial communities in the rhizosphere impact enteric pathogen survival and are impacted by short-term cover cropping. The growth of cover crops is an important agricultural management practice that is widely applied in different cropping systems to improve soil health and plant productivity (165). Numerous types of plants can be used as cover crops; however, legumes and grasses are extensively used to improve nitrogen availability and scavenge for excess soil nutrients, in particular nitrogen, and to reduce soil erosion (166). There is increasing interest in *Fabaceae* (sunn hemp), *Brassica* (rape, mustard, and forage radish), and *Polygonaceae* (buckwheat) cover crops because of their allelopathic effects on weeds and plant diseases (166).

In addition to solarization, short-term cover cropping has been proposed as an alternative mitigation practice to reduce human pathogen contamination from soil because of its ability to be implemented across a wide range of cropping systems and the known allelopathic effects due to the action of secondary plant metabolites on soil disease management (152, 153). The proximity of fresh produce to contaminated soil and enteric pathogen concentrations are two risk factors contributing to produce contamination (167). Generally, produce that is in close proximity to the soil (spinach, lettuce, root crops, melon) has a higher likelihood of contamination compared to produce that develops further above the soil surface (tomatoes, peppers, berries) (16, 17). Contaminated irrigation water and manure are thought to be two important sources of enteric pathogen contamination in fresh produce (128, 129).

Sunn hemp, buckwheat, and mustard greens are three short-term and low-residue cover crops that produce distinct secondary plant metabolites, which have been shown to exhibit bactericidal properties against human pathogens (110, 168, 169). The availability of these plant secondary metabolites to inactivate human pathogens in soil is dependent on the ability to macerate and incorporate plant tissue through disking and tilling after growing the cover crops for 30 to 60 days. Sunn hemp (*Crotalaria juncea*) produces pyrrolizidine alkaloids, which are nitrogen-containing compounds that can be toxic to animals after consumption (170). However, no reports to date have documented their inhibitory effects on enteric pathogen growth in soils or other matrixes.

Buckwheat (*Fagopyrum esculentum*) mainly produces two phenolic compounds: rutin and chlorogenic acid. Rutin in its pure form has not shown inhibitory properties against enteric pathogens (169, 171, 172, 179). Marginal inhibitory effects have been observed when it is mixed with quercetin, suggesting a synergistic inhibitory effect between the substances at a concentration of 100 μg/ml (171). Chlorogenic acid in its pure form has shown inhibitory effects against *S. aureus*, *Salmonella*, and *E. coli* at concentrations of 40 to 80 ppm (169).

Mustard green cover crops (*Brassica juncea*) produce glucosinolates, which are sulfur-containing water-soluble compounds that when hydrolyzed by the myrosinace enzyme produce isothiocyanate compounds known to have inhibitory properties against *E. coli*, *S. enterica* serotype Typhimurium, and *L. monocytogenes* in concentrations ranging from 50 to 200 ppm (117, 168, 173, 174). Allyl isothiocyanate, in a liquid form, has also been shown to inhibit growth of *E. coli*, *S.* Typhimurium, and *L. monocytogenes* in concentrations ranging from 50 to 200 ppm (110). Altogether, combining the use of cover cropping, composting, and solarization could potentially be an alternative and sustainable practice capable of remediating soil contaminated with enteric pathogens. The applicability of this method in commercial organic and conventional fresh produce farming systems is not known.

CITATION

Gutierrez-Rodriguez E, Adhikari A. 2018. Preharvest farming practices impacting fresh produce safety. Microbiol Spectrum 6(2):PFS-0022-2018.

REFERENCES

1. **Institute of Medicine (US) and National Research Council (US) Committee to Ensure Safe Food from Production to Consumption Ensuring Safe Food: From Production to Consumption.** 1998. *Food Safety from Farm to Table: a National Food-Safety Initiative. A Report to the President May 1997.* National Academies Press, Washington, DC.
2. **FDA.** 2016. FDA investigated multistate outbreak of *Salmonella* Poona linked to cucumbers. Available at: http://www.fda.gov/Food/RecallsOutbreaksEmergencies/Outbreaks/ucm461317.htm. Accessed January 2018.
3. **CDC.** 2018. Parasites-cyclosporiasis (Cyclospora infection): outbreak investigations and updates. https://www.cdc.gov/parasites/cyclosporiasis/outbreaks/index.html. Accessed 15 March 2017.
4. **Harris LJ, Farber JN, Beuchat LR, Parish ME, Suslow TV, Garrett EH, Busta FF.** 2003. Outbreaks associated with fresh produce: incidence, growth, and survival of pathogens in fresh and fresh-cut produce. *Compr Rev Food Sci Food Saf* **2**(s1):78–89.
5. **Suslow TV, Oria MP, Beuchat LR, Garrett EH, Parish ME, Harris LJ, Farber JN, Busta FF.** 2003. Production practices as risk factors in microbial food safety of fresh and fresh-cut produce. *Compr Rev Food Sci Food Saf* **2**(s1):38–77.
6. **Bultman MW, Fisher FS, Pappagianis D.** 2013. The ecology of soil-borne human pathogens, p 477–504. *In* Selinus O (ed), *Essentials of Medical Geology*, rev ed. Springer, Dordrecht, The Netherlands.
7. **Delgado-Baquerizo M, Oliverio AM, Brewer TE, Benavent-González A, Eldridge DJ, Bardgett RD, Maestre FT, Singh BK, Fierer N.** 2018. A global atlas of the dominant bacteria found in soil. *Science* **359:**320–325.
8. **Erickson MC.** 2010. Microbial risks associated with cabbage, carrots, celery, onions, and deli salads made with these produce items. *Compr Rev Food Sci Food Saf* **9:**602–619.
9. **Gerba CP.** 2009. The role of water and water testing in produce safety, p 129–142. *In* Fan X, Niemira BA, Doona CJ, Feeherty FE, Gravani RB (ed), *Microbial Safety of Fresh Produce.* Wiley, Indianapolis, IN.
10. **Gorski L, Parker CT, Liang A, Cooley MB, Jay-Russell MT, Gordus AG, Atwill ER, Mandrell RE.** 2011. Prevalence, distribution, and diversity of *Salmonella enterica* in a major produce region of California. *Appl Environ Microbiol* **77:**2734–2748.
11. **Karp DS, Gennet S, Kilonzo C, Partyka M, Chaumont N, Atwill ER, Kremen C.** 2015. Comanaging fresh produce for nature con-

servation and food safety. *Proc Natl Acad Sci U S A* **112:**11126–11131.

12. **Pachepsky Y, Shleton DR, McLain JET, Patel J, Mandrell RE.** 2013. Irrigation water as a source of pathogenic microorganisms in produce: a review. *Adv Agron* **113:**73–137.
13. **van Elsas JD, Chiurazzi M, Mallon CA, Elhottova D, Kristufek V, Salles JF.** 2012. Microbial diversity determines the invasion of soil by a bacterial pathogen. *Proc Natl Acad Sci USA* **109:**1159–1164.
14. **USDANASS.** 2014. 2012 Census of agriculture, United States, summary and state data, **vol. 1.** Geographic area series, part 51. AC-12-A-51. U. S. Department of Agriculture, Washington, DC. http://www.agcensus.usda.gov/Publications/2012/Full_Report/Volume_1,_Chapter_1_US/usv1.pdf. Accessed January 2018.
15. **Painter JA, Hoekstra RM, Ayers T, Tauxe RV, Braden CR, Angulo FJ, Griffin PM.** 2013. Attribution of foodborne illnesses, hospitalizations, and deaths to food commodities by using outbreak data, United States, 1998–2008. *Emerg Infect Dis* **19:**407–415.
16. **Doyle MP, Erickson MC.** 2008. Summer meeting 2007: the problems with fresh produce: an overview. *J Appl Microbiol* **105:**317–330.
17. **Doyle MP, Erickson MC.** 2012. Opportunities for mitigating pathogen contamination during on-farm food production. *Int J Food Microbiol* **152:**54–74.
18. **FDA.** 2017. How did FDA establish requirements for water quality and testing of irrigation water? https://www.fda.gov/downloads/Food/GuidanceRegulation/FSMA/UCM473335.pdf. Accessed 24 November 24 2017.
19. **Beuchat LR, Nail BV, Adler BB, Clavero MR.** 1998. Efficacy of spray application of chlorinated water in killing pathogenic bacteria on raw apples, tomatoes, and lettuce. *J Food Prot* **61:**1305–1311.
20. **Islam M, Doyle MP, Phatak SC, Millner P, Jiang X.** 2004. Persistence of enterohemorrhagic *Escherichia coli* O157:H7 in soil and on leaf lettuce and parsley grown in fields treated with contaminated manure composts or irrigation water. *J Food Prot* **67:**1365–1370.
21. **Natvig EE, Ingham SC, Ingham BH, Cooperband LR, Roper TR.** 2002. *Salmonella enterica* serovar Typhimurium and *Escherichia coli* contamination of root and leaf vegetables grown in soils with incorporated bovine manure. *Appl Environ Microbiol* **68:**2737–2744.
22. **Islam M, Morgan J, Doyle MP, Jiang X.** 2004. Fate of *Escherichia coli* O157:H7 in manure compost-amended soil and on carrots and onions grown in an environmentally controlled growth chamber. *J Food Prot* **67:** 574–578.
23. **Islam M, Morgan J, Doyle MP, Phatak SC, Millner P, Jiang X.** 2004. Fate of *Salmonella enterica* serovar Typhimurium on carrots and radishes grown in fields treated with contaminated manure composts or irrigation water. *Appl Environ Microbiol* **70:**2497–2502.
24. **Alam MJ, Zurek L.** 2006. Seasonal prevalence of *Escherichia coli* O157:H7 in beef cattle feces. *J Food Prot* **69:**3018–3020.
25. **Nielsen LR, Schukken YH, Gröhn YT, Ersbøll AK.** 2004. *Salmonella* Dublin infection in dairy cattle: risk factors for becoming a carrier. *Prev Vet Med* **65:**47–62.
26. **Heaton JC, Jones K.** 2008. Microbial contamination of fruit and vegetables and the behavior of enteropathogens in the phyllosphere: a review. *J Appl Microbiol* **104:**613–626.
27. **Jay-Russell MT.** 2013. What is the risk from wild animals in food-borne pathogen contamination of plants? *Perspect Agric Vet Sci Nutr Nat Resour* **8:**1–16.
28. **Herman KM, Hall AJ, Gould LH.** 2015. Outbreaks attributed to fresh leafy vegetables, United States, 1973–2012. *Epidemiol Infect* **143:** 3011–3021.
29. **Zhao X, Zhong J, Wei C, Lin CW, Ding T.** 2017. Current perspectives on viable but non-culturable state in foodborne pathogens. *Front Microbiol* **8:**580.
30. **Dinu L-D, Bach S.** 2013. Detection of viable but non-culturable *Escherichia coli* O157: H7 from vegetable samples using quantitative PCR with propidium monoazide and immunological assays. *Food Control* **31:**268–273.
31. **Masmoudi S, Denis M, Maalej S.** 2010. Inactivation of the gene *katA* or *sodA* affects the transient entry into the viable but non-culturable response of *Staphylococcus aureus* in natural seawater at low temperature. *Mar Pollut Bull* **60:**2209–2214.
32. **Gião MS, Keevil CW.** 2014. *Listeria monocytogenes* can form biofilms in tap water and enter into the viable but non-cultivable state. *Microb Ecol* **67:**603–611.
33. **Liu J, Zhou R, Li L, Peters BM, Li B, Lin CW, Chuang TL, Chen D, Zhao X, Xiong Z, Xu Z, Shirtliff ME.** 2017. Viable but non-culturable state and toxin gene expression of enterohemorrhagic *Escherichia coli* O157 under cryopreservation. *Res Microbiol* **168:**188–193.
34. **Cappelier JM, Besnard V, Roche SM, Velge P, Federighi M.** 2007. Avirulent viable but non culturable cells of *Listeria monocytogenes* need the presence of an embryo to be recovered in

egg yolk and regain virulence after recovery. *Vet Res* **38:**573–583.

35. **Zeng B, Zhao G, Cao X, Yang Z, Wang C, Hou L.** 2013. Formation and resuscitation of viable but nonculturable *Salmonella* Typhi. *BioMed Res Int* **2013:**907170.
36. **Makino S-I, Kii T, Asakura H, Shirahata T, Ikeda T, Takeshi K, Itoh K.** 2000. Does enterohemorrhagic *Escherichia coli* O157:H7 enter the viable but nonculturable state in salted salmon roe? *Appl Environ Microbiol* **66:** 5536–5539.
37. **FAO, IFAD, UNICEF, WFP, WHO.** 2017. *The State of Food Security and Nutrition in the World 2017. Building Resilience for Peace and Food Security*. FAO, Rome, Italy.
38. **Turner K, Georgiou S, Clark R, Brouwer R.** 2004. Economic valuation of water resources in agriculture: from the sectoral to a functional perspective of natural resource management. FAO, Water Reports 27. http://www.fao.org/docrep/007/y5582e/y5582e00.htm#Contents.
39. **FDA.** 2011. Standards for the growing, harvesting packing and holding of produce for human consumption. Docket No. FDA-2011-N-0921.
40. **UN-Water.** 2015. Compendium of water quality regulatory frameworks: which water for which use? http://www.unwater.org/publications/compendium-water-quality-regulatory-frameworks-water-use/.
41. **CAC (Codex Alimentarius Commission).** 2003. Code of hygienic practice for fresh fruits and vegetables. www.fao.org/input/download/standards/.../CXP_053e_2013.pdf.
42. **CAC (Codex Alimentarius Commission).** 2007. *Fresh Fruits and Vegetables*. www.fao.org/3/a-a1389e.pdf.
43. **Harwood VJ, Levine AD, Scott TM, Chivukula V, Lukasik J, Farrah SR, Rose JB.** 2005. Validity of the indicator organism paradigm for pathogen reduction in reclaimed water and public health protection. *Appl Environ Microbiol* **71:**3163–3170.
44. **Kramer MH, Herwaldt BL, Craun GF, Calderon RL, Juranek DD.** 1996. Surveillance for waterborne-disease outbreaks: United States, 1993–1994. *MMWR CDC Surveill Summ* **45:**1–33.
45. **Partyka ML, Bond RF, Chase JA, Kiger L, Atwill ER.** 2016. Multistate evaluation of microbial water and sediment quality from agricultural recovery basins. *J Environ Qual* **45:**657–665.
46. **Shelton DR, Karns JS, Coppock C, Patel J, Sharma M, Pachepsky YA.** 2011. Relationship between *eae* and *stx* virulence genes and *Escherichia coli* in an agricultural watershed: implications for irrigation water standards and leafy green commodities. *J Food Prot* **74:**18–23.
47. **Meays CL, Broersma K, Nordin R, Mazumder A, Samadpour M.** 2006. Diurnal variability in concentrations and sources of *Escherichia coli* in three streams. *Can J Microbiol* **52:**1130–1135.
48. **McFeters GA, Stuart DG.** 1972. Survival of coliform bacteria in natural waters: field and laboratory studies with membrane-filter chambers. *Appl Microbiol* **24:**805–811.
49. **McLain JET, Williams CF.** 2008. Seasonal variation in accurate identification of *Escherichia coli* within a constructed wetland receiving tertiary-treated municipal effluent. *Water Res* **42:**4041–4048.
50. **Delbeke S, Ceuppens S, Holvoet K, Samuels E, Sampers I, Uyttendaele M.** 2015. Multiplex real-time PCR and culture methods for detection of Shiga toxin-producing *Escherichia coli* and *Salmonella* Thompson in strawberries, a lettuce mix and basil. *Int J Food Microbiol* **193:**1–7.
51. **Pagadala S, Marine SC, Micallef SA, Wang F, Pahl DM, Melendez MV, Kline WL, Oni RA, Walsh CS, Everts KL, Buchanan RL.** 2015. Assessment of region, farming system, irrigation source and sampling time as food safety risk factors for tomatoes. *Int J Food Microbiol* **196:**98–108.
52. **Uyttendaele M, Jaykus LA, Amoah P, Chiodini A, Cunliffe D, Jacxsens L, Holvoet K, Lau M, McClure P, Medema G, Sampers I, Rao Jasti P.** 2015. Microbial hazards in irrigation water: standards, norms, and testing to manage use of water in fresh produce primary production. *Compr Rev Food Sci Food Saf* **14:** 336–356.
53. **Cinotto PJ.** 2005. Occurrence of fecal-indicator bacteria and protocols for identification of fecal contamination sources in selected reaches of the West Branch Brandywine Creek, Chester County, Pennsylvania. U. S. Geological Survey Scientific Investigations Report, 2005-5039.
54. **Solo-Gabriele HM, Wolfert MA, Desmarais TR, Palmer CJ.** 2000. Sources of *Escherichia coli* in a coastal subtropical environment. *Appl Environ Microbiol* **66:**230–237.
55. **Perry KB.** 1998. Basics of frost and freeze protection for horticultural crops. *Horttechnology* **8:**10–15.
56. **Evans RG.** 1999. Frost protection in orchards and vineyards. USDA-Agricultural Research Service. http://www.sidney.ars.usda.gov/personnel/pdfs/Frost%20Protection%20in%20Orchards%20and%20Vineyards.pdf.
57. **Angelo KM, Conrad AR, Saupe A, Dragoo H, West N, Sorenson A, Barnes A, Doyle M, Beal J, Jackson KA, Stroika S, Tarr C, Kucerova Z, Lance S, Gould LH, Wise M, Jackson BR.** 2017.

Multistate outbreak of *Listeria monocytogenes* infections linked to whole apples used in commercially produced, prepackaged caramel apples: United States, 2014–2015. *Epidemiol Infect* **145:**848–856.

58. **Jackson BR, Salter M, Tarr C, Conrad A, Harvey E, Steinbock L, Saupe A, Sorenson A, Katz L, Stroika S, Jackson KA, Carleton H, Kucerova Z, Melka D, Strain E, Parish M, Mody RK, Centers for Disease Control and Prevention (CDC).** 2015. Notes from the field: listeriosis associated with stone fruit: United States, 2014. *MMWR Morb Mortal Wkly Rep* **64:** 282–283.
59. **Evans RG, Kroeger MW, Mahan MO.** 1995. Evaporative cooling of apples by overtree sprinkling. *Appl Eng Agric* **11:**93–99.
60. **Chambers U, Jones VP.** 2015. Effect of over-tree evaporative cooling in orchards on microclimate and accuracy of insect model predictions. *Environ Entomol* **44:**1627–1633.
61. **Guan TT, Blank G, Holley RA.** 2005. Survival of pathogenic bacteria in pesticide solutions and on treated tomato plants. *J Food Prot* **68:** 296–304.
62. **Dobhal S, Zhang G, Royer T, Damicone J, Ma LM.** 2014. Survival and growth of foodborne pathogens in pesticide solutions routinely used in leafy green vegetables and tomato production. *J Sci Food Agric* **94:**2958–2964.
63. **Lopez-Velasco G, Tomas-Callejas A, Diribsa D, Wei P, Suslow TV.** 2013. Growth of *Salmonella enterica* in foliar pesticide solutions and its survival during field production and post-harvest handling of fresh market tomato. *J Appl Microbiol* **114:**1547–1558.
64. **NACMCF.** 1999. Microbiological safety evaluations and recommendations on fresh produce. *Food Control* **10:**117–143.
65. **Santamaría J, Toranzos GA.** 2003. Enteric pathogens and soil: a short review. *Int Microbiol* **6:**5–9.
66. **Griffiths BS, Philippot L.** 2013. Insights into the resistance and resilience of the soil microbial community. *FEMS Microbiol Rev* **37:**112–129.
67. **Vivant AL, Garmyn D, Piveteau P.** 2013. *Listeria monocytogenes*, a down-to-earth pathogen. *Front Cell Infect Microbiol* **3:**87.
68. **Dowe MJ, Jackson ED, Mori JG, Bell CR.** 1997. *Listeria monocytogenes* survival in soil and incidence in agricultural soils. *J Food Prot* **60:**1201–1207.
69. **Ibekwe AM, Watt PM, Shouse PJ, Grieve CM.** 2004. Fate of *Escherichia coli* O157:H7 in irrigation water on soils and plants as validated by culture method and real-time PCR. *Can J Microbiol* **50:**1007–1014.
70. **Gagliardi JV, Karns JS.** 2002. Persistence of *Escherichia coli* O157:H7 in soil and on plant roots. *Environ Microbiol* **4:**89–96.
71. **Locatelli A, Spor A, Jolivet C, Piveteau P, Hartmann A.** 2013. Biotic and abiotic soil properties influence survival of *Listeria monocytogenes* in soil. *PLoS One* **8:**e75969.
72. **Wang H, Ibekwe AM, Ma J, Wu L, Lou J, Wu Z, Liu R, Xu J, Yates SR.** 2014. A glimpse of *Escherichia coli* O157:H7 survival in soils from eastern China. *Sci Total Environ* **476–477:**49–56.
73. **Gutiérrez-Rodríguez E, Gundersen A, Sbodio AO, Suslow TV.** 2012. Variable agronomic practices, cultivar, strain source and initial contamination dose differentially affect survival of *Escherichia coli* on spinach. *J Appl Microbiol* **112:**109–118.
74. **Rousk J, Brookes PC, Bååth E.** 2009. Contrasting soil pH effects on fungal and bacterial growth suggest functional redundancy in carbon mineralization. *Appl Environ Microbiol* **75:** 1589–1596.
75. **van Elsas JD, Semenov AV, Costa R, Trevors JT.** 2011. Survival of *Escherichia coli* in the environment: fundamental and public health aspects. *ISME J* **5:**173–183.
76. **Mathew RP, Feng Y, Githinji L, Ankumah R, Balkcom KS.** 2012. Impact of no-tillage and conventional tillage systems on soil microbial communities. *Appl Environ Soil Sci* **2012:**548620.
77. **Jeffery S, van der Putten WH.** 2011. *Soil Borne Human Diseases.* JRC Scientific and Technical Reports. European Union, EUR 24893 EN. http://publications.jrc.ec.europa.eu/repository/bitstream/111111111/22432/2/lbna24893enn.pdf.
78. **Portnoy DA, Auerbuch V, Glomski IJ.** 2002. The cell biology of *Listeria monocytogenes* infection: the intersection of bacterial pathogenesis and cell-mediated immunity. *J Cell Biol* **158:**409–414.
79. **Fenlon DR.** 1999. *Listeria monocytogenes* in the natural environment, p 21–38. *In* Ryser ET, Marth EH (ed), *Listeria, Listeriosis, and Food Safety*, 2nd ed. Marcel Dekker, New York, NY.
80. **Hamon M, Bierne H, Cossart P.** 2006. *Listeria monocytogenes*: a multifaceted model. *Nat Rev Microbiol* **4:**423–434.
81. **Jiang X, Morgan J, Doyle MP.** 2002. Fate of *Escherichia coli* O157:H7 in manure-amended soil. *Appl Environ Microbiol* **68:**2605–2609.
82. **Islam M, Doyle MP, Phatak SC, Millner P, Jiang X.** 2005. Survival of *Escherichia coli* O157:H7 in soil and on carrots and onions grown in fields treated with contaminated manure composts or irrigation water. *Food Microbiol* **22:**63–70.

83. **van Overbeek LS, Franz E, Semenov AV, de Vos OJ, van Bruggen AHC.** 2010. The effect of the native bacterial community structure on the predictability of *E. coli* O157:H7 survival in manure-amended soil. *Lett Appl Microbiol* **50:** 425–430.
84. **Nicholson FA, Groves SJ, Chambers BJ.** 2005. Pathogen survival during livestock manure storage and following land application. *Bioresour Technol* **96:**135–143.
85. **Kim J, Jiang X.** 2010. The growth potential of *Escherichia coli* O157:H7, *Salmonella* spp. and *Listeria monocytogenes* in dairy manure-based compost in a greenhouse setting under different seasons. *J Appl Microbiol* **109:**2095–2104.
86. **Franz E, van Hoek AHAM, Bouw E, Aarts HJM.** 2011. Variability of *Escherichia coli* O157 strain survival in manure-amended soil in relation to strain origin, virulence profile, and carbon nutrition profile. *Appl Environ Microbiol* **77:**8088–8096.
87. **Erickson MC, Liao J, Ma L, Jiang X, Doyle MP.** 2009. Pathogen inactivation in cow manure compost. *Compost Sci Util* **17:**229–236.
88. **Droffner ML, Brinton WF.** 1995. Survival of *E. coli* and *Salmonella* populations in aerobic thermophilic composts as measured with DNA gene probes. *Zentralbl Hyg Umweltmed* **197:** 387–397.
89. **Erickson MC, Liao J, Boyhan G, Smith C, Ma L, Jiang X, Doyle MP.** 2010. Fate of manure-borne pathogen surrogates in static composting piles of chicken litter and peanut hulls. *Bioresour Technol* **101:**1014–1020.
90. **Calvin L.** 2007. Outbreak linked to spinach forces reassessment of food safety practices. *Amber Waves* **5:**24–31.
91. **Berry ED, Wells JE, Bono JL, Woodbury BL, Kalchayanand N, Norman KN, Suslow TV, López-Velasco G, Millner PD.** 2015. Effect of proximity to a cattle feedlot on *Escherichia coli* O157:H7 contamination of leafy greens and evaluation of the potential for airborne transmission. *Appl Environ Microbiol* **81:**1101–1110.
92. **Jahne MA, Rogers SW, Holsen TM, Grimberg SJ, Ramler IP, Kim S.** 2016. Bioaerosol deposition to food crops near manure application: quantitative microbial risk assessment. *J Environ Qual* **45:**666–674.
93. **Kumar GD, Williams RC, Al Qublan HM, Sriranganathan N, Boyer RR, Eifert JD.** 2017. Airborne soil particulates as vehicles for *Salmonella* contamination of tomatoes. *Int J Food Microbiol* **243:**90–95.
94. **Sivapalasingam S, Barrett E, Kimura A, Van Duyne S, De Witt W, Ying M, Frisch A, Phan Q, Gould E, Shillam P, Reddy V, Cooper T, Hoekstra M, Higgins C, Sanders JP, Tauxe RV, Slutsker L.** 2003. A multistate outbreak of *Salmonella enterica* serotype Newport infection linked to mango consumption: impact of water-dip disinfestation technology. *Clin Infect Dis* **37:**1585–1590.
95. **Walsh KA, Bennett SD, Mahovic M, Gould LH.** 2014. Outbreaks associated with cantaloupe, watermelon, and honeydew in the United States, 1973–2011. *Foodborne Pathog Dis* **11:**945–952.
96. **Pan Y, Breidt F Jr, Kathariou S.** 2006. Resistance of *Listeria monocytogenes* biofilms to sanitizing agents in a simulated food processing environment. *Appl Environ Microbiol* **72:**7711–7717.
97. **Caselli E.** 2017. Hygiene: microbial strategies to reduce pathogens and drug resistance in clinical settings. *Microb Biotechnol* **10:**1079–1083.
98. **Muhterem-Uyar M, Dalmasso M, Bolocan AS, Hernandez M, Kapetanakou AE, Kuchta T, Manios SG, Melero B, Minarovičová J, Nicolau AI, Rovira J, Skandamis PN, Jordan K, Rodríguez-Lázaro D, Stessl B, Wagner M.** 2015. Environmental sampling for *Listeria monocytogenes* control in food processing facilities reveals three contamination scenarios. *Food Control* **51:**94–107.
99. **NMB (National Mango Board).** 2015. Cleaning and sanitizing practices in the mango industry. Food Safety Consulting and Training Solutions. http://www.mangofoodsafety.org/english/Packinghouse/Sanitation/Sanitation_English.pdf.
100. **Zhou B, Luo Y, Turner E, Wang Q, Schneider K.** 2014. Evaluation of current industry practices for maintaining tomato dump tank water quality during packing house operations. *J Food Process Preserv* **38:**2201–2208.
101. **Hedberg CW, Angulo FJ, White KE, Langkop CW, Schell WL, Stobierski MG, Schuchat A, Besser JM, Dietrich S, Helsel L, Griffin PM, McFarland JW, Osterholm MT, The Investigation Team.** 1999. Outbreaks of salmonellosis associated with eating uncooked tomatoes: implications for public health. *Epidemiol Infect* **122:**385–393.
102. **Leff JW, Fierer N.** 2013. Bacterial communities associated with the surfaces of fresh fruits and vegetables. *PLoS One* **8:**e59310.
103. **Maffei DF, Batalha EY, Landgraf M, Schaffner DW, Franco BDGM.** 2016. Microbiology of organic and conventionally grown fresh produce. *Braz J Microbiol* **47**(Suppl 1)**:**99–105.
104. **Hoogenboom LA, Bokhorst JG, Northolt MD, van de Vijver LP, Broex NJ, Mevius DJ, Meijs JAC, Van der Roest J.** 2008. Contaminants and

microorganisms in Dutch organic food products: a comparison with conventional products. *Food Addit Contam Part A Chem Anal Control Expo Risk Assess* **25:**1195–1207.

105. **Williams PRD, Hammitt JK.** 2001. Perceived risks of conventional and organic produce: pesticides, pathogens, and natural toxins. *Risk Anal* **21:**319–330.
106. **Magkos F, Arvaniti F, Zampelas A.** 2006. Organic food: buying more safety or just peace of mind? A critical review of the literature. *Crit Rev Food Sci Nutr* **46:**23–56.
107. **Suslow TV.** 2000. *Postharvest handling for organic crops.* Organic Vegetable Production in California Series. Pub. 7254. University of California Davis. anrcatalog.ucanr.edu/pdf/7254.pdf.
108. **Perricone M, Arace E, Corbo MR, Sinigaglia M, Bevilacqua A.** 2015. Bioactivity of essential oils: a review on their interaction with food components. *Front Microbiol* **6:**76.
109. **Resh HM.** 1995. *Hydroponic Food Production: a Definitive Guidebook for the Advanced Home Gardener and the Commercial Hydroponic Grower,* 5th ed. Woodbridge, Santa Barbara, CA.
110. **Macarisin D, Patel J, Sharma VK.** 2014. Role of curli and plant cultivation conditions on *Escherichia coli* O157:H7 internalization into spinach grown on hydroponics and in soil. *Int J Food Microbiol* **173:**48–53.
111. **Deering AJ, Jack DR, Pruitt RE, Mauer LJ.** 2015. Movement of *Salmonella* serovar Typhimurium and *E. coli* O157:H7 to ripe tomato fruit following various routes of contamination. *Microorganisms* **3:**809–825.
112. **Franz E, Visser AA, Van Diepeningen AD, Klerks MM, Termorshuizen AJ, van Bruggen AH.** 2007. Quantification of contamination of lettuce by GFP-expressing *Escherichia coli* O157:H7 and *Salmonella enterica* serovar Typhimurium. *Food Microbiol* **24:**106–112.
113. **Sharma M, Ingram DT, Patel JR, Millner PD, Wang X, Hull AE, Donnenberg MS.** 2009. A novel approach to investigate the uptake and internalization of *Escherichia coli* O157:H7 in spinach cultivated in soil and hydroponic medium. *J Food Prot* **72:**1513–1520.
114. **Guo X, van Iersel MW, Chen J, Brackett RE, Beuchat LR.** 2002. Evidence of association of salmonellae with tomato plants grown hydroponically in inoculated nutrient solution. *Appl Environ Microbiol* **68:**3639–3643.
115. **Hirneisen KA, Sharma M, Kniel KE.** 2012. Human enteric pathogen internalization by root uptake into food crops. *Foodborne Pathog Dis* **9:**396–405.
116. **Dabney SM, Delgado JA, Reeves DW.** 2001. Using winter cover crops to improve soil and water quality. *Commun Soil Sci Plant Anal* **32:**1221–1250.
117. **Bianchi M, Lowell K.** 2016. Co-management: balancing food safety, the environment, and the bottom line, p 201–206. *In* Jay-Russell MT, Doyle MP (ed), *Food Safety Risks from Wildlife: Challenges in Agriculture, Conservation, and Public Health.* Springer International Publishing, Cham, Switzerland.
118. **Tilman D, Cassman KG, Matson PA, Naylor R, Polasky S.** 2002. Agricultural sustainability and intensive production practices. *Nature* **418:**671–677.
119. **Wu S, Nishihara M, Kawasaki Y, Yokoyama A, Matsuura K, Koga T, Ueno D, Inoue K, Someya T.** 2009. Inactivation of *Escherichia coli* in soil by solarization. *Soil Sci Plant Nutr* **55:**258–263.
120. **Sofi TA, Tewari AK, Razdan VK, Koul VK.** 2013. Long term effect of soil solarization on soil properties and cauliflower vigor. *Phytoparasitica* **42:**1–11.
121. **Stapleton JJ.** 2000. Soil solarization in various agricultural production systems. *Crop Prot* **19:**837–841.
122. **Bianchi M, Lowell K.** 2012. Balancing food safety and sustainability: opportunities for co-management. University of California Agriculture and Natural Resources. http://ucfoodsafety.ucdavis.edu/files/157154.pdf.
123. **Gil MI, Selma MV, Suslow T, Jacxsens L, Uyttendaele M, Allende A.** 2015. Pre- and postharvest preventive measures and intervention strategies to control microbial food safety hazards of fresh leafy vegetables. *Crit Rev Food Sci Nutr* **55:**453–468.
124. **NOP.** 2018. National Organic Program, code of federal regulations: definitions (7-B-I-M-205.3). https://www.ecfr.gov/cgi-bin/text-idx?tpl=/ecfrbrowse/Title07/7cfr205_main_02.tpl.
125. **Muscutt AD, Harris GL, Bailey SW, Davies DB.** 1993. Buffer zones to improve water quality: a review of their potential use in UK agriculture. *Agric Ecosyst Environ* **45:**57–77.
126. **Naiman RJ, Décamps H.** 1997. The ecology of interfaces: riparian zones. *Annu Rev Ecol Syst* **28:**621–658.
127. **Dillaha T, Sherrard J, Lee D, Mostaghimi S, Shanholtz V.** 1988. Evaluation of vegetative filter strips as a best management practice for feed lots. *Water Environ Fed* **60:**1231–1238.
128. **Strawn LK, Gröhn YT, Warchocki S, Worobo RW, Bihn EA, Wiedmann M.** 2013. Risk factors associated with *Salmonella* and *Listeria monocytogenes* contamination of produce fields. *Appl Environ Microbiol* **79:**7618–7627.

129. **Strawn LK, Fortes ED, Bihn EA, Nightingale KK, Gröhn YT, Worobo RW, Wiedmann M, Bergholz PW.** 2013. Landscape and meteorological factors affecting prevalence of three food-borne pathogens in fruit and vegetable farms. *Appl Environ Microbiol* **79:**588–600.
130. **Hoar B, Atwill E, Carlton L, Celis J, Carabez J, Nguyen T.** 2013. Buffers between grazing sheep and leafy crops augment food safety. *Calif Agric* **67:**104–109.
131. **Callahan MT, Micallef SA, Sharma M, Millner PD, Buchanan RL.** 2016. Metrics proposed to prevent the harvest of leafy green crops exposed to floodwater contaminated with *Escherichia coli*. *Appl Environ Microbiol* **82:**3746–3753.
132. **Schmitt TJ, Dosskey MG, Hoagland KD.** 1999. Filter strip performance and processes for different vegetation, widths, and contaminants. *J Environ Qual* **28:**1479–1489.
133. **Natural Resources Conservation Service.** 1999. *Filter strip, national standard no. 393.*
134. **Santhi C, Atwood JD, Lewis J, Potter SR, Srinivasan R.** 2001. *Environmental and economic impacts of reaching and doubling the USDA buffer initiative program on water quality.* ASAE meeting paper no. 01-2068, St. ASAE, Joseph, MI.
135. **Overwash F.** 2014. Natural Resource Conservation Service Conservation Practice Standard. https://efotg.sc.egov.usda.gov/references/public/MN/393mn.pdf.
136. **Coyne MS, Gilfillen RA, Villalba A, Zhang Z, Rhodes R, Dunn L, Blevins RL.** 1998. Fecal bacteria trapping by grass filter strips during simulated rain. *J Soil Water Conserv* **53:**140–145.
137. **Fox GA, Matlock EM, Guzman JA, Sahoo D, Stunkel KB.** 2011. *Escherichia coli* load reduction from runoff by vegetative filter strips: a laboratory-scale study. *J Environ Qual* **40:**980–988.
138. **Chaubey I, Edwards DR, Daniel TC, Moore PA, Nichols DJ.** 1994. Effectiveness of vegetative filter strips in retaining surface-applied swine manure constituents. *Trans ASAE* **37:**845–850.
139. **Schachtman DP, Reid RJ, Ayling SM.** 1998. Phosphorus uptake by plants: from soil to cell. *Plant Physiol* **116:**447–453.
140. **Rahman A, Rahman S, Cihacek L.** 2012. Efficacy of vegetative filter strips (VFS) installed at the edge of feedlot to minimize solids and nutrients from runoff. *Agric Eng Int* **14:**9–21.
141. **Butler DM, Franklin DH, Ranells NN, Poore MH, Green JT Jr.** 2006. Ground cover impacts on sediment and phosphorus export from manured riparian pasture. *J Environ Qual* **35:**2178–2185.
142. **Butler DM, Ranells NB, Franklin DH, Poore MH, Green JT Jr.** 2008. Runoff water quality from manured riparian grasslands with contrasting drainage and simulated grazing pressure. *Agric Ecosyst Environ* **126:**250–260.
143. **Lim, et al.** 1998. Vegetated filter strip removal of cattle manure constituents in runoff. *Trans ASAE* **41:**1375–1381.
144. **Entry JA, Hubbard RK, Thies JE, Fuhrmann JJ.** 2000. The influence of vegetation in riparian filter strips on coliform bacteria. I. Movement and survival in water. *J Environ Qual* **29:**1206–1214.
145. **Goel PK, Rudra RP, Gharabaghi B, Das S, Gupta N.** 2004. Pollutants removal by vegetative filter strips planted with different grasses. ASAE paper no. 042177. ASAE, St. Joseph, MI.
146. **Gharabhagi B, Rudra RR, Whiteley HR, Dickinson WT.** 2001. *Sediment removal efficiency of vegetative filter strips.* ASAE meeting paper no. 012071. ASAE, St. Joseph, MI.
147. **Mersie W, Seybold CA, McNamee C, Lawson MA.** 2003. Abating endosulfan from runoff using vegetative filter strips: the importance of plant species and flow rate. *Agric Ecosyst Environ* **97:**215–223.
148. **Rahman S, Rahman A, Wiederholt RJ.** 2011. *Vegetative Filter Strips Reduce Feedlot Runoff Pollutants.* NDSU Extension Service, North Dakota State University.
149. **Egley GH.** 1983. Weed seed and seedling reductions by soil solarization with transparent polyethylene sheets. *Weed Sci* **31:**404–409.
150. **Standifer LC, Wilson PW, Sorbet RP.** 1984. Effect of solarization on soil weed seed populations. *Weed Sci* **32:**569–573.
151. **Chellemi DO, Olson SM, Mitchell DJ, Secker I, McSorley R.** 1997. Adaptation of soil solarization to the integrated management of soilborne pests of tomato under humid conditions. *Phytopathology* **87:**250–258.
152. **Cohen O, Rubin B.** 2007. Soil solarization and weed management, p 177–200. *In* Upadhyaya MK, Blackshaw RE (ed), *Non-Chemical Weed Management: Principles, Concepts and Technology*. CABI, Wallingford, UK.
153. **Rubin B, Benjamin A.** 1984. Solar heating of the soil: involvement of environmental factors in the weed control process. *Weed Sci* **32:**138–142.
154. **Gamliel A, Austerweil M, Kritzman G.** 2000. Non-chemical approach to soil borne pest management: organic amendments. *Crop Prot* **19:**847–853.
155. **Ishii S, Ksoll WB, Hicks RE, Sadowsky MJ.** 2006. Presence and growth of naturalized *Escherichia coli* in temperate soils from Lake

Superior watersheds. *Appl Environ Microbiol* **72:**612–621.

156. **Adams NE, Mikhailova EA, Schlautman MA, Bridges WC, Newton CH, Nelson LR, Post CJ, Hall KC, Cox SK, Layton PA, Guynn DC.** 2010. Effects of kudzu (*Pueraria montana*) solarization on the chemistry of an Upper Piedmont South Carolina soil. *Soil Sci* **175:**61–71.
157. **Stapleton JJ, Devay JE.** 1984. Thermal components of soil solarization as related to changes in soil and root microflora and increased plant growth response. *Phytopathology* **74:**255.
158. **Berry ED, Wells JE.** 2012. Soil solarization reduces *Escherichia coli* O157:H7 and total *Escherichia coli* on cattle feedlot pen surfaces. *J Food Prot* **75:**7–13.
159. **Holley RA, Arrus KM, Ominski KH, Tenuta M, Blank G.** 2006. *Salmonella* survival in manure-treated soils during simulated seasonal temperature exposure. *J Environ Qual* **35:**1170–1180.
160. **Abd-Elall A, Maysa A.** 2015. Survival and growth behavior of *Salmonella enterica* serovar Typhimurium in lettuce leaves and soil at various temperatures. *Int Food Res J* **22:**1817–1823.
161. **Wessendorf J, Lingens F.** 1989. Effect of culture and soil conditions on survival of *Pseudomonas fluorescens* R1 in soil. *Appl Microbiol Biotechnol* **31:**97–102.
162. **Lau MM, Ingham SC.** 2001. Survival of faecal indicator bacteria in bovine manure incorporated into soil. *Lett Appl Microbiol* **33:**131–136.
163. **Franz E, van Diepeningen AD, de Vos OJ, van Bruggen AHC.** 2005. Effects of cattle feeding regimen and soil management type on the fate of *Escherichia coli* O157:H7 and *Salmonella enterica* serovar Typhimurium in manure, manure-amended soil, and lettuce. *Appl Environ Microbiol* **71:**6165–6174.
164. **Turpin PE, Maycroft KA, Rowlands CL, Wellington EMH.** 1993. Viable but nonculturable salmonellas in soil. *J Appl Bacteriol* **74:**421–427.
165. **Sainju UM, Senwo ZN, Nyakatawa EZ, Tazisong IA, Reddy KC.** 2008. Soil carbon and nitrogen sequestration as affected by long-term tillage, cropping systems, and nitrogen fertilizer source. *Agric Ecosyst Environ* **127:**234–240.
166. **Magdoff F, van Es H.** 2009. *Building Soils for Better Crops*, 3rd ed, chapter 3, cover crops. Sustainable Agriculture Network. Handbook Series No. 4. SARE Outreach, Beltsville, MA.
167. **Jacobsen CS, Bech TB.** 2012. Soil survival of *Salmonella* and transfer to freshwater and fresh produce. *Food Res Int* **45:**557–566.
168. **Wittstock U, Kliebenstein DJ, Lambrix V, Reichelt M, Gershenzon J.** 2003. Glucosinolate hydrolysis and its impact on generalist and specialist insect herbivores, p 101–125. *In* Romeo JT (ed), *Integrative Phytochemistry: From Ethnobotany to Molecular Ecology, Recent Advances in Phytochemistry*. Elsevier Science, Kidlington, UK.
169. **Lou Z, Wang H, Zhu S, Ma C, Wang Z.** 2011. Antibacterial activity and mechanism of action of chlorogenic acid. *J Food Sci* **76:**M398–M403.
170. **Sheahan CM.** 2012. *Plant guide for sunn hemp* (Crotalaria juncea). USDA-Natural Resources Conservation Service, Cape May Plant Materials Center, Cape May, NJ. https://www.nrcs.usda.gov/Internet/FSE_PLANTMATERIALS/publications/njpmcpg11706.pdf.
171. **Arima H, Ashida H, Danno G.** 2002. Rutin-enhanced antibacterial activities of flavonoids against *Bacillus cereus* and *Salmonella enteritidis*. *Biosci Biotechnol Biochem* **66:**1009–1014.
172. **Cetin-Karaca H, Newman MC.** 2015. Antimicrobial efficacy of plant phenolic compounds against *Salmonella* and *Escherichia coli*. *Food Biosci* **11:**8–16.
173. **Halkier BA, Du L.** 1997. The biosynthesis of glucosinolates. *Trends Plant Sci* **2:**425–431.
174. **Gimsing AL, Kirkegaard JA.** 2006. Glucosinolate and isothiocyanate concentration in soil following incorporation of *Brassica biofumigants*. *Soil Biol Biochem* **38:**2255–2264.
175. **Olilo CO, Onyando JO, Moturi WN, Muia AW, Ombui P, Shivoga WA, Roegner AF.** 2016. Effect of vegetated filter strips on transport and deposition rates of *Escherichia coli* in overland flow in the eastern escarpments of the Mau Forest, Njoro River Watershed, Kenya. *Energy Ecol Environ* **1:**157–182.
176. **Douglas-Mankin KR, Okoren CG.** 2011. Field assessment of bacteria and nutrient removal by vegetative filter strips. *Int J Agric Biol Eng* **4:**43–49.
177. **Guber AK, Yakirevich AM, Sadeghi AM, Pachepsky YA, Shelton DR.** 2009. Uncertainty evaluation of coliform bacteria removal from vegetated filter strip under overland flow condition. *J Environ Qual* **38:**1636–1644.
178. **Leonardopoulos J, Papakonstantinou A, Kourti H, Papavassiliou J.** 1980. Survival of shigellae in soil. *Zentralbl Bakteriol Mikrobiol Hyg [B]* **171:**459–465.
179. **Li X, Park NI, Xu H, Woo S-H, Park C-H, Park SU.** 2010. Differential expression of flavonoid biosynthesis genes and accumulation of phenolic compounds in common buckwheat (*Fagopyrum esculentum*). *J Agric Food Chem* **58:**12176–12181.

180. **Ristaino JB, Perry KB, Lumsden RD.** 1991. Effect of solarization and *Gliocladium virens* on sclerotia of *Sclerotium rolfsii,* soil microbiota, and the incidence of southern blight of tomato. *Phytopathology* **81:**1117–1124.

181. **Zhang X, Liu X, Zhang M, Dahlgren RA, Eitzel M.** 2009. A review of vegetated buffers and a meta-analysis of their mitigation efficacy in reducing nonpoint source pollution. *J Environ Qual* **39:**76–84.

3

Preharvest Food Safety Challenges in Beef and Dairy Production

DAVID R. SMITH[1]

INTRODUCTION

> "Food security exists when all people, at all times, have physical and economic access to sufficient, safe and nutritious food that meets their dietary needs and food preferences for an active and healthy life."
>
> World Food Summit, Rome, 1996 (1)

In addition to the need for food to be nutritious, abundant, and available, the term "food security" implies the requirement for food to be safe from health hazards and, at the same time, meet people's food preferences. It is important that considerations about the safety of food begin prior to harvest because some potential food safety hazards introduced at the farm (e.g., chemical residues) cannot be mitigated by subsequent postharvest food processing steps. Also, some people have preferences for consuming food that has not been through postharvest processing (e.g., raw or lightly cooked foods) even though those foods may be unsafe because of microbiological hazards originating from the farm.

Beef and dairy products are important to food security. They are important sources of high-biological-value nutrients. Although less efficient at converting animal feed into muscle proteins than monogastric animals, ruminants have

[1]Mississippi State University, College of Veterinary Medicine, Mississippi State, MS 39762.

Preharvest Food Safety
Edited by Siddhartha Thakur and Kalmia E. Kniel

doi:10.1128/microbiolspec.PFS-0008-2015

the capability to grow and reproduce using resources of limited value for other purposes. For example, cattle and other ruminants can utilize foodstuffs such as grasses and by-product feeds that are not directly usable by people. In addition, they can utilize forages growing on land unsuitable for crop production. However, neither beef nor dairy products are inherently safe. There are chemical, physical, and biological hazards associated with live cattle production settings that may make consumption of beef and dairy products risky.

Assessing Risk

Risk is a function of the probability of something happening and the cost (or benefit) if it does. Risk assessment is a process of evaluating the probability and costs (or benefits) of potential hazards (or opportunities)—termed risk analysis; deciding what actions, at what relative cost, might be taken to mitigate those hazards—termed risk management; and sharing the action plan with members of the team, keeping records to show what was done, and using records to evaluate whether the actions were successful—termed risk communication and documentation (2). Risk management is relative. Each of us faces hazards to our health on a daily basis, and we make decisions about how to act based on our perception of risk. For example, for many people, getting into a car is one of the more dangerous things that they do each day. But most people decide that the benefit of rapidly getting to where they want to go outweighs the risk (probability and cost) of an automobile accident, although they may try to reduce risk by fastening their seatbelts.

One challenge to risk assessment is putting appropriate weight to the cost of a hazard and the probability of its occurrence. For example, consumers' fear of *Escherichia coli* O157:H7 (Shiga-toxin-producing *E. coli* [STEC] O157) reduced demand for beef in the decade that followed a major foodborne outbreak associated with ground beef sandwiches (3). However, perhaps very few people at the time would have guessed that they were 20 times more likely to choke to death on their food than die from a STEC O157 infection, or over 700 times more likely to die in an automobile (National Safety Council, Fig. 1). Thus, there is a need to continually reassess risk and how to manage it. This chapter is about the risks to human health from food safety hazards of beef and dairy products that occur during live animal production.

Cattle Production Systems

Beef and dairy farms differ widely in production practices because of differences in natural, human, and capital resources. Natural resources include things like land and cattle. Beef and dairy production systems are inherently tied to the land by the availability of feed and forage, weather conditions, and geography. Human resources include the availability and skill of labor and manage-

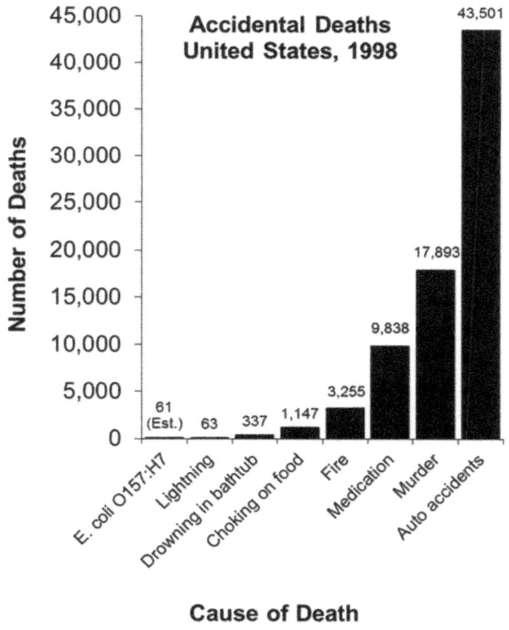

FIGURE 1 Numbers of accidental deaths in the United States in 1998, by cause. From the National Safety Council, http://www.nsc.org.

ment. In many regions of the United States it has become a challenge to hire and retain skilled farm employees. Capital includes the availability of money, credit, and facilities.

In general, beef production can be categorized into phases of cow-calf, stocker-backgrounding, and finishing phases. The goal of cow-calf production is to produce calves by breeding cows and heifers. Cow-calf systems are commonly based on utilizing extensive grass pastures. Beef calves are typically mothered by their dams for 4 to 8 months. Removal of the calf from the dam is termed weaning. After calves are weaned, they may enter a stocker-backgrounding operation for a variable period of time to achieve modest growth, often using inexpensive feed sources. Value may be added to calves during the stocker-backgrounding phase by gathering calves of various body weight and frame from multiple farms, then sorting them into uniform groups based on size or other desirable characteristics, or by marketing calves at a time of the year when they are in demand and prices are higher. Calves may enter finishing feedlots either directly after weaning from cow-calf farms or from stocker-backgrounders. Finishing feedlots often feed nutrient-dense rations to cattle in confinement systems to a target level of muscle and adipose tissue. Feed yard housing may be under a roof on bedded or slatted floors or outside on packed dirt. Most beef in the United States comes from steers or heifers finished in a feed yard; however, in many countries beef may be finished on grass pastures. Beef is also harvested from mature beef or dairy breed cows and bulls marketed when they are past their reproductive potential.

Dairy production includes phases of calf growth, heifer development, and the milking herd. Dairy calves are typically removed from the dam shortly after birth and fed a milk-based diet for six to eight weeks until transitioned onto a solid feed. Bull dairy calves not needed for breeding may be fed to be marketed as veal or fed to a finish weight similar to beef calves. Formula-fed veal are fed a milk-based diet. Dairy cattle may be housed in confinement or spend all or some of their time on pasture. Confinement housing includes dry lots, free stall, bedded pack, or comfort (tie) stalls. Milk is harvested from dairy cows, which typically begin their first lactation at 24 months of age. Typically, dairy cows are bred to calve once a year and have a late gestation dry (not milked) period of approximately 60 days prior to calving.

A Historical Perspective of Preharvest Food Safety

The 20th century efforts to protect human health by controlling tuberculosis in cattle provide some historical perspective on the scientific, political, economic, and practical aspects of making food safe.

Tuberculosis in cattle has been recognized throughout history. Aristotle (ca 400 BC) described the disease in livestock (4). However, following Robert Koch's study of human tuberculosis, the disease in people and cattle was widely recognized as infectious and contagious (5). This led to public health interest in managing exposure risk to the bovine tubercle bacillus. In the early 20th century, tuberculosis was regarded as "the most prevalent and most destructive disease affecting mankind and the domesticated animals" (6). In the first three decades of the 20th century, *Mycobacterium bovis* was isolated from 6 to 30% of human tuberculosis patients in the United States and the United Kingdom. In 1923, it was reported that 10% of all tuberculosis deaths in children under 5 years of age were due to drinking raw cow's milk.

The United States initiated a program for bovine tuberculosis eradication in 1917, using federally funded pre- and postharvest actions including test and cull, controls on animal movement, meat inspection, and pasteurization. In 1918, the U.S. secretary of agriculture reported that "eradication of tuberculosis from live stock means primarily the removal of a constant source of danger to the health of mankind as well as animals" and noted that "progress in eradicating any widespread animal disease depends not only on suitable methods of control, but also in large measure on the desire of live-stock owners to cooperate" (7).

Not all livestock owners wanted to cooperate. Partly, their reluctance was because of the earlier scientific uncertainty about the public health significance of the infection in cattle. They also had other justifiable concerns about the fairness of the process and the cost to their livelihoods. The primary preharvest effort to control bovine tuberculosis was to test and cull infected cattle. Due to poor diagnostic specificity resulting in low positive predictive value, positive test results for bovine tuberculosis sometimes appeared to be spurious. Also, compensation for losses was sometimes perceived to be inadequate. The opposition was so great that law enforcement or the military sometimes had to accompany veterinarians to farms as they tested cattle (8).

In spite of politics and diagnostic uncertainties, the control of bovine tuberculosis in the U.S. cattle population has been a great public health success. Today human infection with *M. bovis* in the United States is exceptionally uncommon because of regulatory efforts to control the disease in live cattle (9). When regulatory efforts to eradicate *M. bovis* from U.S. cattle herds began, 5% of cattle in the United States were infected with *M. bovis*. However, within 25 years the program was achieving success. By 1942 the prevalence of *M. bovis* infection among U.S. cattle was 0.5%, but the program to that point had required the sacrifice of 3,800,000 cattle (10).

Unfortunately, the goal of eradicating bovine tuberculosis from the United States has not been met. Today, prevalence in U.S. cattle is less than 0.001% (9), and in some areas of the country the pathogen has found a reservoir in wild cervids. Even though consumers in the United States have few concerns about acquiring *M. bovis* infection from drinking milk, bovine tuberculosis remains an important infection of cattle, people, and wildlife worldwide.

The history of controlling *M. bovis* in cattle to protect human health illustrates important lessons about preharvest food safety. First, public health statistics represent human suffering with tangible and intangible personal and public costs. We learn from this history that successful food safety programs require (i) effective methods for control based on scientific understanding and clearly defined rules, (ii) acceptance from everyone involved based on a belief that personal or public benefits are achieved through compliance, and (iii) frequently, actions taken at multiple levels of the food system (e.g., pre- and postharvest). Even successful programs incur costs which must be balanced against the benefits. These costs include those incurred by society (public costs) and those incurred by participants (personal costs). Benefits may also be public or personal. Finally, as food becomes incrementally safer, the challenge to provide even greater improvement in safety becomes more difficult and more costly on a marginal basis (i.e., the cost to prevent the next case).

CURRENT PREHARVEST CHALLENGES IN BEEF AND DAIRY PRODUCTS

The on-farm hazards to the safety of beef and dairy product are classified as chemical, physical, or biological, using the terminology of the Hazard Analysis Critical Control Points (HACCP) system (11). Chemical hazards originating on farms may include products used for purposes of agricultural production or environmental contaminants. Physical hazards are various foreign particles or objects. Biological hazards are mainly bacteria that can cause foodborne infections (e.g., from ingesting a sufficient number of pathogenic microorganisms, such as *Salmonella* spp.) or intoxications (e.g., from ingesting preformed bacterial toxins such as staphylococcal enterotoxin).

Chemical Hazards

Chemical hazards are increasingly rare in beef due to effective quality assurance programs, but the potential for harm calls for ongoing vigilance (12). Chemical hazards in beef may occur because of the intentional application or administration of chemicals, such as drugs, vaccines, or pesticides for cattle husbandry purposes or because of unintentional exposures that might typically occur from chemicals present in the environment, including water, feeds, and forages.

Antibiotics are commonly used in beef production for treatment, control, or prevention of bacterial diseases and for growth promotion. Because of these uses, antibiotics are an important source of chemical residues in beef tissue. Residues from other pharmaceuticals and biologics may also contaminate beef, including anthelmintic drugs, vaccine adjuvants, nonsteroidal anti-inflammatory drugs, or hormones such as corticosteroids, anabolic steroids, or prostaglandins. (12). Drugs labeled for use in cattle have established withdrawal periods which define how much time must elapse after drug administration until animals or milk can be marketed for human consumption.

There are many potential environmental contaminants that could result in chemical contamination of beef including lead, dioxins, polychlorinated biphenyls, heavy metals, and radionuclides. The preharvest sources for these contaminants include contaminated water, soils, forages, or harvested feeds. Common sources of lead exposure include automotive batteries and lead-based paint peeling from barns or gates (e.g., from paint marketed in the United States prior to 1978) (13), which calves may lick or consume. Lead bullets or shot can serve as a physical and chemical hazard.

The U.S. Department of Agriculture (USDA) Food Safety Inspection Service (FSIS) monitors beef for chemicals through the U.S. National Residue Program (NRP) for Meat, Poultry, and Egg Products. The NRP is administered by the USDA FSIS in cooperation with the Department of Health and Human Services, the FDA, and the Environmental Protection Agency (EPA). Chemical compounds tested in the program include approved and unapproved veterinary drugs, pesticides, and environmental compounds. The NRP provides a structured process for identifying and evaluating chemical compounds of concern in food animals; analyzing chemical compounds of concern; collecting data, analyzing them, and reporting the results; and identifying the need for regulatory follow-up based on finding violative levels of chemical residues (14). The NRP follows a three-tier sampling system: tier I, scheduled (random) sampling, tier II, targeted sampling at the production or compound class level, and tier III, targeted sampling at the herd/flock or compound class level (14).

The NRP scheduled sampling process is less biased statistically. However, it is difficult to accurately or precisely estimate the prevalence of violative chemical residues in beef carcasses at harvest based on NRP scheduled sampling because the relatively low sample numbers result in few detected residues and 95% confidence estimates that range over a couple of percentage points even when no residue is detected (Fig. 2) (15). However, a greater number of samples are collected when the FSIS public health veterinarians have reason to suspect that a chemical residue might be present, for example, when there is evidence of a disease that may have been treated or the public health veterinarian

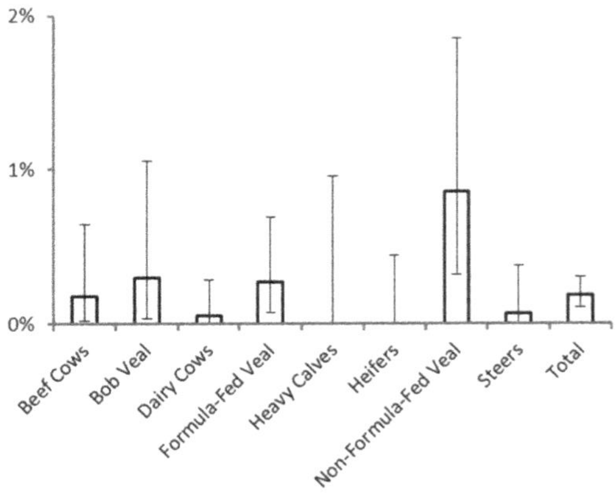

FIGURE 2 Percentage of sampled beef carcasses with violative drug residues based on USDA FSIS statistical sampling in 2012, by class of cattle (15). Error bars represent Clopper-Pearson exact 95% confidence intervals calculated by the author.

suspects the use of a drug (14). Assuming that the targeted inspector-generated sampling process (NRP, tier II) detects almost all carcasses with violative residues, the probability for violative tissue residues in the various cattle production classes can be estimated with relative precision. In 2011, 163,003 inspector-generated samples from cattle carcasses were screened for chemical residues (Fig. 3). Of these, 973 (0.6%) violative residues were detected. Assuming these 973 carcasses represent nearly all of the cattle with chemical residues from among the 34,444,685 cattle harvested, the rate of violative residues in cattle carcasses was 0.0028% (95% confidence interval = 0.0027% to 0.0030%). There were meaningful differences in the probability for chemical residue violations by class of cattle. The greatest rate of violations was among non-formula-fed veal (63 violations/ 14,652 carcasses = 0.4%). In 2011, heifers and steers, typically finished in beef feedlots, represented 28% and 48%, respectively, or

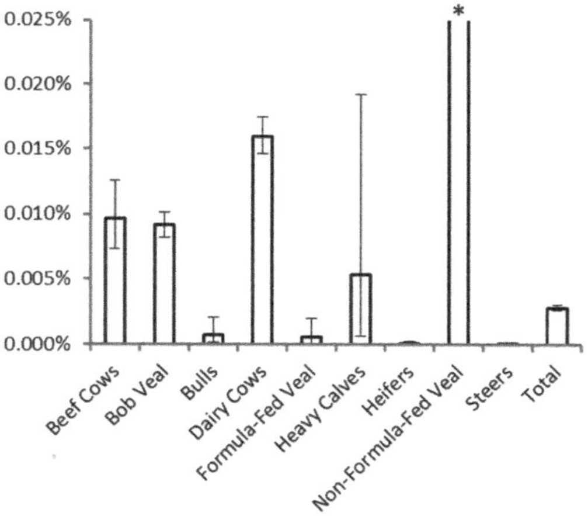

FIGURE 3 Percentage of beef carcasses that had violative drug residues discovered by USDA FSIS targeted inspector-generated sampling (NRP, Tier II) in 2012, by class of cattle (15). Error bars represent Clopper-Pearson exact 95% confidence intervals calculated by the author. *The proportion of non-formula-fed veal samples with violative samples was 0.43% (95% confidence interval = 0.33% to 0.55%).

76% of the total weight of cattle harvested. Collectively, based on targeted sampling, heifers and steers had 30 violative residues detected from 26,280,828 total carcasses for a rate of 1.1 violative residues detected per million head (0.00011%, 95% CI = 0.00008% to 0.00016%) (15).

As with beef cattle, dairy cows may be administered antibiotics, anthelmintics, or other pharmaceuticals or biologics that could leave residues in milk. Besides the oral, parenteral, or topical application of these drugs, dairy cows might also be dosed by the intramammary route to prevent, control, or treat mastitis. In addition to the potential for drug or pesticide residues in milk, various other chemicals used on dairy farms have the potential to contaminate milk before it leaves the farm. For example, cleaning agents and sanitizers used to clean milk pipelines and the bulk tank might be accidentally discharged into the bulk tank when milk is being stored.

Milk may be contaminated with chemicals that originated from environmental sources. As with beef, rare but potential chemical contaminants of milk include lead, dioxins, polychlorinated biphenyls, heavy metals, and radionuclides. In addition, aflatoxins originating from moldy feeds may contaminate milk (16, 17).

The National Conference on Interstate Milk Shipments (NCIMS) has authorized a national program to compile and report the results of milk drug residue testing by industry and state regulatory agencies (18). The Pasteurized Milk Ordinance (PMO) requires that all bulk milk tankers (the tankers that carry bulk raw milk from dairy farms) be sampled and analyzed for drug residues before the milk is processed. Positive-testing bulk milk tankers are rejected for human consumption. Other periodic testing of pasteurized fluid (finished product in packaged form) and milk from the farm bulk tank occurs.

From 1 October 2012 to 30 September 2013, residue testing was performed on 3,761,500 milk samples and 4,115,774 assays for 10 individual drugs or drug groups (18). These

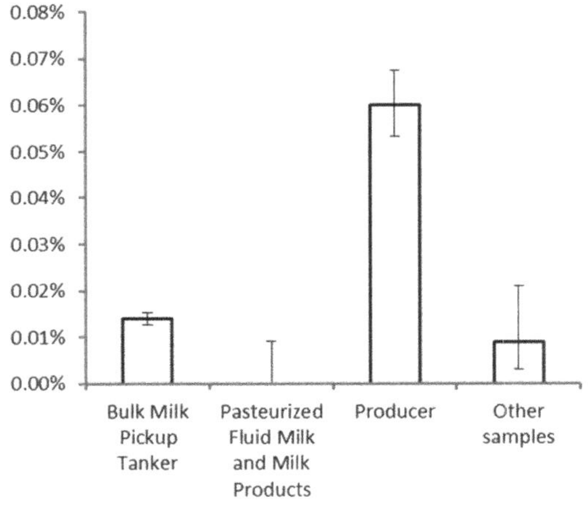

FIGURE 4 Number of violative drug residues detected in U.S. milk samples, by source. Data are from NCIMS, 1 October 2012 to 30 September 2013 (18). Error bars represent Clopper-Pearson exact 95% confidence intervals calculated by the author.

samples came from bulk milk pickup tankers (*n* = 3,198,228), pasteurized fluid milk and milk products (*n* = 40,435); raw milk from farm bulk tanks (*n* = 467,429), and other randomly sampled sources (e.g., from the milk plant, transport tankers; *n* = 55,408). A total of 445 bulk milk tanker samples tested positive for drug residues (0.014%), resulting in 19,553,000 pounds of milk being diverted from human consumption. These results and the results of other samples are shown in Fig. 4. The most commonly used validated tests are for the beta lactam class of antibiotics; 708 of 3,706,466 (0.019%) tests for beta lactam drugs were positive. In addition, 5 of 151,032 (0.003%) tests for sulfonamides and 8 of 134,540 (0.006%) tests for tetracycline drugs were positive for residues (18).

Physical Hazards

The most common physical hazards in beef originating from live cattle are broken hypodermic needles, bullets, arrow tips, and lead or steel shot. These and other items may enter the body of live cattle because of accidents during processing procedures or because of uninformed or malicious acts. For example, needles or darts used for drug administration might break or become imbedded. Some cattlemen have used pistols loaded with shot to round up cattle, and hunters sometimes shoot at cattle. Beef quality assurance programs and other educational programs are helping to reduce these hazards (19–21).

Physical hazards in milk that originate on the farm are rare because milk is filtered. On PMO grade A dairies, those that can supply fluid milk for consumption, the milking system is closed except as the inflation is placed on the teat. Dirt and debris from the teats or udder can enter the system during milking but should be trapped in the milk line filter. There may be more opportunities for dirt and debris to contaminate milk originating from non–grade A (manufacturing grade) dairies because the milking system may be more open to the farm environment, for example, when hand milking into buckets. Even so, the level of milking hygiene required to meet PMO quality standards for bacteria allow little opportunity for gross contamination (16).

Biological Hazards

In the United States, foodborne illness from 31 known pathogens is responsible for an estimated 9.4 million illnesses and 1,351 deaths each year (22). Among the foodborne zoonotic bacterial pathogens associated with beef are STEC, *Salmonella* spp., *Listeria monocytogenes*, and *Campylobacter jejuni*. Milk may be a source of these same agents and, in addition, is more commonly a source of human exposure to *Coxiella burnetii*, *Leptospira* spp., *Staphylococcus aureus* and its enterotoxins, *Bacillus cereus*, and *Clostridia* spp. (16). Bacterial pathogens may contaminate raw milk as it leaves the farm, either because the milk was contaminated from environmental sources during or after milking (e.g., from contaminants on the teat surface) or because of an infection of the mammary gland. Bacterial contamination of milk on the farm is reduced but not eliminated by appropriate milking hygiene and environmental sanitation. Postharvest pasteurization mitigates most preharvest biological hazards in milk. However, farm source pathogens sometimes contaminate postharvest retail products due to system failures that result in cross-contamination with raw product.

Most of the biological hazards that originate with live cattle are rare causes of human illness, but some are common in cattle populations and present ongoing preharvest hazards to food safety. Some zoonotic enteric illnesses, such as STEC O157, salmonellosis, and campylobacteriosis have a strong pattern of seasonality, with incidence of human infection being higher in the summer (23, 24). The explanation for these seasonal patterns of incidence is unclear. A better understanding of this phenomenon might help us understand how to better manage food safety risk. Seasonality may reflect similar environmental effects on pathogen survival or growth, seasonal differences in host population characteristics including the behaviors of individuals (e.g., outdoor activities or changes in diet), or opportunities for pathogen–host interactions from the previous two effects (24).

STEC

STEC is an important cause of human diarrheal illness, including hemorrhagic colitis with occasional severe complications of renal damage and death (25). In the United States, STEC infections are estimated to be responsible for 31 deaths annually, and an estimated 175,905 illnesses and 20 of the 31 deaths each year are attributed to foodborne infections (22). In 1994, the U.S. FSIS declared STEC O157 to be an adulterant in ground beef, then extended the rule to include mechanically tenderized or otherwise nonintact beef. More recently, six non-O157 STECs were added to the list of adulterants (26). The association of STEC O157 with beef has been costly to the industry because of the expense of preventing beef contamination and defending against lawsuits and the decrease in demand for the product (3). Of the estimated $2.7 billion cost of STEC O157 to the beef industry from 1993 to 2003, approximately 60% was due to decreased demand for beef because of consumer fear of foodborne illness associated with ground beef (3).

Even though other pathogens cause more foodborne illnesses (Fig. 5) (27), STEC serves as a model of modern food safety issues attributed to cattle because of the various facets of regulatory and industry efforts to prevent

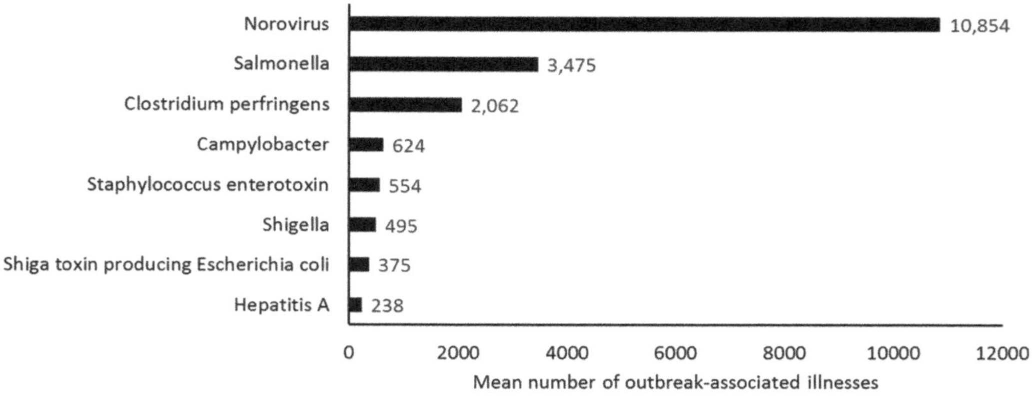

FIGURE 5 Mean annual number of outbreak-associated foodborne illnesses in the United States, by pathogen 2002 to 2006 (27).

food contamination, pre- and postharvest risk factors, and because the organisms have been so widely studied in beef and dairy populations. In contrast to the efforts to control *M. bovis* in live cattle, there is no government-sponsored program in the United States for preharvest prevention or control of STEC.

STEC is transmitted by fecal–oral contact, either by direct contact with contaminated human or animal feces or indirectly through consumption of contaminated food or water (25). Over 200 serotypes of STEC are associated with human illness (28). However, in North America, STEC O157 is the serotype most commonly isolated from human cases and most frequently associated with severe complications (29). The most common non-O157 STEC serotypes associated with human illness in the United States are O26, O111, O103, O121, O45, and O145 (30). Children are particularly at risk for STEC infection and severe complications. In 2012, the annual incidence of laboratory-confirmed STEC O157 infection in the United States was 1.1 cases per 100,000 population (31). However, the incidence in children less than 5 years of age was much greater: 4.7 cases per 100,000 population (31).

Important hazards for human exposure to STECs include day care facilities, nursing homes, children playing with a sick friend, swimming pools, contaminated food and water, and direct exposure to animal environments such as farms, petting zoos, or livestock exhibitions (32–34). The largest outbreaks of STEC O157 infection in the United States and Canada have been because municipal water systems were contaminated with human or animal feces (35–38). However, the greatest total numbers of STEC infections are due to foodborne infection. Many food types have been associated with STEC O157 human infection including beef, fruits, vegetables, leafy greens, and raw milk (33, 34). Some of the early, and most notorious, outbreaks of STEC O157 infection were linked to consumption of undercooked ground beef sandwiches (39–42). The USDA FSIS estimates that one-third of human infections are due to consumption of ground or nonintact beef (43). There may be other farm-associated risks for exposure. People living in or visiting rural areas, particularly those with exposure to cattle or their manure, may have greater risk for STEC infection or serologic evidence of STEC infection (44–48). Other vehicles of fecal–oral transmission of STECs to humans include fomites such as dust (49) and water (35–38) and vectors such as flies (50–54).

Ruminants are particularly prone to STEC colonization, although STEC has been recovered from a variety of animal species (55). In North America, live cattle populations serve as the primary reservoir for human STEC infection (25, 56). Although wild animals appear to be a minor source of human exposure to STEC (57), they have been implicated in notable human STEC outbreaks. For example, a large outbreak of STEC O157 in the United States and Canada was due to consumption of spinach contaminated in the field by feces from feral pigs that also had contact with cattle pastures (58). Strawberries in Oregon were contaminated with STEC O157 by deer feces, resulting in 1 death and at least 14 illnesses (59). It is not clear if wildlife populations serve as reservoirs of STEC infection or merely serve as vehicles of STEC transmission from cattle. Not all STEC strains isolated from cattle are equally pathogenic to humans (56, 60). Regardless, the correlation between seasonal variability in incidence of human STEC O157 illness, prevalence of ground beef contamination, and shedding prevalence of cattle in feedlots (23) and the correlation between the prevalence of carriage of STEC O157 in feces or on hides of live cattle entering the abattoir and subsequent rates of carcass contamination (61, 62) suggest that it is cattle populations that serve as the main reservoir for human exposure.

Numerous studies of the ecology of STEC in beef and dairy production systems and their role in human illness have been conducted (63). The probability for cattle to carry STEC is dependent on both gut and environmental conditions that change over time (63, 64). The factors dictating the probability for cattle to shed STEC are conditions favorable for survival in the bovine gut (65, 66) and the immediate and temporal cattle production setting (67–69).

Cattle are colonized by STEC O157 at the distal rectum (70, 71). Colonization by STEC O157 requires attachment to the intestinal epithelium and induces attaching and effacing lesions, inflammation, and innate and adaptive immune responses in cattle of all ages, indicating that STEC O157 is a bovine pathogen even though infection does not result in clinically observable signs of illness in adult cattle (72, 73). Duration of infection is variable but lasts approximately a month. In field settings reinfection is common (67, 74–76).

Most STEC research in cattle has focused on STEC O157. The ecology of STEC non-O157 serotypes has been less studied, but STEC non-O157 serotypes, including those of public health significance, are found in North American cow–calf, feedlot, and dairy cattle populations (28, 77–80) and in beef processing abattoirs (81, 82). STEC O157 is ubiquitous in beef and dairy cattle populations (63, 67–69, 80, 83–85), with a predilection for calves and yearling animals (85, 86), even in extensive range settings (87–89).

Prevalence of STEC carriage by cattle varies widely by time and place (69, 90, 91). In a longitudinal study of summer-fed feedlot cattle, increased prevalence of STEC O157 fecal shedding was explained by increases in both incidence and duration of shedding (67). This finding suggests that the prevalence of STEC may increase because environmental conditions currently favor survival and subsequent ingestion of the organism (e.g., increased incidence of exposure), that the gut environment currently favors colonization (e.g., increased duration of infection), or both (63).

Greater rates of STEC O157 carriage by feedlot cattle occur in the summer months (68, 92, 93). STEC O157 was cultured from the feces of at least one animal in every pen from a census of 44 pens of summer-fed cattle and 16 of 30 pens of winter-fed cattle from five commercial feedlots in Nebraska, (23, 69). In that study, prevalence within pens of summer-fed cattle ranged from 1 to 80% and 0 to 56% in winter-fed cattle. Also, fecal shedding demonstrated an extra-binomial distribution in both seasons, which suggests that group-level, as opposed to individual-level, factors or characteristics may play an important role in STEC occurrence. Carriage of STEC O157 is greatest in dairy herds during summer months, even if housing is climate controlled (85). The seasonal pattern of STEC carriage by cattle has not been fully explained. Environmental suitability for STEC may explain seasonality (68), but factors associated with length of daylight have also been the subject of study (94–96).

The prevalence of cattle shedding STEC O157 in feces varies widely, even within a season (67, 69, 97). In feedlot cattle, STEC O157 measured on ropes or in rectal feces has been associated with season, ambient temperature within the season, condition of the floor surface, and presence of the organism in other environmental samples, such as fresh fecal pats or drinking water (68, 69, 93). In a 2-year, 2-season longitudinal study of commercial feedlots (68, 98), researchers tied ropes from feed bunk rails to detect high STEC O157 prevalence pens of cattle (90) and pens of cattle with evidence of *Salmonella* spp. STEC O157 was more likely to be recovered from ropes or feces in the summer than winter, even though considerable variation in prevalence was found within each season (68). Within seasons there was significant clustering of rope-positive pens by time, exhibited by runs of weeks when a given pen would test ropes-positive, followed by runs of weeks when the pen would test negative. Feed yards differed in the probability for pens to test rope-positive. This effect was consistent across seasons, suggesting that some management practices or other feed-yard-level characteristics might affect STEC O157 occurrence. The relationship between recovery of STEC O157 from ropes and ambient temperature could be because warmth favors STEC O157 survival in the environment or, at least, its survival on the ropes until they were collected. However, the positive association between recovery of the organism from ropes and its contemporaneous recovery from feces or water supports the conclusion that culture-positive ropes occur in periods when STEC O157 is abundant in the environment (68, 90). In the winter, rope-positive pens were less common, but pens were more likely to test rope-positive if the pen surface was wet and muddy, or dry and dusty, compared to ideal pen surface conditions. The relationship between pen surface condition and recovery of STEC O157 from feedlot cattle has been similarly observed in other longitudinal and cross-sectional studies (69, 99).

Similarly, dairy herds where water was used to flush the pen floor had almost 8 times greater odds of having STEC O157 detected in feces of at least one animal (100). It may be that, compared to ideal conditions, wet and muddy pen surfaces favor STEC survival, or the wet and muddy or dry and dusty pen conditions favor the opportunity for fecal–oral transmission. Collectively, these factors reflect conditions that favor survival of STEC O157 in the environment (e.g., when it's warm and wet) or increased opportunity for fecal–oral transmission to cattle (e.g., when conditions are muddy or dusty, or in the presence of contaminated water). Interestingly, feed yards with the highest probability for ropes to culture STEC O157 positive were the least likely to have ropes culture positive for *Salmonella* spp. (68, 98). This observation may mean that management practices or environmental conditions that favor STEC O157 are not the same as those that favor *Salmonella* spp.

Certain strains of *Lactobacillus acidophilus* have been shown to reduce STEC O157 fecal prevalence when fed continuously to cattle at 10^9 CFU/g (65, 101, 102). The mechanism by which direct-fed microbial products exert an effect on STEC O157 is not understood. Lower levels of inclusion in the diet may not be protective (103). Other approaches to controlling STEC include oral dosing of bacteriophage (104–106) or using antimicrobial chemicals such as sodium chlorate (107–109) or antibiotics (110). Feedlot cattle fed high inclusion rates of distillers grains have been more likely to shed STEC O157 than cattle fed corn-based rations. This may be because of changes in protein, fat, starch, fiber, and water as distillers grains replace corn in the ration (111, 112). It appears that cattle fed diets high in forage or low in starch are more likely to shed STEC O157. However, results have not always been consistent, and the mechanism for this effect is unclear.

Cattle vaccines against STEC O157 are available commercially in the United States and Canada, and their efficacy has been reviewed (113–115). The antigenic targets of these vaccines are either proteins of the type III secreted protein system or molecules responsible for the uptake of iron (66, 116). Vaccination of cattle reduced prevalence of colonization at the terminal rectum (117, 118) and decreased the concentration of organisms in feces (103, 119). The efficacy of vaccination for reducing STEC O157 fecal shedding is about 50 to 60% (114, 115). Computer simulation suggests that vaccination of summer-fed cattle using a 58% effective product would eliminate pens of highest prevalence, resulting in a prevalence distribution similar to what is typically observed in winter-fed cattle (120). When vaccine was applied to all cattle within a region of the feed yard, the efficacy at preventing hide contamination was greater, suggesting that the vaccine efficacy depends on how the products are used to control environmental transmission within groups of cattle or throughout the production system (121). Unfortunately, the efficacy of a vaccine administered to cattle on the farm might be negated by events in subsequent stages of the food system. For example, cross-contamination of cattle hides with STEC O157 may occur during transportation or while cattle are in lairage (62, 122, 123). However, vaccine efficacy has been preserved into the abattoir when efforts were taken to load cattle by treatment groups into clean trucks for transportation to the abattoir (121). Because of serotype specificity, vaccines targeting STEC O157 may offer poor cross-protection against other STEC strains (124). It is not yet clear if STEC O157 vaccines will be widely adopted by cattle feeders because there is not yet an economic signal to indicate that cattle vaccinated against STEC O157 are valued over other cattle.

Salmonella

Salmonella spp. are an important cause of human foodborne illness. Infection with *Salmonella* spp. is estimated to cause 1.2 million human illnesses and 380 deaths in the United States each year, although the number of reported cases of salmonellosis is approximately 42,000 (125). Most foodborne outbreaks of salmonellosis are associated with poultry products, but many foods, including beef and dairy products, may be a source of exposure (126). Cattle infected with *Salmonella* may exhibit clinical signs or, more commonly, are unapparent carriers. Infected cattle can be a source of *Salmonella* exposure to humans through contaminated meat or milk products or because of direct or indirect exposure of people to cattle or their environment. Milk may be contaminated with *Salmonella* from environmental sources during milking or because infected cattle shed the organism in milk (127). *Salmonella* has been recovered from the peripheral lymph nodes of cattle after harvest, presenting a source of beef contamination not amenable to typical postharvest interventions (128, 129).

There are over 2,500 different *Salmonella* serotypes. Some serotypes, such as *Salmonella enterica* Dublin are host adapted to cattle, but most are not. *S. enterica* serovars Typhimurium, Dublin, and Newport are commonly associated with clinical signs of disease in cattle. However, many other serovars may also make cattle sick. Clinical signs in cattle are often associated with stressful events, dietary changes, or concurrent disease. For example, dairy cattle may develop salmonellosis in temporal proximity to parturition (127).

Cattle and other livestock, in a range of production settings, are frequently exposed to *Salmonella* through direct and indirect fecal–oral or oral–oral exposure (127). Sources of *Salmonella* exposure include other infected cattle, contaminated feed, or water. Birds may serve as vectors of transmission within and between farms. *Salmonella* spp. may also be introduced to the farm via contaminated purchased feeds. It is difficult to prevent exposure of cattle to *Salmonella* spp., but vaccines may be effective at minimizing clinical signs (127).

In a survey of U.S. cow–calf farms, *Salmonella* was recovered from 16 of 172 (9.2%) operations and 31 of 5,793 (0.5%) adult cows. Surveillance in beef feedlots indicates that *Salmonella* spp. are endemic to beef feedlots. However, the distribution of serotypes differs meaningfully from the serotypes commonly recovered from human salmonellosis cases (130). In a 2-year study of summer- and winter-fed cattle in five commercial feedlots, *Salmonella* spp. were cultured from feces, water tanks, and ropes that cattle could lick and chew to indicate fecal–oral transmission. *Salmonella* was recovered from ropes in each of the five feedlots and from 50 of 55 (91%) pens of cattle fed in the summer and 14 of 20 (70%) pens of cattle fed in the winter. In that study, researchers found no seasonal difference in the probability of detecting *Salmonella* from ropes, but there were periods in both seasons when *Salmonella* recovery was more likely. Factors associated with fecal–oral transmission were muddy pen surfaces, larger numbers of cattle in the pen (summer-fed), presence of the organism in water tanks, and feed yard (winter-fed) (98).

A national survey of U.S. dairy farms conducted in 2007 included testing the feces from healthy cows for *Salmonella* spp. (131). In the survey, *Salmonella* was cultured from the feces of 523 of 3,804 (14%) healthy adult dairy cattle from 48 of 121 (40%) dairy operations. The percentage of *Salmonella*-positive cows was modified by herd size and region. Cows from dairies in eastern states had a greater probability of testing positive than cattle in western states. In eastern states a greater proportion of cows tested positive on operations with 500 or more cows. There was no significant difference in the prevalence of *Salmonella*-positive cows on dairies of different sizes in western states. Environmental sampling identified Salmonella on 49% of 116 operations. Based on the results of similar surveys conducted on dairies in 1996 and 2002, the proportion of dairy cattle shedding *Salmonella* in feces is increasing. *S. enterica* serotype Montevideo, a common isolate from these studies, is among the top 10 serotypes associated with human illness (131). In the 2007 survey, Salmonella was detected by real time PCR (qPCR) from the milk filter or bulk tank milk of 28% of 538 operations. Dairies with 500 cows or more were twice as likely as dairies with fewer than 100 cows to have *Salmonella* detected on the filter or bulk tank milk (51% and 24%, respectively) (132).

Campylobacter spp.

Campylobacter spp. are a common cause of diarrheal illness in the United States. Campylobacteriosis, characterized by fever, abdominal cramping, and diarrhea, is estimated to affect 1.3 million people in the United States annually. The annual incidence of physician-diagnosed *Campylobacter* infection is about 14 cases per 100,000 people. Human infection is seasonal, with most cases occurring in the summer months. Most people with campylobacteriosis recover completely in 2 to 5 days. However, there are rare complications from infection including arthritis and Guillain-Barré syndrome. About 76 people in the United States die from *Campylobacter* spp. infections each year. *C. jejuni* is the most common cause of human campylobacteriosis (133).

Most cases of campylobacteriosis are associated with eating raw or undercooked poultry products or with cross-contamination of other foods by these products (133). However, live cattle, cattle environments, and raw beef or dairy products can be sources of human foodborne exposure. A national survey of dairy farms in the United States found at least 1 healthy cow shedding *C. jejuni* in feces on 93% of 121 operations. Of 1,885 healthy cows tested, 635 (34%) were positive for *Campylobacter* (131).

L. monocytogenes

L. monocytogenes causes serious and potentially fatal infections in people and cattle (134, 135). Older adults, pregnant women, newborn children, and immunocompromised individuals are at greatest risk for serious illness. Signs in people include headache, fever, muscle aches, stiff neck, confusion, miscarriage, stillbirth, or premature delivery. The incidence of listeriosis in the United States was 0.26 cases per 100,000 individuals in 2013, a decline of 42% from 1996 to 1998. The CDC estimates that 1,600 illnesses and 260 deaths due to listeriosis occur annually in the United States. Human foodborne infections with *L. monocytogenes* are often associated with ready to eat meat products (e.g., deli meats), raw milk, and products manufactured from raw milk (e.g., soft cheeses such as *queso fresco* cheese). The largest outbreak of listeriosis in U.S. history was due to consumption of cantaloupe originating from a single farm in 2011 (134).

Listeria spp. are widespread in the environment, including soil, water, and the digestive tracts of mammals, birds, and fish. *L. monocytogenes* is hardy in the environment and may survive for years in soil and contaminated food. Optimal growth is at 30°C to 37°C, but *L. monocytogenes* can grow at refrigeration temperatures (135). It can become a persistent contaminant of food proocessing facilities, but the organism may also be carried into the plant by live cattle (136).

Asymptomatic carriage of *L. monocytogenes* by cattle is common and presents an important risk to the food safety of beef and milk products (136). *L. monocytogenes* was cultured from either the milk filter or bulk tank milk of 7% of U.S. dairies in a national survey conducted in 2007 (132). Clinical signs of listeriosis in cattle include encephalitis, keratoconjunctivitis, septicemia, mastitis, and abortion. Clinical cases are characterized by unilateral cranial nerve and brainstem lesions. Incidence of clinical *L. monocytogenes* infection in cattle has been estimated at 0.25 and 0.31 cases per 1,000 cows or heifers, respectively. Within a herd, the encephalitic form of listeriosis is usually sporadic, with attack rates of less than 5%. Exposure of cattle to *L. monocytogenes* is typically via contaminated feeds, particularly poorly fermented silage.

Because *L. monocytogenes* is widely distributed in the environment, preventing cattle from exposure is difficult. It is hardy in the environment, so it may recirculate on cattle farms. For example, cattle shed *Listeria* in feces, which is then applied as fertilizer to crops. In turn, during harvest, soil containing *L. monocytogenes* may contaminate crops which are then fed to cattle. Proper silage preparation helps reduce the level of exposure, as does animal and environmental hygiene (132, 135).

C. burnetii

C. burnetii, an obligate intracellular bacterium, is the etiologic agent of Q fever (137). The "Q" stands for "query" and comes from the time when the causal agent was still unknown (138). Clinical signs of Q fever in people range from mild to severe. *C. burnetii* infections may be asymptomatic or mild with spontaneous recovery, and infection is probably underdiagnosed (137, 138). However, when Q fever is diagnosed the illness may be severe, with complications including endocarditis, encephalitis, pneumonia, hepatitis, and splenomegaly. Hospitalization rates for diagnosed infections average 50%. Acute Q fever is typically characterized as a febrile illness with severe headache and, occasionally, atypical pneumonia or hepatitis. The major clinical presentation of chronic Q fever is endocarditis. Other chronic manifestations include vascular infections, osteoarticular infections, chronic hepatitis, chronic pulmonary infections, and chronic fatigue syndrome (137).

The annual incidence of Q fever in the United States is approximately 0.4 cases per million people. About 3% of healthy adults in the United States and 10% of people in high-risk occupations have antibodies against *C. burnetii,* indicating past infection. In 1999, Q fever became a reportable disease in the United States. In 2010, 131 incident cases of Q fever were reported to the CDC. Of these, 106 cases were acute and 25 were chronic. People most commonly become exposed to *C. burnetii* by inhalation of airborne organisms while in barns or other livestock environments. Thus, Q fever is most notable as an occupational hazard of farmers, veterinarians, and farm visitors. In rare circumstances, people become exposed to *C. burnetii* after consuming contaminated raw milk (137). Q fever is more commonly reported from western and plains states. Human infection is seasonal. Cases in the United States peak in April and May, which may coincide with increased human outdoor activity and the birthing season for many domestic livestock species (138).

Cattle, sheep, and goats are the primary reservoirs of human exposure to *C. burnetii,* although dogs and cats may also serve as reservoirs. Various ticks may act as vectors (137). The organism is excreted in milk, urine, and the feces of infected animals. The organisms are shed in high numbers in amniotic fluids and the placenta during parturition. Clinical signs in livestock are uncommon but include abortion in sheep and goats and lower birth weight and infertility in cattle (137). Dairy cattle shedding *C. burnetii* in milk may be at greater risk for subclinical mastitis (139). *C. burnetii* may be ubiquitous to U.S. dairy herds. Of 316 bulk tank milk samples from the United States, 94% were RT-PCR positive for *C. burnetii* (140). *C. burnetii* is hardy in the environment. It is resistant to drying, heat, and common disinfectants (137, 138). The PMO sets the milk pasteurization standards to destroy *C. burnetii.*

REGULATORY AND VOLUNTARY PROGRAMS FOR PREHARVEST SAFETY OF BEEF AND DAIRY PRODUCTS

Much of the success in improving the safety of milk in the United States is due to the PMO, a standardized set of health regulations for milk production that was first published in 1927 and became a model code, adopted by each U.S. state. The PMO regulates all aspects affecting the safety of milk on the farm, during transport, and throughout processing and distribution. For example, the PMO defines standards for dairy cow housing, milking facilities, milking hygiene and sanitation, and milk quality for grade A (fluid) and non–grade A (manufacturing grade) milk.

These state public health regulations have made milk safer. In 1938, milk-borne outbreaks constituted 25% of all disease outbreaks due to infected foods and contaminated water. Currently, milk and fluid milk products continue to be associated with less than 1% of such outbreaks (141). The PMO regulates pre- and postharvest production of milk products because the U.S. Public Health Service recognizes that "effective public health control of milk-borne disease requires the application of sanitation measures throughout the production, handling, pasteurization, and distribution of milk and milk products" (141).

The FDA has regulatory authority over the safety of most foods shipped in interstate commerce, including milk. However, the USDA FSIS has responsibility for the safety of beef at harvest and processing.

The FDA regulates the animal drug approval process, including approving label indications, dosage, frequency of administration, and withdrawal times (the time that must elapse between drug administration and harvest of meat or milk for human consumption). The FDA, under the Federal Food, Drug, and Cosmetic Act, establishes tolerances for veterinary drugs and action levels for food additives and environmental contaminants. Title 21 Code of Federal Regulations includes tolerance levels established by the FDA for chemical residues in milk or meat tissues (14). The FDA and the NCIMS have collaborated through a cooperative, federal–state program to ensure the sanitary quality of grade A milk and milk products shipped in U.S. interstate commerce (18).

The EPA regulates the use of pesticides, including approval of products used for the control of external parasites of cattle. The EPA, under the Federal Insecticide, Fungicide, and Rodenticide Act (as modified by the Food Quality Protection Act), establishes tolerance levels for registered pesticides. Title 40 Code of Federal Regulations includes tolerance levels established by the EPA for chemical residues in milk or meat tissues (14).

Various quality assurance programs have successfully educated beef and dairy producers and their veterinarians on how to improve the safety and quality of beef and dairy products (20, 142). These programs began in earnest in the 1980s to address the risk of injection site blemishes and drug residues in food (12, 21). Today, these programs help producers minimize the likelihood of various preharvest hazards affecting the safety or quality of beef or dairy products as well as protecting the health and well-being of cattle on the farm and during transportation (20, 142). Quality assurance education has been aided by the contributions of various industry stakeholders, including extension and research faculty at U.S. land-grant universities, veterinarians in private practice, industry fieldmen, and allied professional organizations (21).

CONCLUSIONS

Foods of animal origin, including beef and dairy products, are nutritious and important to global food security. However, there are important risks to human health from hazards that are introduced to beef and dairy products on the farm. These hazards may be chemical, physical, or biological in form. Because of various forms of human fallibility and complex microbial ecologies, it is unlikely that many of the preharvest hazards associated with beef and dairy products can be eliminated. However, there is a long history of preventing human foodborne disease by taking action on the farm as well as during harvest and subsequent food processing to reduce the occurrence of such hazards. Notable successes in controlling and preventing on-farm hazards to food safety have occurred because of a combination of voluntary and regulatory efforts.

ACKNOWLEDGMENTS

The author's work is funded in part by the Mikell and Mary Cheek Hall-Davis Endowment for Beef Cattle Health and Reproduction and also by Agriculture and Food Research Initiative Grant No. 2012-68003-30155 from the USDA National Institute of Food and Agriculture, Prevention, Detection and Control of Shiga Toxin-Producing *Escherichia coli* (STEC) from Pre-Harvest Through Consumption of Beef Products Program A4101.

CITATION

Smith DR. 2016. Preharvest food safety challenges in beef and dairy production. Microbiol Spectrum 4(4):PFS-0008-2015.

REFERENCES

1. **Food and Agriculture Organization of the United Nations.** 2014. Food Security Statistics. http://www.fao.org/economic/ess/ess-fs/en/.
2. **Moore PG.** 1977. The manager's struggles with uncertainty. *J R Statist Soc A* **140:**129–165.
3. **Kay S.** 2003. $2.7 billion: the cost of *E. coli* O157:H7. *Meat Poultry* **49:**26–34.
4. **Blancou J.** 2003. Tuberculosis, p 223–241. *In* Blancou J (ed), *History of the Surveillance and Control of Transmissible Animal Diseases.* The International Animal Health Office (OIE), Paris, France.
5. **Koch R.** 1884. Die aetiologie der tuberkulose. *Mitt Kaiserl Gesundheitsamt* **2:**1–88.
6. **Wilson J.** 1902. Report of the Secretary, p 9–116. *Yearbook of Agriculture, 1901.* United States Department of Agriculture, Washington DC.
7. **Kiernan J.** 1919. The accredited-herd plan in tuberculosis eradication, p 215–220. *In Yearbook of Agriculture, 1918.* United States Department of Agriculture, Washington DC.
8. **Welch R.** 2010. The Cedar county cow war of 1931. Iowa State University Center for Agricultural History and Rural Studies.
9. **CDC.** 1990. Epidemiologic notes and reports bovine tuberculosis: Pennsylvania. *Morb Mortal Wkly Rep* **39:**201–203.
10. **Wight A, Lash E, O'Rear H, Crawford A.** 1942. Part 2: important general diseases common to several species; tuberculosis and its eradication, p 237–249. *Yearbook in Agriculture, 1942.*
11. **Anonymous.** 2005. *Meat and Poultry Products Hazards and Control Guide.* U.S. Dept. of Agriculture FSIS.
12. **Fajt V, Griffin D.** 2014. Residue avoidance in beef cattle production systems. *In* Baynes RE, Riviere JE (ed), *Strategies for Reducing Drug and Chemical Residues in Food Animals: International Approaches to Residue Avoidance, Management, and Testing.* John Wiley and Sons, Hoboken, NJ.
13. **Environmental Protection Agency.** 2014. *Lead.* http://www2.epa.gov/lead.
14. **O'keefe M, Muñiz Ortiz JG.** 2014. *United States National Residue Program for Meat, Poultry, and Egg Products 2014 Residue Sampling Plans.* U.S. Department of Agriculture FSIS, Office of Public Health Service, Washington, DC.
15. **Anonymous.** 2013. *United States National Residue Program for Meat, Poultry, and Egg Products. 2011 Residue Sample Results.* U.S. Dept. of Agriculture FSIS, Washington, DC.
16. **Park YW, Albenzio M, Sevi A, Haenlein GFW.** 2013. Milk and dairy products in human nutrition: production, composition, and health, p 261–287. *In* Park YW, Haenlein GFW (ed), *Milk and Dairy Products in Human Nutrition: Production, Composition, and Health,* 1st ed. John Wiley and Sons, Ames, IA.
17. **Mostrom MS, Jacobsen BJ.** 2011. Ruminant mycotoxicosis. *Vet Clin North Am Food Anim Pract* **27:**315–344, viii.
18. **Anonymous.** 2014. National milk drug residue data base fiscal year 2013 annual report: October

1, 2012 - September 30, 2013. GLH, Inc., Lighthouse Point, FL.
19. **Heaton K, Bagley CP.** 2002. *Foriegn Object Contamination in Beef Cattle.* Utah State University Extension.
20. **National Cattlemens Beef Association.** 2014. Beef Quality Assurance National Manual. http://www.bqa.org/Media/BQA/Docs/nationalmanual.pdf. Accessed June 15, 2016
21. **Griffin D.** 2008. Cow-calf operation beef quality assurance, p 587–594. *In* Anderson DE, Rings M (ed), *Current Veterinary Therapy: Food Animal Practice,* 5th ed. Saunders Elsevier, St. Louis, MO.
22. **Scallan E, Hoekstra RM, Angulo FJ, Tauxe RV, Widdowson MA, Roy SL, Jones JL, Griffin PM.** 2011. Foodborne illness acquired in the United States: major pathogens. *Emerg Infect Dis* **17:**7–15.
23. **Williams MS, Withee JL, Ebel ED, Bauer NE, Scholosser WD, Disney WT, Smith DR, Moxley RA.** 2010. Determining relationships between the seasonal occurrence of *Escherichia coli* O157:H7 in live cattle, ground beef, and humans. *Foodborne Path Dis* **7:**1–8.
24. **Lal A, Hales S, French N, Baker MG.** 2012. Seasonality in human zoonotic enteric diseases: a systematic review. *PLoS One* **7:**e31883.
25. **Sargeant JM, Smith DR.** 2003. The epidemiology of *Escherichia coli* O157:H7, p 131–141. *In* Torrence ME, Isaacson RE (ed), *Microbial Food Safety in Animal Agriculture: Current Topics.* Iowa State University Press, Ames, Iowa.
26. **Almanza AV.** 2011. Shiga toxin-producing *Escherichia coli. Fed Reg* **76:**9.
27. **CDC.** 2010. Surveillance for foodborne disease outbreaks: United States, 2007. *MMWR Morb Mortal Wkly Rep* **59:**973–979.
28. **Kalchayanand Na TM, Bosilevac JM, Wheeler TL.** 2011. Non-O157 Shiga toxin-producing *Escherichia coli*: prevalence associated with meat animals and controlling interventions. Abstr., American Meat Science Association 64th Reciprocal Meat Conference, Manhattan, KS, June 19–22.
29. **Mead PS, Slutsker L, Dietz V, McCaig LF, Bresee JS, Shapiro C, Griffin PM, Tauxe RV.** 1999. Food-related illness and death in the United States. *Emerg Infect Dis* **5:**607–625.
30. **Brooks JT, Sowers EG, Wells JG, Greene KD, Griffin PM, Hoekstra RM, Strockbine NA.** 2005. Non-O157 Shiga toxin-producing *Escherichia coli* infections in the United States, 1983-2002. *J Infect Dis* **192:**1422–1429.
31. **Anonymous.** 2013. Incidence and trends of infection with pathogens transmitted commonly through food: Foodborne Diseases Active Surveillance Network, 10 U.S. sites, 1996–2012. *Morb Mortal Wkly Rep* **62:**5.
32. **Feng P.** 1995. *Escherichia coli* serotype O157: H7: novel vehicles of infection and emergence of phenotypic variants. *Emerg Infect Dis* **1.**
33. **Rangel JM, Sparling PH, Crowe C, Griffin PM, Swerdlow DL.** 2005. Epidemiology of *Escherichia coli* O157:H7 outbreaks, United States, 1982–2002. *Emerg Infect Dis* **11:**603–609.
34. **Sparling PH.** 1998. *Escherichia coli* O157:H7 outbreaks in the United States, 1982–1996. *J Am Vet Med Assoc* **213:**1733–1733.
35. **Swerdlow DL, Woodruff BA, Brady RC, Griffin PM, Tippen S, Donnell HD Jr, Geldreich E, Payne BJ, Meyer A Jr, Wells JG.** 1992. A waterborne outbreak in Missouri of *Escherichia coli* O157:H7 associated with bloody diarrhea and death. *Ann Intern Med* **117:**812–819.
36. **Kondro W.** 2000. *E. coli* outbreak deaths spark judicial inquiry in Canada. *Lancet* **355:**2058.
37. **Kondro W.** 2000. Canada reacts to water contamination. *Lancet* **355:**2228.
38. **Anonymous.** 1999. Outbreak of *Escherichia coli* O157:H7 and *Campylobacter* among attendees of the Washington County Fair: New York, 1999. *Morb Mortal Wkly Rep* **48:**803.
39. **Kassenborg HD, Hedberg CW, Hoekstra M, Evans MC, Chin AE, Marcus R, Vugia DJ, Smith K, Ahuja SD, Slutsker L, Griffin PM.** 2004. Farm visits and undercooked hamburgers as major risk factors for sporadic *Escherichia coli* O157:H7 infection: data from a case-control study in 5 FoodNet sites. *Clin Infect Dis* **38** (Suppl 3)**:**S271–S278.
40. **Ryan CA, Tauxe RV, Hosek GW, Wells JG, Stoesz PA, McFadden HW Jr, Smith PW, Wright GF, Blake PA.** 1986. *Escherichia coli* O157:H7 diarrhea in a nursing home: clinical, epidemiological, and pathological findings. *J Infect Dis* **154:**631–638.
41. **Riley LW, Remis RS, Helgerson SD, McGee HB, Wells JG, Davis BR, Hebert RJ, Olcott ES, Johnson LM, Hargrett NT, Blake PA, Cohen ML.** 1983. Hemorrhagic colitis associated with a rare *Escherichia coli* serotype. *N Engl J Med* **308:**681–685.
42. **Slutsker L, Ries AA, Maloney K, Wells JG, Greene KD, Griffin PM.** 1998. A nationwide case-control study of *Escherichia coli* O157:H7 infection in the United States. *J Infect Dis* **177:** 962–966.
43. **Withee J, Williams M, Schlosser W, Bauer N, Ebel E.** 2009. Streamlined analysis for evaluating the use of preharvest interventions intended to prevent *Escherichia coli* O157:H7 illness in humans. *Foodborne Path Dis* **6:**817–825.
44. **Valcour JE, Michel P, McEwen SA, Wilson JB.** 2002. Associations between indicators of

livestock farming intensity and incidence of human Shiga toxin-producing *Escherichia coli* infection. *Emerg Infect Dis* **8:**252–257.

45. **Trevena WB, Willshaw GA, Cheasty T, Wray C, Gallagher J.** 1996. Vero cytotoxin-producing *E coli* O157 infection associated with farms. *Lancet* **347:**60–61.
46. **Reymond D, Johnson RP, Karmali MA, Petric M, Winkler M, Johnson S, Rahn K, Renwick S, Wilson J, Clarke RC.** 1996. Neutralizing antibodies to *Escherichia coli* Vero cytotoxin 1 and antibodies to O157 lipopolysaccharide in healthy farm family members and urban residents. *J Clin Microbiol* **34:**2053–2057.
47. **Rahn K, Renwick SA, Johnson RP, Wilson JB, Clarke RC, Alves D, McEwen SA, Lior H, Spika J.** 1998. Follow-up study of verocytoxigenic *Escherichia coli* infection in dairy farm families. *J Infect Dis* **177:**1138–1139.
48. **Wilson J, Spika J, Clarke R, McEwen S, Johnson R, Rahn K, Renwick S, Karmali M, Lior H, Alves D, Gyles C, Sandhu K.** 1998. Verocytotoxigenic *Escherichia coli* infection in dairy farm families. *Can Commun Dis Rep* **24:**17–20.
49. **Varma JK, Greene KD, Reller ME, DeLong SM, Trottier J, Nowicki SF, DiOrio M, Koch EM, Bannerman TL, York ST, Lambert-Fair MA, Wells JG, Mead PS.** 2003. An outbreak of *Escherichia coli* O157 infection following exposure to a contaminated building. *JAMA* **290:**2709–2712.
50. **Alam MJ, Zurek L.** 2004. Association of *Escherichia coli* O157:H7 with houseflies on a cattle farm. *Appl Environ Microbiol* **70:**7578–7580.
51. **Hancock DD, Besser TE, Rice DH, Ebel ED, Herriott DE, Carpenter LV.** 1998. Multiple sources of *Escherichia coli* O157 in feedlots and dairy farms in the northwestern USA. *Prev Vet Med* **35:**11–19.
52. **Janisiewicz WJ, Conway WS, Brown MW, Sapers GM, Fratamico P, Buchanan RL.** 1999. Fate of *Escherichia coli* O157:H7 on fresh-cut apple tissue and its potential for transmission by fruit flies. *Appl Environ Microbiol* **65:**1–5.
53. **Kobayashi M, Sasaki T, Saito N.** 1999. Houseflies: not simple mechanical vectors of enterohemorrhagic *Escherichia coli* O157:H7. *Am J Trop Med Hyg* **61:**625–629.
54. **Moriya K, Fujibayashi T, Yoshihara T, Matsuda A, Sumi N, Umezaki N, Kurahashi H, Agui N, Wada A, Watanabe H.** 1999. Verotoxin-producing *Escherichia coli* O157:H7 carried by the housefly in Japan. *Med Vet Entomol* **13:**214–216.
55. **Beutin L, Geier D, Steinruck H, Zimmermann S, Scheutz F.** 1993. Prevalence and some properties of verotoxin (Shiga-like toxin)-producing *Escherichia coli* in seven different species of healthy domestic animals. *J Clin Microbiol* **31:**2483–2488.
56. **Karmali MA, Gannon V, Sargeant JM.** 2010. Verocytotoxin-producing *Escherichia coli* (VTEC). *Vet Microbiol* **140:**360–370.
57. **Ferens WA, Hovde CJ.** 2011. *Escherichia coli* O157:H7: animal reservoir and sources of human infection. *Foodborne Pathog Dis* **8:**465–487.
58. **Jay MT, Cooley M, Carychao D, Wiscomb GW, Sweitzer RA, Crawford-Miksza L, Farrar JA, Lau DK, O'Connell J, Millington A, Asmundson RV, Atwill ER, Mandrell RE.** 2007. *Escherichia coli* O157:H7 in feral swine near spinach fields and cattle, central California coast. *Emerg Infect Dis* **13:**1908–1911.
59. **Anonymous.** 2012. *Strawberries, Deer and Other Investigations.* Oregon Health Authority, Portland, Oregon.
60. **Kim J, Nietfeldt J, Benson A.** 1999. Octamer based genome scanning distinguishes a subpopulation of *Escherichia coli* O157:H7 strains in cattle. *Proc Natl Acad Sci USA* **96:**13288–13293.
61. **Elder RO, Keen JE, Siragusa GR, Barkocy-Gallagher GA, Koohmaraie M, Laegreid WW.** 2000. Correlation of enterohemorrhagic *Escherichia coli* O157 prevalence in feces, hides, and carcasses of beef cattle during processing. *Proc Natl Acad Sci USA* **97:**2999–3003.
62. **Arthur TM, Bosilevac JM, Nou X, Shackelford SD, Wheeler TL, Kent MP, Jaroni D, Pauling B, Allen DM, Koohmaraie M.** 2004. *Escherichia coli* O157 prevalence and enumeration of aerobic bacteria, *Enterobacteriaceae*, and *Escherichia coli* O157 at various steps in commercial beef processing plants. *J Food Prot* **67:**658–665.
63. **Smith DR.** 2014. Cattle production systems: ecology of existing and emerging *Escherichia coli* types related to foodborne illness. *Annu Rev Anim Biosci* **2:**23.
64. **Savageau MA.** 1983. *Escherichia coli* habitats, cell types, and molecular mechanisms of gene control. *Am Naturalist* **122:**732–744.
65. **Peterson RE, Klopfenstein TJ, Erickson GE, Folmer J, Hinkley S, Moxley RA, Smith DR.** 2007. Effect of *Lactobacillus acidophilus* strain NP51 on *Escherichia coli* O157:H7 fecal shedding and finishing performance in beef feedlot cattle. *J Food Prot* **70:**287–291.
66. **Potter AA, Klashinsky S, Li Y, Frey E, Townsend H, Rogan D, Erickson G, Hinkley S, Klopfenstein T, Moxley RA, Smith DR, Finlay BB.** 2004. Decreased shedding of *Escherichia coli* O157:H7 by cattle following vaccination with type III secreted proteins. *Vaccine* **22:**362–369.

67. **Khaitsa ML, Smith DR, Stoner JA, Parkhurst AM, Hinkley S, Klopfenstein TJ, Moxley RA.** 2003. Incidence, duration, and prevalence of *Escherichia coli* O157:H7 fecal shedding by feedlot cattle during the finishing period. *J Food Prot* **66:**1972–1977.
68. **Smith DR, Moxley RA, Clowser SL, Folmer JD, Hinkley S, Erickson GE, Klopfenstein TJ.** 2005. Use of rope devices to describe and explain the feedlot ecology of *Escherichia coli* O157:H7 by time and place. *Foodborne Pathog Dis* **2:**50–60.
69. **Smith DR, Blackford MP, Younts SM, Moxley RA, Gray JT, Hungerford LL, Milton CT, Klopfenstein TJ.** 2001. Ecological relationships between the prevalence of cattle shedding *Escherichia coli* O157:H7 and characteristics of the cattle or conditions of the feedlot pen. *J Food Prot* **64:**1899–1903.
70. **Naylor SW, Low JC, Besser TE, Mahajan A, Gunn GJ, Pearce MC, McKendrick IJ, Smith DG, Gally DL.** 2003. Lymphoid follicle-dense mucosa at the terminal rectum is the principal site of colonization of enterohemorrhagic *Escherichia coli* O157:H7 in the bovine host. *Infect Immun* **71:**1505–1512.
71. **Grauke LJ, Kudva IT, Yoon JW, Hunt CW, Williams CJ, Hovde CJ.** 2002. Gastrointestinal tract location of *Escherichia coli* O157:H7 in ruminants. *Appl Environ Microbiol* **68:**2269–2277.
72. **Baehler AA, Moxley RA.** 2000. *Escherichia coli* O157:H7 induces attaching-effacing lesions in large intestinal mucosal explants from adult cattle. *FEMS Micribiol Lett* **185:**239–242.
73. **Moxley RA.** 2004. *Escherichia coli* O157:H7: an update on intestinal colonization and virulence mechanisms. *Anim Health Res Rev* **5:**15–33.
74. **Besser TE, Hancock DD, Pritchett LC, McRae EM, Rice DH, Tarr PI.** 1997. Duration of detection of fecal excretion of *Escherichia coli* O157:H7 in cattle. *J Infect Dis* **175:**729.
75. **Sanderson MW, Besser TE, Gay JM, Gay CC, Hancock DD.** 1999. Fecal *Escherichia coli* O157:H7 shedding patterns of orally inoculated calves. *Vet Microbiol* **69:**199–205.
76. **Rice DH, Sheng HQ, Wynia SA, Hovde CJ.** 2003. Rectoanal mucosal swab culture is more sensitive than fecal culture and distinguishes *Escherichia coli* O157:H7-colonized cattle and those transiently shedding the same organism. *J Clin Microbiol* **41:**4924–4929.
77. **Dargatz DA, Bai J, Lubbers BV, Kopral CA, An B, Anderson GA.** 2013. Prevalence of *Escherichia coli* O-types and Shiga-toxin genes in fecal samples from feedlot cattle. *Foodborne Pathog Dis* **10:**392–296.
78. **Renter DG, Morris JG Jr, Sargeant JM, Hungerford LL, Berezowski J, Ngo T, Williams K, Acheson DW.** 2005. Prevalence, risk factors, O serogroups, and virulence profiles of Shiga toxin-producing bacteria from cattle production environments. *J Food Prot* **68:**1556–1565.
79. **Wells JG, Shipman LD, Greene KD, Sowers EG, Green JH, Cameron DN, Downes FP, Martin ML, Griffin PM, Ostroff SM.** 1991. Isolation of *Escherichia coli* serotype O157:H7 and other Shiga-like-toxin-producing *E. coli* from dairy cattle. *J Clin Microbiol* **29:**985–989.
80. **Hussein HS, Sakuma T.** 2005. Prevalence of Shiga toxin-producing *Escherichia coli* in dairy cattle and their products. *J Dairy Sci* **88:**450–465.
81. **Arthur TM, Barkocy-Gallagher GA, Rivera-Betancourt M, Koohmaraie M.** 2002. Prevalence and characterization of non-O157 Shiga toxin-producing *Escherichia coli* on carcasses in commercial beef cattle processing plants. *Appl Environ Microbiol* **68:**4847–4852.
82. **Barkocy-Gallagher GA, Arthur TM, Rivera-Betancourt M, Nou X, Shackelford SD, Wheeler TL, Koohmaraie M.** 2003. Seasonal prevalence of Shiga toxin-producing *Escherichia coli*, including O157:H7 and non-O157 serotypes, and *Salmonella* in commercial beef processing plants. *J Food Prot* **66:**1978–1986.
83. **Hancock DD, Rice DH, Thomas L, Dargatz DA, Besser TE.** 1997. Epidemiology of *Escherichia coli* O157 in feedlot cattle. *J Food Prot* **60:** 462–465.
84. **Hancock D, Besser TE, Rice DH, Herriot DE, Tarr PI.** 1997. A longitudinal study of *Escherichia coli* O157:H7 in fourteen cattle herds. *Epidemiol Infect* **118:**193–195.
85. **Stanford K, Croy D, Bach SJ, Wallins GL, Zahiroddini H, McAllister TA.** 2005. Ecology of *Escherichia coli* O157:H7 in commercial dairies in southern Alberta. *J Dairy Sci* **88:** 4441–4451.
86. **Herriott DE, Hancock DD, Ebel ED, Carpenter LV, Rice DH, Besser TE.** 1998. Association of herd management factors with colonization of dairy cattle by Shiga toxin-positive *Escherichia coli* O157. *J Food Prot* **61:**802–807.
87. **Laegreid WW, Elder RO, Keen JE.** 1999. Prevalence of *Escherichia coli* O157:H7 in range beef calves at weaning. *Epidemiol Infect* **123:**291–298.
88. **Dunn JR, Keen JE, Del VR, Wittum TE, Thompson RA.** 2004. *Escherichia coli* O157:H7 in a cohort of weaned, preconditioned range beef calves. *J Food Prot* **67:**2391–2396.
89. **Renter DG, Sargeant JM, Hungerford LL.** 2004. Distribution of *Escherichia coli* O157:H7

within and among cattle operations in pasture-based agricultural areas. *Am J Vet Res* **65:** 1367–1376.

90. **Smith DR, Gray JT, Moxley RA, Younts-Dahl SM, Blackford MP, Hinkley S, Hungerford LL, Milton CT, Klopfenstein TJ.** 2004. A diagnostic strategy to determine the Shiga toxin-producing *Escherichia coli* O157 status of pens of feedlot cattle. *Epidemiol Infect* **132:**297–302.
91. **Stanford K, Bach SJ, Marx TH, Jones S, Hansen JR, Wallins GL, Zahiroddini H, McAllister TA.** 2005. Monitoring *Escherichia coli* O157:H7 in inoculated and naturally colonized feedlot cattle and their environment. *J Food Prot* **68:**26–33.
92. **Van Donkersgoed J, Berg J, Potter A, Hancock D, Besser T, Rice D, Lejeune J, Klashinsky S.** 2001. Environmental sources and transmission of *Escherichia coli* O157 in feedlot cattle. *Can Vet J* **42:**714–720.
93. **Renter DG, Smith DR, King R, Stilborn R, Berg J, Berezowski J, McFall M.** 2008. Detection and determinants of *Escherichia coli* O157:H7 in Alberta feedlot pens immediately prior to slaughter. *Can J Vet Res* **72:**217–227.
94. **Edrington TS, Callaway TR, Hallford DM, Chen L, Anderson RC, Nisbet DJ.** 2008. Effects of exogenous melatonin and tryptophan on fecal shedding of *E. coli* O157:H7 in cattle. *Microb Ecol* **55:**553–560.
95. **Edrington TS, Callaway TR, Ives SE, Engler MJ, Looper ML, Anderson RC, Nisbet DJ.** 2006. Seasonal shedding of *Escherichia coli* O157:H7 in ruminants: a new hypothesis. *Foodborne Pathog Dis* **3:**413–421.
96. **Edrington TS, Farrow RL, MacKinnon KM, Callaway TR, Anderson RC, Nisbet DJ.** 2012. Influence of vitamin D on fecal shedding of *Escherichia coli* O157:H7 in naturally colonized cattle. *J Food Prot* **75:**314–319.
97. **Moxley RA, Smith DR, Luebbe M, Erickson GE, Klopfenstein TJ, Rogan D.** 2009. *Escherichia coli* O157:H7 vaccine dose-effect in feedlot cattle. *Foodborne Pathog Dis* **6:**879–884.
98. **Smith DR, Moxley RA, Clowser SL, Folmer JD, Hinkley S, Erickson GE, Klopfenstein TJ.** 2005. Use of rope devices to describe and explain the feedlot ecology of *Salmonella* by time and place. *Foodborne Pathog Dis* **2:**61–69.
99. **Smith DR, Moxley RA, Peterson RE, Klopfenstein T, Erickson GE, Clowser SL.** 2008. A two-dose regimen of a vaccine against *Escherichia coli* O157:H7 type III secreted proteins reduced environmental transmission of the agent in a large-scale commercial beef feedlot clinical trial. *Foodborne Pathog Dis* **5:**589–598.
100. **Garber L, Wells S, Schroeder-Tucker L, Ferris K.** 1999. Factors associated with fecal shedding of verotoxin-producing *Escherichia coli* O157 on dairy farms. *J Food Prot* **62:**307–312.
101. **Younts-Dahl SM, Galyean ML, Loneragan GH, Elam NA, Brashears MM.** 2004. Dietary supplementation with *Lactobacillus*- and *Propionibacterium*-based direct-fed microbials and prevalence of *Escherichia coli* O157 in beef feedlot cattle and on hides at harvest. *J Food Prot* **67:**889–893.
102. **Younts-Dahl SM, Osborn GD, Galyean ML, Rivera JD, Loneragan GH, Brashears MM.** 2005. Reduction of *Escherichia coli* O157 in finishing beef cattle by various doses of *Lactobacillus acidophilus* in direct-fed microbials. *J Food Prot* **68:**6–10.
103. **Cull CA, Paddock ZD, Nagaraja TG, Bello NM, Babcock AH, Renter DG.** 2012. Efficacy of a vaccine and a direct-fed microbial against fecal shedding of *Escherichia coli* O157:H7 in a randomized pen-level field trial of commercial feedlot cattle. *Vaccine* **30:**6210–6215.
104. **Bach SJ, Johnson RP, Stanford K, McAllister TA.** 2009. Bacteriophages reduce *Escherichia coli* O157:H7 levels in experimentally inoculated sheep. *Can J Anim Sci* **89:**285–293.
105. **Callaway TR, Edrington TS, Brabban AD, Anderson RC, Rossman ML, Engler MJ, Carr MA, Genovese KJ, Keen JE, Looper ML, Kutter EM, Nisbet DJ.** 2008. Bacteriophage isolated from feedlot cattle can reduce *Escherichia coli* O157:H7 populations in ruminant gastrointestinal tracts. *Foodborne Pathog Dis* **5:**183–191.
106. **Stanford K, McAllister TA, Niu YD, Stephens TP, Mazzocco A, Waddell TE, Johnson RP.** 2010. Oral delivery systems for encapsulated bacteriophages targeted at *Escherichia coli* O157:H7 in feedlot cattle. *J Food Prot* **73:**1304–1312.
107. **Callaway TR, Anderson RC, Genovese KJ, Poole TL, Anderson TJ, Byrd JA, Kubena LF, Nisbet DJ.** 2002. Sodium chlorate supplementation reduces *E. coli* O157:H7 populations in cattle. *J Anim Sci* **80:**1683–1689.
108. **Callaway TR, Edrington TS, Anderson RC, Genovese KJ, Poole TL, Elder RO, Byrd JA, Bischoff KM, Nisbet DJ.** 2003. *Escherichia coli* O157:H7 populations in sheep can be reduced by chlorate supplementation. *J Food Prot* **66:**194–199.
109. **Edrington TS, Callaway TR, Andersen RC, Genovese KJ, Jung YS, McReynolds JL, Bischoff KM, Nisbet DJ.** 2003. Reduction of *E. coli* O517:H7 populations in sheep by supplementation of an experimental sodium chlorate product. *Small Ruminant Res* **49:**173–181.

110. **Loneragan GH, Brashears MM.** 2005. Preharvest interventions to reduce carriage of *E. coli* O157 by harvest-ready feedlot cattle. *Meat Sci* **71:**72–78.
111. **Wells JE, Shackelford SD, Berry ED, Kalchayanand N, Guerini MN, Varel VH, Arthur TM, Bosilevac JM, Freetly HC, Wheeler TL, Ferrell CL, Koohmaraie M.** 2009. Prevalence and level of *Escherichia coli* O157:H7 in feces and on hides of feedlot steers fed diets with or without wet distillers grains with solubles. *J Food Prot* **72:**1624–1633.
112. **Jacob ME, Paddock ZD, Renter DG, Lechtenberg KF, Nagaraja TG.** 2010. Inclusion of dried or wet distillers' grains at different levels in diets of feedlot cattle affects fecal shedding of *Escherichia coli* O157:H7. *Appl Environ Microbiol* **76:**7238–7242.
113. **Smith DR, Vogstad AR.** 2012. Vaccination as a method of *E. coli* O157:H7 reduction in feedlot cattle. *In* Callaway TR, Edrington TS (ed), *On Farm Strategies to Control Foodborne Pathogens.* Nova Science Publishers, Hauppauge, NY.
114. **Snedeker KG, Campbell M, Sargeant JM.** 2012. A systematic review of vaccinations to reduce the shedding of *Escherichia coli* O157 in the faeces of domestic ruminants. *Zoonoses Public Health* **59:**126–138.
115. **Varela NP, Dick P, Wilson J.** 2012. Assessing the existing information on the efficacy of bovine vaccination against *Escherichia coli* O157:H7: a systematic review and meta-analysis. *Zoonoses Public Health* **60:**253–268.
116. **Fox JT, Thomson DU, Drouillard JS, Thornton AB, Burkhardt DT, Emery DA, Nagaraja TG.** 2009. Efficacy of *Escherichia coli* O157:H7 siderophore receptor/porin proteins-based vaccine in feedlot cattle naturally shedding *E. coli* O157. *Foodborne Pathog Dis* **6:**893–899.
117. **Peterson RE, Klopfenstein TJ, Moxley RA, Erickson GE, Hinkley S, Bretschneider G, Berberov EM, Rogan D, Smith DR.** 2007. Effect of a vaccine product containing type III secreted proteins on the probability of *Escherichia coli* O157:H7 fecal shedding and mucosal colonization in feedlot cattle. *J Food Prot* **70:** 2568–2577.
118. **Smith DR, Moxley RA, Peterson RE, Klopfenstein TJ, Erickson GE, Bretschneider G, Berberov EM, Clowser S.** 2009. A two-dose regimen of a vaccine against type III secreted proteins reduced *Escherichia coli* O157:H7 colonization of the terminal rectum in beef cattle in commercial feedlots. *Foodborne Pathog Dis* **6:**155–161.
119. **Thomson DU, Loneragan GH, Thornton AB, Lechtenberg KF, Emery DA, Burkhardt DT, Nagaraja TG.** 2009. Use of a siderophore receptor and porin proteins-based vaccine to control the burden of *Escherichia coli* O157:H7 in feedlot cattle. *Foodborne Pathog Dis* **6:**871–877.
120. **Vogstad AR.** 2012. *Modeling the efficacy and effectiveness of* Escherichia coli *O157:H7 preharvest interventions. Masters thesis,* University of Nebraska-Lincoln, Lincoln, NE.
121. **Smith DR, Moxley RA, Klopfenstein TJ, Erickson GE.** 2009. A randomized longitudinal trial to test the effect of regional vaccination within a cattle feedyard on *Escherichia coli* O157:H7 rectal colonization, fecal shedding, and hide contamination. *Foodborne Pathog Dis* **6:**885–892.
122. **Miller MF, Loneragan GH, Harris DD, Adams KD, Brooks JC, Brashears MM.** 2008. Environmental dust exposure as a factor contributing to an increase in *Escherichia coli* O157:H7 and *Salmonella* populations on cattle hides in feedyards. *J Food Prot* **71:**2078–2081.
123. **Reicks AL, Brashears MM, Adams KD, Brooks JC, Blanton JR, Miller MF.** 2007. Impact of transportation of feedlot cattle to the harvest facility on the prevalence of *Escherichia coli* O157:H7, *Salmonella,* and total aerobic microorganisms on hides. *J Food Prot* **70:**17–21.
124. **Asper DJ, Sekirov I, Finlay BB, Rogan D, Potter AA.** 2007. Cross reactivity of enterohemorrhagic *Escherichia coli* O157:H7-specific sera with non-O157 serotypes. *Vaccine* **25:**8262–8269.
125. **CDC.** 2014. *Salmonella.* http://www.cdc.gov/salmonella/.
126. **Gould LH, Walsh KA, Vieira AR, Herman K, Williams IT, Hall AJ, Cole D, Centers for Disease Control and Prevention.** 2013. Surveillance for foodborne disease outbreaks: United States, 1998-2008. *MMWR Surveill Summ* **62:**1–34. http://www.cdc.gov/mmwr/preview/mmwrhtml/ss6202a1.htm.
127. **Mohler VL, House J.** 2009. Salmonellosis in ruminants, p 106–111. *In* Anderson DE, Rings DM (ed), *Current Veterinary Therapy: Food Animal Practice,* 5th ed. Saunders, Elsevier, St. Louis, MO.
128. **Gragg SE, Loneragan GH, Brashears MM, Arthur TM, Bosilevac JM, Kalchayanand N, Wang R, Schmidt JW, Brooks JC, Shackelford SD, Wheeler TL, Brown TR, Edrington TS, Brichta-Harhay DM.** 2013. Cross-sectional study examining *Salmonella enterica* carriage in subiliac lymph nodes of cull and feedlot cattle at harvest. *Foodborne Pathog Dis* **10:**368–374.
129. **Gragg SE, Loneragan GH, Nightingale KK, Brichta-Harhay DM, Ruiz H, Elder JR, Garcia LG, Miller MF, Echeverry A, Ramirez**

Porras RG, Brashears MM. 2013. Substantial within-animal diversity of *Salmonella* isolates from lymph nodes, feces, and hides of cattle at slaughter. *Appl Environ Microbiol* **79:**4744–4750.

130. **Anonymous.** 2014. *Salmonella in U.S. Cattle Feedlots.* USDA, APHIS, Veterinary Services, Center for Epidemiology and Animal Health, Fort Collins, CO.
131. **Anonymous.** 2009. *Salmonella and Campylobacter on U.S. Dairy Operations, 1996–2007.* USDA, APHIS, Veterinary Services, Center for Epidemiology and Animal Health, Fort Collins, CO.
132. **Anonymous.** 2009. *Prevalence of Salmonella and Listeria in Bulk Tank Milk and Inline Filters on U.S. Dairies, 2007.* USDA, APHIS, Veterinary Services, Center for Epidemiology and Animal Health, Fort Collins, CO.
133. **CDC.** 2014. Campylobacter. http://www.cdc.gov/foodsafety/diseases/campylobacter/index.html.
134. **CDC.** July 29, 2014 2014. Listeria (*Listeriosis*). http://www.cdc.gov/listeria/.
135. **Francoz D.** 2009. Cranial nerve abnormalities, p 299–306. *In* Anderson DE, Rings DM (ed), *Current Veterinary Therapy: Food Animal Practice,* 5th ed. Saunders, Elsevier, St. Louis, MO.
136. **Koohmaraie M, Arthur TM, Bosilevac JM, Guerini M, Shackelford SD, Wheeler TL.** 2005. Post-harvest interventions to reduce/eliminate pathogens in beef. *Meat Sci* **71:**79–91.
137. **Maurin M, Raoult D.** 1999. Q fever. *Clin Microbiol Rev* **12:**518–553.
138. **CDC.** 2014. *Q Fever.* http://www.cdc.gov/qfever/.
139. **Barlow J, Rauch B, Welcome F, Kim SG, Dubovi E, Schukken Y.** 2008. Association between *Coxiella burnetii* shedding in milk and subclinical mastitis in dairy cattle. *Vet Res* **39:**23.
140. **Kim SG, Kim EH, Lafferty CJ, Dubovi E.** 2005. *Coxiella burnetii* in bulk tank milk samples, United States. *Emerg Infect Dis* **11:**619–621.
141. **Anonymous.** 2009. Grade “A” pasteurized milk ordinance, 2007 revision. US Food and Drug Administration.
142. **Anonymous.** 2013. Milk and dairy beef drug residue prevention, p 1–64. National Milk Producers Federation, Arlington, VA.

4

Preharvest Food Safety in Broiler Chicken Production

WALID Q. ALALI[1] and CHARLES L. HOFACRE[2]

INTRODUCTION

The goal of preharvest broiler food safety is to minimize opportunities for the introduction, persistence, and transmission of foodborne pathogens (mostly *Salmonella* and *Campylobacter*) in the bird flock. Current intervention strategies to combat foodborne pathogens in preharvest broiler production target *Salmonella* and *Campylobacter* in general, or more specifically target certain serotypes that are most frequently associated with human disease.

OVERVIEW OF *SALMONELLA* AND *CAMPYLOBACTER* INFECTIONS IN HUMANS THAT ARE LINKED TO POULTRY PRODUCTS

Salmonella and *Campylobacter* infections are a significant public health concern in the United States and worldwide (1–3). In the United States, the Centers for Disease Control and Prevention (CDC) estimated that 1.03 million and 0.8 million cases of *Salmonella* and *Campylobacter* infections, respectively, occur annually (3). Both pathogens are frequently associated with the consumption of contaminated poultry products. It is estimated that 10 to 29% of

[1]College of Public Health, Hamad bin Khalifa University, Doha-Qatar; [2]Department of Population Health, Poultry Diagnostic and Research Center, University of Georgia, Athens, GA 30602.

Preharvest Food Safety
Edited by Siddhartha Thakur and Kalmia E. Kniel

doi:10.1128/microbiolspec.PFS-0002-2014

Salmonella infections and 43% of *Campylobacter* infections are associated with poultry products (4–7).

Poultry meat production and consumption is of significant economic importance in many countries worldwide. Poultry is the second most consumed meat (33% of all meats) globally, after pork (37%) (8). However, poultry is the fastest growing meat sector as reflected by the increase in world consumption. The consumption of poultry meat increased by 14.4% from 2004 to 2008 (from 80 million tons in 2004 to 93 million tons in 2008). While the consumption increased slightly in the United States (2.9%), it increased by 25% in China. The United States is ranked number one in poultry meat production, followed by China and Brazil. However, the largest exporters of poultry meat are Brazil and the United States.

With such increasing poultry meat production and consumption worldwide, the safety of poultry products is a priority to consumers, producers, and government regulators. Poultry meat has been identified as one of the most significant food vehicles for *Salmonella* and *Campylobacter*. For instance, in the United States, an estimated 100,000 to 300,000 cases of human salmonellosis each year have been attributed to the consumption of contaminated poultry products. Despite the control and intervention strategies established to control *Salmonella* contamination in processed poultry, the U.S. Department of Agriculture (USDA) Food Safety and Inspection Service (FSIS) reported a 4.3% *Salmonella* prevalence on raw broiler carcasses in poultry processing plants in 2012 (9). Since the United States produces an estimated 8.9 billion broiler birds a year, around 382 million carcasses can be contaminated each year. The latest estimate in 2011 from a USDA-FSIS microbiological baseline data collection survey revealed that *Campylobacter* contamination prevalence on broilers was 9.4% (10).

The per capita consumption of poultry products has increased 6.5-fold since 1910 (11). An increase in consumption of poultry increases the potential risk for exposure to *Salmonella* and *Campylobacter* through contaminated poultry products unless steps are taken to improve the microbiological safety of the products.

The most commonly identified *Salmonella* serovars associated with human infections in the United States were *Salmonella enterica* Enteritidis, Typhimurium, Newport, Javiana, I 4,[5],12:i:-, Montevideo, Heidelberg, Muenchen, Infantis, and Braenderup (12). According to a USDA-FSIS report in 2011, the top serovars associated with broiler carcasses were *S. enterica* Kentucky, Enteritidis, Typhimurium, Infantis, Heidelberg, Johannesburg, 4,[5],12:I:-, 8,20:-:z6, Mbandaka, and Berta (13). Although it's not commonly associated with human infection, *S.* Kentucky has been identified as the most common serotype isolated from broilers (51.8% of *Salmonella* detected on carcasses) and retail chicken breasts (25.3%) (13, 14). *S.* Kentucky was recently identified as a potential emerging multidrug-resistant human pathogen in Europe and North Africa because the numbers of cases have risen significantly over the last decade (15). In the United States, *S.* Kentucky increased from averaging about 62 cases per year in 1996 to 2004 before climbing to 123 cases in 2006 (12). The prevalence of multidrug-resistant *S.* Kentucky in poultry is also alarming; according to National Antimicrobial Resistance Monitoring System data for 2007, 50% of isolates from broiler meat that were resistant to ≥5 antimicrobials were identified as *S.* Kentucky (14).

Unlike *Salmonella*, the State Public Health laboratories that participate in the CDC FoodNet surveillance program do not serotype *Campylobacter* isolates (16). Based on the scientific literature, the most commonly identified *Campylobacter* serovars associated with human infections in the United States were *Campylobacter jejuni* and *Campylobacter coli* (17–19). *C. jejuni* is the most common serovar associated with poultry products, followed by *C. coli* (20–23).

STRUCTURE OF INTEGRATED POULTRY PRODUCTION SYSTEM IN THE UNITED STATES

It is necessary to understand the integrated broiler production system to better understand the transmission dynamics of *Salmonella* and *Campylobacter* as well as the development and application of management strategies and intervention methods. The poultry industry in the United States and in many countries worldwide is a vertically integrated production, processing, and distribution system. Most poultry companies in the United States have contract agreements with farmers (growers) that bind farmers to grow birds (breeder or broilers) in exchange for payment. Operating expenses such as repairs and maintenance, clean-up cost, manure, and dead bird disposal are the responsibility of the farmer. Most integrators require houses to be built and equipped according to strict specifications. The poultry companies control the rest of the production processes, which include providing chicks to the farmers growing breeders (parent stock) or broiler flocks; producing and distributing feed; providing veterinary and management services to the farms; hatching the breeder-broiler eggs; transporting broiler birds to the processing plant; and processing, production, and marketing of broiler meat.

USDA-FSIS NEW STANDARDS FOR *SALMONELLA* AND *CAMPYLOBACTER*

The USDA-FSIS implemented new performance standards for *Salmonella*, and for the first time, standards for *Campylobacter* on broiler carcasses at slaughter/processing plants became effective 1 July 2011. The number of samples collected (n = 51) for broiler pathogen verification remains the same as before; however, each carcass is now analyzed not only for *Salmonella*, but also for *Campylobacter*. A processing plant passes the *Salmonella* and *Campylobacter* standards if no more than 5 and 8 positive samples, respectively, in a 51-sample set for broilers are detected. If a plant fails to meet these standards, the FSIS issues a warning and collects a follow-up 51 samples and analyzes them for both pathogens.

The New *Salmonella* Standards for Broilers

Based on the latest microbiological baseline survey of broiler carcasses (2011–2012), FSIS estimates of *Salmonella* prevalence on postchilled carcasses was 4.3% (9). With a goal of further reduction of human illness associated with chicken products, the FSIS issued these new standards to improve pathogen control in processing plants. The FSIS believes that the standard of no more than 5 positives out of 51 samples will lead to an 80% probability that a processing plant will meet a 7.5% performance standard. The qualitative (presence/absence) *Salmonella* testing method will continue as before, using the 400-ml carcass rinse to determine which carcasses are positive. The FSIS predicts that in the first two years of implementing the new standards for *Salmonella* for broilers, a reduction of approximately 26,000 human cases will occur compared to the old standards (i.e., no more than 13 positives out of a 51-sample set).

The New *Campylobacter* Standards for Broilers

In the microbiological survey in 2011, the FSIS collected quantitative and qualitative data for *Campylobacter* on broiler carcasses. The FSIS estimate of *Campylobacter* prevalence on postchilled carcasses was 9.4%. The agency uses a direct plating method (1 ml of carcass rinsate) to quantify (enumerate) *Campylobacter*, as well as a qualitative (enrichment) method using 30 ml of rinsate to detect the organism when direct plating is negative. The 1-ml direct plating test procedure is less sensitive than the 30-ml procedure, but it detects a high level of contamination.

The performance standards for *Campylobacter* will be based on the results from the two detection methods: the number of samples that are positives based on the 1-ml method and the combined percentage of positive results of either the 1-ml and/or the 30-ml methods.

The performance standard for the 1-ml method is no more than 8 positive samples of the 51-sample set, whereas the performance standard for results of both methods is no more than 27 positive of 51 samples. The FSIS predicts that in the first two years of implementing the new standards for *Campylobacter* for broilers, a reduction of approximately 39,000 human cases will occur compared to prior implementation of the standards.

CONTROL OF FOODBORNE PATHOGENS IN BROILER PRODUCTION SYSTEMS AT THE PREHARVEST LEVEL

Several factors can affect *Salmonella* and *Campylobacter* colonization in broilers, including the age and genetic susceptibility of the birds, stress due to overcrowding or underlying illness, level of pathogen exposure and virulence of the organism to facilitate colonization and/or evade host defenses, and competition with gut microbiota for colonization sites (19, 24, 25). Broilers are most susceptible to pathogen colonization during the first few days of hatch by vertical transmission from infected parents or by horizontal transmission at the hatcheries during feeding, handling, and transportation (24, 26, 27). Additionally, birds are highly susceptible to *Salmonella* and *Campylobacter* infections during feed withdrawal prior to slaughter due to stress, consumption of contaminated litter, and alterations of the crop pH and intestinal microbiota (28–30). Preharvest interventions to control *Salmonella* and *Campylobacter* in broilers should be directed largely at the most critical points for pathogen introduction to the flock. In this section, we discuss the different routes of *Salmonella* and *Campylobacter* entry to broiler flocks and possible intervention methods.

Feed Contamination and Recontamination

Feed in general is considered a vehicle for foodborne pathogen transmission in food animal production, and in poultry production in particular. Poultry birds that consume contaminated feed could get infected and colonized with pathogens, resulting in pathogen colonization and fecal shedding in the farm environment. Although the poultry industry understands the role and importance of feed as a vehicle of foodborne pathogen transmission in poultry production, contamination of feed at the mill continues to be a problem since control and/or elimination of contaminants is considered a daunting task. There are several challenges involved with detection of *Salmonella* and/or *Campylobacter* in poultry feed: (i) organisms are not uniformly distributed in feed; (ii) organisms might be damaged or injured, making them difficult to isolate/detect; (iii) microbiological and molecular laboratory methods have detection limits, especially for dry sample matrix; and (iv) deciding on the number of samples to collect for microbial analysis as well as the sample analytical volume is a challenge. That being said, the goal of pathogen control in feed should be to ensure that feed contaminants are below a certain level (i.e., threshold) which will pose minimal risk to human and bird health. This can be achieved by following "sound and practical" microbial sampling plans, good manufacturing practices, and application of intervention strategies such as chemical and physical treatments to control pathogens in feed.

It has been reported that poultry feed is often contaminated with a variety of pathogens such as *Salmonella* spp. and *Enterococcus* (31, 32). Feed is mostly free of *Campylobacter* since feed is dried and pelleted under high temperature; therefore, the survival of this organism in feed is unlikely (33, 34). Animal feed manufacturing and distribution is a multistage process as follows: (i) in the field (preharvest) with plant crop (e.g., corn and soybean) processing as well as production of protein meals (e.g., animal, bone, and fish); (ii) at the feed mill (i.e., harvest), where feed goes through grinding, mixing, pelleting, and then storage at the mill; and (iii) in post-harvest, which includes transportation of feed to animal farms, storage at the farms, and dispensing of feed to animals for consumption. Bacterial contamination of animal feed can occur at various manufacturing stages. Control of pathogens in raw feed ingredients and during production has the most significant impact on feed contamination. At the feed mill, poultry feed can become contaminated during manufacture/processing as a result of cross-contamination. The most critical point for microbial contamination and growth at the feed mill is the pellet cooler. The combination of moisture and heat is a favorable environment for microbial survival and growth. The heating process during production is required to pellet the feed, which usually kills most pathogens in the produced feed. Inadequate temperature for the pelleting equipment and feed conditioner would increase the likelihood of microbial survival.

Contamination of feed before and after the heating point is common and can be attributed to many factors within the feed mill facilities. These include unclean receiving and unloading area, unclean intake pits, dust generated by the feed ingredients, dirty conveyers with leftover feed from previous loads, inadequate feed storage conditions, and presence of visitors in unsanitary cloths (35). Feed ingredients should be inspected prior to unloading for signs of rodent contamination, bird droppings, and insect infestation. Inspection is particularly important given the difficulty of testing large amounts of incoming ingredients and of testing the feed during and after production. This due to the cost, time, and labor associated with the microbial analysis. Therefore, feed mills should consider investigating which suppliers have been consistent and reliable in delivering ingredients without macro and/or micro levels of contamination. The poultry industry may avoid using high-risk *Salmonella*-contaminated ingredients such as animal by-products to reduce the likelihood of contaminated feed. Microbial sampling plans designed to determine the minimum number of sampling units that represent the microbiological quality of feed should be considered when needed.

Several chemical and physical interventions have been applied to feed to prevent and control *Salmonella* contamination. These methods include treating feed with acids (e.g., organic and propionic acids) and formaldehyde, heat treatments, and irradiation (36–39). Combinations of acids and heat treatments are common in the poultry feed industry because they have been proven to effectively reduce the level of *Salmonella* in the end product (40).

Breeder Flocks (Parent Stock)

Vaccination programs for breeder flocks are becoming part of the regular food safety control system in most poultry companies in the United States and the European Union. The goal of vaccination in broiler breeder farms is to reduce the vertical transmission of *Salmonella* spp. to the hatching progeny and ultimately to reduce the incidence of the organisms being carried into the processing plant. Jones et al. (41) surveyed *Salmonella* contamination in broiler production and processing systems in the United States. They observed similar rates of contamination in fecal droppings collected from breeder houses, day-old chicks, and live birds in broiler houses. They suggested that the breeder house environment is an important vehicle for the transmission of contamination to the final product.

For *Salmonella*, the two types of vaccinations used for breeder flocks are inactivated (killed) and live-attenuated. Both types of vaccines have demonstrated effectiveness in controlling *Salmonella* spp. transmission to broiler chicks (42–45). Live-attenuated vaccines are used to provide hens with long-term immunity against *Salmonella* spp. colonization, whereas killed vaccines provide immunity that passes from the hen into the eggs. Some poultry companies use both types of vaccine for better immunity against *Salmonella* in breeder broiler birds.

A recent study compared poultry companies that vaccinated their breeders with live-attenuated *S.* Typhimurium followed by a booster dose of killed *Salmonella* vaccine (*S.* Berta and *S.* Kentucky) to companies that did not vaccinate (45). The authors reported that vaccinated breeder birds had significantly lower prevalence of *Salmonella* in the ceca (38.3% versus 64.2%) and the reproductive tracts (14.22% versus 51.7%) when compared to nonvaccinated breeders. Furthermore, the authors found that broiler progeny chicks from vaccinated breeders had significantly lower *Salmonella* prevalence (18.1%) compared to chicks from nonvaccinated breeders (33.5%). Moreover, breeder vaccination was a significant factor associated with lower environmental *Salmonella* contamination of broiler chicken farms. The authors also followed the birds to processing plants, where they found a significantly lower *Salmonella* prevalence on broilers (from vaccinated breeders) entering the processing plants (23.4%) compared to nonvaccinated flocks (33.5%). In another study, Berghaus et al. (44) compared breeder flocks that were vaccinated with a killed *Salmonella* vaccine (three serotypes: Typhimurium, Enteritidis, and Kentucky) to breeder flocks that were not vaccinated. They reported that vaccinated breeder flocks had significantly higher humoral immunity, and their progeny had significantly lower *Salmonella* prevalence and numbers compared to their nonvaccinated counterparts. However, vaccination of breeders did not significantly reduce the environmental *Salmonella* contamination of breeder farms. Poultry companies are encouraged to vaccinate their breeder flocks with live-attenuated and/or killed vaccine, especially if the company has a history of high prevalence of *Salmonella* in their broiler flocks and/or at the processing plant.

Regarding *Campylobacter* vaccine programs for breeders, researchers revealed that development of live-attenuated vaccine has been hampered by the genomic and phenotypic instability of this organism (46, 47). Researchers have used *S.* Typhimurium engineered genes that delivered immunity to *Campylobacter*, but with limited success (48–50). Furthermore, other types of *Campylobacter* vaccines have been investigated. It has been demonstrated that killed and flagellin-based *Campylobacter* vaccines do not possess an effective immune response to protect chickens from colonization with this organism (51–55). *Campylobacter* resides in the mucosal layer of the chicken intestine, so it might not stimulate enough immune response to provide protection (46). To date, no *Campylobacter* vaccine that is effective in commercial poultry settings has been developed for breeder or broiler flocks.

Hatchery

Broiler eggs can be contaminated at the breeder farm and at the hatchery (56). The hatching chick might be infected with *Salmonella* spp. through vertical and/or horizontal transmission. Vertical transmission occurs when breeder hens with contaminated reproductive tracts lay fertile broiler eggs that harbor *Salmonella* (57, 58). Horizontal transmission can occur when contaminated eggs are hatched and *Salmonella* spreads among newly hatched chicks (59). At the farm, laid eggs are exposed to environmental contaminants (e.g., feces and dust) which may contain *Salmonella.* This pathogen can contaminate the outer surface of the egg shell and may penetrate the shell and multiply inside the egg (60).

Commercial hatcheries are considered a major source of *Salmonella* infection of young chicks (61–63). This is mainly because *Salmonella*-negative chicks are exposed to *Salmonella*-positive chicks in a confined environment. The dust generated from contaminated eggs and the fluff from newly hatched chicks can widely spread *Salmonella* in the hatchery cabinet (61–63). Researchers reported that *Salmonella* was detected on

71% of eggshell fragments, on 80% of swabs from the chick conveyor belt, and on 74% of chick tray liners that contained feces (61). In another study, Byrd et al. (64) reported a 12% *Salmonella* prevalence (n = 700 chick tray liners tested) in hatcheries. Epidemiological studies have demonstrated a link between *Salmonella* isolates from the hatchery and the isolates found on processed chicken carcasses. Bhatia and McNabb (65) followed *Salmonella* contamination from the hatchery to the processing plant and found that the same *Salmonella* serotype was detected in fluff and meconium samples collected at the hatchery, in litter samples at the broiler farm, and on carcasses following processing.

Interestingly, hatcheries are considered a minor source of *Campylobacter* transmission and infection of chicks (66). This organism was detected in 0.75% (n = 2,000) of tray liners of day-of-hatch broiler chicks collected from eight commercial hatcheries (66). Vertical transmission of *Campylobacter* from breeder hens to their progeny is not well understood. A number of experimental studies revealed that breeder hens can potentially transmit *Campylobacter* vertically through eggs (67–71). Furthermore, Cox et al. (72) found that *Campylobacter* isolates from breeder hens' fecal samples had identical ribotype patterns to those isolates from their offspring's fecal droppings. These authors suggested that hatching cabinets and the hatchery environment can be a source of *Campylobacter* contamination and transmission to newly hatched birds.

Hatcheries employ a number of interventions to control the transmission of these pathogens. These include disinfection of the incubator and hatching cabinet surfaces as well as disinfection of the hatching eggs using UV light, ozone, electrostatic charging, and electrostatic spraying (73–77).

Farm and Farm Environment

There is a wide array of potential routes of *Salmonella* and *Campylobacter* infection of broilers as well as intervention methods to control these pathogens at broiler farms. In this section we will discuss (i) the role of poultry litter in pathogen transmission and control methods, (ii) the association between the feed withdrawal period and pathogen colonization as well as control methods, (iii) biosecurity measures at the farm to control infection of broiler flocks, (iv) antimicrobial use and pathogen prevalence in flocks, and (v) prebiotic and probiotic use to control these pathogens in flocks.

Litter

Litter is the absorbent bedding material used to cover the floor of a broiler chicken house. Depending on the availability of materials, litter can be made of wood shavings, pine straw, peanut hulls, and other dry, absorbent, low-cost materials. Used litter consists of a large amount of feces, feathers, and spilled feed and high levels of moisture. In the United States, the litter is generally used for several broiler flock cycles. A fresh layer of bedding is often added to the litter prior to placement of a new flock. In other countries, the litter is completely removed, and then the broiler house is cleaned and disinfected before placement of a new flock. Studies have investigated the epidemiological link between isolates detected on litter and contamination of broiler carcasses. Bhatia et al. (78) reported that *Salmonella* found on broiler litter was a reliable indication of both flock infection and carcass contamination. Furthermore, Bhatia and McNabb (65) revealed that the same *Salmonella* serotypes in litter samples at broiler farms were found on carcasses during processing. Furthermore, Corry et al. (79) found that *Salmonella* serotypes detected in litter samples during grow-out were usually detected in fewer birds at the end of the grow-out as well as in carcasses during processing. Goren et al. (80) surveyed broiler production systems in The Netherlands and found a positive correlation between the *Salmonella* prevalence from litter at 5-weeks of age and that from ceca at processing. None of these studies determined the molecular genetic similarity of the isolates using subtyping techniques (e.g., pulsed field gel electrophoresis) from litter, birds, and processing plants.

Several types of amendments can be applied to litter to reduce the emission of ammonia by modifying the litter pH. The effect of these amendments on *Salmonella* prevalence and other pathogens in the litter has been the subject of a number of studies (81–85). These studies examined the effect of litter amendment products such as organic acids, formalin, sodium bisulfate, aluminum sulfate, and sulfuric acid on *Salmonella* and other pathogens such as *Campylobacter*. In an experimental study, the authors reported a significant reduction in *Salmonella* prevalence in litter treated with sulfuric acid (85). Furthermore, Williams (84) found that the use of 4 or 6% formalin to spray litter led to elimination of *Salmonella* from the litter surface. The effect of citric, tartaric, and salicylic acids on pathogens in poultry litter was assessed in an experimental trial (81). The authors reported a significant reduction in *S.* Enteritidis prevalence, with salicylic acid being the most effective. However, other studies reported less effect of litter amendments on pathogen prevalence. Line's (82) experimental study revealed that two products (sodium bisulfate and aluminum sulfate) did not significantly reduce *Salmonella* prevalence in poultry litter, but they did for *Campylobacter*. Similarly, Pope and Cherry (83) did not find a significant *Salmonella* reduction when they applied sodium bisulfate to litter.

Interestingly although one would expect that prolonged use of litter with multiple broiler flocks will increase pathogen contamination in the litter (85), other studies have demonstrated the contrary. Corrier et al. (86) found in an experimental trial that birds raised on used litter had higher cecal concentrations of volatile fatty acid and lower prevalence of *Salmonella* cecal colonization compared to birds raised on new litter. In another study, Roll et al. (87) followed *Salmonella* prevalence on boot covers at broiler farms at one poultry company in Brazil for 2 years. The authors reported that *Salmonella* prevalence on reused litter declined over time.

It is recommended that farmers keep their broiler litter as dry as possible, replace the top layer of the litter with a new one, and apply litter amendments that are acid-based.

Feed withdrawal

A common practice in the poultry industry is to withdraw feed before broilers are shipped to the processing plant (88, 89). Feed withdrawal (8 to 12 h preslaughter) reduces the amount of feces in the birds' crop and gut and therefore reduces the potential risk of carcass contamination with *Salmonella* and *Campylobacter* during processing (90–92). However, feed withdrawal has been shown to increase the populations of *Salmonella* and *Campylobacter* in the ceca and crops of broilers (92–96). As the birds search for feed during the withdrawal period, they consume contaminated litter, which may result in crop and ceca pathogen contamination. Furthermore, feed withdrawal has been associated with an increase in crop pH due to reduction in *Lactobacillus* fermentation and lactic acid. The rise in pH has been linked to increased *Salmonella* survival in the crop (94, 95). Furthermore, crops are 85 times more likely to break open in a processing plant compared to ceca, which could lead to carcass contamination with these pathogens (93, 94, 96). Interestingly, crops were found to be 3.5 times more likely to be contaminated with *Salmonella* compared to ceca after feed withdrawal (93, 94). As for *Campylobacter*, feed withdrawal increases this pathogen prevalence in the crop, but it did not significantly alter the *Campylobacter* prevalence in the ceca (92).

A number of studies have examined the effect of various interventions prior to or during the feed withdrawal period to control pathogen contamination of the crop and ceca.

These interventions included the acidification of drinking water (97, 98), treating the drinking water with chlorate (99), feeding broilers a glucose-based cocktail (100), acidification of feed (101–104), and using essential oils in drinking water (105, 106). Parker et al. (98), who determined that a commercial organic acid blend, when administered at 0.04% and 0.08%, significantly reduced the prevalence of *Salmonella* in crops and ceca (by 1.5 logs) of broiler birds compared to controls. Byrd et al. (97) also reported a significant reduction in the percentage of *Salmonella* culture-positive crops from broilers treated with lactic acid in drinking water compared to controls. In addition, the authors reported that 0.5% lactic acid in water did not statistically lower the *Salmonella* prevalence in ceca compared to controls. In recent studies by Alali (105, 106), a commercial essential oil blend used in drinking water significantly reduced *Salmonella* colonization in crops of broilers, compared to 0.44% lactic acid (97) and a 0.4% organic acid blend (98). Another product (i.e., chlorate) was used in broiler drinking water (24 to 48 h prior to slaughter) and reduced the percentage of *Salmonella*-contaminated crops and ceca compared to controls (99). It is important that the product used in water or feed to control pathogens in crops and/or ceca both be cost-effective and not negatively impact bird health, performance (weight gain and feed conversation ratio), and water consumption. Treated drinking water consumption by the broilers is important to measure. For instance, it was reported that birds consumed less water treated with organic acid compared to untreated water (99).

Biosecurity

Biosecurity at the poultry farm can be defined as the management action plans designed by the poultry company to prevent and control the introduction of pathogens into the flock. This includes action plans to prevent and control *Salmonella* and *Campylobacter* from entering the breeder and broiler houses, which include strategies to prevent and control wildlife, rodents, insects, unauthorized human visitors, and fomites (e.g., feed truck and farm equipment), because they are all potential *Salmonella* and *Campylobacter* vectors/carriers.

Researchers have shown that darkling beetles, mites, and flies, commonly found in the poultry litter and poultry farm environment, can be vectors for *Salmonella* and *Campylobacter* (107–114). In a broiler house, the infestation of darkling beetles can be massive (up to 1,000 insects per square meter of litter) (115). Birds can become infected with pathogens by ingesting contaminated beetles, mites, and fly larvae. Wild animals (e.g., dogs and cats) and wild birds (e.g., waterfowl shorebirds, and gulls) can also be vectors of foodborne pathogens (116). Rodents (rats and mice) can be a significant source of *Salmonella* and *Campylobacter* in farms and the surrounding areas (113, 117, 118).

Fomites such as dirt and soil were found to be contaminated with pathogens. Bailey et al. (113) found that soil samples collected from the entrance of broiler houses were *Salmonella*-positive (6.1%, n = 131). Other biosecurity risk factors associated with pathogen prevalence at the farm were reported. In an epidemiological study in France, the authors revealed that the removal of mobile equipment inside the broiler house prior to cleaning and disinfection and using specific containers to collect dead birds decreased the risk of *Salmonella* contamination in broilers (119). In another study, by Gradel et al. (120) in The Netherlands, the authors reported a link between reduced *Salmonella* prevalence in broiler farms and the following factors: workers washing their hands with antiseptic soap prior to entering the broiler house, proper disposal of dead birds, using gravel around the broiler house, regularly checking on rodent baits in the house, and the use of combined surface and pulse-fogging disinfection methods. Cleaning and disinfection of broiler houses after removal of litter was considered a significant factor that reduced *Salmonella* contamination in the next placement flock (121). However, cleaning and disinfection was not significantly associated with the prevalence of *Campylobacter* fecal shedding (122). In agreement with Gradel et al. (120), Renwick et al. (123) reported that disposal of dead chickens by incineration or complete removal from the farm premises was associated with reduced *Salmonella* prevalence in the farm. The practice of workers dipping their boots in disinfectant solution prior to entering broiler houses, and the frequency of changing the solution, was found to reduce colonization of birds with *Campylobacter* (124, 125).

Poultry housing should be designed to prevent entry of any vector such as rodents and wild birds. Pest-control measures such as traps and baits should be used inside and outside of the house and checked regularly. Human access to the house and to the farm should be regulated. Workers' sanitation and hygiene measures such as footbaths, hand washing, and the use of disposable coveralls and overshoes should be implemented. It is recommended that sharing equipment and transport vehicles with other farms should be minimal. Equipment and vehicles should be effectively decontaminated before entering the farm environment. Stricter biosecurity measures should be employed for breeder farms/houses such as shower-in–shower-out (i.e., shower prior to and after entry to the breeder farm), or at least disposable footwear, coveralls, and hair covering should be used.

Salmonella- and *Campylobacter*-contaminated water poses a low risk of pathogen transmission to poultry flocks. *Salmonella* was detected in 1.4% (n = 731) of water samples collected from water lines in broiler farms (113), whereas *Campylobacter* was detected in 9% (n = 100) of water line samples in broiler farms (126). Water contamination with *Salmonella* was associated with fecal contamination at broiler farms (123). *Campylobacter* is very sensitive to chlorination of water (127, 128). However, the presence of organic materials might reduce the effectiveness of chlorination on *Campylobacter* as well as *Salmonella*. Studies have revealed that the addition of sanitizers (e.g., chlorine) to drinking water might reduce the risk of a broiler flock being positive for *Campylobacter* (125, 129). Interestingly, Stern et al. (130) found that chlorination of water at 2 to 5 ppm had no effect on *Campylobacter* colonization in broilers' guts. Therefore, alternative sanitizers or possibly higher chlorine concentrations in drinking water should be considered; however, bird water consumption should be monitored and compared to untreated water. Cleaning the water lines between flocks with sanitizers to destroy biofilms is important to improve drinking water quality.

Antibiotic use

In the past, antibiotics were used widely in the industry at subtherapeutic levels as growth promoters and as prophylactics to minimize the risk of colonization by enteric pathogens. The relationship between antibiotic use and prevalence/levels of pathogens in poultry production is not clear. Most studies of antibiotic-resistant bacteria have focused on the relationship between antibiotic use and the emergence, development, and transmission of antibiotic-resistant bacteria. Currently, there is no scientific answer to the following question: Does subtherapeutic antibiotic use in poultry farms impact the prevalence/level of *Salmonella* and *Campylobacter* in broilers? To be able to answer this question, longitudinal epidemiological studies are needed to compare the prevalence and levels of *Salmonella* and *Campylobacter* in broiler farms before and after discontinuation of antibiotics.

Few studies have compared the *Salmonella* prevalence in conventional and antibiotic-free (i.e., organic/free-range) broiler flocks at the farm level. Siemon et al. (131) and Alali et al. (132) reported significantly lower prevalence of *Salmonella* at pasture and USDA-certified organic farms, respectively, compared to conventional farms. Other studies of pasture broiler farms reported no *Salmonella* detection

(133), low *Salmonella* prevalence of 2.9% (135), and 25% *Salmonella* prevalence (155). For *Campylobacter* prevalence on pasture and organic farms (antibiotic-free farms), the overall prevalence was high, ranging from 61 to 90.4% (133–136), except for Hanning et al. (137), who found that *Campylobacter* prevalence was 30%. However, at retail, *Salmonella* and *Campylobacter* prevalence on organic and antibiotic free broilers meat has been shown to be higher and lower than conventionally reared broiler meat (138–141).

Prebiotics and probiotics

Prebiotics and probiotics are added to feed to reduce pathogen colonization and subsequent shedding in poultry. Prebiotics are nondigestible carbohydrates that stimulate the growth and/or activity of one or more bacteria in the gastrointestinal tract (142). Probiotics are single or multiple defined microorganisms that act as protective microflora in the gut to prevent colonization by pathogens (143, 144). Competitive exclusion (CE) is a form of probiotic culture but is composed of both defined and undefined microorganisms from adult chicken intestines and is given via spray or water to newly hatched chicks (145). Young chicks are highly susceptible to infection with *Salmonella* and *Campylobacter*, largely due their underdeveloped intestinal microbial populations. Therefore, administration of CE enhances the composition of the chick microbial populations and exclude pathogens from colonizing the intestine (144, 146, 147). Exclusion of pathogens is thought to result from both direct and indirect pathways. Direct effects include the production of volatile fatty acids, competition for colonization sites and nutrients, and production of bacteriocins by natural microbes; indirect effects include stimulation of the host immune system and increased peristalsis (148). Studies have demonstrated the efficacy of CE in laboratories and field trials (144, 149, 150). However, under commercial conditions, exclusion of pathogens such as *Salmonella* has been reported to be highly variable depending on the organism status in the chick gut and the hatchery environment (80, 151, 152). Various CE products are on the market for use in poultry to control *Salmonella* inection. Due to the difference in colonization location of *Campylobacter* and *Salmonella*, a different composition of competitive bacteria may be needed to control *Campylobacter* (153). According to Schneitz (154), the most effective products are those that contain both large numbers and a diversity of bacterial strains.

CONCLUSIONS

Foodborne pathogen transmission in preharvest poultry production is complex and therefore requires a comprehensive food safety control and prevention program to minimize broiler carcass contamination. A wide range of management strategies and intervention measures exist to control *Salmonella* and *Campylobacter* in broiler preharvest production. No one single intervention method or management strategy has been identified to control *Salmonella* and/or *Campylobacter* in preharvest broiler production. Instead, a collection of measures that are shown to be effective and cost-effective need to be applied to control these pathogens. We believe that vaccination of breeder flocks for *Salmonella*, strong biosecurity programs for breeders and broiler flocks, well-timed feed withdrawal with interventions, and the use of feed free of animal by-products are keys to reducing the prevalence of *Salmonella* and *Campylobacter* in broiler flocks.

It is clear that interventions that are effective in experimental or field trial settings might not necessarily be as effective in commercial poultry settings. Although longitudinal studies are expensive to conduct at commercial settings, they are needed to investigate the effectiveness of interventions at farms. Furthermore, the economic feasibility of any intervention application in a commercial setting should be assessed.

CITATION

Alali WQ, Hofacre CL. 2016. Preharvest food safety in broiler chicken production. Microbiol Spectrum 4(4):PFS-0002-2014.

REFERENCES

1. **World Health Organization.** 2005. Drug-resistant *Salmonella*. http://www.who.int/mediacentre/factsheets/fs139/en/.
2. **Flint JA, Van Duynhoven YT, Angulo FJ, DeLong SM, Braun P, Kirk M, Scallan E, Fitzgerald M, Adak GK, Sockett P, Ellis A, Hall G, Gargouri N, Walke H, Braam P.** 2005. Estimating the burden of acute gastroenteritis, foodborne disease, and pathogens commonly transmitted by food: an international review. *Clin Infect Dis* **41:**698–704.
3. **Scallan E, Hoekstra RM, Angulo FJ, Tauxe RV, Widdowson MA, Roy SL, Jones JL, Griffin PM.** 2011. Foodborne illness acquired in the United States: major pathogens. *Emerg Infect Dis* **17:**7–15.
4. **FDA.** 2005. Human health impact of fluoroquinolone resistant *Campylobacter* associated with the consumption of chicken, 2001. http://www.fda.gov/downloads/AnimalVeterinary/SafetyHealth/RecallsWithdrawals/UCM152308.pdf.
5. **USDA.** 2008. U.S. Department of Agriculture, Food Safety and Inspection Service. Improvements for poultry slaughter inspection. Appendix A. Attribution of human salmonellosis to chicken in the USA.
6. **Braden CR.** 2006. *Salmonella enterica* serotype Enteritidis and eggs: a national epidemic in the United States. *Clin Infect Dis* **43:**512–517.
7. **Painter JA, Hoekstra RM, Ayers T, Tauxe RV, Braden CR, Angulo FJ, Griffin PM.** 2013. Attribution of foodborne illnesses, hospitalizations, and deaths to food commodities by using outbreak data, United States, 1998-2008. *Emerg Infect Dis* **19:**407–415.
8. **Food and Agriculture Organization of the United Nations.** 2010. *Poultry meat and eggs.* http://www.fao.org/docrep/012/al175e/al175e.pdf.
9. **USDA.** 2012. *Progress report on* Salmonella *and* Campylobacter *testing of raw meat and poultry products, CY 1998-2012.* http://www.fsis.usda.gov/wps/portal/fsis/topics/data-collection-and-reports/microbiology/annual-progress-reports/2012-annual-report.
10. **USDA.** 2011. *FSIS HACCP verification* Campylobacter *results: July-December, 2011.* http://www.fsis.usda.gov/wps/portal/fsis/topics/data-collection-and-reports/microbiology/campylobacter/ct_index.
11. **Buzby JC, Farah HA.** 2006. Chicken consumption continues longrun rise. *Amber Waves* **4:**5.
12. **CDC.** 2011. *Foodborne Diseases Active Surveillance Network (FoodNet) 2011 surveillance report.* http://www.cdc.gov/foodnet/PDFs/2011_annual_report_508c.pdf.
13. **USDA, Food Safety and Inspection Service.** 2011. *Serotypes profile of* Salmonella *isolates from meat and poultry products: January 1998 through December 2011.* http://www.fsis.usda.gov/wps/wcm/connect/26c0911b-b61e-4877-a630-23f314300ef8/salmonella-serotype-annual-2011.pdf?MOD=AJPERES.
14. **FDA.** 2010. *National Antimicrobial Resistance Monitoring System: Enteric Bacteria (NARMS): 2007 Executive Report.* U.S. Department of Health and Human Services, Food and Drug Administration.
15. **Le Hello S, Hendriksen RS, Doublet B, Fisher I, Nielsen EM, Whichard JM, Bouchrif B, Fashae K, Granier SA, Jourdan-Da Silva N, Cloeckaert A, Threlfall EJ, Angulo FJ, Aarestrup FM, Wain J, Weill FX.** 2011. International spread of an epidemic population of *Salmonella enterica* serotype Kentucky ST198 resistant to ciprofloxacin. *J Infect Dis* **204:**675–684.
16. **Olson CK, Ethelberg S, Van Pelt ERVT.** 2008. Epidemiology of *Campylobacter jejuni* infections in the United States and other industrialized nations, p 163–189. *In* Nachamkin I, Szymanski CM, Blaser MJ (ed), *Campylobacter jejuni: Current and Future Trends.* American Society for Microbiology Press, Washington, DC.
17. **Swaminathan B, Barrett TJ, Hunter SB, Tauxe RV, CDC PulseNet Task Force.** 2001. PulseNet: the molecular subtyping network for foodborne bacterial disease surveillance, United States. *Emerg Infect Dis* **7:**382–389.
18. **Altekruse SF, Stern NJ, Fields PI, Swerdlow DL.** 1999. *Campylobacter jejuni*: an emerging foodborne pathogen. *Emerg Infect Dis* **5:**28–35.
19. **Lee MD, Newell DG.** 2006. *Campylobacter* in poultry: filling an ecological niche. *Avian Dis* **50:**1–9.
20. **Wong TL, Hollis L, Cornelius A, Nicol C, Cook R, Hudson JA.** 2007. Prevalence, numbers, and subtypes of *Campylobacter jejuni* and *Campylobacter coli* in uncooked retail meat samples. *J Food Prot* **70:**566–573.
21. **Madden RH, Moran L, Scates P, McBride J, Kelly C.** 2011. Prevalence of *Campylobacter* and *Salmonella* in raw chicken on retail sale in

the republic of Ireland. *J Food Prot* **74:**1912–1916.

22. **Hoang KV, Stern NJ, Saxton AM, Xu F, Zeng X, Lin J.** 2011. Prevalence, development, and molecular mechanisms of bacteriocin resistance in *Campylobacter. Appl Environ Microbiol* **77:**2309–2316.
23. **Hardy CG, Lackey LG, Cannon J, Price LB, Silbergeld EK.** 2011. Prevalence of potentially neuropathic *Campylobacter jejuni* strains on commercial broiler chicken products. *Int J Food Microbiol* **145:**395–399.
24. **Bailey JS.** 1988. Integrated colonization control of *Salmonella* in poultry. *Poult Sci* **67:**928–932.
25. **Cox NA, Berrang ME, Cason JA.** 2000. *Salmonella* penetration of egg shells and proliferation in broiler hatching eggs: a review. *Poult Sci* **79:**1571–1574.
26. **Lahellec C, Colin P.** 1985. Relationship between serotypes of *Salmonellae* from hatcheries and rearing farms and those from processed poultry carcases. *Br Poult Sci* **26:**179–186.
27. **Opitz HM, El-Begearmi M, Flegg P, Beane D.** 1993. Effectiveness of five feed additives in chicks infected with *Salmonella* Enteritidis phage type 13A. *Appl Poult Sci.* **2:**147–153.
28. **Ramirez GA, Sarlin LL, Caldwell DJ, Yezak CR Jr, Hume ME, Corrier DE, Deloach JR, Hargis BM.** 1997. Effect of feed withdrawal on the incidence of *Salmonella* in the crops and ceca of market age broiler chickens. *Poult Sci* **76:**654–656.
29. **Corrier DE, Byrd JA, Hargis BM, Hume ME, Bailey RH, Stanker LH.** 1999. Presence of *Salmonella* in the crop and ceca of broiler chickens before and after preslaughter feed withdrawal. *Poult Sci* **78:**45–49.
30. **Dunkley KD, McReynolds JL, Hume ME, Dunkley CS, Callaway TR, Kubena LF, Nisbet DJ, Ricke SC.** 2007. Molting in *Salmonella* Enteritidis-challenged laying hens fed alfalfa crumbles. II. Fermentation and microbial ecology response. *Poult Sci* **86:**2101–2109.
31. **Cox NA, Bailey JS, Thomson JE, Juven BJ.** 1983. *Salmonella* and other *Enterobacteriaceae* found in commercial poultry feed. *Poult Sci* **62:**2169–2175.
32. **Davies RH, Wray C.** 1997. Distribution of *Salmonella* contamination in ten animal feedmills. *Vet Microbiol* **57:**159–169.
33. **Doyle MP, Roman DJ.** 1982. Sensitivity of *Campylobacter jejuni* to drying. *J Food Prot* **45:**507–510.
34. **Humphrey TJ, Henley A, Lanning DG.** 1993. The colonization of broiler chickens with *Campylobacter jejuni*: some epidemiological investigations. *Epidemiol Infect* **110:**601–607.
35. **Jones FT, Richardson KE.** 2004. *Salmonella* in commercially manufactured feeds. *Poult Sci* **83:**384–391.
36. **Ha SD, Maciorowski KG, Kwon YM, Jones FT, Ricke SC.** 1998. Indigenous feed microflora and *Salmonella typhimurium* marker strain survival in poultry mash diets containing varying levels of protein. *Anim Feed Sci Tech* **76:**23–33.
37. **Maciorowski KG, Jones FT, Pillai SD, Ricke SC.** 2004. Incidence, sources, and control of foodborne *Salmonella* spp. in poultry feeds. *Worlds Poult Sci J* **60:**446–457.
38. **Ricke SC.** 2003. Perspectives on the use of organic acids and short chain fatty acids as antimicrobials. *Poult Sci* **82:**632–639.
39. **Leeson S, Marcotte M.** 1993. Irradiation of poultry feed. I. Microbial status and bird response. *Worlds Poult Sci J* **49:**19–33.
40. **Matlho G, Himathongkham S, Riemann H, Kass P.** 1997. Destruction of *Salmonella enteritidis* in poultry feed by combination of heat and propionic acid. *Avian Dis* **41:**58–61.
41. **Jones FT, Axtell RC, Rives DV, Scheideler SE, Tarver FR Jr, Walker RL, Wineland MJ.** 1991. A survey of *Salmonella* contamination in modern broiler production. *J Food Prot* **54:**502–507.
42. **Chambers JR, Lu X.** 2002. Probiotics and maternal vaccination for *Salmonella* control in broiler chickens. *J Appl Poult Res* **11:**320–327.
43. **Hassan JO, Curtiss R III.** 1996. Effect of vaccination of hens with an avirulent strain of *Salmonella typhimurium* on immunity of progeny challenged with wild-type *Salmonella* strains. *Infect Immun* **64:**938–944.
44. **Berghaus RD, Thayer SG, Maurer JJ, Hofacre CL.** 2011. Effect of vaccinating breeder chickens with a killed *Salmonella* vaccine on *Salmonella* prevalences and loads in breeder and broiler chicken flocks. *J Food Prot* **74:**727–734.
45. **Dórea FC, Cole DJ, Hofacre C, Zamperini K, Mathis D, Doyle MP, Lee MD, Maurer JJ.** 2010. Effect of *Salmonella* vaccination of breeder chickens on contamination of broiler chicken carcasses in integrated poultry operations. *Appl Environ Microbiol* **76:**7820–7825.
46. **de Zoete MR, van Putten JPM, Wagenaar JA.** 2007. Vaccination of chickens against *Campylobacter. Vaccine* **25:**5548–5557.
47. **Ridley AM, Toszeghy MJ, Cawthraw SA, Wassenaar TM, Newell DG.** 2008. Genetic instability is associated with changes in the colonization potential of *Campylobacter jejuni*

in the avian intestine. *J Appl Microbiol* **105:** 95–104.
48. **Wyszynska A, Raczko A, Lis M, Jagusztyn-Krynicka EK.** 2004. Oral immunization of chickens with avirulent *Salmonella* vaccine strain carrying *C. jejuni* 72Dz/92 cjaA gene elicits specific humoral immune response associated with protection against challenge with wild-type *Campylobacter*. *Vaccine* **22:**1379–1389.
49. **Sizemore DR, Warner B, Lawrence J, Jones A, Killeen KP.** 2006. Live, attenuated *Salmonella typhimurium* vectoring *Campylobacter* antigens. *Vaccine* **24:**3793–3803.
50. **Buckley AM, Wang J, Hudson DL, Grant AJ, Jones MA, Maskell DJ, Stevens MP.** 2010. Evaluation of live-attenuated *Salmonella* vaccines expressing *Campylobacter* antigens for control of *C. jejuni* in poultry. *Vaccine* **28:** 1094–1105.
51. **Quéré P, Girard F.** 1999. Systemic adjuvant effect of cholera toxin in the chicken. *Vet Immunol Immunopathol* **70:**135–141.
52. **Vervelde L, Janse EM, Vermeulen AN, Jeurissen SHM.** 1998. Induction of a local and systemic immune response using cholera toxin as vehicle to deliver antigen in the lamina propria of the chicken intestine. *Vet Immunol Immunopathol* **62:**261–272.
53. **Widders PR, Thomas LM, Long KA, Tokhi MA, Panaccio M, Apos E.** 1998. The specificity of antibody in chickens immunised to reduce intestinal colonisation with *Campylobacter jejuni*. *Vet Microbiol* **64:**39–50.
54. **Logan SM, Kelly JF, Thibault P, Ewing CP, Guerry P.** 2002. Structural heterogeneity of carbohydrate modifications affects serospecificity of *Campylobacter* flagellins. *Mol Microbiol* **46:**587–597.
55. **Doig P, Kinsella N, Guerry P, Trust TJ.** 1996. Characterization of a post-translational modification of *Campylobacter* flagellin: identification of a sero-specific glycosyl moiety. *Mol Microbiol* **19:**379–387.
56. **Cox NA, Berrang ME, Mauldin JM.** 2001. Extent of *Salmonellae* contamination in primary breeder hatcheries in 1998 as compared to 1991. *J Appl Poult Res* **10:**202–205.
57. **Timoney JF, Shivaprasad HL, Baker RC, Rowe B.** 1989. Egg transmission after infection of hens with *Salmonella* Enteritidis phage type 4. *Vet Rec* **125:**600–601.
58. **Shivaprasad HL, Timoney JF, Morales S, Lucio B, Baker RC.** 1990. Pathogenesis of *Salmonella* Enteritidis infection in laying chickens. I. Studies on egg transmission, clinical signs, fecal shedding, and serologic responses. *Avian Dis* **34:**548–557.
59. **Cason JA, Cox NA, Bailey JS.** 1994. Transmission of *Salmonella typhimurium* during hatching of broiler chicks. *Avian Dis* **38:**583–588.
60. **Rizk SS, Ayres JC, Kraft AA.** 1966. Effect of holding condition on the development of *Salmonellae* in artificially inoculated hens' eggs. *Poult Sci* **45:**825–829.
61. **Cox NA, Bailey JS, Mauldin JM, Blankenship LC.** 1990. Presence and impact of *Salmonella* contamination in commercial broiler hatcheries. *Poult Sci* **69:**1606–1609.
62. **Cox NA, Bailey JS, Mauldin JM, Blankenship LC, Wilson JL.** 1991. Extent of *Salmonellae* contamination in breeder hatcheries. *Poult Sci* **70:**416–418.
63. **Cox NA, Bailey JS, Berrang ME.** 1996. Alternative routes for *Salmonella* intestinal tract colonization of chicks. *J Appl Poult Res* **5:** 282–288.
64. **Byrd JA, DeLoach JR, Corrier DE, Nisbet DJ, Stanker LH.** 1999. Evaluation of *Salmonella* serotype distributions from commercial broiler hatcheries and grower houses. *Avian Dis* **43:**39–47.
65. **Bhatia TR, McNabb GD.** 1980. Dissemination of *Salmonella* in broiler-chicken operations. *Avian Dis* **24:**616–624.
66. **Byrd J, Bailey RH, Wills R, Nisbet D.** 2007. Recovery of *Campylobacter* from commercial broiler hatchery trayliners. *Poult Sci* **86:**26–29.
67. **Cox NA, Bailey JS, Richardson LJ, Buhr RJ, Cosby DE, Wilson JL, Hiett KL, Siragusa GR, Bourassa DV.** 2005. Presence of naturally occurring *Campylobacter* and *Salmonella* in the mature and immature ovarian follicles of late-life broiler breeder hens. *Avian Dis* **49:** 285–287.
68. **Buhr RJ, Cox NA, Stern NJ, Musgrove MT, Wilson JL, Hiett KL.** 2002. Recovery of *Campylobacter* from segments of the reproductive tract of broiler breeder hens. *Avian Dis* **46:** 919–924.
69. **Buhr RJ, Musgrove MT, Richardson LJ, Cox NA, Wilson JL, Bailey JS, Cosby DE, Bourassa DV.** 2005. Recovery of *Campylobacter jejuni* in feces and semen of caged broiler breeder roosters following three routes of inoculation. *Avian Dis* **49:**577–581.
70. **Hiett KL, Cox NA, Buhr RJ, Stern NJ.** 2002. Genotype analyses of *Campylobacter* isolated from distinct segments of the reproductive tracts of broiler breeder hens. *Curr Microbiol* **45:**400–404.
71. **Doyle MP.** 1984. Association of *Campylobacter jejuni* with laying hens and eggs. *Appl Environ Microbiol* **47:**533–536.

72. **Cox NA, Stern NJ, Hiett KL, Berrang ME.** 2002. Identification of a new source of *Campylobacter* contamination in poultry: transmission from breeder hens to broiler chickens. *Avian Dis* **46:**535–541.
73. **Dunn J.** 1996. Pulsed light and pulsed electric field for foods and eggs. *Poult Sci* **75:**1133–1136.
74. **Coufal CD, Chavez C, Knape KD, Carey JB.** 2003. Evaluation of a method of ultraviolet light sanitation of broiler hatching eggs. *Poult Sci* **82:**754–759.
75. **Mitchell BW, Buhr RJ, Berrang ME, Bailey JS, Cox NA.** 2002. Reducing airborne pathogens, dust and *Salmonella* transmission in experimental hatching cabinets using an electrostatic space charge system. *Poult Sci* **81:**49–55.
76. **Rodriguez-Romo LA, Yousef AE.** 2005. Inactivation of *Salmonella enterica* serovar Enteritidis on shell eggs by ozone and UV radiation. *J Food Prot* **68:**711–717.
77. **Russell SM.** 2003. Effect of sanitizers applied by electrostatic spraying on pathogenic and indicator bacteria attached to the surface of eggs. *J Appl Poult Res* **12:**183–189.
78. **Bhatia TRS, McNabb GD, Wyman H, Nayar GPS.** 1979. *Salmonella* isolation from litter as an indicator of flock infection and carcass contamination. *Avian Dis* **23:**838–847.
79. **Corry JE, Allen VM, Hudson WR, Breslin MF, Davies RH.** 2002. Sources of *Salmonella* on broiler carcasses during transportation and processing: modes of contamination and methods of control. *J Appl Microbiol* **92:**424–432.
80. **Goren E, de Jong WA, Doornenbal P, Bolder NM, Mulder RWAW, Jansen A.** 1988. Reduction of *Salmonella* infection of broilers by spray application of intestinal microflora: a longitudinal study. *Vet Q* **10:**249–255.
81. **Ivanov IE.** 2001. Treatment of broiler litter with organic acids. *Res Vet Sci* **70:**169–173.
82. **Line JE.** 2002. *Campylobacter* and *Salmonella* populations associated with chickens raised on acidified litter. *Poult Sci* **81:**1473–1477.
83. **Pope MJ, Cherry TE.** 2000. An evaluation of the presence of pathogens on broilers raised on poultry litter treatment-treated litter. *Poult Sci* **79:**1351–1355.
84. **Williams JE.** 1980. Formalin destruction of *Salmonellae* in poultry litter. *Poult Sci* **59:**2717–2724.
85. **Vicente JL, Higgins SE, Hargis BM, Tellez G.** 2007. Effect of poultry guard litter amendment on horizontal transmission of *Salmonella* Enteritidis in broiler chicks. *Int J Poult Sci* **6:**314–317.
86. **Corrier DE, Hinton A Jr, Hargis B, DeLoach JR.** 1992. Effect of used litter from floor pens of adult broilers on *Salmonella* colonization of broiler chicks. *Avian Dis* **36:**897–902.
87. **Roll VF, Dai Prá MA, Roll AP.** 2011. Research on *Salmonella* in broiler litter reused for up to 14 consecutive flocks. *Poult Sci* **90:**2257–2262.
88. **Rigby CE, Pettit JR.** 1981. Effects of feed withdrawal on the weight, fecal excretion and *Salmonella* status of market age broiler chickens. *Can J Comparat Med* **45:**363–365.
89. **Rigby CE, Pettit JR.** 1980. Changes in the *Salmonella* status of broiler chickens subjected to simulated shipping conditions. *Can J Comparat Med* **44:**374–381.
90. **Lyon CE, Papa CM, Wilson RL Jr.** 1991. Effect of feed withdrawal on yields, muscle pH, and texture of broiler breast meat. *Poult Sci* **70:**1020–1025.
91. **Buhr RJ, Northcutt JK, Lyon CE, Rowland GN.** 1998. Influence of time off feed on broiler viscera weight, diameter, and shear. *Poult Sci* **77:**758–764.
92. **Byrd JA, Corrier DE, Hume ME, Bailey RH, Stanker LH, Hargis BM.** 1998. Effect of feed withdrawal on *Campylobacter* in the crops of market-age broiler chickens. *Avian Dis* **42:** 802–806.
93. **Ramirez GA, Sarlin LL, Caldwell DJ, Yezak CR Jr, Hume ME, Corrier DE, Deloach JR, Hargis BM.** 1997. Effect of feed withdrawal on the incidence of *Salmonella* in the crops and ceca of market age broiler chickens. *Poult Sci* **76:**654–656.
94. **Humphrey TJ, Baskerville A, Whitehead A, Rowe B, Henley A.** 1993. Influence of feeding patterns on the artificial infection of laying hens with *Salmonella* Enteritidis phage type 4. *Vet Rec* **132:**407–409.
95. **Corrier DE, Byrd JA, Hargis BM, Hume ME, Bailey RH, Stanker LH.** 1999. Survival of *Salmonella* in the crop contents of market-age broilers during feed withdrawal. *Avian Dis* **43:**453–460.
96. **Hargis BM, Caldwell DJ, Brewer RL, Corrier DE, Deloach JR.** 1995. Evaluation of the chicken crop as a source of *Salmonella* contamination for broiler carcasses. *Poult Sci* **74:**1548–1552.
97. **Byrd JA, Hargis BM, Caldwell DJ, Bailey RH, Herron KL, McReynolds JL, Brewer RL, Anderson RC, Bischoff KM, Callaway TR, Kubena LF.** 2001. Effect of lactic acid administration in the drinking water during preslaughter feed withdrawal on *Salmonella* and *Campylobacter* contamination of broilers. *Poult Sci* **80:**278–283.
98. **Parker D, Hofacre C, Mathis GF, Quiroz MA, Dibner J, Knight C.** 2007. *Organic acid water*

treatment reduced Salmonella *horizontal transmission in broiler chickens.* http://www.cabi.org/Uploads/animal-science/worlds-poultry-science-association/WPSA-italy-2006/10272.pdf.

99. **Byrd JA, Anderson RC, Callaway TR, Moore RW, Knape KD, Kubena LF, Ziprin RL, Nisbet DJ.** 2003. Effect of experimental chlorate product administration in the drinking water on *Salmonella* Typhimurium contamination of broilers. *Poult Sci* **82:**1403–1406.
100. **Hinton A Jr, Buhr RJ, Ingram KD.** 2002. Carbohydrate-based cocktails that decrease the population of *Salmonella* and *Campylobacter* in the crop of broiler chickens subjected to feed withdrawal. *Poult Sci* **81:**780–784.
101. **Hinton M, Linton AH.** 1988. Control of *Salmonella* infections in broiler chickens by the acid treatment of their feed. *Vet Rec* **123:** 416–421.
102. **Van Immerseel F, Boyen F, Gantois I, Timbermont L, Bohez L, Pasmans F, Haesebrouck F, Ducatelle R.** 2005. Supplementation of coated butyric acid in the feed reduces colonization and shedding of *Salmonella* in poultry. *Poult Sci* **84:**1851–1856.
103. **Heres L, Engel B, Urlings HA, Wagenaar JA, van Knapen F.** 2004. Effect of acidified feed on susceptibility of broiler chickens to intestinal infection by *Campylobacter* and *Salmonella*. *Vet Microbiol* **99:**259–267.
104. **Cox NA, McHan F, Bailey JS, Shotts EB.** 1994. Effect of butyric or lactic acid on the *in vivo* colonization of *Salmonella* Typhimurium. *J Appl Poult Res* **3:**315–318.
105. **Alali WQ, Hofacre CL, Mathis GF, Faltys G.** 2013. Effect of essential oil compound on shedding and colonization of *Salmonella enterica* serovar Heidelberg in broilers. *Poult Sci* **92:**836–841.
106. **Alali WQ, Hofacre CL, Mathis GF, Faltys G, Ricke SC, Doyle MP.** 2013. Effect of non-pharmaceutical compounds on shedding and colonization of *Salmonella enterica* serovar Heidelberg in broilers. *Food Contr* **31:**125–128.
107. **Roche AJ, Cox NA, Richardson LJ, Buhr RJ, Cason JA, Fairchild BD, Hinkle NC.** 2009. Transmission of *Salmonella* to broilers by contaminated larval and adult lesser mealworms, *Alphitobius diaperinus* (*Coleoptera: tenebrionidae*). *Poult Sci* **88:**44–48.
108. **Bates C, Hiett KL, Stern NJ.** 2004. Relationship of *Campylobacter* isolated from poultry and from darkling beetles in New Zealand. *Avian Dis* **48:**138–147.
109. **McAllister JC, Steelman CD, Skeeles JK.** 1994. Reservoir competence of the lesser mealworm (*Coleoptera: Tenebrionidae*) for *Salmonella* Typhimurium (*Eubacteriales: Enterobacteriaceae*). *J Med Entomol* **31:**369–372.
110. **Holt PS, Geden CJ, Moore RW, Gast RK.** 2007. Isolation of *Salmonella enterica* serovar Enteritidis from houseflies (*Musca domestica*) found in rooms containing *Salmonella* serovar Enteritidis-challenged hens. *Appl Environ Microbiol* **73:**6030–6035.
111. **Olsen AR, Hammack TS.** 2000. Isolation of *Salmonella* spp. from the housefly, *Musca domestica* L., and the dump fly, *Hydrotaea aenescens* (Wiedemann) (*Diptera: Muscidae*), at caged-layer houses. *J Food Prot* **63:**958–960.
112. **Kinde H, Castellan DM, Kerr D, Campbell J, Breitmeyer R, Ardans A.** 2005. Longitudinal monitoring of two commercial layer flocks and their environments for *Salmonella enterica* serovar Enteritidis and other *Salmonellae*. *Avian Dis* **49:**189–194.
113. **Bailey JS, Stern NJ, Fedorka-Cray P, Craven SE, Cox NA, Cosby DE, Ladely S, Musgrove MT.** 2001. Sources and movement of *Salmonella* through integrated poultry operations: a multistate epidemiological investigation. *J Food Prot* **64:**1690–1697.
114. **Rosef O, Kapperud G.** 1983. House flies (*Musca domestica*) as possible vectors of *Campylobacter fetus* subsp. *jejuni*. *Appl Environ Microbiol* **45:**381–383.
115. **Arends J.** 1997. External parasites and poultry pests, p 785–813. *In* Calneck B (ed), *Diseases of Poultry*, **vol 10**. Iowa State University Press, Ames, IA.
116. **Skov MN, Madsen JJ, Rahbek C, Lodal J, Jespersen JB, Jørgensen JC, Dietz HH, Chriél M, Baggesen DL.** 2008. Transmission of *Salmonella* between wildlife and meat-production animals in Denmark. *J Appl Microbiol* **105:**1558–1568.
117. **Davies RH, Wray C.** 1995. Mice as carriers of *Salmonella* Enteritidis on persistently infected poultry units. *Vet Rec* **137:**337–341.
118. **Stern NJ, Fedorka-Cray P, Bailey JS, Cox NA, Craven SE, Hiett KL, Musgrove MT, Ladely S, Cosby D, Mead GC.** 2001. Distribution of *Campylobacter* spp. in selected U.S. poultry production and processing operations. *J Food Prot* **64:**1705–1710.
119. **Le Bouquin S, Allain V, Rouxel S, Petetin I, Picherot M, Michel V, Chemaly M.** 2010. Prevalence and risk factors for *Salmonella* spp. contamination in French broiler-chicken flocks at the end of the rearing period. *Prev Vet Med* **97:**245–251.
120. **Gradel KO, Rattenborg E.** 2003. A questionnaire-based, retrospective field study

of persistence of *Salmonella* Enteritidis and *Salmonella* Typhimurium in Danish broiler houses. *Prev Vet Med* **56:**267–284.

121. **Marin C, Balasch S, Vega S, Lainez M.** 2011. Sources of *Salmonella* contamination during broiler production in Eastern Spain. *Prev Vet Med* **98:**39–45.
122. **Wedderkopp A, Gradel KO, Jørgensen JC, Madsen M.** 2001. Pre-harvest surveillance of *Campylobacter* and *Salmonella* in Danish broiler flocks: a 2-year study. *Int J Food Microbiol* **68:** 53–59.
123. **Renwick SA, Irwin RJ, Clarke RC, McNab WB, Poppe C, McEwen SA.** 1992. Epidemiological associations between characteristics of registered broiler chicken flocks in Canada and the *Salmonella* culture status of floor litter and drinking water. *Can Vet J* **33:**449–458.
124. **Humphrey TJ, Henley A, Lanning DG.** 1993. The colonization of broiler chickens with *Campylobacter jejuni*: some epidemiological investigations. *Epidemiol Infect* **110:**601–607.
125. **Evans SJ, Sayers AR.** 2000. A longitudinal study of *Campylobacter* infection of broiler flocks in Great Britain. *Prev Vet Med* **46:**209–223.
126. **Gregory E, Barnhart H, Dreesen DW, Stern NJ, Corn JL.** 1997. Epidemiological study of *Campylobacter* spp. in broilers: source, time of colonization, and prevalence. *Avian Dis* **41:**890–898.
127. **Blaser MJ, Smith PF, Wang WL, Hoff JC.** 1986. Inactivation of *Campylobacter jejuni* by chlorine and monochloramine. *Appl Environ Microbiol* **51:**307–311.
128. **Wang WL, Powers BW, Leuchtefeld NW, Blaser MJ.** 1983. Effects of disinfectants on *Campylobacter jejuni*. *Appl Environ Microbiol* **45:**1202–1205.
129. **Kapperud G, Skjerve E, Vik L, Hauge K, Lysaker A, Aalmen I, Ostroff SM, Potter M.** 1993. Epidemiological investigation of risk factors for *Campylobacter* colonization in Norwegian broiler flocks. *Epidemiol Infect* **111:**245–255.
130. **Stern NJ, Robach MC, Cox NA, Musgrove MT.** 2002. Effect of drinking water chlorination on *Campylobacter* spp. colonization of broilers. *Avian Dis* **46:**401–404.
131. **Siemon CE, Bahnson PB, Gebreyes WA.** 2007. Comparative investigation of prevalence and antimicrobial resistance of *Salmonella* between pasture and conventionally reared poultry. *Avian Dis* **51:**112–117.
132. **Alali WQ, Thakur S, Berghaus RD, Martin MP, Gebreyes WA.** 2010. Prevalence and distribution of *Salmonella* in organic and conventional broiler poultry farms. *Foodborne Pathog Dis* **7:**1363–1371.
133. **Hoogenboom LA, Bokhorst JG, Northolt MD, van de Vijver LP, Broex NJ, Mevius DJ, Meijs JA, Van der Roest J.** 2008. Contaminants and microorganisms in Dutch organic food products: a comparison with conventional products. *Food Addit Contam Part A Chem Anal Control Expo Risk Assess* **25:**1195–1207.
134. **Luangtongkum T, Morishita TY, Ison AJ, Huang S, McDermott PF, Zhang Q.** 2006. Effect of conventional and organic production practices on the prevalence and antimicrobial resistance of *Campylobacter* spp. in poultry. *Appl Environ Microbiol* **72:**3600–3607.
135. **Esteban JI, Oporto B, Aduriz G, Juste RA, Hurtado A.** 2008. A survey of foodborne pathogens in free-range poultry farms. *Int J Food Microbiol* **123:**177–182.
136. **Colles FM, Jones TA, McCarthy ND, Sheppard SK, Cody AJ, Dingle KE, Dawkins MS, Maiden MCJ.** 2008. *Campylobacter* infection of broiler chickens in a free-range environment. *Environ Microbiol* **10:**2042–2050.
137. **Hanning I, Biswas D, Herrera P, Roesler M, Ricke SC.** 2010. Prevalence and characterization of *Campylobacter jejuni* isolated from pasture flock poultry. *J Food Sci* **75:**M496–M502.
138. **Bailey JS, Cosby DE.** 2005. *Salmonella* prevalence in free-range and certified organic chickens. *J Food Prot* **68:**2451–2453.
139. **Cui S, Ge B, Zheng J, Meng J.** 2005. Prevalence and antimicrobial resistance of *Campylobacter* spp. and *Salmonella* serovars in organic chickens from Maryland retail stores. *Appl Environ Microbiol* **71:**4108–4111.
140. **Lestari SI, Han F, Wang F, Ge B.** 2009. Prevalence and antimicrobial resistance of *Salmonella* serovars in conventional and organic chickens from Louisiana retail stores. *J Food Protect* **72:**1165–1172.
141. **Hanning I, Biswas D, Herrera P, Roesler M, Ricke SC.** 2010. Prevalence and characterization of *Campylobacter jejuni* isolated from pasture flock poultry. *J Food Sci* **75:**M496–M502.
142. **Gibson GR, Roberfroid MB.** 1995. Dietary modulation of the human colonic microbiota: introducing the concept of prebiotics. *J Nutr* **125:**1401–1412.
143. **Ricke SC, Pillai SD.** 1999. Conventional and molecular methods for understanding probiotic bacteria functionality in gastrointestinal tracts. *Crit Rev Microbiol* **25:**19–38.
144. **Nisbet D.** 2002. Defined competitive exclusion cultures in the prevention of enteropathogen colonisation in poultry and swine. *Antonie van Leeuwenhoek* **81:**481–486.

145. **Impey CS, Mead GC, Hinton M.** 1987. Influence of continuous challenge via the feed on competitive exclusion of *Salmonellas* from broiler chicks. *J Appl Bacteriol* **63:**139–146.
146. **Zhang G, Ma L, Doyle MP.** 2007. Potential competitive exclusion bacteria from poultry inhibitory to *Campylobacter jejuni* and *Salmonella*. *J Food Prot* **70:**867–873.
147. **Zhang G, Ma L, Doyle MP.** 2007. *Salmonellae* reduction in poultry by competitive exclusion bacteria *Lactobacillus salivarius* and *Streptococcus cristatus*. *J Food Prot* **70:**874–878.
148. **Doyle MP, Erickson MC.** 2006. Reducing the carriage of foodborne pathogens in livestock and poultry. *Poult Sci* **85:**960–973.
149. **Corrier DE, Nisbet DJ, Byrd JA II, Hargis BM, Keith NK, Peterson M, Deloach JR.** 1998. Dosage titration of a characterized competitive exclusion culture to inhibit *Salmonella* colonization in broiler chickens during growout. *J Food Prot* **61:**796–801.
150. **McReynolds JL, Moore RW, McElroy AP, Hargis BM, Caldwell DJ.** 2007. Evaluation of a competitive exclusion culture and megan vac 1 on *Salmonella* Typhimurium colonization in neonatal broiler chickens. *J Appl Poult Res* **16:**456–463.
151. **Patterson JA, Burkholder KM.** 2003. Application of prebiotics and probiotics in poultry production. *Poult Sci* **82:**627–631.
152. **Revolledo L, Ferreira AJP, Mead GC.** 2006. Prospects in *Salmonella* control: competitive exclusion, probiotics, and enhancement of avian intestinal immunity. *J Appl Poult Res* **15:**341–351.
153. **Beery JT, Hugdahl MB, Doyle MP.** 1988. Colonization of gastrointestinal tracts of chicks by *Campylobacter jejuni*. *Appl Environ Microbiol* **54:**2365–2370.
154. **Schneitz C.** 2005. Competitive exclusion in poultry: 30 years of research. *Food Contr* **16:** 657–667.
155. **Melendez SN, Hanning I, Han J, Nayak R, Clement AR.** 2010. *Salmonella enterica* isolates from pasture raised poultry exhibit antimicrobial resistance and class I integrons. *J Appl Microbiol* **109:**1957–1966.

5

Egg Safety in the Realm of Preharvest Food Safety

MANPREET SINGH[1] and JAGPINDER BRAR[1]

INTRODUCTION

Eggs are one of the basic components of our food system. Their easy availability and being an inexpensive source of proteins make them attractive to consumers. The average annual consumption of eggs is around 250 eggs per capita in the United States, and according to statistical data, in 2013 the average per unit consumption was 252.6 eggs per person (1). Shell eggs can either be sold directly to consumers or be further processed and added as an ingredient in other food products. Due to the markets where eggs are sold, be it directly to consumers, as an ingredient in other food products, or as a processed product such as dried egg powder, liquid egg, etc., it is appropriate to decontaminate eggs to prevent foodborne illnesses. In most egg recipes half-cooked eggs are used, thus allowing certain pathogenic bacteria such as *Salmonella*, *Campylobacter*, and *Clostridium perfringens* to survive the low-temperature cooking. As a result, emphasis should be put on the decontamination techniques prior to packaging shell eggs.

Shell eggs can be contaminated with various pathogenic bacteria, but the prevalence of *Salmonella* is exceptionally high (2). *Salmonella* can penetrate the outer shell of eggs and infect the egg content; however, it can also be transferred from the ovaries of a hen to the egg content (3). Among different

[1]Department of Food Science, Purdue University, West Lafayette, IN 47907.

Preharvest Food Safety
Edited by Siddhartha Thakur and Kalmia E. Kniel

doi:10.1128/microbiolspec.PFS-0005-2014

serotypes, *Salmonella enterica* serotypes Enteritidis and Typhimurium are the most problematic for the egg industry. Therefore, the U.S. Food and Drug Administration (FDA) has developed a guideline to control the prevalence of *S.* Enteritidis in laying hen houses and processing plants. *Salmonella* infection causes salmonellosis, a gastrointestinal disease that causes diarrhea, fever, and even death among immunocompromised patients. The FDA estimates that around 142,000 cases of salmonellosis are associated with the consumption of infected eggs each year (4). *C. perfringens* is also one of the main pathogens in poultry production. The prevalence of *C. perfringens* is not very high, but it can cause necrotic enteritis, an acute enterotoxemia disease in chickens, hence reducing the egg quality. *Campylobacter* is highly prevalent in chickens, but it cannot penetrate eggshells as effectively as *Salmonella*, and ovarian transmission is not very common in cases of *Campylobacter* (5). To reduce pathogens from shell eggs, various types of decontamination techniques are used in the egg industry. These include washing eggs in disinfectants, pasteurization, UV radiation, gas plasma, etc. The details of these techniques will be discussed later in the article.

CONTAMINATION OF SHELL EGGS

Eggs contain natural protective layers for protection against bacterial invasion; the main components of these protective layers are the eggshell and the cuticle. Eggshell has two membranes: an outer membrane of 50 to 70 μm thickness and an inner membrane of 15 to 25 μm thickness. The hard outer membrane is made of calcium carbonate and protects the egg content from mechanical damage (6). The eggshell contains pores that help with gaseous exchange, and an egg can contain 10,000 to 20,000 pores with a size of more than 10 μm in diameter (7). The larger pores of an eggshell provide potential hiding spots for pathogens; to prevent bacteria from hiding in these pores eggs contain an outer protein layer called the cuticle. This layer ranges from 10 to 15 μm and can protect eggs against bacterial invasion as well as maintain moisture inside the shell (8). The cuticle is very sensitive and can easily be removed by common industry practices such as washing, chlorine treatment, etc. Samiullah et al. (9) studied the effect of washing on penetration of *Salmonella enterica* serotype Infantis in shell eggs. They found that washing eggs in hypochlorite solution removed the cuticle layer. The inoculation of whole eggs caused a higher penetration of *S.* Infantis in the cuticle-less shell eggs observed in this study. Therefore, it is very important to protect shell eggs from recontamination after the cuticle layer is removed.

Along with the outer layers, eggs contain bactericidal enzymes such as ovomucin and lysozyme in the albumen that can hydrolyze the outer cell membrane and kill some pathogenic bacteria. Albumen is a protein-rich component of shell eggs and also kills invading bacteria (10). However, the nutrient-rich egg content makes it very favorable for the growth of pathogens, and if bacteria invade the egg content, they can multiply at refrigerated temperatures as well.

S. Enteritidis is the biggest concern in the egg industry, and the FDA has stated that eggs infected with *S.* Enteritidis cause 79,000 illness and 30 deaths every year in the United States (4). The contamination of eggs mainly occurs in two ways, the first being contamination of eggshells by coming into direct contact with infected birds or bird feces containing pathogenic microbes, known as horizontal transmission. The vent is the same exit for eggs and feces in a hen's body. This can cause fecal contamination of the outer shell of an egg. Liong et al. (11) studied the penetration of *S.* Enteritidis from the outer shell via confocal scanning laser microscopy. The researchers measured the thickness of the inner membrane in the range of 15 to 26 μm and the outer membrane in the range of 50 to 70 μm. Egg membranes were suspended into 9 log CFU/ml solution of *S.* Enteritidis, and it was observed that *S.* Enteritidis penetrated 25 to 30 μm into the membrane in 24 hours. Chen et al. (12) also reported penetration of *S.* Enteritidis from eggshell to inside the egg. In this study, eggs were dipped into 6, 7, and 8 log CFU/ml suspensions of *S.* Enteritidis for 16 h at 40°C and 3 days at 21°C. After 10 weeks of storage at 21°C, all of the eggs dipped into the 8 log suspension had *S.* Enteritidis–positive content, whereas 82% of the same exposed to 6 to 7 log CFU/ml were positive.

The other means of contamination of egg content is ovarian transmission. *Salmonella* can be transferred from the intestines of an infected hen to the reproductive tract and the egg content. This type of contamination is dangerous, because it cannot be removed by using surface decontamination techniques such as washing or UV irradiation. Humphrey et al. (3) studied the ovarian transmission of *Salmonella* from naturally infected hens to their eggs. The authors found that the egg content of 11 eggs out of 1,119 was positive with *Salmonella.*

Housing systems, feed components, antibiotics, vaccination, water quality, and flock size are preharvest factors that can impact the load of pathogenic bacteria on hens and hence on the egg shells. It is important to raise laying hens in a clean environment and maintain the air and water quality to reduce the pathogenic load on shell eggs. The details of different factors affecting microbial load in laying hens are given below.

Feed Components and Additives

Feed components affect the internal organs and bacterial colonization in chickens. Chicken feed mainly consists of corn, wheat, and soybeans along with small quantities of animal fat, sugar, starch, fishmeal, amino acids, and some feed additives (13). Corn-based feed is most common in the United States because of its availability, while in Europe, a wheat- and barley-based chicken diet is commonly used. Bjerrum et al. (14) studied the effect of a wheat-based diet on the *S.* Typhimurium count in internal organs and found that the use of wheat-based chicken feed reduced the pH in the gizzard as well as *S.* Typhimurium counts in the gizzard and ileum, but there was no difference in *S.* Typhimurium counts in the ceca and rectum of chickens. On the other hand, various researchers have shown a higher prevalence of necrotic enteritis caused by *C. perfringens* type A in chickens fed wheat- and barley-based diets (15–17). Keeping these things in mind, a corn-based chicken diet is usually recommended for chickens. However, in addition to feed components, the physical properties of feed also affect the *C. perfringens* load and the digestive organs of chickens. Engberg et al. (13) found that a wheat-based diet in pellet form is more helpful than in mash form in reducing C. *perfringens* load from the intestines of chickens. A lower number of *C. perfringens* organisms was found in the ceca and rectum of pellet-fed chickens compared to mash-fed chickens.

Along with main feed components, various other feed additives are added to chicken feed to enhance the feed quality. Different feed additives such as antibiotics, prebiotics, and probiotics are used as a preventive measure to control the prevalence of pathogenic bacteria in the chicken intestine (18). Antibiotics have successfully been used by various researchers in reducing *Salmonella* and *Campylobacter* shedding from chicken feces (19, 20). Bolder et al. (19) studied the effect of antibiotics on shedding of pathogens in chicken feces and observed that in young chicks inoculated with 8 log/ml of *S.* Enteritidis, *C. perfringens,* and *Campylobacter jejuni* and feeding with 9 mg/kg of flavophospholipol and 120 g/kg of salinomycin sodium significantly reduced *S.* Enteritidis and *C. perfringens* shedding from feces. Furthermore, the presence of salinomycin sodium in chicken feed reduced the prevalence of *S.* Enteritidis in chickens. However, there are some concerns regarding antibiotic residues in meat and the development of antibiotic-resistant bacteria resulting from feed additive antibiotics. Therefore, the European Union banned the use of antibiotics in 1999, with some exceptions (18).

Competitive exclusion is a method of reducing pathogens in the intestinal tract of chickens by feeding them nonpathogenic microorganisms or probiotics, which would compete with the pathogens for nutrients. This method is commonly used in the United States and proved to be very effective in reducing antibiotic-resistant *Salmonella* in chickens. Line et al. (21) studied the effect of yeast components in chicken feed in reducing *Salmonella* and *Campylobacter* load from chickens. *Saccharomyces boulardii*, a nonpathogenic yeast, was given to chickens at the rate of 1 g/kg and 100 g/kg of feed. The chickens were infected with an oral dose of 8 log CFU/ml of *S.* Typhimurium and *C. jejuni*. The yeast component in feed helped in significantly reducing *S.* Typhimurium load. Only 20% and 5% of chickens were colonized with *Salmonella* fed with 1 g/kg and 100 g/kg of the yeast component, respectively, compared to 70% of birds without a yeast component in their feed. However, the colonization of *C. jejuni* in chicken remains unaffected. Al-Zenki et al. (22) also found competitive exclusion to be effective in reducing *Salmonella* load in chicken carcasses as well as in the interior organs. A lower number of *Salmonella* organisms was found when 1 g/kg of *Saccharomyces cerevisiae* and 100 mg/kg of *Pediococcus acidilactici* were mixed with the chicken feed.

Another important feed additive is prebiotics. These are nondigestive food ingredients that stimulate growth of probiotics and help reduce the population of pathogenic bacteria in chicken intestines (23). Mannan-oligosaccharide (MOS) is a mannose-based carbohydrate that is widely used as a prebiotic in chicken feed. Spring et al. (24) found that MOS is effective in reducing *S.* Typhimurium load in chickens. In this study, 3-day-old chicks were inoculated with 4 log CFU/ml of *S.* Typhimurium and were fed a 4,000 ppm of MOS supplement along with their feed. The cecal population of *S.* Typhimurium was reduced from 5.40 to 4.07 log CFU/g. Addition of the prebiotic helped significantly reduce *Salmonella* load from chickens.

Vaccination

Chickens are vaccinated to protect against the colonization of pathogenic microorganisms—specifically, *Salmonella.* The FDA's egg safety rule recommends that producers vaccinate pullets before bringing the flocks into a production house, but this is not mandatory in the United States (25). Vaccination of weak strains helps in making antibodies against the targeted strain and hence reduces its colonization in different body parts. Vaccination has been proven effective in reducing *Salmonella* counts in organs and shell eggs on a lab scale in various research studies. Miyamoto et al. (26) studied the effect of oil-emulsion *S.* Enteritidis bacterin vaccine in reducing its load on shell eggs. Hens were vaccinated at the age of 38 weeks, and a booster was given 4 weeks later. The vaginal vent of the hens was inoculated with 7 log CFU/ml of *S.* Enteritidis after 2 weeks of booster vaccination, and results indicated that 37% of the eggs from unvaccinated hens were positive for *S.* Enteritidis, whereas only 19% from vaccinated hens were positive suggesting significant reductions of S. Enteritidis in eggs obtained from vaccinated hens. In another study, Gast et al. (27) found that oil-emulsion *S.* Enteritidis bacterin helps reduce *S.* Enteritidis colonization in the organs of the hens. However, the populations of *S.* Enteritidis remained unchanged in the feces.

Vaccination can be given in two forms: live and inactivated/killed. Live vaccine contains live cells of the targeted organism, causes illness in the vaccinated chickens, and stimulates immune responses. The advantages of using live vaccine are that it can be given via eye drops, in drinking water, or by spraying it on a chicken flock (28). These advantages make it applicable on a larger industrial scale. Eye drops of live vaccine are very effective because the vaccine passes to the Harderian glands, which are the most important organs for developing immune responses in chickens (29). The other application methods of live vaccination, such as drinking water, feed additive, and spraying, are not very effective and should be used at a higher frequency.

The other type of vaccination is inactivated vaccine, in which cells of a targeted organism are inactivated/killed and then injected into a chicken's body. The only method of application is injecting this vaccine into the intramuscular or subcutaneous layer of a chicken's body (28). This is very labor-intensive but is effective in inducing an immune response against the targeted organism. Researchers have shown that live vaccines are more effective in inducing immunity in chickens than are inactivated vaccines (30, 31). Babu et al. (31) studied the effect of live and killed vaccine against *Salmonella* in chickens and found that live vaccine was more effective in increasing the lymphocyte population against *S.* Enteritidis antigen. However, the efficacy of a particular type of vaccine mainly depends on the virulence factors of the targeted microorganism. The vaccination program for commercial layers is presented in Table 1.

Housing System

The housing system is one of the major factors that affect pathogenic bacterial load on chickens. The manure-handling system, air quality, and area per unit bird vary with different housing systems, and the growing systems can broadly be divided into two main groups: battery cage and alternative methods. In a battery cage system, birds are confined in multistoried, small cages with an egg conveyer belt moving underneath as shown in Fig. 1. This is the most common housing system used in the egg industry, because it is economical and less labor intensive. In the United States, the average space per hen should range from 460 to 540 cm^2 (32). The main advantages of a cage system are that it reduces the cost of egg production, increases livability for hens, and makes it easy to collect eggs on a moving conveyer belt (33). However, Singh et al. (34) found that the birds

TABLE 1 Vaccination program for commercial layers[a]

Age	Vaccine	Route	Type
1 day	Marek's disease	Subcutaneous	Turkey herpesvirus and SB-1
14–21 days	Newcastle/infectious bronchitis	Water	B1/Mass
14–21 days	Infectious bursal disease	Water	Intermediate
5 wk	Newcastle/infectious bronchitis	Water or coarse spray	B1/Mass
8–10 wk	Newcastle/infectious bronchitis	Water or coarse spray	B1 or LaSota/Mass
10–12 wk	Encephalomyelitis	Wing web	Live, chick-embryo origin
10–12 wk	Fowlpox	Wing web	Modified live
10–12 wk	Laryngotracheitis	Intraocular	Modified live
10–14 wk	*Mycoplasma gallisepticum*[b]	Intraocular or spray	Mild live strain
Or 18 wk		Parenteral	Inactivated
12–14 wk	Newcastle/infectious bronchitis	Water or aerosol	B1 or LaSota/Mass
16–18 wk	Newcastle/infectious bronchitis	Water or aerosol	B1 or LaSota/Mass
Every 60–90 days or 18 wk	Newcastle/infectious bronchitis	Parenteral	Inactivated

[a]This is an example of a vaccination program. Individual programs are highly variable and reflect local conditions, disease prevalence, severity of challenge, and individual preferences.

[b]The use of *M. gallisepticum* vaccine is regulated or prohibited in some states. SB-1 or MDV301 may be combined with turkey herpesvirus in some areas. Vaccination for infectious bursal disease, laryngotracheitis, and fowlpox depends on local requirements. Other strains of infectious bronchitis (Connecticut, Arkansas 99, Florida 88, etc.) are included in some areas. *M. gallisepticum* and *Haemophilus gallinarum* (coryza) are used only on infected, multiage premises in some areas. Source: *The Merck Veterinary Manual* (81).

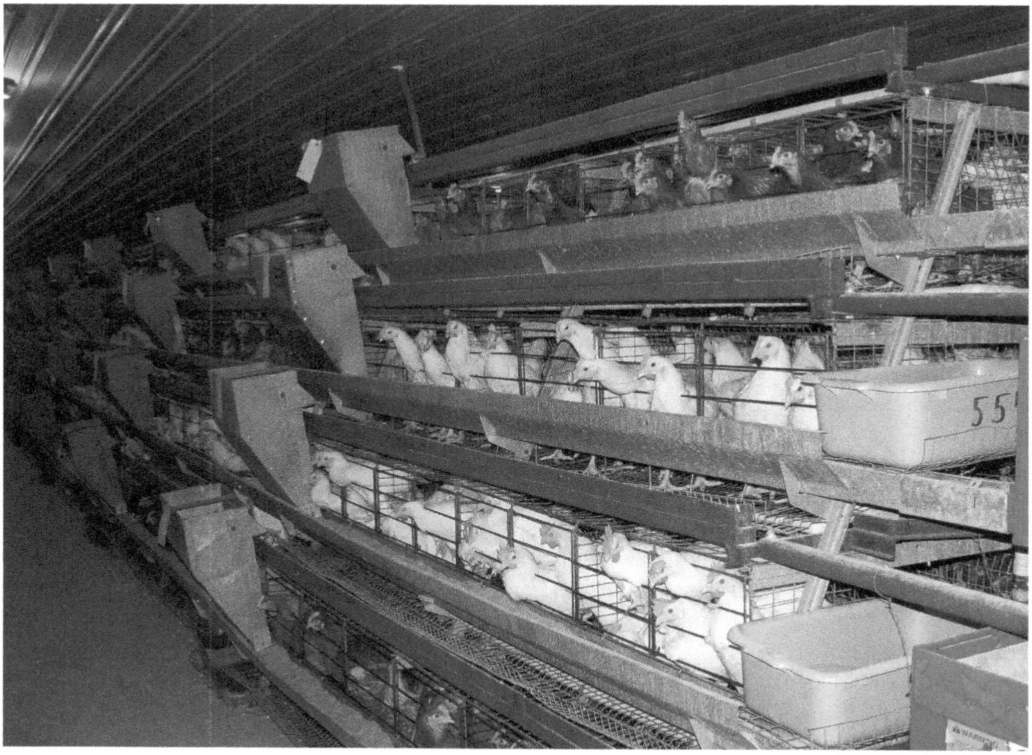

FIGURE 1 A battery cage housing system for laying hens. Credit: Dr. Kenneth E. Anderson, Director, North Carolina Layer Performance and Management Program, North Carolina State University.

were under more stress in a cage system during the initiation of the laying cycle and that the stress decreased as the birds acclimatized to the environment. For comparison, the stress period was short for a floor pen system because the pullets were grown in the same system. The cage system has been banned in Europe since 2012 due to public concerns regarding less space per unit hen (35).

Alternative housing systems include the free-range system, where birds are grown in an open environment and are free to move in an open area as shown in Fig. 2. According to the animal husbandry guidelines for egg laying hens in the United States, a minimum of 1,390 cm^2 area is required per unit bird to maintain their natural behavior in a free-range system, and 15% of the total space should be provided for litter. Feeders should be distributed in the entire area in such a way that birds don't need to move more than 26 feet to reach a feeder (32). In comparing the egg quality of the caged and free-range birds, Van Den Brand et al. (36) found that birds took longer to grow and remained lighter in a free-range environment during the observational period. Free-range birds produced eggs with a darker yolk than those of caged birds. The authors presumed that consumption of grass and herbs could be the reason for the darker yolk and concluded that due to interaction with the outside environment it is difficult to control the internal and external egg quality in a free-range system.

Various research studies show the prevalence of *Salmonella* and other pathogenic bacteria in both systems. Methner et al. (37) found that the prevalence of *Salmonella*

FIGURE 2 **A free-range housing system for laying hens. Credit: Dr. Kenneth E. Anderson, Director, North Carolina Layer Performance and Management Program, North Carolina State University.**

was higher in a battery cage system than in alternative housing systems such as free range, floor management with free range, and floor management without free range. A total of 329 flocks from different housing systems were tested for the presence of *Salmonella*, and it was found that 42.3% of flocks with battery cage systems were positive. In contrast, De Vylder et al. (38) found higher horizontal transmission of *S.* Enteritidis in an alternative housing system because more birds come in direct contact with each other. In this study, 129 birds were infected with *S.* Enteritidis at the age of 16 weeks and then placed with an equal number of *Salmonella*-free birds in a different housing system. After 4 weeks, a higher rate of horizontal transmission was observed in the aviary and floor-based system compared to the cage system. However, these studies are not conclusive because other factors such as air quality and flock size are not included in the results.

Other Preharvest Factors

Various other factors such as flock size, water quality, air quality, stress, and the presence of rodents also affect *Salmonella* infection in a laying house. The chances of infection increase with a higher number of pullets in a flock (39), because infected pullets can come into direct contact with others and cause horizontal transmission of pathogenic bacteria. Higher flock density can increase the spread of diseases or pathogenic microorganisms. Moreover, multiple flocks at one farm can further increase the chances of cross-contamination as different flocks share water and feed lines. Hens from one laying house can also be infected from other houses via egg conveying belts. Therefore, it is very important to disinfect the water lines, feed lines, and egg conveying belts before replacing the flocks (4). Water and air quality should also be tested frequently to comply with the FDA rule. Other environmental factors such as rodents, insects, and flies are also carriers of pathogenic bacteria, and they can increase cross-contamination inside a flock. The FDA egg rule recommends that growing houses be set up with mechanical traps, glue boards, and other biosecurity measures for trapping rodents and insects in the vicinity of the house. The egg rule also recommends removing debris and vegetation from inside and nearby areas of a growing house (4).

Stress affects the immune system of chickens and makes them prone to infections. Mashaly et al. (40) studied the effect of heat stress on egg quality and the immune system of laying hens. They found that hens subjected to higher temperature and humidity had lower white blood cell and antibody counts compared to the control group. Induced molting of hens also causes stress. Induced molting is a process in which feed is removed from the aged hens until they lose a certain amount of weight. Molted hens do not lay eggs for some weeks, and lost nutrients are replenished. Molting helps increase the efficiency of a producing plant because it recycles hens for another laying cycle. However, the stress caused by induced molting weakens the hens' immune system. Holt (41) found that the colonization of *S.* Enteritidis is higher in feces as well as the internal organs of the hens that have experienced induced molting.

PREVENTION CONTROLS FOR EGG SAFETY

The contamination of shell eggs can be reduced by following good management practices and precautionary measures. Some good management practices include increasing biosecurity in the processing plant, disinfecting flocks, and vaccinations (42). Pathogen carriers such as humans, insects, and rodents are the potential sources of contamination. People other than employees should not be allowed inside a poultry house. The chances of cross-contamination can be reduced further by forbidding employees from working in other processing plants or keeping their own birds at home (25). Insects such as mites, lice, ticks, flies, and mosquitoes can spread contamination in hens as well as in eggs. Preventive measures such as spraying dust pesticides, cleaning frequently, and avoid standing water should be taken to reduce insect populations. Various types of rodent traps, glue boards, and baits should be installed in different areas to further fortify biosecurity in processing plants (43). The conveying belts, used for moving feed and eggs, should be decontaminated frequently because they are a potential source of cross-contamination between the flock houses (4). Fecal contamination from one flock can infect the second flock in a laying house. It is very important to decontaminate all equipment, including cages and belts, before introducing new flocks to a poultry house to reduce contamination (4). Environmental samples including air samples and swabs should be taken frequently at different locations in the poultry house to keep in compliance with the FDA egg safety rule.

DECONTAMINATION TECHNOLOGIES FOR SHELL EGGS

Once eggs are harvested from a laying house system, they are decontaminated in a processing plant before packing. In the United States, the egg surface is decontaminated by washing with alkaline detergents, chlorinated water, or U.S. Department of Agriculture–approved chemical disinfectants. In Europe, egg washing is banned because it removes the outer cuticle of the shell. Also, washing all eggs together can cause cross-contamination from infected eggs to clean eggs. However, immediate drying of shell eggs after washing can significantly reduce the chances of microbial infection (44).

Various other surface and egg content decontamination techniques are used in the egg industry. Researchers have developed techniques that do not include washing eggshells. A summary of various decontamination strategies used in the egg industry, their efficacy in reducing pathogen load from shell eggs, and their benefits and limitations are presented in Table 2.

Chemical Disinfectants

Various types of disinfecting chemical solutions are used in the egg industry to reduce microbial loads on shell eggs. Chlorine is the most common disinfectant used in the food industry. Eggs are washed with chlorinated water for decontamination of pathogenic microorganisms. Along with chlorine, other chemicals such as chlorhexidiene, sodium dodecyl sulfate, Lugol's solution (mixture of iodine, potassium iodiate, and ethanol), and hydrogen peroxide are used in the egg industry. However, various authors have reported their inefficacy for reducing the entire bacterial load from shell eggs. Himathongkham et al. (45) found that dipping infected eggs for 30 min at 20°C in 200 ppm chlorine resulted in about 13% disinfection, whereas chlorhexidine treatment for 2 min at 20°C reduced about 40% of *S.* Enteritidis from shell eggs. Lugol's solution on shell eggs for 5 min at 20°C disinfected only 2 out of 85 infected eggs. Dipping eggs in boiling water for 3 s disinfected 100% of the eggs, but it also resulted in breakage of the outer membrane. Padron (46) found that dipping shell eggs inoculated with *S.* Typhimurium in a 6% hydrogen peroxide solution killed 95% of the total bacteria, but only 55% of the eggs were disinfected by this process. In another study, Gast (47) found that dipping shell eggs infected with *S.* Enteritidis in a 2% solution of tincture of iodine for 5 s disinfected only

TABLE 2 Decontamination processes and their effect on reduction of pathogens from shell eggs

Process	Reduction of pathogens
Chemical disinfectants	200 ppm chlorine disinfects 13% of eggs from *S.* Enteritidis in 30 min at 20°C; 0.5% chlorhexidiene treatment reduced only 40% of *S.* Enteritidis; dipping in boiling water for 3 s disinfects 100% of the shell eggs, but it breaks the shell membrane (44)
Pasteurization of eggs	Pasteurization at 57°C for 25 min reduced 3 log CFU/shell egg of *S.* Enteritidis in a water bath; at 55°C for 180 min in a hot air oven 5 log CFU/shell egg reduction was observed (51)
Thermoultrasonication	Pasteurization at 60°C for 3.5 min combined with ultrasonication increased the bacterial load reduction by 100-fold (56); pasteurization at 54°C for 5 min with 24 kHz ultrasonic waves reduced *S.* Enteritidis by 5.83 log CFU/shell egg (57)
Ozone treatment	5.9 log CFU/eggshell reduction of *S.* Enteritidis by exposing to ozone at 15 lb/in^2 for 10 min (61); 1.4 mg of ozone/liter of water reduced 2.47 log CFU of *S.* Enteritidis in 90 s at 4°C; 1.37 log CFU reduction at 22°C (62); ozone treatment at 140 g/m^3 combined with heating at 57°C for 21 min resulted in reduction of 4.2 out of 5.39 log CFU/shell egg of *S.* Enteritidis (60)
Nonthermal gas plasma treatment	4.7 log CFU reduction of *S.* Enteritidis and *S.* Typhimurium after exposing shell eggs to gas plasma for 90 min at 25°C (67)
Acidic electrolyzed water	Spraying electrolyzed oxidized water for 15 s per h for 24 h decontaminated 11 out of 15 eggs containing *S.* Typhimurium and *S. aureus* (71); dipping eggs in acidic and alkaline electrolyzed water containing 41 mg/liter available chlorine for 1 min reduced *S.* Enteritidis by 3.66 log CFU/shell egg (70)
Pulsed light technology	3.6 log CFU/shell egg of *S.* Enteritidis organisms were killed when 12 J/cm^2 of pulsed light was applied; washing eggs with warm water reduced the efficacy of this process (76)
UV irradiation	*Salmonella* was reduced by 5.4 log CFU/g when exposed to 23.6 J/cm^2 of UV light for 20 s from 9.5 m (78); 4.3 log CFU/g of *S.* Enteritidis was reduced by exposure of 2500 W/cm^2 of UV for 5 min at 25°C (61)
Formaldehyde fumigation	Fumigation of contaminated shell eggs with formaldehyde for 1 h decontaminated 32% of unwashed eggs (80)

45% of the shell eggs. From these studies, it can be concluded that chemical disinfectants are not very effective for egg safety and should be used with other disinfecting techniques.

Pasteurization of Eggs

Pasteurization is one of the oldest and most well-established methods for decontaminating milk, eggs, and other food products. Shell eggs are heated at a mild temperature in the range of 52 to 60°C for 2 to 30 min for pasteurization (48). The egg pasteurization temperature recommended by the Food Safety and Inspection Service (FSIS) of U.S. Department of Agriculture is 57.5°C for egg white and 61.1°C for egg yolk (49). The main purpose of this process is not only to reduce microbial load from the surface of shell eggs but also to preserve their color and functional properties, strength, and quality.

The heating system used for pasteurization should have low come-up time (time taken to reach the desired temperature), high efficiency, and high temperature-holding capacity. Various types of heating systems including water bath, hot air ovens, and microwave ovens are used for pasteurizing shell eggs in the food industry. Stadelman et al. (50) compared the come-up times and heating efficiency of the water bath with the hot air oven to reach an optimum temperature of 55 to 57°C. The time required for heating the egg yolk to 56°C with a water bath and hot air oven are 20 min and 60 min, respectively. The lower come-up time of water baths makes them more effective for pasteurization of shell eggs. In another study, Hou et al. (51) looked at the effectiveness of different types of heating systems in reducing bacterial load. They found that pasteurizing shell eggs inoculated with *S.* Enteritidis in a water bath at 57°C for 25 min reduced bacterial load by 3 log CFU/shell egg, while a hot air oven at 55°C for 180 min killed 5 log CFU/shell egg of bacterial population. The authors also found that pasteurization is most effective with the combination of water bath at 57°C for 25 min followed by hot air oven at 55°C for 25 min. A total 7 log CFU/shell egg reduction in bacterial load was observed when the processes were combined, without denaturing the egg proteins.

Despite the higher efficiency of water bath heating, there is a chance of overheating eggs that causes denaturing and loss of functional properties due to longer exposure to heat. Microwave heating can help in further reducing the come-up time and hence reduce the chances of overheating. Dev et al. (52) studied pasteurization of shell eggs using three microwave power densities: 0.75 W/g, 1 W/g, and 2 W/g. The authors also compared the come-up times for egg albumen and egg yolk in a beaker and in an eggshell. Egg yolk in a beaker took 9 min to reach 60°C, compared to 3.5 min in shell egg with 2.0 W/g power density of microwave. The lower come-up time for yolk in shell egg is because its spherical geometry converges microwaves toward the center where yolk is located and increases the heating of the yolk (53). Lower come-up times and heating duration make microwave heating suitable for egg pasteurization.

Thermoultrasonication

Ultrasound waves with frequency in the range of 20 to 100 kHz cause thinning of cell membranes and produce micromechanical shocks that damage the cell components and therefore cause cell death (54). However, bacterial spores and some other pathogenic bacteria are resistant to ultrasound wave treatments. In order to enhance the efficacy of the ultrasound waves, it is used in combination with thermal or pressure treatments (55). The process of combining the effects of ultrasonic waves and heat is known as thermoultrasonication. Cabeza et al. (56) studied the efficacy of heat treatment and ultrasonic waves in reducing *S. enterica* serotype Senftenberg load from shell eggs. Shell eggs artificially inoculated with *S.* Senftenberg were pasteurized at

57.8 to 67°C with and without ultrasonic waves. It was found that the decimal reduction time (D-value) for *S.* Senftenberg was decreased by 65.2% at 57.8°C with the effect of ultrasonic waves. At 67°C, the D-value was reduced by 11.1% with the combined effect of both processes. The authors also found that thermoultrasonication helped in increasing bacterial load destruction by 100-fold for pasteurization of shell eggs at 60°C for 3.5 min.

Thermoultrasonication is not only effective in microbial decontamination but also maintains the physical and nutritional properties of shell eggs (57). In another study, Cabeza et al. (57) found a reduction of 5.83 log CFU/shell egg in the *S.* Enteritidis population by using thermoultrasonication. Eggs were inoculated with *S.* Enteritidis and then pasteurized at 54°C for 5 min with 24 kHz of ultrasonic waves. The authors also measured the effect of storage time and conditions on the thickness, density, and strength of the shell with or without the decontamination treatment via thermoultrasonication. The treated and untreated shell eggs were stored at 4°C, 10°C, 20°C, and 32°C with 85%, 80%, 43%, and 30% relative humidity, respectively, for 15 days. There was no significant difference found in physical properties such as thickness, density, and strength of the shell eggs with or without treatment after storage. However, some changes in cuticle morphology were observed in the treated eggs, but no bacterial growth was observed in any storage condition after 15 days of storage. The authors concluded that thermoultrasonication is an effective process for egg safety and can effectively replace pasteurization of shell eggs.

Ozone Treatment

Ozone (O_3) is an oxidizing agent and can potentially be used to kill pathogenic microorganisms in food products. The presence of a third oxygen atom makes ozone a highly oxidizing agent, and it oxidizes important components of a cell such as the phospholipid bilayer, lipoproteins, and unsaturated fatty acids and causes cell death. Ozone can also cause destruction of the nucleic acid (58). The FDA has approved the use of ozone as a sanitizer in the food industry (59). Various research studies have shown that ozone treatment significantly reduces *Salmonella* populations in shell eggs (60–62). Rodriguez-Romo and Yousef (61) used ozone gas treatment for decontamination of artificially inoculated shell eggs with an *S.* Enteritidis load of 5.9 to 6.6 log CFU/eggshell. The authors observed a reduction of 5.9 log CFU/eggshell of *S.* Enteritidis by exposing inoculated shell eggs to ozone gas at 15 lb/in^2 for 10 min.

Along with ozone gas, ozonated water treatment is a nonthermal technique that can also help decontaminate shell eggs. Ozone gas is dissolved in water and shell eggs are washed with this water for decontamination. Koidis et al. (62) used ozonated water at 1.4 mg of ozone/liter of water to decontaminate eggs and observed a reduction of 2.47 log CFU/eggshell of *S.* Enteritidis concentration after treatment of 90 s at 4°C. At a higher temperature, ozonated water tends to lose dissolved ozone and hence its potency decreases. The authors found a decrease in decontamination of *S.* Enteritidis at 22°C because only 1.37 log CFU/eggshell was reduced.

Although ozone exposure reduces the bacterial load on shell eggs, exposure at a higher concentration of ozone gas can alter the egg components. Fuhrmann et al. (63) studied the effect of ozone exposure on egg components and found that the free SH groups of egg white protein were reduced significantly when eggs were exposed to 50 ml/liter of ozone, which reduces the nutritional value of the eggs. Also, even a small concentration of ozone can remove the cuticle of eggs by damaging the cuticle proteins. Moreover, using ozone alone is not very effective in reducing the bacterial load of shell eggs. To increase its efficacy as a disinfectant and reduce higher exposure to ozone, this process is combined with other treatments. Perry et al. (60) studied the combination of ozone and heat treatment to kill *S.* Enteritidis in shell eggs. They found that combing these treatments yielded significantly more reduction of *S.* Enteritidis in shell eggs. Heat treatment at 57°C for 21 min, ozone exposure at 140 g/m^3 of ozone at 180 to 190 Kpa pressure, and combination of both treatments was used on eggs inoculated with 2.4 x 10^5 CFU/ml of *S.* Enteritidis. Reductions of 0.11 and 3.1 log CFU/ml were calculated with ozone treatment and heat treatment, respectively. The combination of both processes produced 4.2 log CFU/shell egg cycle reductions in the population of *S.* Enteritidis. The authors recommended using a combination of these treatments for better results.

Nonthermal Gas Plasma Treatment

Ionizing gas molecules by applying an electric field generates gas plasma. The ionized gas molecules move with high speed and lose their energy by colliding with other bigger particles/bacterial cells (64). It is believed that the collision between ions and cells induces perforation in the cytoplasmic membrane of microorganisms, which causes cell lysis (65, 66). This property makes gas plasma effective in killing microorganisms in food products. Gas plasma can be generated at low temperatures and pressures (65). Nonthermal plasma is desirable for egg decontamination, because low temperatures do not affect the nutritional qualities or the physical properties of the eggshell.

Gas composition and relative humidity of gas plasma play an important role in determining the germicidal property of gas plasma. Ragni et al. (67) studied the efficacy of gas plasma in reducing *S.* Enteritidis and *S.* Typhimurium load on shell eggs at different relative humidity. Gas plasma was generated by resistive barrier discharge prototype. Shell eggs were inoculated with 5.5 to 6.5 log CFU/eggshell of *S.* Enteritidis and *S.* Typhimurium and then exposed to gas plasma for 0 to 90 min at 25°C with 35% and 65% relative humidity. At higher relative humidity more OH^- ions were produced due to higher water content of the gas that causes oxidation of cell components, and hence higher microbial reduction of 4.7 log CFU/eggshell was observed. The microbial reduction of 35% relative humidity gas was 2.5 log CFU/eggshell. It was also found that the gas plasma treatment did not affect the cuticle and the inner layer of the eggs. The authors concluded that nonthermal plasma at higher relative humidity could reduce *Salmonella* load from shell eggs without altering the nutritional quality of the eggs.

Acidic Electrolyzed Water

Usage of alkaline detergent and then chlorine treatment is the most common technique in the egg industry for cleaning and disinfecting shell eggs, but higher chorine exposure can degrade the egg's shell and quality (68). Many researchers have found that electrolyzed oxidized water (EOW) or acidic electrolyzed water (AEW) can replace alkaline detergent and chlorinated water washing and is less destructive to the egg shell and egg quality. Different researchers have recently shown the capability of AEW for deactivating *S.* Enteritidis (69, 70). Electrolyzed water is produced by the electrolysis of a weak sodium chloride solution. The presence of chlorine and a lower pH in the range of 2 to 2.5 of the AEW makes it an effective way to disinfect shell eggs (68).

Russell (71) studied the effect of EOW on the pathogenic bacteria present on shell eggs. In this study, shell eggs were inoculated with *S.* Typhimurium, *Staphylococcus aureus,* and *Listeria monocytogenes,* and then EO water was sprayed on the eggs for 15 s each hour for 24 h. The EOW completely eradicated *S. aureus* and *L. monocytogenes* from more than 11 out of 15 shell eggs in four repetitions. Also, 4.75 eggs out of 15 were disinfected from *S.* Typhimurium after four repetitions. In another study, Bialka et al. (68) compared the efficacy of EOW with the standard alkaline detergent wash and chlorine treatment for microbial safety. Shell eggs inoculated with *S.* Enteritidis were submerged in alkaline EOW and then in acidic EOW. The EOW treatment showed a significantly higher reduction (2.1 log) in the *S.* Enteritidis population compared to standard alkaline detergent washing methods (1.7 log). Also, using EOW water did not result in any degradation of the strength of the eggshells. Park et al. (70) also found that the alkaline EOW followed by acidic EOW treatment is equivalent to washing with chlorinated water in reducing *S.* Enteritidis and *L. monocytogenes* populations from shell eggs. The mean reduction in *S.* Enteritidis population after 5 min of washing with 200 mg/liter chlorinated water was 3.81 log CFU/shell egg. In contrast, the combination of alkaline and acidic EOW treatment for 1 min each with 41 mg/liter available chlorine reduced the *S.* Enteritidis population by 3.66 log CFU/shell egg. Similar results were found for *L. monocytogenes* with chlorinated water treatment (4.01 log CFU/shell egg) and EOW treatment (4.39 log CFU/shell egg).

However, the lower pH of EOW could volatilize the available chlorine content, which would not only reduce its efficacy as an antimicrobial but would also make it harmful to humans and the environment (72). This drawback limits the use of AEW on an industrial scale. To overcome this problem, Cao et al. (73) investigated the effect of slightly acidic electrolyzed water (SAEW) for killing *S.* Enteritidis in shell eggs. SAEW is collected from the anode side of electrolysis of dilute hydrochloric acid, and it contains hypochlorous acid, which has very strong antimicrobial activity (74). Washing inoculated shell eggs for 3 min with SAEW that has 15 mg/liter available chlorine resulted in a reduction of 6.5 log cycles of *S.* Enteritidis. Furthermore, the wastewater did not contain any *S.* Enteritidis. The authors concluded that SAEW could reduce pathogenic bacteria load from shell eggs without causing environmental pollution.

Pulsed Light Technology

Pulsed light technology is a nonthermal technology for killing pathogenic bacteria in food products. With this technology, microorganisms are killed by photochemical and photothermal effects of pulsed light with wavelengths ranging from 200 to 1000 nm and an exposure time of 10^{-2} to 10^{-3} ms. Thymine and cytosine in DNA structures form pyrimidine dimers in the presence of UV light and therefore cause disruption of the DNA double strand and cell death. This phenomenon is called the photochemical effect of UV. Takeshita et al. (75) confirmed the presence of pyrimidine dimers and DNA rupture when yeast cells were exposed to UV radiation. The other effect of pulsed light that causes cell death is the photothermal effect. The absorption of light by microorganisms causes overheating, which in turn causes vaporization of water from the cytoplasm and rupture of the cell membrane (75).

Hierro et al. (76) studied the efficacy of pulsed light technology for inactivation of *S.* Enteritidis on agar media and shell eggs. Pulsed light treatment with 0.7 J/cm^2 showed inactivation of 6.7 log CFU/cm^2 of *S.* Enteritidis from an agar media plate. The inactivation of *S.* Enteritidis was higher (3.6 log CFU/egg) when the unwashed eggs were pulsed compared to washed eggs (1.8 log CFU/g). This is mainly because washing eggs with warm water may damage the cuticle, which prevents loss of water and acts as a barrier to bacterial invasion (44). With a damaged cuticle, microorganisms can hide in the pores of the eggshell and do not come into direct contact of pulsed light, and hence a lower deactivation rate was observed. However, the authors observed reactivation of *S.* Enteritidis with exposure to light and recommended that the eggs should be kept in a dark place after pulsed light treatment.

UV Irradiation

UV irradiation is one of the most common nonthermal techniques used to kill pathogenic microorganisms on the surface of food products. A UV wavelength in the range of 220 to 300 nm is considered highly lethal to the microorganism because it disrupts the double-stranded DNA by forming pyrimidine dimers (77). The efficacy of UV radiation for eliminating pathogenic microorganisms from food products is well documented (77). This technique also helps protect the nutritional value as well as the strength of shell eggs. The height of albumen and Haugh unit is measured to quantify the egg quality. Keklik et al. (78) studied the effectiveness of UV light for eliminating *S.* Enteritidis from shell eggs. A UV light pulse ranging from 1 to 30 s with a distance of 9.5 and 14.5 cm from the UV light source was used. A reduction of 5.4 log CFU/cm^2 in the *Salmonella* population was observed when inoculated eggs were exposed to 23.6 J/cm^2 of UV light for 20 s from 9.5 cm, without any significant change in the albumen height and Haugh unit of the eggs. The force required to break the eggs did not change, suggesting that the egg strength was also not affected by the UV treatment. However, an increase of 16.3°C in temperature was observed with 30 s exposure time from 9.5 cm. Therefore, the authors recommended using a 20 s exposure of UV light for decontaminating shell eggs. In a different study, Rodriguez-Romo and Yousef (61) found 4.3 log CFU/g of egg shell reduction in *S.* Enteritidis population after treating shell eggs with 2500 μ-W/cm^2 for 5 min at 25°C. These authors also found a synergetic effect of combining UV light with ozone treatment and observed more than 4.6 log CFU/g of egg reduction after 2 min treatment time.

Although UV irradiation is very effective in reducing the surface pathogenic load of shell eggs, UV irradiation is not very effective in reducing bacteria from inside the eggs. De Reu et al. (79) studied the effectiveness of UV irradiation in reducing the surface microbial load and internal bacterial egg contamination. The artificially inoculated shell eggs containing 5.5 x 10^4 CFU/shell egg of *Escherichia coli* and 4.6 x 10^4 of *S. aureus* were exposed to 253.7 nm UV light with an intensity of 10 mW/cm^2 for 4.7 and 18.8 s. A reduction of 3 and 4 log CFU/shell egg was observed for both exposure times, respectively, for both bacteria species. However, for the internally contaminated eggs, the bacterial load of untreated eggs was 4.37 log CFU/ml, and the same for UV irradiated eggs was 4.07 log CFU/ml. There was no significant reduction in internal bacterial load of the eggs from using UV irradiation.

Formaldehyde Fumigation

Bacterial infection of newly hatched chickens is highly likely when the eggs are infected with pathogenic bacteria. It is very important to decontaminate eggs before hatching to reduce infection of young chicks. The shell eggs are fumigated with formaldehyde for decontamination before hatching. Formalin, containing formaldehyde dissolved in water, is reacted with potassium permanganate for the formation of formaldehyde fumes. Whistler and Sheldon (80) studied the effect of formaldehyde fumigation on the bacterial load of the shell surface. A batch of 10 shell eggs was washed with warm water at 40°C to remove the dirt particles, and one set remained unwashed. For formaldehyde production, 40 ml of formalin was mixed with 20 g of potassium permanganate, and the eggs were exposed to it for 1 h. The washed eggs did not show any bacteria after fumigation, but growth in an enrichment broth revealed the presence of some bacteria on these eggs. Also, 68% of the unwashed eggs were still positive for bacteria after fumigation. The authors also found that a 1 h exposure to formaldehyde fumigation did not change the hatchability of the eggs.

FDA RULE ON SHELL EGG SAFETY

The final FDA egg rule has guidelines to reduce contamination in shell eggs. It is mandatory for a processing plant with more than 3,000 laying hens to follow the egg rule and to register with the FDA. However, producers who sell shell eggs directly to consumers or who do not produce eggs for the table market are exempt from the rule. This rule makes it mandatory for the producers to obtain pullets which are grown under *S.* Enteritidis–monitored conditions; those include following "SE Clean" standards of the National Poultry Improvement Plan, taking environmental samples when pullets are 14 to 16 weeks old, and cleaning and disinfecting the pullet house if an environmental sample is positive for *S.* Enteritidis. It recommends using vaccination against *Salmonella* in pullets, but the vaccination is not mandatory. Egg producers are also required to have *S.* Enteritidis control plans including a biosecurity program, equipment cleaning, and provision for monitoring flies and preventing insects and stray animals from entering the poultry house. It is mandatory to take environmental samples in laying houses to test for *S.* Enteritidis when laying hens are 40 to 45 weeks old and also 4 to 6 weeks after molting. If an environmental sample is positive with *S.* Enteritidis, removing all the visible manure, dust, feathers, and old feed is required. A sample of 1000 intact eggs from one day's production must be tested for *S.* Enteritidis after finding a positive environmental sample, and the results must be obtained within 10 calendar days. The testing must be repeated four times at 2 week intervals. The prevalence of infected eggs cannot be more than 1 from a sample size of 1400 eggs. The infected eggs cannot be sold as table eggs and must be diverted for further processing that insures 5 log reduction for *S.* Enteritidis. The shipping container with infected eggs must be labeled. The shell eggs must be refrigerated to less than 45°F 36 h after laying. The producers are required to keep records documenting compliance for every flock and to keep them for at least 1 year after the flock has been taken out of production (25).

CONCLUSIONS

With increased demands and consumption of shell eggs and egg products, the safety of this commodity is imperative. Although there are several practices and technologies (as discussed in this chapter), the harmonious functioning of all aspects from farm to table is essential to ensure the safety of eggs and egg products. Due to documented evidence of vertical transmission that leads to internalization of *Salmonella* in eggs, good practices need to be followed even before the chicks arrive on the farms at the hatchery. Given the available information and the steps discussed in this article, external contamination of shell eggs is the primary source of cross-contamination in shell eggs and egg products, so a combination of good practices along with decontamination strategies is essential to limit contact between the shell exterior and the egg contents to enhance their safety.

CITATION

Singh M, Brar J. 2016. Egg safety in the realm of preharvest food safety. Microbiol Spectrum 4(4):PFS-0005-2014.

REFERENCES

1. **American Egg Board.** 2014. About the US egg industry. http://www.aeb.org/farmers-and-marketers/industry-overview.
2. **Howard ZR, O'Bryan CA, Crandall PG, Ricke SC.** 2012. *Salmonella* Enteritidis in shell eggs: current issues and prospects for control. *Food Res Int* **45:**755–764.
3. **Humphrey TJ, Baskerville A, Mawer S, Rowe B, Hopper S.** 1989. *Salmonella* Enteritidis phage type 4 from the contents of intact eggs: a study involving naturally infected hens. *Epidemiol Infect* **103:**415–423.
4. **Food and Drug Administration.** 2014. FDA improves egg safety. http://www.fda.gov/ForConsumers/ConsumerUpdates/ucm170640.htm.
5. **Sahin O, Kobalka P, Zhang Q.** 2003. Detection and survival of *Campylobacter* in chicken eggs. *J Appl Microbiol* **95:**1070–1079.

6. **Hunton P.** 2005. Research on eggshell structure and quality: an historical overview. *Rev Bras Cienc Avic* **7:**67–71.
7. **Baxter-Jones C.** 1991. Egg hygiene: microbial contamination, significance and control. *Avian Incubation* **22:**269–276.
8. **Board R, Halls N.** 1973. The cuticle: a barrier to liquid and particle penetration of the shell of the hen's egg. *Br Poult Sci* **14:**69–97.
9. **Samiullah, Chousalkar KK, Roberts JR, Sexton M, May D, Kiermeier A.** 2013. Effects of egg shell quality and washing on *Salmonella* Infantis penetration. *Int J Food Microbiol* **165:**77–83.
10. **Sugino H, Nitoda T, Juneja LR.** 1996. General chemical composition of hen eggs, p 13–24. *In* Yamamoto T, Juneja LR, Hatta H, Kim M (ed), *Hen Eggs: Basic and Applied Science.* CRC Press, Boca Raton, FL.
11. **Liong JW, Frank JF, Bailey S.** 1997. Visualization of eggshell membranes and their interaction with *Salmonella* Enteritidis using confocal scanning laser microscopy. *J Food Prot* **60:**1022–1028.
12. **Chen J, Clarke R, Griffiths MW.** 1996. Penetration of *Salmonella* Enteritidis through whole-shell eggs and survival of the bacteria in whole-liquid egg during refrigeration storage. *J Food Prot* **59:**915–921.
13. **Engberg RM, Hedemann MS, Jensen BB.** 2002. The influence of grinding and pelleting of feed on the microbial composition and activity in the digestive tract of broiler chickens. *Br Poult Sci* **43:**569–579.
14. **Bjerrum L, Pedersen AB, Engberg RM.** 2005. The influence of whole wheat feeding on *Salmonella* infection and gut flora composition in broilers. *Avian Dis* **49:**9–15.
15. **Annett CB, Viste JR, Chirino-Trejo M, Classen HL, Middleton DM, Simko E.** 2002. Necrotic enteritis: effect of barley, wheat and corn diets on proliferation of *Clostridium perfringens* type A. *Avian Pathol* **31:**598–601.
16. **Branton SL, Reece FN, Hagler WM Jr.** 1987. Influence of a wheat diet on mortality of broiler chickens associated with necrotic enteritis. *Poult Sci* **66:**1326–1330.
17. **Kaldhusdal M, Hofshagen M.** 1992. Barley inclusion and avoparcin supplementation in broiler diets. 2. Clinical, pathological, and bacteriological findings in a mild form of necrotic enteritis. *Poult Sci* **71:**1145–1153.
18. **Van Immerseel F, Cauwerts K, Devriese LA, Haesebrouck F, Ducatelle R.** 2002. Feed additives to control *Salmonella* in poultry. *Worlds Poult Sci J* **58:**501–513.
19. **Bolder NM, Wagenaar JA, Putirulan FF, Veldman KT, Sommer M.** 1999. The effect of flavophospholipol (Flavomycin) and salinomycin sodium (Sacox) on the excretion of *Clostridium perfringens, Salmonella* Enteritidis, and *Campylobacter jejuni* in broilers after experimental infection. *Poult Sci* **78:**1681–1689.
20. **Johansen CH, Bjerrum L, Pedersen K.** 2007. Impact of salinomycin on the intestinal microflora of broiler chickens. *Acta Vet Scand* **49:**30.
21. **Line JE, Bailey JS, Cox NA, Stern NJ, Tompkins T.** 1998. Effect of yeast-supplemented feed on *Salmonella* and *Campylobacter* populations in broilers. *Poult Sci* **77:**405–410.
22. **Al-Zenki SF, Al-Nasser AY, Al-Saffar AE, Abdullah FK, Al-Bahouh ME, Al-Haddad AS, Alomirah H, Mashaly M.** 2009. Effects of using a chicken-origin competitive exclusion culture and probiotic cultures on reducing *Salmonella* in broilers. *J Appl Poult Res* **18:**23–29.
23. **Gibson GR, Roberfroid MB.** 1995. Dietary modulation of the human colonic microbiota: introducing the concept of prebiotics. *J Nutr* **125:**1401–1412.
24. **Spring P, Wenk C, Dawson KA, Newman KE.** 2000. The effects of dietary mannaoligosaccharides on cecal parameters and the concentrations of enteric bacteria in the ceca of *Salmonella*-challenged broiler chicks. *Poult Sci* **79:**205–211.
25. **Food and Drug Administration.** 2009. Egg Safety Final Rule. http://www.fda.gov/Food/GuidanceRegulation/GuidanceDocuments RegulatoryInformation/Eggs/ucm170615.htm. Accessed 14 June 2016.
26. **Miyamoto T, Kitaoka D, Withanage GS, Fukata T, Sasai K, Baba E.** 1999. Evaluation of the efficacy of *Salmonella* Enteritidis oil-emulsion bacterin in an intravaginal challenge model in hens. *Avian Dis* **43:**497–505.
27. **Gast RK, Stone HD, Holt PS, Beard CW.** 1992. Evaluation of the efficacy of an oil-emulsion bacterin for protecting chickens against *Salmonella* Enteritidis. *Avian Dis* **36:**992–999.
28. **Alders R, Spradbrow P.** 2001. *Controlling Newcastle Disease in Village Chickens: A Field Manual.* Australian Center for International Agricultural Research, Canberra, Australia.
29. **Davelaar FG, Noordzij A, Vanderdonk JA.** 1982. A study on the synthesis and secretion of immunoglobulins by the Jarderian gland of the fowl after eyedrop vaccination against infectious bronchitis at 1-day-old. *Avian Pathol* **11:**63–79.

30. **Silva EN, Snoeyenbos GH, Weinack OM, Smyser CF.** 1981. Studies on the use of 9R strain of *Salmonella* Gallinarum as a vaccine in chickens. *Avian Dis* **25:**38–52.
31. **Babu U, Scott M, Myers MJ, Okamura M, Gaines D, Yancy HF, Lillehoj H, Heckert RA, Raybourne RB.** 2003. Effects of live attenuated and killed *Salmonella* vaccine on T-lymphocyte mediated immunity in laying hens. *Vet Immunol Immunopathol* **91:**39–44.
32. **United Egg Producers.** 2010. Animal Husbandry Guidelines for U.S. Egg Laying Flocks. http://www.uepcertified.com/pdf/2010-uep-animal-welfare-guidelines.pdf.
33. **Hannah JF, Wilson JL, Cox NA, Cason JA, Bourassa DV, Musgrove MT, Richardson LJ, Rigsby LL, Buhr RJ.** 2011. Comparison of shell bacteria from unwashed and washed table eggs harvested from caged laying hens and cage-free floor-housed laying hens. *Poult Sci* **90:**1586–1593.
34. **Singh R, Cheng KM, Silversides FG.** 2009. Production performance and egg quality of four strains of laying hens kept in conventional cages and floor pens. *Poult Sci* **88:**256–264.
35. **European Communities.** 1999. Council Directive 1999/74/EC of 19 July 1999: minimum standards for the protection of laying hens. *Off J* **L203:**53–57.
36. **Van Den Brand H, Parmentier HK, Kemp B.** 2004. Effects of housing system (outdoor vs cages) and age of laying hens on egg characteristics. *Br Poult Sci* **45:**745–752.
37. **Methner U, Diller R, Reiche R, Böhland K.** 2006. Occurence of salmonellae in laying hens in different housing systems and inferences for control. *Berl Munch Tierarztl Wochenschr* **119:**467–473. (In German.)
38. **De Vylder J, Dewulf J, Van Hoorebeke S, Pasmans F, Haesebrouck F, Ducatelle R, Van Immerseel F.** 2011. Horizontal transmission of *Salmonella* Enteritidis in groups of experimentally infected laying hens housed in different housing systems. *Poult Sci* **90:**1391–1396.
39. **Carrique-Mas JJ, Davies RH.** 2008. *Salmonella* Enteritidis in commercial layer flocks in Europe: legislative background, on-farm sampling and main challenges. *Rev Bras Cienc Avic* **10:**1–9.
40. **Mashaly MM, Hendricks GL III, Kalama MA, Gehad AE, Abbas AO, Patterson PH.** 2004. Effect of heat stress on production parameters and immune responses of commercial laying hens. *Poult Sci* **83:**889–894.
41. **Holt PS.** 2003. Molting and *Salmonella enterica* serovar Enteritidis infection: the problem and some solutions. *Poult Sci* **82:**1008–1010.
42. **Garber L, Smeltzer M, Fedorka-Cray P, Ladely S, Ferris K.** 2003. *Salmonella enterica* serotype enteritidis in table egg layer house environments and in mice in U.S. layer houses and associated risk factors. *Avian Dis* **47:**134–142.
43. **Kuney DR.** 2002. External parasites, insects, and rodents, p 169–184. *In* Bell DD, Weaver WD, Jr (ed), *Commercial Chicken and Egg Production*, 5th ed. Springer, New York, NY.
44. **European Food Safety Authority.** 2005. Opinion of the scientific panel on biological hazards on the request from the commission related to the microbiological risks on washing of table eggs. *EFSA J* **269:**1–39.
45. **Himathongkham S, Riemann H, Ernst R.** 1999. Efficacy of disinfection of shell eggs externally contaminated with *Salmonella* Enteritidis. Implications for egg testing. *Int J Food Microbiol* **49:**161–167.
46. **Padron M.** 1995. Egg dipping in hydrogen peroxide solution to eliminate *Salmonella* Typhimurium from eggshell membranes. *Avian Dis* **39:**627–630.
47. **Gast RK.** 1993. Immersion in boiling water to disinfect egg shells before culturing egg contents for *Salmonella* Enteritidis. *J Food Prot* **56:**533–535.
48. **Koonz CH, Kauffman FL.** 1951. U.S. Patent No. 2,565,311. Washington, DC: U.S. Patent and Trademark.
49. **Coleman M, Ebel E, Golden N, Hogue A, Kadry A, Kause J, Latimer H, Marks H, Quiring N, Schlosser W, Schroeder C.** 2005. *Risk assessments for* Salmonella *Enteritidis in shell eggs and* Salmonella *spp. in egg products.* FSIS-USDA. http://www.fsis.usda.gov/shared/PDF/SE_Risk_Assess_Oct2005.pdf.
50. **Stadelman WJ, Singh RK, Muriana PM, Hou H.** 1996. Pasteurization of eggs in the shell. *Poult Sci* **75:**1122–1125.
51. **Hou H, Singh RK, Muriana PM, Stadelman WJ.** 1996. Pasteurization of intact shell eggs. *Food Microbiol* **13:**93–101.
52. **Dev SRS, Raghavan GSV, Gariepy Y.** 2008. Dielectric properties of egg components and microwave heating for in-shell pasteurization of eggs. *J Food Eng* **86:**207–214.
53. **Datta AK, Almeida M.** 2005. Properties relevant to infrared heating of foods. *In* Rao MA, Rizvi SSH, Datta AK (ed), *Engineering Properties of Foods*, 3rd ed. CRC Press.
54. **Butz P, Tauscher B.** 2002. Emerging technologies: chemical aspects. *Food Res Int* **35:**279–284.
55. **Piyasena P, Mohareb E, McKellar RC.** 2003. Inactivation of microbes using ultrasound: a review. *Int J Food Microbiol* **87:**207–216.

56. **Cabeza MC, García ML, de la Hoz L, Cambero I, Ordóñez JA.** 2005. Destruction of *Salmonella* Senftenberg on the shells of intact eggs by thermoultrasonication. *J Food Prot* **68:**841–844.
57. **Cabeza MC, Cambero MI, De la Hoz L, Garcia ML, Ordonez JA.** 2011. Effect of the thermoultrasonic treatment on the eggshell integrity and their impact on the microbial quality. *Innov Food Sci Emerg Technol* **12:**111–117.
58. **Kim JG, Yousef AE, Dave S.** 1999. Application of ozone for enhancing the microbiological safety and quality of foods: a review. *J Food Prot* **62:**1071–1087.
59. **Food and Drug Administration.** 2006. *Code of Federal Regulations, Title 21, Section 73.368, issued April 2006.* Washington, DC: Government Printing Office. http://www.grokfood.com/regulations.
60. **Perry JJ, Rodriguez-Romo LA, Yousef AE.** 2008. Inactivation of *Salmonella enterica* serovar Enteritidis in shell eggs by sequential application of heat and ozone. *Lett Appl Microbiol* **46:**620–625.
61. **Rodriguez-Romo LA, Yousef AE.** 2005. Inactivation of *Salmonella enterica* serovar Enteritidis on shell eggs by ozone and UV radiation. *J Food Prot* **68:**711–717.
62. **Koidis P, Bori M, Vareltzis K.** 2000. Efficacy of ozone treatment to eliminate *Salmonella* Enteritidis from eggshell surface. *Arch Lebensmittelhyg* **51:**4–6.
63. **Fuhrmann H, Rupp N, Büchner A, Braun P.** 2010. The effect of gaseous ozone treatment on egg components. *J Sci Food Agric* **90:**593–598.
64. **Laroussi M, Leipold F.** 2004. Evaluation of the roles of reactive species, heat, and UV radiation in the inactivation of bacterial cells by air plasmas at atmospheric pressure. *Int J Mass Spectrum* **233:**81–86.
65. **Moreau M, Orange N, Feuilloley MGJ.** 2008. Non-thermal plasma technologies: new tools for bio-decontamination. *Biotechnol Adv* **26:** 610–617.
66. **Fernández A, Noriega E, Thompson A.** 2013. Inactivation of *Salmonella enterica* serovar Typhimurium on fresh produce by cold atmospheric gas plasma technology. *Food Microbiol* **33:**24–29.
67. **Ragni L, Berardinelli A, Vannini L, Montanari C, Sirri F, Guerzoni E, Garunieri A.** 2010. Non-thermal atmospheric gas plasma device for surface decontamination of shell eggs. *J Food Eng* **100:**125–132.
68. **Bialka KL, Demirci A, Knabel SJ, Patterson PH, Puri VM.** 2004. Efficacy of electrolyzed oxidizing water for the microbial safety and quality of eggs. *Poult Sci* **83:**2071–2078.
69. **Fabrizio KA, Sharma RR, Demirci A, Cutter CN.** 2002. Comparison of electrolyzed oxidizing water with various antimicrobial interventions to reduce *Salmonella* species on poultry. *Poult Sci* **81:**1598–1605.
70. **Park CM, Hung YC, Lin CS, Brackett RE.** 2005. Efficacy of electrolyzed water in inactivating *Salmonella* Enteritidis and *Listeria monocytogenes* on shell eggs. *J Food Prot* **68:** 986–990.
71. **Russell SM.** 2003. The effect of electrolyzed oxidative water applied using electrostatic spraying on pathogenic and indicator bacteria on the surface of eggs. *Poult Sci* **82:**158–162.
72. **Len SV, Hung YC, Erickson M, Kim C.** 2000. Ultraviolet spectrophotometric characterization and bactericidal properties of electrolyzed oxidizing water as influenced by amperage and pH. *J Food Prot* **63:**1534–1537.
73. **Cao W, Zhu ZW, Shi ZX, Wang CY, Li BM.** 2009. Efficiency of slightly acidic electrolyzed water for inactivation of *Salmonella* Enteritidis and its contaminated shell eggs. *Int J Food Microbiol* **130:**88–93.
74. **Yoshifumi H.** 2003. Improvement of the electrolysis equipment and application of slightly acidic electrolyzed water for dairy farming. *J Jpn Soc Agric Mach* **65:**27–29.
75. **Takeshita K, Shibato J, Sameshima T, Fukunaga S, Isobe S, Arihara K, Itoh M.** 2003. Damage of yeast cells induced by pulsed light irradiation. *Int J Food Microbiol* **85:**151–158.
76. **Hierro E, Manzano S, Ordóñez JA, de la Hoz L, Fernández M.** 2009. Inactivation of *Salmonella enterica* serovar Enteritidis on shell eggs by pulsed light technology. *Int J Food Microbiol* **135:**125–130.
77. **Morgan R.** 1989. UV 'green' light disinfection. *Dairy Ind Int* **54:**33–35.
78. **Keklik NM, Demirci A, Patterson PH, Puri VM.** 2010. Pulsed UV light inactivation of *Salmonella* Enteritidis on eggshells and its effects on egg quality. *J Food Prot* **73:**1408–1415.
79. **De Reu K, Grijspeerdt K, Herman L, Heyndrickx M, Uyttendaele M, Debevere J, Putirulan FF, Bolder NM.** 2006. The effect of a commercial UV disinfection system on the bacterial load of shell eggs. *Lett Appl Microbiol* **42:**144–148.
80. **Whistler PE, Sheldon BW.** 1989. Bactericidal activity, eggshell conductance, and hatchability effects of ozone versus formaldehyde disinfection. *Poult Sci* **68:**1074–1077.
81. **Aiello SE.** 2015. *The Merck Veterinary Manual,* 10th ed. Merck and Co., Inc., Kenilworth, NJ.

6

Nuts and Grains: Microbiology and Preharvest Contamination Risks

PARDEEPINDER K. BRAR[1] and MICHELLE D. DANYLUK[1]

INTRODUCTION

Almonds, walnuts, and cashews are the major tree nuts produced in the world. The United States is the largest producer of tree nuts (1). Peanuts, also known as groundnuts, are produced largely in China, followed by India (1). Almonds are the major nuts produced by the United States, followed by peanuts, walnuts, and pistachios in this order (1). According to the 2014–2015 Global Statistical Review and world nut and dried fruit map, the world's tree nut production increased by 5.4% compared to the previous season (1). By definition, nuts can be divided into several categories, including drupes (e.g., almonds, coconuts, pecans, pistachios, walnuts), legumes (e.g., peanuts), nuts (e.g., acorns, chestnuts, hazelnuts), and seeds (e.g., brazil nuts, cashews, pumpkin, pine nuts, sesame, sunflower) (2). Nuts are consumed as a whole or as an ingredient in confectionary, bakery, and snack products. Nuts can be consumed raw or can be processed by thermal treatments such as oil roasting, dry roasting, blanching, and others.

Cereal grains are an important food resource worldwide (3). Rice, wheat, and corn are the major grains produced around the world. Sorghum and millet are also produced in several regions, with the production being much lower

[1]Department of Food Science and Human Nutrition, Citrus Research and Education Center, Institute of Food and Agricultural Sciences, University of Florida, Lake Alfred, FL 33850.

Preharvest Food Safety
Edited by Siddhartha Thakur and Kalmia E. Kniel

doi:10.1128/microbiolspec.PFS-0023-2018

than the three major crops (3). Global cereal production is expected to decline by 1.2 million tons (0.6%) in 2017 compared to the 2016 crop, with the production in 2017 predicted to be 2,593 million tons (4). Approximately 60% of the calories consumed in developing countries and 30% of calories consumed in developed countries are derived from cereals (3). Grains are consumed as a whole or in refined form. Many whole grains are a rich source of dietary fiber (5).

NUT PRODUCTION, MICROBIOLOGY, AND CONTAMINATION SOURCES

Tree nuts take years to bear fruit after planting. The production, harvesting, and post-harvest practices vary among different types of tree nuts. With most of the production and harvesting conducted mechanically, the degree of mechanization can vary (6).

Almond production in the United States primarily occurs in California (1). Almond production includes irrigation of crops using furrow, sprinkler, and drip irrigation. Nitrogen, potassium, and zinc are the major nutrients applied to the crop (7). Almond harvesting is a well-mechanized process involving mechanical tree shakers and other equipment. Almonds are allowed to dry on ground for a few days to 2 weeks after knocking the nuts (8). Nuts are swept into windrows, placed onto trailers, and transported to hullers. When rainy conditions prevail, nuts should be dried prior to hulling. Drying can be done using forced hot air at 49 to 54°C in a batch dryer or at 82°C in a continuous flow dryer. After removal of the hulls, nuts are placed in bulk storage and transported to the processing plant (8).

Hazelnuts, or filberts, are native tree nuts of Europe and adjacent areas (9), and in the United States the majority of production occurs in Oregon (1). About 70% of the world production of hazelnuts occurs in Turkey (9). Hazelnuts are best suited for well-drained soils with high environmental humidity conditions (10). The U.S. production (99%) of hazelnuts is concentrated in Oregon due to suitable weather conditions (10). Oregon rainfall is the major source of water for hazelnuts due to the good water-holding capacity of the clay soil (10). Nitrogen and potassium are the main fertilizers applied to hazelnuts (10). Mechanical tree shakers are used for harvesting of hazelnuts, followed by arranging the nuts in the center of rows using a sweeper. Nuts are then transported to the nut dryer or processor.

Macadamias are native tree nuts of Australia (9). U.S. macadamia production is primarily concentrated in Hawaii (1). Water and fertilizer requirements of macadamia nuts vary based on the production regions. During harvest, macadamia nuts are allowed to fall onto the ground naturally and are picked from the ground. The flowering of macadamia nuts takes place over several months, which makes it difficult to predict the exact harvest month of nuts. Mechanical tree shakers are not very useful for macadamia nuts (6). Upon harvesting, macadamia nuts are dried using a two-stage process. The first stage reduces the moisture content of nuts to 14%, and the second stage brings the moisture content down to 4%.

Pecans are the only tree nuts native to the United States and are produced in several states including, Georgia, New Mexico, Texas, Oklahoma, and Arizona (1). Pecans are a high-water crop, and depending on the region, different irrigation methods are used (11). Routinely, nitrogen, phosphorus, potassium, and zinc are applied to pecans. Nitrogen is applied through irrigation, and zinc is applied through foliar application (11). Pecans are harvested mechanically by shaking trees and scraping nuts from the ground. If pecans remain on the ground for too long, nuts can become rancid and the seed coat of the nut will darken (12). During processing, pecans are conditioned in cold or hot water to soften the shells prior to cracking or shelling. After separating the inedible portions of nuts, pecans are dried to reduce the moisture content.

Iran is the leading producer of pistachios, followed by the United States (9). In the United

States, about 98% of pistachio production occurs in California (13). Among the tree nuts, pistachios are the most tolerant to drought-related stress conditions. Copper, zinc, and boron are the major nutritional deficiencies in pistachios (14). The high moisture content of pistachios (40 to 50%) makes them susceptible to contamination and mechanical injury if nuts drop onto the orchard floor during harvesting. Pistachios are harvested with shake-catch mechanical harvesters, followed by placing the nuts in bins or trucks and transporting them to the huller or dehydration plants (14). Pistachios must be processed within 12 to 24 h of harvesting (13) to prevent quality issues. After the removal of hull, nuts are dried to a moisture content of 5 to 6% using a single-stage or two-stage system. Unhulled pistachios are kept in cold and low relative humidity conditions to minimize losses (0°C and <70% relative humidity) (8).

Two main varieties of walnuts are grown in the United States, English walnuts and black walnuts. Black walnuts are used as rootstocks for English walnuts (13). Virtually all walnut production in the United States occurs in California. Among the different types of walnuts, English walnuts are primarily grown in California (1). Walnuts are generally produced in areas with little to no rainfall to prevent foliar diseases, and the crop is generally dependent on irrigation for its water requirements. Like almonds, nitrogen, potassium, and zinc are the major nutrients applied to the crop (15). Walnuts are swept into windrows immediately after mechanical harvesting and placed on carts, bins, or trailers. They are transported to hullers to remove the hull, then dried at 43°C to reduce the moisture content to 8%. Walnuts can become rancid if dried at temperatures higher than 110°C (8). The drying period can range from 4 h to 2 days depending on the moisture content of the nuts. In-shell nuts can then be transported to the processing plants (8).

Peanuts are native to South America. In the United States, Georgia is the major producer of peanuts in terms of acreage as well as production (16). Peanuts (also called groundnuts), unlike tree nuts, are grown below the ground and take 4 to 5 months to reach maturity. Peanuts are usually inverted using a specialized machine before harvesting and allowed to dry in windrows in open air. They are usually harvested at approximately 35 to 50% moisture and dried to prevent aflatoxin growth (17). Too quick or too slow drying of peanuts can increase the splitting of kernels during shelling (18). The moisture content of dried-in-shell peanuts is maintained around 7 to 10% to prevent mold growth, and shelled peanuts are dried to 7% for maximum shelf life (18). Peanuts are generally not exposed to water during processing. Dirt, stones, plant debris, and other foreign materials are removed from peanuts with the help of screens and blowers. Cleaned peanuts are sized and graded to prevent crushing of kernels while shelling. They are passed through rollers to crack the pods and separate kernels. Shaker screens are used to separate shells from kernels. Following separation, kernels are graded and either packed in bags or shipped in bulk (17). The majority of the peanut crop gets processed, but there is demand for fresh peanuts as well (16).

Natural Microflora on Nuts

Both nuts and grains are low in moisture content and water activity, which prevents the growth of most microorganisms on their surface. Close proximity of nuts to soil during production (peanuts), harvesting (almonds, pecans, walnuts), and postharvest handling can influence the type and population of microorganisms present on nuts (19). *Alternaria* spp., *Aspergillus* spp., *Cunninghamella* spp., *Cladosporium* spp., *Fusarium* spp., *Penicillium* spp., *Rhizopus* spp., *Trichoderma* spp., and *Verticillium* spp. have been identified from tree nuts; some of these strains can produce mycotoxins (19, 20). When fungal organisms are present superficially on the surface of shells or nuts, they can be removed; however, some fungal species such as *Aspergillus* spp.,

Penicillium spp., and *Rhizopus* spp. have been obtained from inside almonds, brazil nuts, pistachios, and walnuts, indicating that these organisms can internalize and close contaminate in-shell nuts (19). The significant differences in the type of fungal organisms present on nuts have been observed between field-collected and store-bought almonds, brazil nuts, pistachios, and walnuts, where *Aspergillus flavus, Aspergillus niger,* and *Rhizopus* spp. were associated with store-bought tree nuts, and *Penicillium* spp. were associated with field-harvested tree nuts (19). Peanuts are grown beneath the soil and can be contaminated with the aflatoxin-producing fungi *A. flavus* and *Aspergillus parasiticus.* Other genera commonly associated with peanuts include *Fusarium* spp., *Macrophomina phaseolina, Botryodiplodia theobromae, Penicillium* spp., and *Rhizopus* spp. (21).

Fungal spores can be transmitted to nuts from air, insects, and soil. Physical, chemical, and biological factors can impact the production of aflatoxin by fungus on food (22). Environmental factors such as temperature, relative humidity, composition of gaseous compounds, and moisture content of stored nuts can influence the growth of microorganisms. Chemical factors such as the use of fungicides and insecticides can alter mold colonization and aflatoxin production (22). Insect or bird damage of nuts amplifies the chances of invasion of nuts by molds and concomitant aflatoxin production (23). Aflatoxin is a potent carcinogen and teratogen to humans and mainly affects peanuts, corn, corn seed, and tree nut crops (23). In Europe, raw and ready-to-eat almonds have been required to not to exceed 15 and 10 ppb of total aflatoxin, respectively, since 2010 (24). The stringent threshold levels of aflatoxin set for trading purposes are due to growing food safety concerns at an international level.

Enterococcus and coliform microorganisms are the predominantly identified bacteria on nut surfaces (25). The major bacterial genera identified are *Streptococcus, Staphylococcus, Bacillus, Xanthomonas, Achromobacter, Pseudomonas, Micrococcus,* and *Brevibacterium* (26, 27).

Bacterial Pathogens on Nuts

Nuts can become contaminated with foodborne pathogens at any stage of production, harvesting, processing, distribution, or consumption. Proper food safety measures should be adopted to prevent contamination issues. *Salmonella* is considered the target organism for dry foods, including tree nuts and peanuts, because of its long-term persistence and high heat resistance on dry foods (28–31). The presence of high levels of fat contributes to the enhanced resistance of pathogens on nuts (32). Infiltration of *Salmonella* into in-shell pecans is higher at 21°C and 35°C than at 4°C and −20°C (33). Infiltrated pathogens demonstrate higher thermal resistance than pathogens present on the surface (34). Other bacterial foodborne pathogens associated with dry foods, including tree nuts and peanuts, are *Bacillus cereus, Clostridium botulinum, Clostridium perfringens, Cronobacter, Escherichia coli* O157:H7, *L. monocytogenes,* and *Staphylococcus aureus* (32).

Preharvest Contamination Sources for Nuts

Soil

Preharvest contamination of tree nuts can occur from contaminated orchard soil (35). The 2000–2001 almond outbreak associated with *Salmonella enterica* serovar Enteritidis PT 30 is an example of contamination acquired from an orchard, but the exact source of contamination of orchard soil has not been identified (31, 35). A 5-year study was conducted to evaluate the persistence of *S.* Enteritidis PT 30 in orchard soil associated with the outbreak (31). A total of 53 (23%) *Salmonella*-positive samples were obtained; all isolates (100%) were identified as *S.* Enteritidis PT 30. *Salmonella* was more frequently isolated during the almond-harvesting months of August to October, indicating the potential transfer of *Salmonella* from orchard soil to almonds as they fall to the ground during harvest or the wide distribution of the organism through an orchard due to dust production during harvest (31). Rainfall during harvesting periods can increase the amount of soil adhered to almonds and can impact the frequency of *Salmonella* being isolated. The probable reasons for the long-term survival of *Salmonella* in orchard soil have also been investigated by researchers. Almond orchard soil type, moisture content, surface and subsurface soil temperature, and air temperature were studied to determine their influence on *Salmonella* survival in orchard soil (36). Almond hulls that may get wet in the orchard due to rainfall or irrigation events and temperature are significant factors influencing the survival, and potential growth, of *Salmonella* in almond orchard soil (36, 37). Almond hulls contain significantly more soluble sugars than almond shells (36), and the liquid generated when dry hulls get wet supports the growth of *Salmonella* both in the hulls themselves and in soils the liquid may leach into (36, 37). Pecan packing tissue, the tissue that is inside a pecan shell in addition to the nut, consists of tannins and polyphenols and is toxic to human pathogens, which may provide some degree of natural defense against human pathogens (38). Walnut tannins are also known to have an antibacterial effect (39). Tree nuts come in contact with soil after harvesting, whereas peanuts are grown beneath the soil surface. The presence of soilborne pathogens in peanuts is inevitable, emphasizing the need to adopt proper postharvest handling and processing practices to avoid contamination of the final product.

Animals

Repetitive mowing, application of herbicides, and grazing are some of the practices adopted by orchard growers to manage cover crops (40). Sheep, goats, and cattle are used for grazing in orchards with the aim of getting better economic returns (40). Orchard grazing was commonly practiced until the 1950s, when concerns such as overgrazing, bark damage, and pathogen contamination issues reduced this practice (40). Taking pets out for walks in orchards is also practiced in some areas (41). Animal intrusion in the orchard can lead to several food safety concerns. Foodborne pathogens harbored by animals can be deposited in orchard soil by animal feces. Research conducted in the 1970s (42) analyzed the food safety concerns associated with grazing animals in pecan orchards. Two harvest seasons, 1970 (wet) and 1971 (dry), were considered in the study. Results indicated that 6 times higher *E. coli* contamination of pecans was found when orchards were grazed compared to ungrazed orchards, with higher positive samples obtained in the wet year (1970) compared to the dry year (1971). The hard shells of tree nuts such as almonds, pecans, and walnuts protect their kernels from dust, discoloration, and microorganisms. When the shell was intact, *E. coli* could not penetrate to the inside of in-shell pecans (42). Water absorption by in-shell pecans for 48 h can lead to cracking of shells along the suture line, which can further increase the chances of *E. coli* invasion (42).

Nut orchards can attract several kind of wild animals including deer, hogs, mice, rabbits, squirrels, and turkeys (43). Rats and mice are considered the main cause of damage in nut and fruit orchards in the United States. Accurate and timely identification of damage from rats and mice is necessary to develop effective management programs. The use of snap traps, baits, and remote-triggered game cameras are some of the techniques used to catch rats and mice in orchards. Extensive information on the design of bait stations, bait application, and cost analysis of baiting programs is available (43). Birds can also cause considerable damage to agricultural crops. Ravens, crows, scrub jays, magpies, and black birds are some of the birds responsible for damaging nut crops (44).

Harvest

Until harvesting, tree nuts do not typically come into contact with humans, soil, or equipment, which can be critical sources of contamination. If nuts are harvested to the ground, the orchard floor should be prepared prior to harvest to ensure that there are no potential sources of contamination, such as animal feces, for nuts to be harvested into. Early splitting of pistachio shells can make the kernel more vulnerable to contamination during harvest and postharvest handling (45). Mechanical harvesting of tree nuts consists of equipment such as tree shakers, nut sweepers, vacuum harvesters, trash separators, and others. Peanuts are mechanically harvested using peanut digger and harvesting combines. The use of contaminated harvesting equipment can contaminate nuts upon contact, and care should be taken to ensure that harvesting equipment is cleaned. Mechanical harvesting should be conducted in dry weather to minimize the potential for cross-contamination. In wet weather, leaves, soil, and debris can adhere strongly to nut surfaces and can affect the cleaning process and increase the chances of contamination. Historically, almonds were harvested onto canvas tarps; these nuts had lower aerobic plate and yeast and mold counts than almonds harvested from the orchard floor (27), but this practice is no longer in use. All workers involved in harvesting activities should be trained in good health and hygiene practices.

Handling

Many nuts are mechanically dried immediately after harvesting to reduce the moisture content to less than 7% with the use of forced dry heat (8). Drying nuts to an adequate moisture level is critical to prevent microbial growth on nuts that may result in aflatoxin formation. The moisture content of the nuts should be determined at several spots in each load to ensure accurate measurement. Certain nuts such as macadamias, pecans, and walnuts are exposed to water conditions to soften the outer shell prior to shelling (6). This process is called conditioning. Almonds are exposed to water during the blanching process to remove the outer skin. Seeds such as melon, sesame, and pumpkin are also sometimes exposed to water treatment for the hulling process (46). Proper drying after water treatment is crucial for preventing the growth of microflora that are present on nuts. Beuchat and Mann (28) demonstrated the inactivation of *Salmonella* on inoculated in-shell pecans during conditioning treatment. Immersion-inoculated in-shell pecans immersed in chlorinated water at concentrations of 0, 100, 200, and 400 μg/ml did not reduce *Salmonella* populations significantly even after 24 h of treatment. *Salmonella* populations were significantly higher after 16 h and then after 24 h of treatment in chlorinated water. The increase in *Salmonella* concentration was attributed to the release of cells attached to the shell and internal tissues (47). The use of chlorine to prevent cross-contamination during conditioning or hulling may be challenging due to high organic loads present in the water. The use of hot water (80 to 95°C) was effective in reducing *Salmonella* cells by ca. 5 log CFU/g, but the use of higher temperatures (85 to 95°C) can compromise the organoleptic properties of nuts after 15 to 20 min. of exposure (47).

Nuts may be processed prior to consumption; the reduction of pathogens on the surface of dry foods depends on several factors such as the type of organism, process parameters, type of heat treatment, and type of food item. The time and temperature used for nut processing are critical parameters to ensure the desired lethality of foodborne pathogens. Failure to comply with the process parameters can lead to food safety issues. Dry heat is less lethal than moist heat for *Salmonella* on dry products. Average *D* values for *Salmonella* cocktails inoculated onto wheat, sesame seeds, walnuts, pecan nuts, pumpkin seeds, and brazil nuts and exposed to dry heat at 105°C were 132.5, 102.5, 174.6, 170.9, 235.3, and 242.1 min (48). During wet heat treatment such as almond blanching, *D* values for *Salmonella* were 2.6, 1.5, 0.75, and 0.39 min at 60, 70, 80, and 88°C water, respectively (49). Heat treatments conducted using appropriate process parameters are capable of ensuring significant and adequate reduction of pathogens on nuts.

Storage

Following harvest, nuts should be stored in a facility that is dry and provides protection from rain, groundwater drainage, rodents, and pest infestation. The storage facility should have minimum fluctuations in temperature to prevent microbial growth and insect infestation. Storage temperature should be monitored at fixed intervals. Nut handlers may store their product under controlled conditions to maintain quality. The shelf life and quality of nuts are influenced by the storage conditions (50). Almonds can be stored by nut handlers at ambient, refrigerated (4°C), or frozen (−20°C) conditions for 12 months or longer (30). The common storage temperatures (and suggested storage times) for pecans, both in-shell and kernels, include −18°C (for up to 6 to 8 years), 0°C (for about 12 to 18 months), or ambient temperature (for about 3 to 6 months) (51). Common storage temperatures (and suggested times) for peanuts, both in-shell and kernels, include −18°C (for 2 to 10 years), 1 to 5°C (for approximately 1 year), or ambient temperature (up to 6 months) (52). After processing and packaging, nuts are typically stored under ambient conditions at the retail level; consumers may store nuts under ambient, refrigerated, or freezer conditions for extended periods (53).

Salmonella survives on nut surfaces at lower temperatures (i.e., refrigeration or freezer conditions) without significant declines in populations over time, whereas slow, but significant, reductions are typically observed at ambient temperatures (29–31, 33, 54). The survival of *Salmonella*, *E. coli* O157:H7, and *L. monocytogenes* has been documented on almond kernels at −19, 4, 24, and 35°C (30), on peanuts at −24, 4, and 22°C (29), on pecans at −24, 4, and 22°C (29), on pistachios at −19, 4, and 24°C (30), and on walnut kernels at −20, 4, and 23°C (54) for at least 365 days. *Salmonella* can survive in peanut butter and peanut butter spreads for at least 168 days at 5 and 21°C (55). At −20, 4, 21, and 37°C, *Salmonella* can survive on pecan halves and pieces for 365 days and on in-shell pecans for 550 days (33). *Salmonella* survives better than *E. coli* O157:H7 and *L. monocytogenes* on the surface of nuts (29, 30, 54), making it the target organism for low-water-activity foods. Under dry conditions, *Salmonella* produces cell division inhibitors and accumulates compatible solutes to maintain the turgor and osmotic pressure that helps it survive in the environment (56).

Prevalence and Concentration of Pathogens on Nuts

Limited information is available on the prevalence and concentration of human pathogens on nuts. The prevalence of *Salmonella* on raw almond kernels, sampled over a period of 7 years, was 0.98 ± 0.32%, where 137 almond samples out of 13,972 almonds were positive for *Salmonella* (57–59). The concentration of *Salmonella* determined using a three-tube most probable number (MPN) method on raw almond kernels ranged from 0.0044 to 0.15 MPN/100 g. Out of 81 positive samples, 35 serotypes of *Salmonella* were identified (58). Natural *Salmonella* prevalence on raw in-shell almonds was 1.5% of 455 100-g samples tested (57). Using the MPM method, 44 (0.95%) out of 4,641 in-shell pecans were positive for *Salmonella*, and the levels obtained were 0.47 to 39 MPN/100 g with a mean of 2.4 MPN/100 g (60). A total of 31 serotypes of *Salmonella* were obtained from 42 *Salmonella*-positive pecan samples (60). The prevalence of naturally present *Salmonella* on raw shelled peanuts in the United States was 2.3% of 944 peanut samples (375 g each) from three crop years (61), and the corresponding concentration of *Salmonella* on peanuts as determined by an MPN assay was <0.030 to 2.4 MPN/g (3 to 240 MPN/100 g). Another study of raw shelled peanuts found *Salmonella* and enterohemorrhagic *E. coli* in 0.67% and 0.030% of 10,162 peanut samples (350 g each), respectively, averaged over three crop years (62);

the calculated *Salmonella* levels were 0.74 to 5.25 MPN/350 g (0.21 to 1.5 MPN/100 g).

Processing practices reduce the microbial load on nuts. The presence of pathogens on processed ready-to-eat products is unacceptable and makes the product unfit for consumption. A survey of edible roasted nuts conducted in the U.K. retail market resulted in one *S. enterica* serovar Havana-contaminated pistachio sample out of a total of 727 samples (0.13%) of different nuts tested (63). Edible brazil nut kernels analyzed for the presence of *Salmonella* in the U.K. retail market resulted in 0.40% of 469 positive samples contaminated with *S. enterica* serovars Seftenberg and Tennessee at the levels of <0.010 to 0.23 MPN/g (<1 to 23 MPN/100 g) using a 10-tube MPN method (63). The same group of researchers (63) also found *S. enterica* serovar Anatum from a sample of mixed nuts (almond, Brazil nut, cashew, peanut, walnut) out of 329 samples tested during their survey. The levels of *S.* Anatum obtained from a mixed nut sample was <0.010 MPN/g (1 MPN/100g) using a 10-tube MPN method (63). A study conducted to determine the microbiological safety of edible seeds in the U.K. retail market observed *Salmonella* in alfalfa seeds (1.7% of 58 samples), linseeds (0.40% of 284), melon seeds (8.5% of 47), sesame seeds (1.7% of 771), and sunflower seeds (0.10% of 976), with the overall prevalence of *Salmonella* in edible seeds observed to be 0.60% (64). The presence of pathogens on ready-to-eat nuts is a serious concern and, if not identified in a timely manner, can result in foodborne illnesses.

Recalls and Outbreaks

Almonds, cashews, coconuts, hazelnuts, pine nuts, pecans, pistachios, and peanuts, as well as several nut products, have been associated with foodborne outbreaks and/or recalls (65, 66). The majority of these outbreaks and recalls have been associated with *Salmonella*. The foodborne illness cases from multiple U.S. states (67–73) and/or from other countries (35, 74–83) are highlighted in these outbreaks. Outbreaks and recalls due to *E. coli* O157:H7 (84, 85) and recalls due to *L. monocytogenes* contamination (86) are documented for some nuts and nut products. Due to the long shelf life and the long-term survival of human pathogens on nuts, most of the nut-associated outbreaks have lasted for a period of several months. The cases from an almond and peanut butter outbreak were identified over 8 to 9 months and 4 to 10 months, respectively (35, 67, 68, 71, 74). Nut outbreak investigations rarely link the outbreak strain to its original source. The 2000–2001 outbreak of salmonellosis caused by *S.* Enteritidis PT 30 associated with raw almonds was a rare outbreak where the investigation identified an almond orchard to be the source of contamination, but the source of contamination of the orchard soil was not identified. The *S.* Enteritidis PT 30 strain was subsequently isolated from the almond orchard floor over a period of 5 years (31).

Seeds have also been identified as the source of foodborne illnesses and recalls in the recent past. Sesame seed products (tahini, halva, and hummus) have been involved in seven foodborne disease outbreaks in different parts of the world since 2001 (87–91). In all the cases, the product was procured from either Turkey or Lebanon. Epidemiological investigations identified *Salmonella* as the causative agent in all these outbreaks. In 2014, chia seeds and chia powder were involved in a multistate outbreak of salmonellosis in the United States (92). The outbreak caused 73 infections and 8 hospitalizations in the United States and Canada (92).

GRAIN MICROBIOLOGY AND CONTAMINATION SOURCES

By definition, grains are the fruiting bodies of various grasses (93). Grains belong to the family *Poaceae* and are similar in structure and biochemical properties. Typically, grains are dense in energy and vary in their starch content from 50 to 80% depending upon their origin and environmental conditions. Grains include wheat, rice, barley, oats, maize, buckwheat, sorghum, millet, and mixed grains. Grains are used for food, feed, and silage production in all parts of the world (93).

Fungi Associated with Grains

Fungal pathogens are more tolerant to dry environments than bacterial pathogens and are considered a significant problem for cereal products. Air, soil, human contact, animals, insects, and pests can be possible sources of contamination of grains. Environmental conditions such as warm temperature, high humidity, rainfall, drought, sunlight, frost, and wind can lead to proliferation of fungal contaminants on grains. Contamination of cereal grains with various spoilage- and mycotoxin-producing fungi can occur prior to harvesting (94). Fungi associated with cereals can be divided into "field fungi" and "storage fungi." Field fungi primarily consist of *Alternaria*, *Cladosporium*, *Fusarium*, and *Helminthosporium*. Field fungi infect grains in the field at high moisture content (18 to 30%), high water activity (>0.9), and high humidity conditions (90 to 100%) (95). Field fungi can cause seed discoloration, shriveling, blemishing, loss of germination, and mycotoxin production and have the potential to survive on grains for long periods of time, depending on the environmental conditions of storage (95). Storage fungi can infect and invade cereal grain after harvesting. *Aspergillus*, *Rhizopus*, *Mucor*, *Wallemia*, and *Penicillium* are the genera of fungi infecting grains during storage (96). Damaged grains are more prone to fungal infestation and mycotoxin production than undamaged grains (97).

Various postharvest techniques, such as physical separation of damaged kernels, separation through filtration, solvent extraction, milling, and inactivation of mycotoxins by heat can reduce the microbial contaminants from grains and enhance the product safety (94). However, the complete elimination of mycotoxins from grain is impossible to achieve (98), emphasizing the requirement to have better prevention strategies in place. Various processing techniques such as baking, cooking, brewing, roasting, and flaking have variable effects on mycotoxin concentration (98). Utilization of proper harvesting, handling, and storage practices can prevent contamination to a large extent. Lowering the moisture content of grains to 12% from 14% and maintaining low humidity conditions in the storage environment can prevent fungal growth during storage. The use of antifungal agents such as propionic acid and acetic acid has also been practiced to prevent fungal contamination of grains (96). Regulatory limits set by the U.S. FDA for mycotoxin in grains helps keep the grain supply safe for consumption in the United States.

Bacteria Associated with Grains

A wide variety of spoilage and pathogenic bacteria can grow on cereal grains, including lactic acid bacteria, coliforms, *Enterococci*, *B. cereus*, *Clostridium* spp., *Salmonella*, and *S. aureus* (99). Most of the microorganisms are present on the pericarp layer of the grain. Removal of the outer layer of grains can significantly reduce the microbial load. The removal of 4% of total weight of wheat using abrasive milling can reduce microbial populations by ca. 87% (100); simple agitation in a liquid medium did not result in the removal of the strongly adhered microorganisms from the surface of wheat grains (100).

As with nuts, grains can acquire contamination from soil, harvesting equipment, animal feces, or inadequate storage conditions. Postprocess handling can also cause serious food safety issues. Spore-forming bacteria can withstand the adverse processing conditions and can be carried over to the cereal products, which can lead to spoilage issues such as ropiness in bread caused by *Bacillus subtilis*. Bacteriostatic agents such as propionates can assist in the prevention of rope formation in bread and other quality issues (101). *B. cereus* is an important foodborne pathogen of concern for grains, especially rice. The bacterium is present on the outer casing of rice and can withstand high cooking temperatures and improper refrigeration conditions. Germinated spores can be killed with adequate cooking conditions. Immediate cooling of cooked rice or maintaining the cooked rice at a hot temperature is pertinent to prevent foodborne illnesses associated with *B. cereus* (102). *C. botulinum* is a soilborne spore-forming organism and can come in contact with grains during production. The growth of bacteria can be prevented by keeping the grains dry and cool. Water activity of <0.93 can prevent the production of toxin by *C. botulinum* (103).

Food safety recalls and outbreaks have been linked to cereal products due to pathogen contamination. In 1998, *S. enterica* serovar Agona was involved in the disease outbreak linked to toasted oats prepared at a Minnesota cereal manufacturing facility. Ten years later, in 2008, Malt-O-Meal unsweetened puffed rice and puffed wheat cereal prepared at the same facility was also implicated in a multistate disease outbreak caused by *S.* Agona (104). The original source of this outbreak was not identified; however, the pulsed-field gel electrophoresis pattern of the *S.* Agona strain obtained from the puffed cereal product was indistinguishable from the one obtained from toasted oats, leading to permanently shutting down that section of the facility. The presence of *S.* Agona in the same facility after a span of 10 years indicates the long-term persistence of this pathogen in the dry food production environment. *Cronobacter* is also commonly associated with dry food environments; it has been isolated from dry products such as rice seeds (105), rice starch and flour, and brown rice (106), as well as from a dry processing plant during environmental sampling (107). *Cronobacter* can survive and grow in infant rice cereal at various temperature conditions (108, 109). The survival of *Cronobacter* spp. has been documented for the period of 12 months at 4, 21, and 30°C conditions, typically observed during retail and consumer storage at home (109).

GOOD AGRICULTURAL PRACTICES FOR NUTS AND GRAINS

The use of safe practices can help in reducing the natural microbial loads of nuts. Good agricultural practices are the various farming practices that have the potential to reduce contamination in food. Specific guidelines for individual crops vary, but general guidelines recommend prevention of contamination from soil, water, animals, and humans. Various recommended practices for nuts and grain producers are as follows.

Site History

Prior use of agriculture land for animal husbandry, pasture purposes, or animal feeding operations can cause food safety issues in nuts, mainly peanuts, which are grown beneath the soil surface (110). Industrial waste dumped into the soil consists of high levels of heavy metals such as lead, arsenic, and mercury. The lack of mobility of heavy metals in plants prevents their uptake by tree nuts (110), but peanuts can be exposed to heavy metals during their growth below the ground level. Agricultural land used for nut and grain cultivation should not have a history of being an industrial dump site. Even the proximity of agricultural land to animals can lead to the contamination of nuts through water runoff from animal operations to farm land and/or air flow from livestock to farm land. Vegetative buffers can help in preventing water runoff from adjacent lands and prevent the introduction of airborne pathogens into orchard land by blocking the wind coming from nearby areas (110).

Water Source

Water is used on the farm for irrigation, for mixing pesticides and fertilizer, and for foliar application. At shelling facilities, water is used for conditioning tree nuts to soften the outer shell and facilitate the shelling process. At any step where water contacts the harvestable portion of the crop, care should be taken that the water is safe and of adequate quality for its intended use. In harvest and postharvest applications, potable water should be used, and if practical, sanitizer levels should be maintained in water to avoid cross-contamination. Contaminated water used for conditioning purposes has the potential to contaminate nuts (33). If hot water conditioning is considered as a "kill step," proper validation of the treatment is required.

Agricultural Land Management

The orchard floor used for tree nut harvesting should be properly managed for weeds and grasses to expedite the harvesting process (111). Grazing of animals in tree nut orchards should either not be practiced or animals should be removed a few months prior to harvesting to allow manure to cake and dry. Wildlife (e.g., rodents, birds) can contaminate the nuts in the trees and should be kept out of the orchard site as much as possible. Although it may be difficult to completely eliminate grazing animals and wildlife from orchards, effective management strategies can help in minimizing the risk associated with them (43, 44), such as preharvest inspections for animal feces and its removal prior to harvest. Insect pests can also damage the nut crop and should be controlled using integrated pest management programs.

Agricultural Chemical Use

Spraying agricultural land and crops with insecticides, pesticides, and herbicides protects the crop from damage. Spraying should be conducted by trained personnel and according to the instructions provided by the manufacturer. Only Environmental Protection Agency (EPA)-registered chemicals should be used for spraying, and they should be used according to label requirements. Individuals should have personal protective equipment including gloves, masks, and spray suits to prevent exposure to pesticide. All chemicals should be stored in locked, well-ventilated rooms away from the food source. Only authorized personnel should have access to the chemical storage room.

Manure Application

Human pathogens such as *E. coli* O157:H7 and *Salmonella* reside in the gastrointestinal tract of animals such as cattle, pigs, deer, sheep, and others. Animals shed these pathogens in their feces. Improperly treated compost or manure used in the field can be a potential source of contamination for crops, mainly peanuts growing beneath the surface of the soil. Growers should be aware of contamination and environmental issues associated with using improperly treated manure on farms (112). Manure storage sites close to orchards or agricultural land can also be a source of contamination of nuts. With an inadequate slope, runoff from the storage site to the orchard during rainfall or dust due to prevailing winds can increase the chances of contamination, so manure storage sites, if present, should be contained with physical barriers.

Worker Hygiene

Workers should be trained to handle the product with care and avoid any chances of cross-contamination upon handling. Proper restroom facilities, hand-washing facilities, and drinking water facilities should be available for workers. Workers should be provided with proper sick leave policies, or management strategies should be in place preventing ill workers from coming into contact with nuts, grains, or food contact surfaces to avoid cross-contamination.

Harvesting

Every effort should be made to prevent contamination during harvesting. Harvesting equipment should be in good working order, should be kept clean, and should not introduce pathogens to nuts. Equipment and transport vehicles should be inspected for obvious dirt and debris and should be cleaned prior to use. Harvesting bins should not be used for storing or transporting toxic chemicals, fertilizers, or pesticides but should be designated for nut or grain storage only.

Plant Cleaning and Sanitation

Effective cleaning and sanitation practices are critical for product safety. Facilities handling nuts and grains where no water is used in the process should be kept dry, and specific dry cleaning and sanitizing procedures should be followed, unless there is adequate time for equipment to dry after cleaning. Cereal grain and nut processing plants can have varying amounts of oil, carbohydrate, and protein deposits on surfaces, depending on the processed product (113). Dry cleaning of dry facilities minimizes moisture conditions, which subsequently prevents the proliferation of microorganisms. Processes such as scraping, wiping, sweeping, blowing, vacuuming, purging, and flushing are used for dry cleaning purposes (113). The design of equipment can also influence the effectiveness of the cleaning and sanitation practices used. Alcohol-based sanitizers (114) are more effective in controlling microbial populations in dust than water-based sanitizers and are recommended for sanitation of dry food facilities.

Documentation and Record Keeping

Records about the history of the farmland, fertilizers, pesticide application, and worker training and information on water sources, water testing records, and harvest dates should be kept. Trace-back is an important tool to identify the exact source of food items. It plays an important role in foodborne recall and outbreak situations, where consumer health could be at risk. Adequate information about the load, variety, and farm should be recorded to track-back the product to the farm, when required.

CITATION

Brar PK, Danyluk MD. 2018. Nuts and grains: microbiology and preharvest contamination risks. Microbiol Spectrum 6(2):PFS-0023-2018.

REFERENCES

1. **International Nut and Dried Fruit Council.** 2015. 2014/2015 world nuts and dried fruits trade map. http://www.nutfruit.org/what-we-do/industry/statistics. Accessed 2 August 2017.
2. **Alden L.** 2005. Cook's thesaurus: Nuts. http://www.foodsubs.com/Nuts.html. Accessed 17 October 2014.
3. **Awika JM.** 2011. Major cereal grains production and use around the world, p 1–13. *In* Awika JM, Piironen V, Bean S (ed), *Advances in Cereal Science: Implications to Food Processing and Health Promotion.* ACS Symposium Series ACS, Washington, DC.
4. **Food and Agriculture Organization.** 2017. FAO Cereal supply and demand brief. http://www.fao.org/worldfoodsituation/csdb/en/. Accessed 31 July 2017.
5. **American Heart Association (AHA).** 2016. Eat 3 or more whole grain foods every day. http://www.heart.org/HEARTORG/HealthyLiving/HealthyEating/Nutrition/Eat-3-or-More-Whole-Grain-Foods-Every-Day_UCM_320264_Article.jsp#.WYJq0FEpCpo. Accessed 2 August 2017.
6. **Wells ML.** 2013. Agricultural practices to reduce microbial contamination of nuts, p 1–21. *In* Harris LJ (ed), *Improving the Safety and Quality of Nuts.* Woodhead Publishing, Cambridge, UK.
7. **Micke WC.** 1994. Almond orchard management. University of California Division of Agricultural Science. Publication no. 3364.
8. **Kader AA, Thompson JF.** 2002. Post-harvest handling systems: tree nuts, p 399–406. *In* Kader AA (ed), *Post-harvest technology of horticultural crops,* 3rd edition. University of California Division of Agriculture & Natural Resources, Oakland, CA.
9. **Janick J, Paul RE.** 2008. *Encyclopedia of Fruits and Nuts.* CABI, Oxfordshire, UK.
10. **Olson J.** 2002. Growing hazelnuts in the Pacific Northwest. Oregon State University, extension circular 1219.
11. **Wells ML, Hawkins G.** 2007. Best management practices of poultry litter in pecan orchards. University of Georgia Cooperative Extension circular no. 939.

12. **Woodruff JG.** 1979. *Tree Nuts: Production, Processing, Products.* AVI Publishing, Westport, CT.
13. **Boriss H.** 2015. *Pistachios.* http://www.agmrc.org/commodities-products/nuts/pistachios/. Accessed 3 August 2017.
14. **Crane JC, Maranto J.** 1988. Pistachio production. University of California. Agricultural and Natural Resources. publication no. 2279.
15. **Ramos DE.** 1998. Walnut production manual. University of California, Division of Agricultural Science. Publication number 3373.
16. **Marzolo G.** 2016. *Peanuts.* http://www.agmrc.org/commodities-products/nuts/peanut-profile/. Accessed 3 August 2017.
17. **EPA.** 2010. Peanut processing. http://www.epa.gov/ttnchiel/ap42/ch09/final/c9s10-2b.pdf. Accessed 26 August 2014.
18. **Tillman BL, Wright DL.** 2012. Producing quality peanut seed. University of Florida Institute of Food and Agricultural Sciences.
19. **Bayman P, Baker JL, Mahoney NE.** 2002. *Aspergillus* on tree nuts: incidence and associations. *Mycopathologia* **155:**161–169.
20. **Beuchat LR.** 1975. Incidence of molds on pecan nuts at different points during harvesting. *Appl Microbiol* **29:**852–854.
21. **McDonald D.** 1970. Fungal infection of groundnut fruit before harvest. *Trans Br Mycol Soc* **54:** 453–460.
22. **Gurses M.** 2006. Mycoflora and aflatoxin content of hazelnuts, walnuts, peanuts, almonds and roasted chickpeas (leblebi) sold in Turkey. *Int J Food Prop* **9:**395–399.
23. **Campbell BC, Molyneux RJ, Schatzki TF.** 2003. Current research on reducing pre- and post-harvest aflatoxin contamination of U.S. almond, pistachio and walnuts. *J Toxicol* **22:** 225–266.
24. **Zivoli R, Gambacorta L, Perrone G, Solfrizzo M.** 2014. Effect of almond processing on levels and distribution of aflatoxins in finished products and byproducts. *J Agric Food Chem* **62:**5707–5715.
25. **Hyndman JB.** 1963. Comparison of enterococci and coliform microorganisms in commercially produced pecan nut meats. *Appl Microbiol* **11:** 268–272.
26. **Al-Moghazy M, Boveri S, Pulvirenti A.** 2014. Microbiological safety in pistachios and pistachio containing products. *Food Control* **36:**88–93.
27. **King AD Jr, Miller MJ, Eldridge LC.** 1970. Almond harvesting, processing, and microbial flora. *Appl Microbiol* **20:**208–214.
28. **Beuchat LR, Mann DA.** 2011. Inactivation of *Salmonella* on pecan nutmeats by hot air treatment and oil roasting. *J Food Prot* **74:** 1441–1450.
29. **Brar PK, Proano LG, Friedrich LM, Harris LJ, Danyluk MD.** 2015. Survival of *Salmonella, Escherichia coli* O157:H7, and *Listeria monocytogenes* on raw peanut and pecan kernels stored at −24, 4, and 22°C. *J Food Prot* **78:** 323–332.
30. **Kimber MA, Kaur H, Wang L, Danyluk MD, Harris LJ.** 2012. Survival of *Salmonella, Escherichia coli* O157:H7, and *Listeria monocytogenes* on inoculated almonds and pistachios stored at −19, 4, and 24°C. *J Food Prot* **75:**1394–1403.
31. **Uesugi AR, Danyluk MD, Mandrell RE, Harris LJ.** 2007. Isolation of *Salmonella* Enteritidis phage type 30 from a single almond orchard over a 5-year period. *J Food Prot* **70:**1784–1789.
32. **Beuchat LR, Komitopoulou E, Beckers H, Betts RP, Bourdichon F, Fanning S, Joosten HM, Ter Kuile BH.** 2013. Low-water activity foods: increased concern as vehicles of foodborne pathogens. *J Food Prot* **76:**150–172.
33. **Beuchat LR, Mann DA.** 2010. Factors affecting infiltration and survival of *Salmonella* on in-shell pecans and pecan nutmeats. *J Food Prot* **73:** 1257–1268.
34. **Beuchat LR, Mann DA.** 2011. Inactivation of *Salmonella* on in-shell pecans during conditioning treatments preceding cracking and shelling. *J Food Prot* **74:**588–602.
35. **Isaacs S, Aramini J, Ciebin B, Farrar JA, Ahmed R, Middleton D, Chandran AU, Harris LJ, Howes M, Chan E, Pichette AS, Campbell K, Gupta A, Lior LY, Pearce M, Clark C, Rodgers F, Jamieson F, Brophy I, Ellis A, Salmonella Enteritidis PT30 Outbreak Investigation Working Group.** 2005. An international outbreak of salmonellosis associated with raw almonds contaminated with a rare phage type of *Salmonella* enteritidis. *J Food Prot* **68:**191–198.
36. **Danyluk MD, Nozawa-Inoue M, Hristova KR, Scow KM, Lampinen B, Harris LJ.** 2008. Survival and growth of *Salmonella* Enteritidis PT 30 in almond orchard soils. *J Appl Microbiol* **104:**1391–1399.
37. **Uesugi AR, Harris LJ.** 2006. Growth of *Salmonella* Enteritidis phage type 30 in almond hull and shell slurries and survival in drying almond hulls. *J Food Prot* **69:**712–718.
38. **Beuchat LR, Heaton EK.** 1975. *Salmonella* survival on pecans as influenced by processing and storage conditions. *Appl Microbiol* **29:**795–801.
39. **Amarowicz R, Dykes GA, Pegg RB.** 2008. Antibacterial activity of tannin constituents from *Phaseolus vulgaris, Fagoypyrum esculentum, Corylus avellana* and *Juglans nigra*. *Fitoterapia* **79:**217–219.

40. **Wilson LM, Hardestry LH.** 2006. Targeted grazing with sheep and goats in orchard setting, p 99–106. *In* Launchbaugh K (ed), *Targeted Grazing: A Natural Approach to Vegetation Management and Landscape Enhancement.* American Sheep Industry Association.
41. **Tarrant B.** 1994. *How to Hunt Birds with Gun Dogs*, p 26–43. Stackpole Books, Mechanicberg, PA.
42. **Marcus KA, Amling HJ.** 1973. *Escherichia coli* field contamination of pecan nuts. *Appl Microbiol* **26:**279–281.
43. **Quinn N, Balwin RA.** 2014. Managing roof rats and deer mice in nut and fruit orchards. University of California, ANR publication 8513. http://anrcatalog.ucanr.edu. Accessed 30 March 2015.
44. **Crabb AC, Salmon TP, Marsh RE.** 1986. Bird problems in California pistachio production. Proceedings of the 12th Vertebrate Pest Conference. http://digitalcommons.unl.edu/cgi/viewcontent.cgi?article=1016&context=vpc12. Accessed 11 February 2015.
45. **Sommer NF, Buchanon JR, Fortlage RJ.** 1986. Relation of early splitting and tattering of pistachio nuts to aflatoxin in the orchard. *Phytopathology* **76:**692–694.
46. **Bankole S, Osha A, Joda A, Enikumehim O.** 2005. Effect of drying method on the quality and storability of 'egusi'melon seeds (*Colocynthis citrullus* L.). *Afr J Biotechnol* **4:**799–803.
47. **Beuchat LR, Mann DA.** 2011. Inactivation of *Salmonella* on in-shell pecans during conditioning treatments preceding cracking and shelling. *J Food Prot* **74:**588–602.
48. **Limburn R, Gaze JE.** 2013. Heat resistance determination of *Salmonella* in low Aw foods. Campden BRI. Chipping Campden, Gloucestershire, UK.
49. **Harris LJ, Uesugi AR, Abd SJ, McCarthy KL.** 2012. Survival of *Salmonella* Enteritidis PT 30 on inoculated almond kernels in hot water treatments. *Food Res Int* **45:**1093–1098.
50. **Srichamnong W, Wootton M, Srzednicki G.** 2010. Effect of nut in-shell storage conditions on volatile profile in macadamia nuts. *Julius-Kühn-Arch* **425:**270–274.
51. **Picha D, Pyzner J.** 2013. Storage hints for pecans. Louisiana State University Agricultural Center. http://www.lsuagcenter.com/NR/rdonlyres/B4948FCF-C7FA-41F2-9536-8F450B7AF058/20289/STORAGEHINTSFORPECANS.pdf. Accessed 19 August 2014.
52. **Maness N.** 2005. Peanut. *In* Gross K, Saltveit M, Wang CY (ed), *The Commercial Storage of Fruits, Vegetables, and Florist and Nursery Stocks*, 4th ed. USDA, ARS Agricultural Handbook no. 66.
53. **Lee LE, Metz D, Giovanni M, Bruhn CM.** 2011. Consumer knowledge and handling of tree nuts: food safety implications. *Food Prot Trends* **31:**18–27.
54. **Blessington T, Mitcham EJ, Harris LJ.** 2012. Survival of *Salmonella enterica, Escherichia coli* O157:H7, and *Listeria monocytogenes* on inoculated walnut kernels during storage. *J Food Prot* **75:**245–254.
55. **Burnett SL, Gehm ER, Weissinger WR, Beuchat LR.** 2000. Survival of *Salmonella* in peanut butter and peanut butter spread. *J Appl Microbiol* **89:**472–477.
56. **Winfield MD, Groisman EA.** 2003. Role of nonhost environments in the lifestyles of *Salmonella* and *Escherichia coli*. *Appl Environ Microbiol* **69:**3687–3694.
57. **Bansal A, Jones TM, Abd SJ, Danyluk MD, Harris LJ.** 2010. Most-probable-number determination of *Salmonella* levels in naturally contaminated raw almonds using two sample preparation methods. *J Food Prot* **73:**1986–1992.
58. **Danyluk MD, Harris LJ, Sperber WH.** 2007. Nuts and cereals, p 171–183. *In* Doyle MP, Beuchat LR (ed), *Food Microbiology: Fundamentals and Frontiers*. ASM Press, Washington, DC.
59. **Lambertini E, Danyluk MD, Schaffner DW, Winter CK, Harris LJ.** 2012. Risk of salmonellosis from consumption of almonds in the North American market. *Food Res Int* **45:**1166–1174.
60. **Brar PK, Strawn LK, Danyluk MD.** 2016. Prevalence, level and types of *Salmonella* isolated from North American in-shell pecans over four harvest years. *J Food Prot* **79:**352–360.
61. **Calhoun S, Post L, Warren B, Thompson S, Bontempo AR.** 2013. Prevalence and concentration of *Salmonella* on raw shelled peanuts in the United States. *J Food Prot* **76:**575–579.
62. **Miksch RR, Leek J, Myoda S, Nguyen T, Tenney K, Svidenko V, Greeson K, Samadpour M.** 2013. Prevalence and counts of *Salmonella* and enterohemorrhagic *Escherichia coli* in raw, shelled runner peanuts. *J Food Prot* **76:**1668–1675.
63. **Little CL, Rawal N, de Pinna E, McLauchlin J.** 2010. Survey of *Salmonella* contamination of edible nut kernels on retail sale in the UK. *Food Microbiol* **27:**171–174.
64. **Willis C, Little CL, Sagoo S, de Pinna E, Threlfall J.** 2009. Assessment of the microbiological safety of edible dried seeds from retail premises in the United Kingdom with a focus on *Salmonella* spp. *Food Microbiol* **26:**847–852.
65. **Harris LJ, Palumbo M, Beuchat LR, Danyluk MD.** 2014. Outbreaks of foodborne illness

associated with the consumption of tree nuts, peanuts and sesame seeds [table and references]. *In Outbreaks from tree nuts, peanuts and sesame seeds.* Available at: http://ucfoodsafety.ucdavis.edu/files/169530.pdf. Accessed 25 August 2014.

66. **Palumbo M, Beuchat LR, Danyluk MD, Harris LJ.** 2014. Recalls of tree nuts and peanuts in the U.S., 2001 to present [table and references]. In U.S. recalls of nuts. Available at: http://ucfoodsafety.ucdavis.edu/files/162415.pdf. Accessed 7 June 2014.
67. **Centers for Disease Control and Prevention (CDC).** 2007. Multistate outbreak of *Salmonella* serotype Tennessee infections associated with peanut butter: United States, 2006–2007. *MMWR Morb Mortal Wkly Rep* **56:**521–524.
68. **Centers for Disease Control and Prevention (CDC).** 2009. Multistate outbreak of *Salmonella* infections associated with peanut butter and peanut butter-containing products: United States, 2008–2009. *MMWR Morb Mortal Wkly Rep* **58:**85–90.
69. **Centers for Disease Control and Prevention (CDC).** 2009. *Salmonella* in pistachio nuts, 2009. http://www.cdc.gov/salmonella/pistachios/update.html. Accessed 19 April 2014.
70. **Centers for Disease Control and Prevention (CDC).** 2011. *Multistate outbreak of human* Salmonella *Enteritidis infections linked to Turkish pine nuts.* http://www.cdc.gov/salmonella/pinenuts-enteriditis/111711/index.html. Accessed 19 August 2013.
71. **Centers for Disease Control and Prevention (CDC).** 2012. Multistate outbreak of *Salmonella* Bredeney infections linked to peanut butter manufactured by Sunland, Inc. (final update). http://www.cdc.gov/salmonella/bredeney-09-12/index.html. Accessed 19 August 2014.
72. **Centers for Disease Control and Prevention (CDC).** 2014. Multistate outbreak of *Salmonella* Stanley infections linked to raw cashew cheese (final update). http://www.cdc.gov/salmonella/stanley-01-14/index.html. Accessed 25 August 2014.
73. **Center for Disease Control and Prevention (CDC).** 2017. Multistate outbreak of Shiga toxin producing *Escherichia coli* O157:H7 infections linked to I.M. Healthy Brand SoyNut Butter (Final Update). https://www.cdc.gov/ecoli/2017/o157h7-03-17/index.html. Accessed 1 August 2017.
74. **Centers for Disease Control and Prevention (CDC).** 2004. Outbreak of *Salmonella* serotype Enteritidis infections associated with raw almonds: United States and Canada, 2003–2004. *MMWR Morb Mortal Wkly Rep* **53:**484–487.
75. **Centers for Disease Control and Prevention (CDC).** 2014. Multistate outbreak of *Salmonella* Braenderup infections linked to nut butter manufactured by nSpired Natural Foods, Inc. (final update). http://www.cdc.gov/salmonella/braenderup-08-14/index.html.
76. **Chou JH, Hwang PH, Malison MD.** 1988. An outbreak of type A foodborne botulism in Taiwan due to commercially preserved peanuts. *Int J Epidemiol* **17:**899–902.
77. **Kirk MD, Little CL, Lem M, Fyfe M, Genobile D, Tan A, Threlfall J, Paccagnella A, Lightfoot D, Lyi H, McIntyre L, Ward L, Brown DJ, Surnam S, Fisher IST.** 2004. An outbreak due to peanuts in their shell caused by *Salmonella* enterica serotypes Stanley and Newport: sharing molecular information to solve international outbreaks. *Epidemiol Infect* **132:**571–577.
78. **Ledet Müller L, Hjertqvist M, Payne L, Pettersson H, Olsson A, Plym Forshell L, Andersson Y.** 2007. Cluster of *Salmonella* Enteritidis in Sweden 2005–2006: suspected source: almonds. *Euro Surveill* **12:**E9–E10. http://www.eurosurveillance.org/content/10.2807/esm.12.06.00718-en.
79. **Semple AB, Parry WH, Graham AJ.** 1961. Paratyphoid fever traced to desiccated coconut. *Lancet* **2:**364–365.
80. **Scheil W, Cameron S, Dalton C, Murray C, Wilson D.** 1998. A South Australian *Salmonella* Mbandaka outbreak investigation using a database to select controls. *Aust N Z J Public Health* **22:**536–539.
81. **OzFoodNet Working Group.** 2010. OzFoodNet quarterly report, 1 April to 30 June 2010. *Commun Dis Intell Q Rep* **34:**345–354. http://www.health.gov.au/internet/main/publishing.nsf/Content/cda-cdi3403o.htm.
82. **Ward L, Brusin S, Duckworth G, O'Brien S.** 1999. *Salmonella* java phage type Dundee: rise in cases in England: update. *Eurosurveillance* **3:**1435. http://www.eurosurveillance.org/ViewArticle.aspx?ArticleId=1435.
83. **Wilson MM, MacKenzie EF.** 1955. Typhoid fever and salmonellosis due to the consumption of infected desiccated coconut. *J Appl Bacteriol* **18:**510–521.
84. **Centers for Disease Control and Prevention (CDC).** 2011. Investigation update: multistate outbreak of *E. coli* O157:H7 infections associated with in-shell hazelnuts. http://www.cdc.gov/ecoli/2011/hazelnutsO157/index.html. Accessed 30 April 2014.
85. **Canadian Food Inspection Agency.** 2011. Health hazard alert: certain walnuts and walnut products may contain *E. coli* O157:H7 bacteria.

http://www.inspection.gc.ca/about-the-cfia/newsroom/food-recall-warnings/complete-listing/2011-04-03/eng/1359548340192/1359548340223
86. **Food and Drug Administration (FDA).** 2014. Parkers Farm Acquisition, LLC issues voluntary recall of products due to listeria contamination. Press release available at https://content.govdelivery.com/accounts/MIDARD/bulletins/acfc1c.
87. **Aavitsland P, Alvseike O, Guérin PJ, Stavnes TL.** 2001. International outbreak of *Salmonella* Typhimurium DT104: update from Norway. *Eurosurveillance* **5:**1701. http://www.eurosurveillance.org/ViewArticle.aspx?ArticleId=1701.
88. **Centers for Disease Control and Prevention (CDC).** 2012. Multistate outbreak of *Salmonella* serotype Bovismorbificans infections associated with hummus and tahini: United States, 2011. *MMWR Morb Mortal Wkly Rep* **61:**944–947.
89. **Centers for Disease Control and Prevention (CDC).** 2013. Multistate outbreak of *Salmonella* Montevideo and *Salmonella* Mbandaka infections linked to tahini sesame paste (final update). http://www.cdc.gov/salmonella/montevideo-tahini-05-13/index.html. Accessed 19 August 2014.
90. **Paine S, Thornley C, Wilson M, Dufour M, Sexton K, Miller J, King G, Bell S, Bandaranayake D, Mackereth G.** 2014. An outbreak of multiple serotypes of *salmonella* in New Zealand linked to consumption of contaminated tahini imported from Turkey. *Foodborne Pathog Dis* **11:**887–892.
91. **Unicomb LE, Simmons G, Merritt T, Gregory J, Nicol C, Jelfs P, Kirk M, Tan A, Thomson R, Adamopoulos J, Little CL, Currie A, Dalton CB.** 2005. Sesame seed products contaminated with *Salmonella*: three outbreaks associated with tahini. *Epidemiol Infect* **133:**1065–1072.
92. **Centers for Disease Control and Prevention.** 2014. Multistate outbreak of *Salmonella* infections linked to organic sprouted chia powder. http://www.cdc.gov/salmonella/newport-05-14/index.html. Accessed 12 February 2015.
93. **Wrigley CW.** 2010. An introduction to the cereal grains: major providers for mankind's food needs. *In* Wrigley CW, Bate IL (ed), *Cereal Grains: Assessing and Managing Quality*. Woodhead Publishing, Cambridge, UK.
94. **Choudhary AK, Kumari P.** 2010. Management of mycotoxin contamination in pre-harvest and post-harvest crops: present status and future prospects. *J Phytol* **2:**37–52.
95. **International Commission of Microbiological Specification for Foods (ICMSF).** 2005. *Microorganisms in food 6: Microbial Ecology of Food Commodities*, 2nd ed, p 392–439. Kluwer Academic, New York, NY.
96. **Harris LJ, Shebuski JR, Danyluk MD, Palumbo MS, Beuchat LR.** 2013. Nuts, seeds and cereals. *In* Doyle MP, Buchanan RL (ed), *Food Microbiology: Fundamentals and Frontiers*, 4th ed. ASM Press, Washington, DC.
97. **Fandohan P, Ahouansou R, Houssou P, Hell K, Marasas WFO, Wingfield MJ.** 2006. Impact of mechanical shelling and dehulling on *Fusarium* infection and fumonisin contamination in maize. *Food Addit Contam* **23:**415–421.
98. **Milani J, Maleki G.** 2014. Effects of processing on mycotoxin stability in cereals. *J Sci Food Agric* **94:**2372–2375.
99. **Bullerman LB, Bianchini A.** 2011. The microbiology of cereal and cereal products. Food Quality Safety. http://www.foodqualityandsafety.com/article/the-microbiology-of-cereals-and-cereal-products/. Accessed 9 September 2014.
100. **Laca A, Mousia Z, Diaz M, Webb C, Pendiella SS.** 2006. Distribution of microbial contamination within cereal grains. *J Food Eng* **72:**332–338.
101. **Soumalainen TH, Mayra-Makinen AM.** 1999. Propionic acid bacteria as protective cultures in fermented milks and breads. *Lait* **79:**165–174.
102. **Gilbert RJ, Stringer MF, Peace TC.** 1974. The survival and growth of *Bacillus cereus* in boiled and fried rice in relation to outbreaks of food poisoning. *J Hyg (Lond)* **73:**433–444.
103. **Silva FVM, Gibbs PA.** 2010. Non-proteolytic *Clostridium botulinum* spores in low acid cold distributed foods and design of pasteurization processes. *Trends Food Sci Technol* **21:** 95–105.
104. **Russo ET, Biggerstaff G, Hoekstra RM, Meyer S, Patel N, Miller B, Quick R, Salmonella Agona Outbreak Investigation Team.** 2013. A recurrent, multistate outbreak of *Salmonella* serotype Agona infections associated with dry, unsweetened cereal consumption, United States, 2008. *J Food Prot* **76:**227–230.
105. **Cottyn B, Regalado E, Lanoot B, De Cleene M, Mew TW, Swings J.** 2001. Bacterial populations associated with rice seed in the tropical environment. *Phytopathology* **91:**282–292.
106. **Jung MK, Park JH.** 2006. Prevalence and thermal stability of *Enterobacter sakazakii* from unprocessed ready to eat agricultural products and powdered infant formulas. *Food Sci Biotechnol* **15:**152–157.
107. **Kandhai MC, Reij MW, Gorris LGM, Guillaume-Gentil O, van Schothorst M.** 2004. Occurrence of *Enterobacter sakazakii* in food production environments and households. *Lancet* **363:**39–40.

108. **Richards GM, Gurtler JB, Beuchat LR.** 2005. Survival and growth of *Enterobacter sakazakii* in infant rice cereal reconstituted with water, milk, liquid infant formula, or apple juice. *J Appl Microbiol* **99:**844–850.
109. **Lin LC, Beuchat LR.** 2007. Survival of *Enterobacter sakazakii* in infant cereal as affected by composition, water activity, and temperature. *Food Microbiol* **24:**767–777.
110. **California Pistachio Research Board.** 2009. Good agricultural practices manual. Guidelines for California pistachio growers. http://www.gmaonline.org/downloads/technical-guidance-and-tools/Addendum_2_GAP_for_Pistachio_Growers.pdf. Accessed 6 October 2014.
111. **Vossen P, Ingals C.** 2002. Orchard floor management. http://cesonoma.ucanr.edu/files/27205.pdf. Accessed 20 February 2015.
112. **Kumar RR, Park BJ, Cho JY.** 2013. Application and environmental risks of livestock manure. *J Korean Soc Appl Bio Chem* **56:**497–503.
113. **Dirksen J.** 2011. Sanitation for cereal product processing plants begins with a plan. *Cereal Foods World* **56:**148–150.
114. **Du W-X, Danyluk MD, Harris LJ.** 2010. Efficacy of aqueous and alcohol-based quaternary ammonium sanitizers for reducing *Salmonella* in dusts generated in almond hulling and shelling facilities. *J Food Sci* **75:**M7–M13.

Risks Associated with Fish and Seafood

7

SAILAJA CHINTAGARI,[1] NICOLE HAZARD,[2] GENEVIEVE EDWARDS,[2]
RAVI JADEJA,[3] and MARLENE JANES[3]

INTRODUCTION

In the United States, the most popular fish and seafood products consumed per capita in 2015 were shrimp (4.00 lb), salmon (2.88 lb), tuna (2.20 lb), tilapia (1.38 lb), pollack (0.97 lb), pangasius (0.74 lb), cod (0.60 lb), crab (0.56 lb), catfish (0.52 lb), and clams (0.35 lb) (1). Fresh fish and seafood are highly perishable, and microbiological spoilage is one of the important factors that limit shelf life and safety. Fresh seafood can be contaminated at any point from rearing or harvesting to processing to transport or due to cross-contamination by consumer mishandling at home.

Fish and seafood have been the cause of many foodborne diseases and outbreaks in the United States and worldwide. The main microorganisms that cause foodborne outbreaks associated with fish and seafood include bacteria, viruses, and parasites (Fig. 1). The potential risk associated with fish and seafood can often be directly related to the environmental conditions and microbial quality of the water from which it was caught (2). Water qualities such as temperature, salt content, distance between location of catch and

[1]Department of Food Science and Technology, University of Georgia, Griffin, GA 30223; [2]School of Nutrition and Food Sciences, Louisiana State University, Baton Rouge, LA 70894; [3]Department of Animal Science, Oklahoma State University, Stillwater, OK 74078.

Preharvest Food Safety
Edited by Siddhartha Thakur and Kalmia E. Kniel

doi:10.1128/microbiolspec.PFS-0013-2016

polluted areas, and natural occurrence of bacteria in the water can affect the microbial quality of the fish and seafood (2). Polluted waters have been associated with bacterial contamination of fish and seafood, and the sources of pollution have included overboard sewage discharge into harvest areas, illegal harvesting from sewage-contaminated waters, and sewage runoff from points inland after heavy rains or flooding (3). Additionally, the season during which the fish and seafood are harvested, the method of catch, chilling conditions, and how they are handled, prepared, and served can increase the risk of contamination (2, 3). Fish and seafood may also become contaminated as a result of storage or transportation at improper temperatures and contamination by food handlers or through cross-contamination.

Crustaceans contaminated with bacteria are often the culprit in foodborne disease outbreaks. Many of these outbreaks are due to improperly prepared or mishandled crustaceans (Fig. 1) (4). Detailed investigations of bacteria-related foodborne illnesses have frequently focused on oysters, since they are often consumed uncooked (5). However, a significant number of bacterial foodborne disease outbreaks are caused by eating crustaceans (Fig. 1). There is usually no apparent reason for crustaceans to cause foodborne illness, since they are typically cooked before being consumed (6).

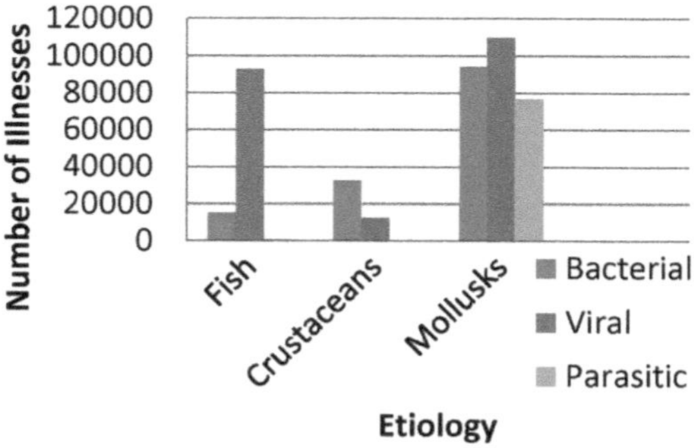

FIGURE 1 Foodborne illness cases associated with fish and shellfish (crustaceans and mollusks) in the United States, 1998 to 2008 (4).

There are incidents of fish and seafood becoming contaminated with foodborne pathogens such as *Listeria* (7–9), *Salmonella* (10), and *Vibrio* spp. (11, 12); hepatitis A (13); and norovirus (13). Some of these pathogens are naturally found in the environment (*Vibrio* spp., *Listeria monocytogenes*) and can pose serious health risks if the food is not cooked adequately to kill them. Another factor that contributes to foodborne outbreaks is that 20% of the fish and shellfish consumed in the United States are derived from recreational or subsistence fishing and are not subject to health-based control. Usually consumers rely on the color of the flesh as a parameter for doneness during cooking, which does not ensure the safety of these products. In 2007, the National Advisory Committee on Microbiological Criteria for Foods was asked to provide scientific guidance to the FDA and the National Marine Fisheries on cooking procedures for fish and seafood for consumers (14). The report found that most of the consumer cooking methods for fish and seafood were based on quality and not on scientific information to ensure destruction of foodborne pathogens.

BACTERIAL RISKS ASSOCIATED WITH FISH AND SEAFOOD

Outbreaks associated with the consumption of tainted seafood have been documented for many centuries. However, in the early 20th century, some major outbreaks gained the attention of public officials. One such outbreak occurred in 1925 on the East Coast. An outbreak of typhoid fever was eventually traced back to oysters contaminated by sewage (15). It was so severe that it prompted officials to petition the surgeon general of the United States to draw up formal guidelines for the safety of the public to replace the loose recommendations the shellfish industry followed at the time. Over time, small outbreaks continued around the country, and there were still many advances in sanitation, hygiene, and general processing that had yet to be discovered or implemented as regular practice.

Things started changing in the 1970s with advances in fields such as bacteriology and microbiology. For example, in 1978, when over 1,100 people became ill with *Vibrio parahaemolyticus* at a shrimp dinner in Port Allen, LA, a thorough investigation of this foodborne disease outbreak was conducted. Upon further examination, it was discovered that the food was grossly mishandled: not only were the shrimp cross-contaminated after cooking, but they were held unrefrigerated for 8 hours in the middle of the Louisiana summer before being served (16). Those are two extreme examples, and since then, federal agencies such as the FDA and the CDC have instituted strict shellfish industry regulations, have developed sanitization practices that minimize cross-contamination, Hazard Analysis Critical Control Point (HACCP), and in general, have tried to ensure safe food handling and consumer safety. However, recent statistics have shown that foodborne disease outbreaks due to shellfish are still a concern for consumers (17). Currently, conditions in processing plants are very strict, but there is still typically at least one outbreak per year that leads to a product recall. Recently, fish and seafood-associated bacterial or viral illnesses have typically been caused by cross-contamination or mishandling, either in private residences or in restaurant/delicatessen settings (18).

Salmonella

Reports to the CDC of outbreaks associated with fish and seafood have shown that *Salmonella* was the leading cause of these outbreaks. Of these *Salmonella* outbreaks, 10, with 224 cases, were caused by crustaceans; 14 outbreaks, with 852 cases, were caused by fish; and 2 outbreaks, with 13 cases, were caused by molluscan shellfish (17). *Salmonella* has a prevalence rate of 7.4% along the U.S. coastlines, with *Salmonella enterica* serovar Newport occurring at a higher percentage (75%) than other serovars (19). Acute gastroenteritis caused by *Salmonella* spp. continues to be a worldwide public health concern (19). In humans, salmonellosis is usually due to the consumption of contaminated food or water. The fecal wastes from infected animals and humans are important sources of bacterial contamination of the environment and the food chain (20). During a 9-year study (1990 to 1998), the FDA noted an overall incidence of *Salmonella* in 7.2% of 11,312 samples from imported and 1.3% of 768 samples from domestic U.S. seafood (10). *Salmonella* has also been detected in U.S. market oysters (19) and in other imported seafood from different countries (21). The incidence of *Salmonella* in seafood is highest in the central Pacific and African countries and lowest in Europe, including Russia, and North America (12% versus 1.6%) (10). The presence of *Salmonella* spp. in seafood has been reported in Vietnam (22), India (20, 23, 24), Sri Lanka (25), Thailand (26), Taiwan (27), and Japan (28). All of these countries export seafood into the United States, and the FDA is concerned about possible contamination of these products with *Salmonella*.

The FDA has conducted studies which showed that aquacultured seafood was more likely than wild-caught seafood to contain *Salmonella* (29). Many researchers also have evaluated the presence of *Salmonella*, fecal coliforms, and *Escherichia coli* in shrimp aquaculture ponds (6, 30–32). The relationship between the occurrence of *Salmonella* in shrimp from aquaculture operations, and the concentration of fecal bacteria in the source and grow-out pond water has been described by Koonse et al. (29). These could be the possible routes of contamination of shrimp and other seafood with this pathogen.

Salmonella infection presents as either enteric syndrome, also called typhoid, or as gastroenteritis, which is more common (5). Many species of *Salmonella* can be found on seafood because *Salmonella* is present in estuaries. However, it is still debated if *Salmo-*

nella is in waters because it is a part of the natural marine flora, if it is due to contamination from sewage run-off, or both (33). When a pathogen is commonly found on surfaces, the environment, and the food product, it is almost impossible to tell if the contamination is from a processing failure, improper storage conditions, unsanitary workers, or some combination of factors (10).

It is generally believed that the gastroenteritis syndrome of *Salmonella* is one of the most underreported foodborne illnesses because it is self-limiting. *Salmonella* has been implicated in many foodborne outbreaks involving seafood mixtures such as crab cakes. In these cases, it is not clear which component of the crab mixture is responsible for contributing the *Salmonella*. Many outbreaks are caused by eggs used as a binding agent in the seafood mixtures, but some have no obvious cause. For example, a 10-person outbreak in Ohio caused by crab cakes, lobster cakes, and crab-stuffed lobster served in a restaurant in July 2001 was eventually traced back to contaminated eggs (34). However, that same year, another 10-person outbreak caused by crab cakes served in a Washington, DC, restaurant could not be traced to any egg contamination source (34). Therefore, more investigation into the possible source of contamination of *Salmonella* in fish and seafood should be considered because it is one of the major foodborne bacterial pathogens.

L. monocytogenes

L. monocytogenes is prevalent in nature and can be found in soil, foliage, and the feces of animals and humans (2). This species is indigenous to the marine and estuarine environments, so its association with fish and seafood should be expected (35). It has also been known to establish itself as an in-house bacterium in a processing facility. It can create a biofilm on stainless steel surfaces and can be isolated from equipment, cold stores, and floors, enabling it to recontaminate products in the production environment (2). In-house reservoirs of *L. monocytogenes* have been reported from fish-processing establishments (2), and the bacterium has been isolated from domestic and imported fresh, frozen, and processed seafood products, including crustaceans, molluscan shellfish, and fish (36).

L. monocytogenes has many serotypes, but it is serotype ½b that is associated with seafood contamination (37). *L. monocytogenes* has unique survival properties: it is psychrotrophic (able to grow at refrigeration temperatures) (38), can survive irradiation (39), can grow in high salt concentration, and can survive a wide range of pH (37). The greatest threat from *L. monocytogenes* is through ready-to-eat products such as processed crabmeat. A study by Farber (40) demonstrated that *L. monocytogenes* grows better on crabmeat than on other seafood. Since it can grow to high concentrations in refrigerated, vacuum-packed, ready-to-eat foods that will not be subjected to further processing such as heating, there is a serious health risk associated with this organism (41). This has caused some safety and regulation issues, since there is no established infectious dose (36, 37).

Raw fish and shrimp have been linked to an outbreak of *L. monocytogenes* which caused nine deaths in New Zealand (7). Since this outbreak, the seafood industry has been concerned with the ability of *L. monocytogenes* to grow to high levels in shrimp when stored at refrigerated temperatures (7–9). Consumers buy raw fish and seafood from their local grocery store and cook it at home, which greatly reduces the risk of outbreaks, but the extent of cooking depends on various factors such as the size of the seafood product and the type of cuisine or dish. The shelf life of seafood is greatly influenced by microbial load, added to the fact that these are highly perishable commodities (8). The National Advisory Committee on Microbiological Criteria for Foods report concluded that *L. monocytogenes* is a pathogen of concern in ready-to-eat seafood products but not for raw fish and seafood products that will be cooked by consumers (14).

Vibrio cholerae

V. cholerae is a bacterium that many believe is no longer a threat, especially to U.S. citizens. Cholera is thought to be an issue for countries with questionable sanitation practices. Unfortunately, there are approximately one to two cases of cholera reported per week in the United States (42). While many of these cases of cholera are associated with travel to places with endemic cholera outbreaks, it should be remembered that the Gulf Coast has a long history of cholera. The first confirmed case was in Louisiana in 1832, and the last case was in 1873.

There were no more reported cases of cholera in the Gulf Coast states until 1973. Then a 4-case cluster of 11 people infected with *V. cholerae* O1 biotype El Tor serotype Inaba was reported in Abbeville, LA, in 1978, the result of eating contaminated crabs that had been boiled between 10 and 20 minutes or steamed up to 35 minutes (43). *V. cholerae* has been repeatedly isolated in blue crabs from the Gulf Coast. The Gulf Coast has been a reservoir of naturally occurring environmental toxigenic *V. cholerae*, and the crabs harvested from that area remain a risk to consumers, especially during the warmer months (44). In 2005 there were two confirmed cases of toxigenic *V. cholerae* O1, serotype Inaba, biotype El Tor, isolated from a couple from Louisiana after they ate locally caught crabs and shrimp (45).

V. cholerae O1, due to its higher resistance compared to other *Vibrio* bacteria, could lead to safety hazards in seafood products (46). A majority of fish and seafood products are cooked at home by the consumer or in commercial/institutional settings. This reduces the number of microorganisms in shrimp (47), but the extent of cooking plays a large role in destruction of this bacteria. *V. cholerae* O1 is recognized as a bacterium that is indigenous in the marine environment (48). Marine foods have been identified as vehicles for the transmission of cholera (49). The factors and mechanisms that affect the bacterium's survival in the aquatic environment are not completely understood (50). However, some research groups have stated that this pathogen is able to attach to abiotic surfaces, to zooplankton and phytoplankton, and to the carapaces of crustaceans such as shrimp and crab (12, 51, 52). *Vibrio* bacteria are generally considered to be heat-sensitive, but some reports show that *V. cholerae* O1 has some resistance in hot foods (53).

V. parahaemolyticus

In the Unites States, *V. parahaemolyticus* was first noted in the Chesapeake Bay in association with dead and dying blue crabs in 1969, whereas the first fully documented case occurred in Maryland in 1971, which was also associated with steamed crab (54). The ecology of Chesapeake Bay was examined, and it was found that the incidence of *V. parahaemolyticus* was correlated with water temperatures. The organism was not found in sea waters during the winter months, but it could be isolated from sediments in small numbers. It was later established that *V. parahaemolyticus* could survive the winter months by attaching to planktons and then proliferate inside the planktons (55). With the increase in water temperatures, *V. parahaemolyticus* is released from the planktons and can be easily detected in the waters.

V. parahaemolyticus illness is mainly associated with consumption of contaminated raw or undercooked shellfish. Oysters, clams, and mussels and cooked crustaceans such as shrimp and crab have also been implicated in *V. parahaemolyticus* infections (6). *V. parahaemolyticus* can increase to high levels in oysters because of its ability to enhance its concentration and survival in oysters. Oysters are filter feeders, and *V. parahaemolyticus* is associated with zooplanktons, which results in enhancing the bioconcentration of the bacteria in oysters (56).

Oysters which were harvested from the Gulf of Mexico, in particular, showed high *Vibrio* counts. A survey of retail oysters in the Gulf of Mexico, North Atlantic, Pacific, and Mid-Atlantic regions found high *V. parahaemolyticus* and *Vibrio vulnificus* counts during the summer months (57), when *V. parahaemolyticus* counts in freshly harvested oysters could exceed 10^4 most probable number/g (57).

V. parahaemolyticus is the *Vibrio* spp. most associated with blue crabs (58). Due to the halophilic nature of the *Vibrio* spp., *V. parahaemolyticus* grows very well in the same high-salinity habitat necessary for blue crabs to complete their life cycle (59). Both *V. cholerae* and *V. parahaemolyticus* have been found to bio-accumulate in the gut and gills of blue crabs, most likely due to the crabs' omnivorous diet (60). *V. parahaemolyticus* was recognized as an emerging foodborne illness in 1950 (61). In 1971, the United States experienced its first major *V. parahaemolyticus* foodborne outbreak, which was associated with crabs in Maryland. The outbreak caused approximately 425 people to become ill and was traced to improperly steamed crabs (54, 62). In 1998, the CDC received a report that 13 people in Florida became ill with *V. parahaemolyticus* from eating crabs (58). In New York in 2006, 80 people were diagnosed with *V. parahaemolyticus* after eating crab in a restaurant (63). Approximately 25 serotypes of *V. parahaemolyticus* are being monitored by the CDC (64). In addition, emerging research has determined that a specific gene, the *tdh* gene, was responsible for a virulence factor capable of causing the hemolytic syndrome when *V. parahaemolyticus* colony counts were well below the FDA-accepted *V. parahaemolyticus* limits (5). Less than 5% of environmental isolates produce *tdh* (65).

V. vulnificus

V. vulnificus is considered the most serious of all the pathogenic *Vibrio* spp. because it has been identified as being the leading cause of seafood-related fatalities (61). The infectious dose of *V. vulnificus* is 10^3 bacteria/g of food, but it is one of the more heat-sensitive bacteria and is easily destroyed with proper cooking (66). Of the thermal death times listed in the FDA fish and fishery products hazards and control guidance manual, those for the most virulent strains of *V. vulnificus* were much lower than those for *L. monocytogenes* (2). The danger with foodborne illness associated with *V. vulnificus* is its propensity to progress into severe necrotizing wound infections or fatal septicemia in patients with pre-existing conditions such as hemochromatosis or cirrhosis (67, 68). Liver disease plays a particular factor in the virulence of *V. vulnificus* due to the availability of free iron in the patient's serum (67, 68). Of the *Vibrio* cases that occur, *V. vulnificus* has the highest mortality rate: approximately 50% of the cases result in death approximately 48 hours postconsumption (64). Interestingly, it is very common for only one member of a family to show symptoms of *V. vulnificus*, mainly due to only specific family members having compromised immune systems (61). *V. vulnificus* and *V. parahaemolyticus* are regularly isolated together in crabs sampled for bacterial titers (69).

Clostridium botulinum

C. botulinum is found in marine sediments mainly as spores and can contaminate fish and seafood. If the conditions are right, the spores can germinate into the vegetative state and start producing neurotoxins that cause botulism. The optimal conditions for *C. botulinum* to produce toxins are anaerobic conditions and at pH above 4.6.

Each year about 150 confirmed cases of botulism occur in the United States for all food categories (70). Controlling the growth of *C. botulinum* in fish and seafood can be achieved by reducing the pH below 4.6, using salt or sodium nitrite, by lowering the moisture content, and by lowering the temperature (71).

VIRAL RISKS ASSOCIATED WITH FISH AND SEAFOOD

Seawater can become contaminated with enteric viruses and pose a health risk to people. These viruses enter source waterways through the direct or indirect discharge of treated and untreated human and animal waste into rivers, streams, and estuaries (72). In general, waterborne human enteric viruses pose a greater health risk than enteric bacteria due to the low infectious dose, which may be as little as one virion (73).

Enteric viruses replicate in the gastrointestinal tract and are shed in the feces of infected individuals. Most enteric viruses are morphologically similar and consist of an icosahedral, nonenveloped capsid, which surrounds a single-stranded RNA (e.g., norovirus, hepatitis A virus) molecule. Noroviruses and hepatitis A are the most common enteric viruses transmitted by fish and seafood (13).

Studies have shown that sediments may entrap enteric viruses and that when disturbed they can release viruses into the body of water (74, 75). Some aquatic organisms such as bivalve molluscan shellfish are filter feeders and are able to accumulate microorganisms from the surrounding water to a high titer with a concentration factor up to 99-fold (76). Unlike bacteria, viruses can be retained in oyster tissue for a long period of time and make the depuration process relatively ineffective (77). Consequently, even after shellfish are considered safe from a bacteriological standpoint (e.g., following conventional depuration), the risk to consumers' health may still exist.

The National Shellfish Sanitation Program (NSSP) routinely determines the absence of bacterial pathogens from shellfish growing areas by using coliform bacteria as indicators of water quality. However, when growing areas are implicated as the source of shellfish causing illness consistent with viral etiology, the NSSP requires closure for a minimum of 21 days.

Since viruses persist longer than bacteria in growing waters and in shellfish (78), it takes considerably longer for shellfish to eliminate viruses (78), and, while the persistence of these coliform bacteria in shellfish growing areas is comparable to that of bacterial pathogens, the relationship between bacterial indicators and the presence of enteric viruses such as noroviruses is poor (78, 79). Therefore, if shellfish harvest areas become unexpectedly contaminated, the likelihood exists that viral pathogens may remain viable in shellfish long after growing waters appear safe according to the NSSP bacteriological criteria. Recognizing these facts, and lacking an alternative viral indicator or any other reasonable way to judge, the NSSP originally stipulated a 3-week closing period as the criterion for achieving safe shellfish when viral pathogens are known or suspected to be involved.

NOROVIRUS

A majority of seafood outbreaks result from the consumption of raw or undercooked bivalve molluscan shellfish contaminated with enteric viruses (80). One of the major enteric viruses associated with shellfish is norovirus, which is one of the causative agents of viral gastroenteritis in humans and has caused several outbreaks throughout the world (81). Norovirus is associated with up to 10% of hospitalizations in the United States, up to 200,000 deaths in children under 5 years old in developing countries, and mortality in the elderly (82). According to the CDC, norovirus was the number-one foodborne illness in the United States in 2011 and had the highest death rate (83).

Norovirus outbreaks are frequently reported in the United States. According to the FDA, many norovirus outbreaks have been linked to consumption of oysters obtained from commercial harvesting areas along the Gulf Coast: Louisiana (Port Sulphur, 2010, and Calcasieu Basin, 2013), Mississippi (Pass

Christian, 2009), and Texas (San Antonio Bay, 2007). On 28 and 29 April 2012, 14 people became ill with norovirus after consuming oysters at a restaurant in New Orleans, LA. The oysters were traced back to oyster harvesting area 23, Terrebonne Parish, off the coast of Louisiana, which resulted in the temporary closure of the harvesting area (84).

In 2013 scientists investigated the occurrence of norovirus and microbial indicators of fecal contamination in eastern oysters (*Crassostrea virginica*) and water from commercial harvesting areas along the Louisiana Gulf Coast (January to November of 2013). All the sampling locations (harvesting areas 9 through 13) were among the most active commercial oyster harvesting areas and remained open during the sampling period. The results of this study found that the microbial fecal indicators (aerobic plate count, enterococci, fecal coliforms, *E. coli*, male-specific coliphages, and somatic coliphages) were detected at levels lower than public health concerns. Despite low levels of fecal contamination in the open areas for oyster and harvesting water collection, norovirus GII was detected in oysters collected from area 12 in June 2013 (85).

An outbreak of norovirus occurred in January 2013 in Cameron Parish, Louisiana. The individuals who became ill had eaten oysters collected from Louisiana Gulf Coast area 30. A stool specimen was obtained from an infected individual, and the suspected oysters were also analyzed. The norovirus strain in the stool belonged to GII.4 Sydney; however, the oysters were negative and could not be linked to the outbreak (85).

Current regulations regarding the safety of seafood rely on the levels of fecal coliforms and/or *E. coli* present in seafood and/or harvest waters. Studies have found that norovirus can be detected in seafood even when microbial indicators are low and in compliance with the U.S. federal standards. As such, there could be a potential health hazard to seafood consumers. This emphasizes the need for regular monitoring of norovirus in commercial fish and seafood harvesting areas to reduce the risk of viral outbreaks.

HEPATITIS A

Enteric viruses such as hepatitis A virus are responsible for a large proportion of food- and waterborne illnesses. These viruses are transmitted to humans via the fecal-oral route, usually from contaminated water or foods such as raw shellfish (86). Hepatitis A infection is the leading worldwide cause of acute viral hepatitis, and outbreaks have occurred among consumers of shellfish harvested from fecally polluted waters. Outbreaks of hepatitis A virus caused by the consumption of raw shellfish have been reported regularly since 1962, and raw or undercooked clams or oysters were implicated as the most frequent vehicles of infection (75, 87–89). In 2005 a multistate outbreak of hepatitis A occurred in Alabama, Florida, South Carolina, and Tennessee due to consumption of oysters obtained from Louisiana harvesting sites (89). There is a greater likelihood of shellfish harboring hepatitis A virus in autumn and winter, coinciding with the seasons that shellfish are among the most popular dining choices (86).

Hepatitis A virus appears to be extremely stable in the environment, with only a 100-fold decline in infectivity over 4 weeks at room temperature, and up to 12 months in fresh or salt water (6). Hepatitis A virus appears to be relatively resistant to free chlorine, especially when the virus is associated with organic matter.

PARASITIC RISKS ASSOCIATED WITH FISH AND SEAFOOD

The most common parasites associated with outbreaks in fish and seafood include round worms (*Anisakis* spp., *Pseudoterranova* spp., *Eustrongylides* spp., and *Gnathostoma* spp.), tape worms (*Diphyllobothrium* spp.), and flukes (*Clonorchis sinensis*, *Opisthorchis* spp., *Heterophyes* spp., *Metagonimus* spp., *Nanophyetus salmincola*, and *Paragonimus* spp.).

Parasitic outbreaks in fish and seafood are rare in the United States. Most outbreaks have been associated with eating raw or undercooked fish and seafood. The heat process used to control bacterial pathogens will also kill parasites (71).

AQUACULTURE PRODUCTION RISKS ASSOCIATED WITH FISH AND SEAFOOD

Aquaculture production has greatly increased outside and inside the United States, and now supplies 46% of the world's seafood supply (90). This trend could lead to increased risks of safety hazards related to fish and seafood consumption due to biological contamination in farm culture waters compared to natural seawater as a result of the proximity of culture farms to urban areas (2). A number of factors have been shown to influence the safety of aquaculture products including location, farmed species, husbandry practices, postharvest processing, and cultural habits of food preparation and consumption (91).

Salmonella contamination in cultured shrimp products is a problem, and contamination of culture environments can occur through the following routes: run-off of organic water into ponds during rainfall; animal waste introduced directly (bird droppings or frogs) or indirectly (runoff); fertilization of ponds with noncomposted manures; integrated farming systems with animals housed in proximity to ponds; toilets discharging into ponds; contaminated source water; unsanitary ice, water, containers, and poor hygienic handling practices; and contaminated feed (90, 91). A study by the FDA showed that aquacultured seafood was more likely to contain *Salmonella* than wild-caught seafood (29). Additionally, several reports have been made on the prevalence of *Salmonella* in shrimp culture environments (91).

Another major issue facing the aquaculture industry is the fact that extensive use of antibiotics in agricultural animal production can result in the development of antibiotic-resistant pathogens and that these pathogens can infect and transfer resistance to humans (92). The development of resistant pathogens in aquaculture environments has been well documented, and the transfer of resistance-encoding plasmids between aquaculture environments and humans has been reported (21, 92). The term "resistance" refers to the microorganism's ability to adapt and survive antimicrobials (91). The public health consequences of resistance include failure of treatment, increased severity and duration of infections, hospitalization, and mortality (93). Antibiotics that are authorized for use in aquaculture include oxytetracycline, florfenicol, chorionic gonadotropin, formalin solutions, tricaine methanesulfate, sulfadimethoxine/ormetropin, and hydrogen peroxide (71).

The concern about antibiotic-resistant pathogens has been spotlighted by an increased prevalence of antimicrobial-resistant *Salmonella* in shrimp and other claims that ready-to-eat shrimp is an international vehicle of antibiotic-resistant bacteria (35). Both the European Food Safety Authority and the National Antimicrobial Resistance Monitoring System have reported on resistant and multiresistant *Salmonella* isolates (91). A *Salmonella* strain resistant to extended-spectrum beta-lactamase has been recognized worldwide (93). Furthermore, a study of antibiotic use in shrimp farming in Thailand revealed that the use of antibiotics among farmers in that area could result in a severe risk of development of antibiotic-resistant bacterial strains (92). Of particular concern in this study was the prophylactic use of antibiotics at subtherapeutic levels. The study claimed that 74% of the farmers used antibiotics in farm management, and a minimum of 13 antibiotics were used. Additionally, the farmers either used higher doses or what they considered more potent antibiotics for treatment rather than prevention. It was discovered that many of the farmers studied did not have enough information on the efficient use of the antibiotics. A main cause of concern was the farmers' widespread use of fluoroquinolones, which are important due to their treatment of a broad range of human pathogens (92).

CAUSES OF SEAFOOD RISKS

Food safety concerns among consumers have been growing over time. A survey conducted by the International Food Information Council Foundation in 2008 asked consumers the following question: "To what extent, if at all, do you feel confident in the safety of the U.S. food supply?" Over 50% of the respondents felt confident about the safety of the U.S. food supply (n = 1,000). Food safety practices do not always match confidence. This was evident when the same survey asked the following question: "Which of the following actions do you perform regularly when cooking, preparing, and consuming food products? 1) Wash my hands with soap and water, 2) Cook to required temperature (such as 165 degrees F for poultry), 3) Use different cutting boards for each product (such as raw meat or poultry or produce), 4) Use a food thermometer to check the doneness of meat and poultry items." The survey revealed that most consumers washed their hands before cooking (79%) and cooked food to the proper temperature (68%). However, consumers did not use different cutting boards for each food product (71%) or use a thermometer to check temperature (29%). It is unclear how the consumers determined if they had cooked the food product to the proper temperature if they had not used a thermometer. Furthermore, the survey found that most consumers felt that the science-based information related to food and health is confusing and conflicting (94).

In 1996 Buzby and Ready (95) reported on >1,000 respondents to a national survey about where they obtained their food safety information, whether they trusted this information, and their major concerns about food safety. Thirty seven percent obtained their information from newspapers and magazines, and only 16.5% from government publications. Over 40% of the survey respondents did not trust the accuracy of food safety information in any form, including government publications and food labeling.

The Cooperative Extension System in several states has included food safety education and a food behavior checklist in their Food Stamp Nutrition Education Program. Texas Agriculture Extension completed a behavior checklist phone survey with 459 participants. The self-reported behavior survey was used to identify changes in the amount of time food was left out at room temperature and hand washing for 20 seconds before handling food (96). The University of California Cooperative Extension's Food Stamp Nutrition Education Program reported that participants did improve safe food-handling practices after participating in the "Be Food Safe" curriculum with 1,900 people in 19 counties (97).

Raw or Minimally Processed Fish and Seafood

During the past several decades, researchers have continuously emphasized foodborne infection cases in humans which were caused by consuming contaminated fresh, raw fish and seafood. *Vibrio* spp. have been identified as the most significant cause of foodborne hospitalizations that can lead to death from eating raw seafood. The magnitude of the risk increases when food preparation and consumption trends do not change regarding eating raw or undercooked seafood.

Ready-to-Eat Fish and Seafood

Ready-to-eat seafood items are potentially high-risk foods, and regulatory agencies in the United States have adopted a zero-tolerance policy toward contamination of these products with *L. monocytogenes*, which can grow easily on the surface of ready-to-eat food products. The salt content, pH, addition of less preservative, and water activity of smoked fish permits the growth of undesirable microorganisms on the surface of this product. A major concern about smoked fish products is *L. monocytogenes* and *Salmonella* spp. (98). The highest incidence of *L. monocytogenes* is associated with cold-smoked fish rather than hot-smoked fish because this pathogen does not survive the hot-smoke process (98). Contamination of smoked fish with *L. monocytogenes* ranges from 17.9 to 22.3% (99). Studies have shown that cold-smoked salmon is a good substrate for *L. monocytogenes*, which grows well at refrigerated temperatures on the surface of this product even under vacuum conditions (100). *Salmonella* spp. have been associated with smoked fish outbreaks (10). The incidence of *Salmonella* in smoked fish is 3.9%, and *Salmonella* Newport and *Salmonella* Anatum are the most prominent serotypes isolated from smoked fish in the United States (98). Reducing the risk of ready-to-eat fish and seafood can be achieved through sanitation practices and HACCP plans in the processing plants.

Fully Cooked Fish and Seafood

The freshness and safety of fish and seafood vary depending on many factors such as contamination during farming, harvesting, and handling and other postharvest activities. Consumers can store fresh raw fish and seafood for days at refrigerated temperatures ranging from 0.5 to 4.5°C. However, at this temperature some pathogenic bacteria such as *L. monocytogenes* have the potential to multiply. Heat treatment such as proper cooking has an important role in the safety and sensory acceptance of fish and seafood. The definition of cooking is as follows: "The application of heat to a food to modify raw product properties in order to meet sensory expectations of consumers and to reduce its microbial load, which improves its safety and may extend its shelf life" (101). For this definition, it is important to determine consumer acceptance of cooked fish and seafood, which influences consumers' willingness to employ proper cooking methods such as using a thermometer during cooking to monitor the internal temperature of products.

However, cooking does not eliminate certain hazards such as heat-stable natural or microbial toxins and biogenic amines (e.g., histamine) if they are already present. Such hazards are generally controlled through effective use of Good Manufacturing Practices (GMPs) and HACCP plans. The various cooking methods for fish and seafood include baking *en papillote* (in a folded pouch) for 10 minutes to reach an internal temperature of 400 to 450°F, barbequing, deep-fat-frying at an oil temperature of 375°F for 3 to 5 minutes, grilling, broiling, microwaving, oven frying at 500 to 550°F, pan frying, sautéing at 375°F, planking at 400°F for 10 minutes, smoking at 245°F for 30 minutes in 2.5 to 3.5% NaCl, steaming 1 to 2 inches above water, and stir frying at 375°F.

A number of foodborne outbreaks occur due to improper cooking of food by consumers. Inadequate cooking and storage of food is considered to be the main cause of foodborne infection (102). It has been suggested that domestic household conditions and inadequate cooking account for 11% and inappropriate storage for up to 50% of outbreaks. This leaves room for extensive research in food safety for developing guidelines at the consumer or domestic household level that will aid in reducing the number of outbreaks. There clearly is a need for consumer-friendly guidelines to ensure the maximum possible food safety.

Thermal resistance in microorganisms can vary from one genus to other and also between species. Chintagari (103) inoculated three *Vibrio* spp., *Salmonella* spp., and *Listeria* spp. on the surface of shrimp and then stored the samples at 3°C. On days 0, 1, and 2 the shrimp samples were placed in a boiling water bath and were removed at different internal temperatures. *Vibrio* spp. were the least resistant to heat, with bacterial counts reaching nondetectable levels at 55°C (Fig. 2). *Salmonella* reached nondetectable levels on shrimp at 75°C (Fig. 3), and *Listeria* spp.

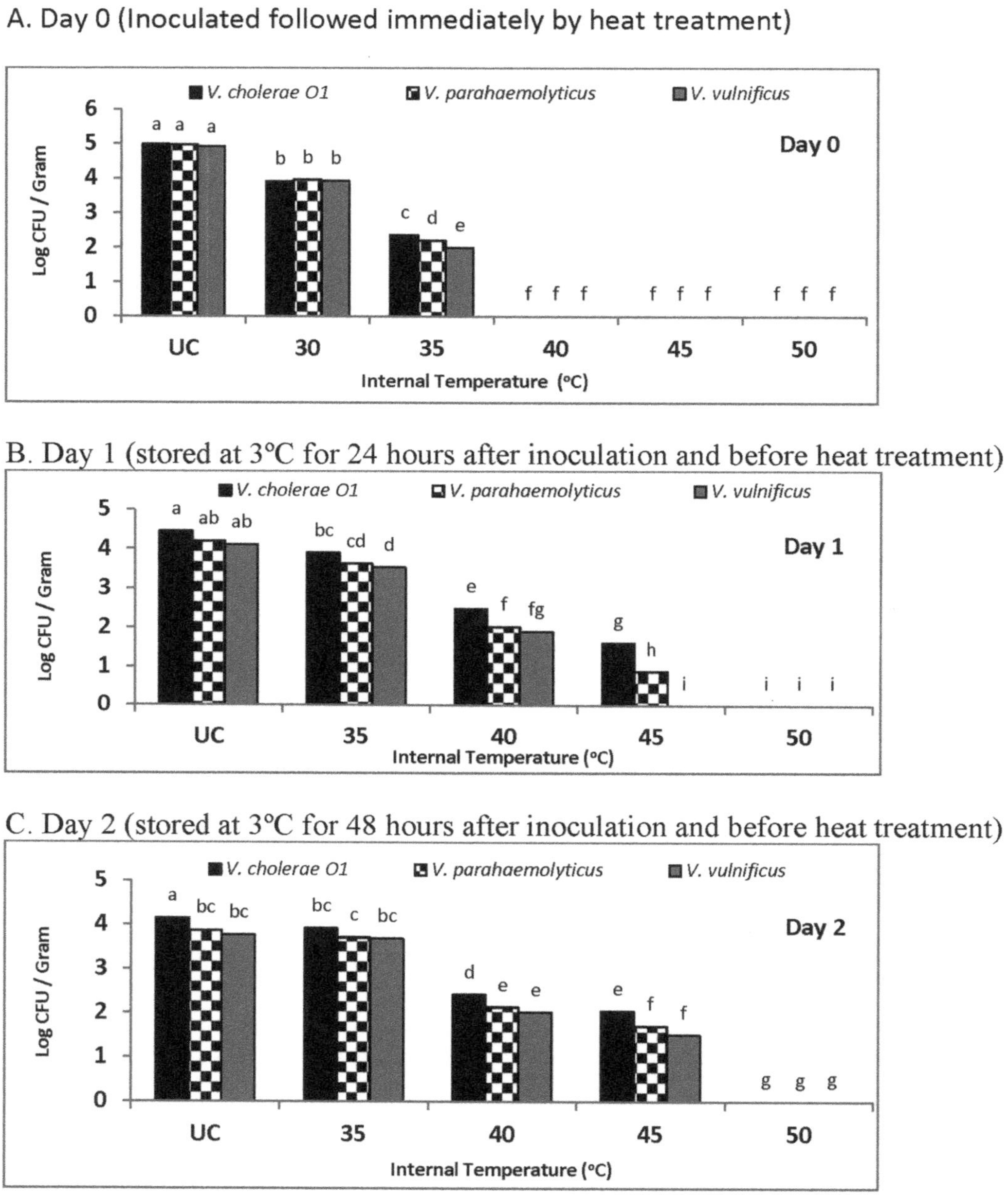

FIGURE 2 Thermal resistance of *V. vulnificus*, *V. parahaemolyticus*, and *V. cholerae* O1 at different internal temperatures in shrimp when subjected to boiling. UC, Uncooked shrimp sample. Data presented in the bar diagram are the mean of three different experiments, and the bars with different letters are significantly different from each other ($P < 0.05$).

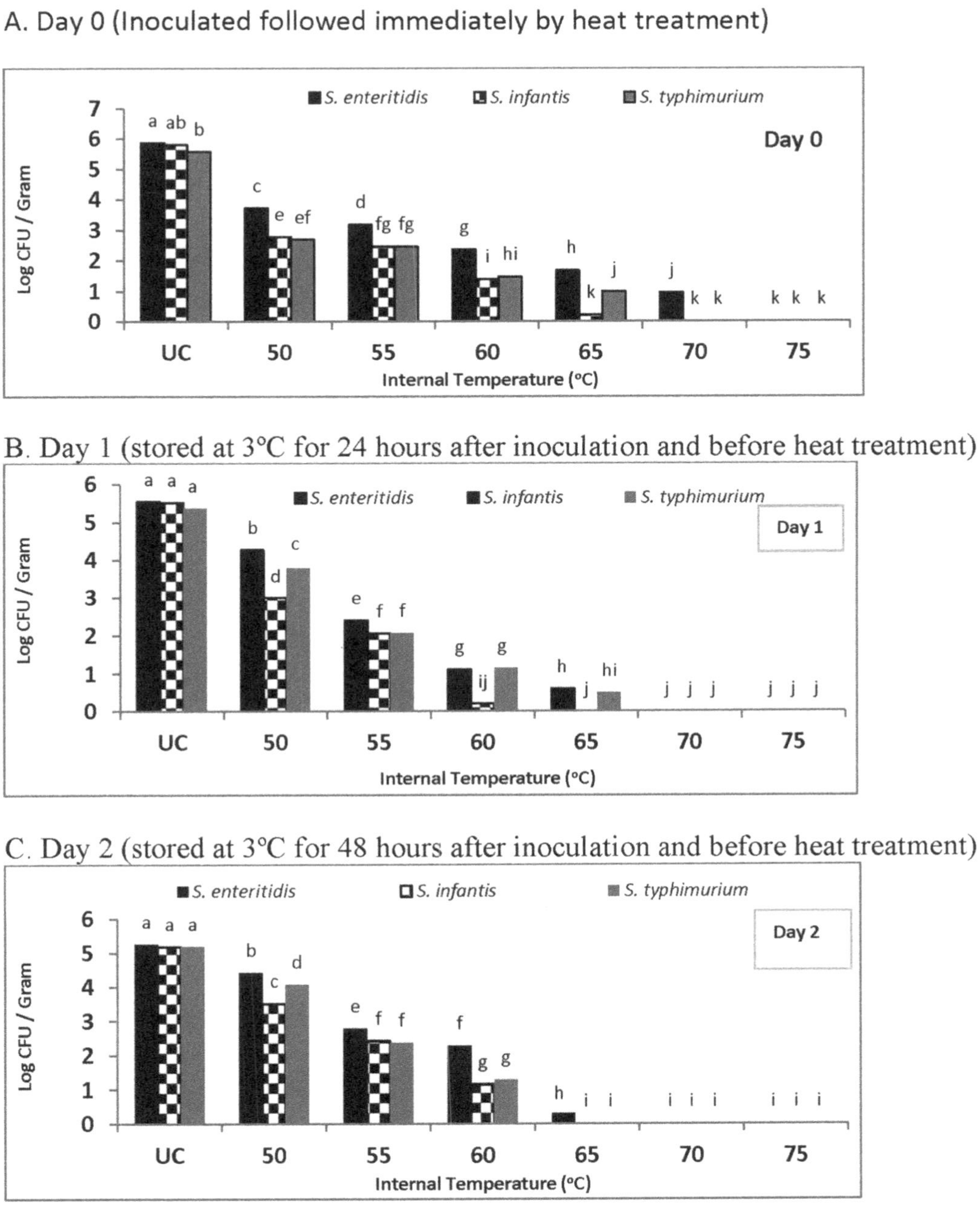

FIGURE 3 Thermal resistance of *Salmonella enterica* serovars Enteritidis (*S. enteritidis* above), Infantis (*S. infantis* above), and Typhimurium (*S. Typhimurium* above) at different internal temperatures in shrimp when subjected to boiling. UC, Uncooked shrimp sample. Data presented in the bar diagram are the mean of three different experiments, and the bars with different letters are significantly different from each other ($P < 0.05$).

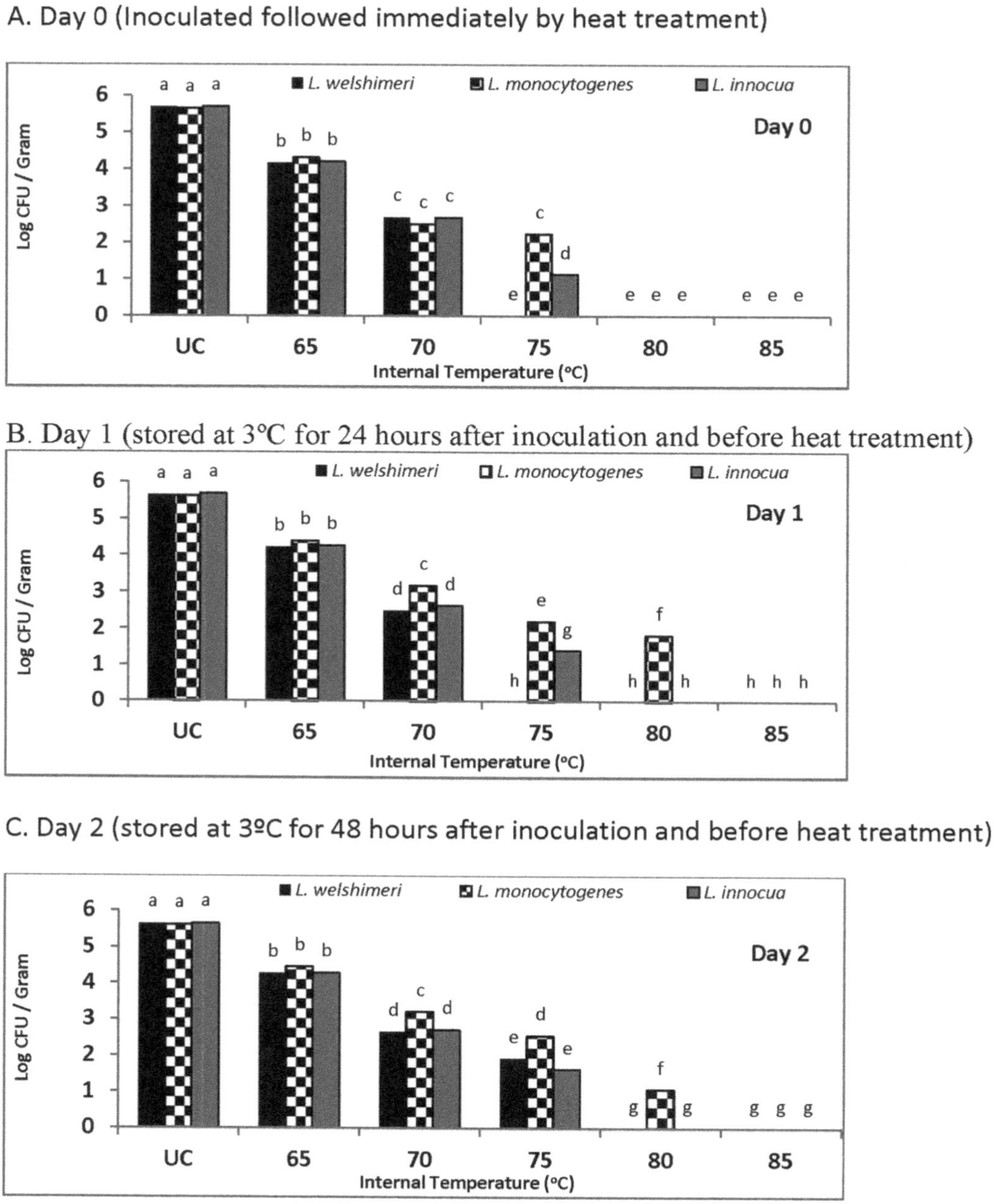

FIGURE 4 Thermal resistance of *Listeria welshimeri*, *Listeria monocytogenes*, and *Listeria innocua* at different internal temperatures in shrimp when subjected to boiling. UC, Uncooked shrimp sample. Data presented in the bar diagram are the mean of three different experiments, and the bars with different letters are significantly different from each other ($P < 0.05$).

showed the highest resistance, reaching nondetectable levels at 85°C (Fig. 4).

Thermal resistance can also be affected by factors such as the conditions under which foods contaminated with pathogens are stored. Consumers generally store fish and seafood at refrigerated temperatures for several days before using. This could promote biofilm formation and increase the heat resistance of some foodborne pathogens. Comparatively high resistance of *V. cholerae* O1 has been linked to its ability to form colonies on the shells of shrimps. Several studies have demonstrated that once *V. cholerae* O1 has attached to chitin particles or crustacean external surfaces, the microorganism is able to initiate a process of colonization (51, 104). This process can be associated with increased resistance to various stresses such as temperature but also to those caused by chemical disinfectants (105), low temperatures (106), and low pH levels (104). Chintagari (103) found that three species of *Vibrio* were more resistant to heat treatment when surface-inoculated shrimp were stored at refrigerated temperatures before boiling (Fig. 2). Shultz et al. (107) conducted a similar kind of experiment with cockles and concluded that cockles should be cooked until the slowest-heating cockles reach 71°C, for 1 minute. The difference in temperature requirements for cockles and shrimp may be due to the differences in sizes and composition.

A popular method of cooking shrimp is by boiling. Consumers often boil shrimp until they float to the surface of the water and change to a red color. Edwards et al. (108) found that boiling shrimp until they float reduces *Listeria* and *Salmonella* spp. to nondetectable levels, but color change was not a good indication of the reduction of foodborne pathogens due to color variation. Baking and broiling are two popular methods of cooking catfish. Consumers often follow the 10-minute rule as a general rule of thumb when cooking catfish. This rule says to measure the dressed fish, fillet, or steak at its thickest part and then allow 10 minutes of cooking time per inch of thickness. For fish that are less than 1 inch thick, the cooking time is shortened proportionally. Additionally, a common method used by consumers to determine doneness when cooking is to observe the color change of the cooked product. For example, when boiling crab, the color change from gray to red is associated with it being thoroughly cooked. For a serving size of crab (4 crabs), to reduce foodborne pathogens to nondetectable levels, it is recommended to boil the crabs for 10 minutes and cool 5 additional minutes for an internal temperature of at least 85°C and a total cooking time of 15 minutes; steam four crabs for 15 minutes and cool 5 additional minutes to reach an internal temperature of at least 85°C, with a total cooking time of 20 minutes (109) (Fig. 5 and Fig. 6). For catfish, consumers look for the meat to turn opaque when cooked. This color change is used for safety evaluation and is also associated with sensory acceptance. Further scientific research needs to be done to verify that color change is a good indication of safety in fish and seafood products.

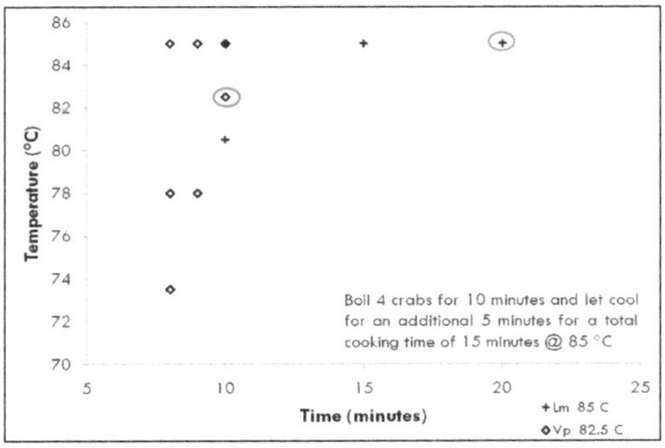

FIGURE 5 Graph of below-detection-limit/nondetectable level reached for boiling a serving size of four crabs at each time point and the optimum temperature each achieved for *L. monocytogenes* (Lm) and *V. parahaemolyticus* (Vp).

CONCLUSION

Risks associated with fish and seafood will continue to be a problem unless consumers'

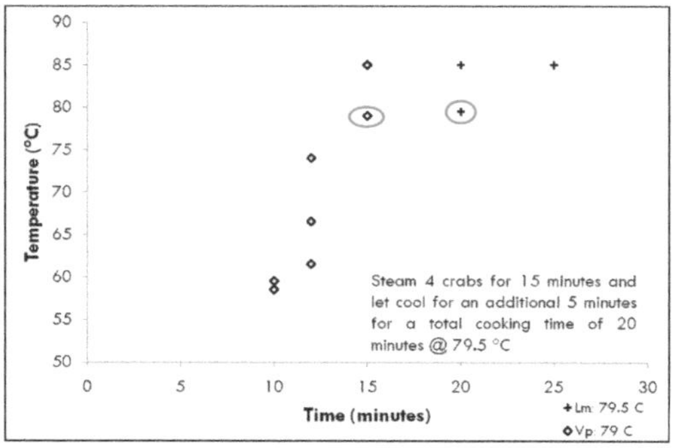

FIGURE 6 Graph of below-detection-limit/non-detectable level reached for steaming a serving size of four crabs at each time point and the optimum temperature each achieved for *L. monocytogenes* (Lm) and *V. parahaemolyticus* (Vp).

attitudes change toward eating these products raw. Most consumers know of the risks associated with eating raw oysters but continue to eat them. With the increase in the demand for fish and seafood, aquaculture production is increasing. Aquaculture production of fish and seafood could lead to new risks that will need to be addressed in the future to control foodborne pathogens.

CITATION

Chintagari S, Hazard N, Edwards G, Jadeja R, Janes M. 2017. Risks associated with fish and seafood. Microbiol Spectrum 5(1):PFS-0013-2016.

REFERENCES

1. **National Fishery Institution.** 2013. Top 10 list for seafood consumption. http://www.aboutseafood.com/about.
2. **Feldhusen F.** 2000. The role of seafood in bacterial foodborne diseases. *Microbes Infect* **2:** 1651–1660.
3. **Iwamoto M, Ayers T, Mahon BE, Swerdlow DL.** 2010. Epidemiology of seafood-associated infections in the United States. *Clin Microbiol Rev* **23:**399–411.
4. **Painter LA, Hoekstra RM, Ayers T, Tauxe RV, Braden CR, Angulo FJ, Griffin PM.** 2013. Attribution of foodborne illnesses, hospitalizations, and deaths to food commodities by using outbreak data, United States, 1998–2008. *Emerg Infect Dis* **19:**407–415.
5. **Butt AA, Aldridge KE, Sanders CV.** 2004. Infections related to the ingestion of seafood. Part I. viral and bacterial infections. *Lancet Infect Dis* **4:**201–212.
6. **Sattar SA, Tetro J, Bidawid S, Farber J.** 2000. Foodborne spread of hepatitis A: Recent studies on virus survival, transfer and inactivation. *Can J Infect Dis* **11:**159–163.
7. **Lennon D, Lewis B, Mantell C, Becroft D, Dove B, Farmer K, Tonkin S, Yeates N, Stamp R, Mickleson K.** 1984. Epidemic perinatal listeriosis. *Pediatr Infect Dis* **3:**30–34.
8. **Mu DYW, Huang KW, Gates WH, Wu WH.** 1997. Effect of trisodium phosphate on *Listeria monocytogenes* attached to rainbow trout (*Oncorhynchus Mykiss*) and shrimp (*Penaeus* spp.) during refrigerated storage. *J Food Safety* **17:**37–46.
9. **Weagant SD, Sado PN, Colburn KG, Torkelson JD, Stanley FA, Krane MH, Shields SE, Thayer CF.** 1988. The incidence of *Listeria* species in frozen seafood products. *J Food Prot* **51:**655–657.
10. **Heinitz ML, Ruble RD, Wagner DE, Tatini SR.** 2000. Incidence of *Salmonella* in fish and seafood. *J Food Prot* **63:**579–592.
11. **Fatma AC, Sarmasik A, Koseoglu B.** 2006. Occurrence of *Vibrio* spp. and *Aeromonas* spp. in shellfish harvested off Dardanelles cost of Turkey. *Food Control* **17:**648–652.
12. **Hood MA, Ness GE, Rodrick GE.** 1981. Isolation of *Vibrio cholerae* serotype O1 from the eastern oyster, *Crassostrea virginica*. *Appl Environ Microbiol* **41:**559–560.
13. **Koopmans M, von Bonsdorff CH, Vinjé J, de Medici D, Monroe S.** 2002. Foodborne viruses. *FEMS Microbiol Rev* **26:**187–205.
14. **National Advisory Committee on Microbiological Criteria For Foods.** 2008. Response to the questions posed by the Food and Drug Administration and the National Marine Fisheries Service regarding determination of cooking parameters for safe seafood for consumers. *J Food Prot* **71:**1287–1308.
15. **Redfield HW.** 1925. Food and drugs. *Am J Public Health (NY)* **15:**660–663.
16. **Blythe DG, Hack E, Washington G.** 2001. Seafood and environmental toxins, p 680. *In* Hui YH, Kitts D, Stanford PS (ed), *Foodborne Disease Handbook*, 2nd ed, vol 4. Marcel Dekker, New York, NY.
17. **Lynch M, Painter J, Woodruff R, Braden C, Centers for Disease Control and Prevention.** 2006. Surveillance for foodborne-disease

outbreaks: United States, 1998–2002. *MMWR Surveill Summ* **55:**1–42.

18. **Centers for Disease Control and Prevention.** 2015. *2013. Surveillance for Foodborne-Disease Outbreaks: United States, 2013 Annual Report.* U.S. Department of Health and Human Services, CDC, Atlanta, Georgia.
19. **Brands DA, Billington SJ, Levine JF, Joens LA.** 2005. Genotypes and antibiotic resistance of *Salmonella* Newport isolates from U.S. market oysters. *Foodborne Pathog Dis* **2:**111–114.
20. **Varma PRG, Mathen C, Mathew A.** 1985. Bacteriological quality of frozen seafoods for export with special references to *Salmonella,* p 483–484. *In* Rivindran K, Nair NU, Perigreen PA, Madhavan P, Pillai AGG, Panicker PA, Thomas M (ed), *Harvest and Post-Harvest Technology of Fish.* Society of Fisheries Technologists (India). Cochin, India.
21. **Khan AA, Cheng CM, Van KT, West CS, Nawaz MS, Khan SA.** 2006. Characterization of class 1 integron resistance gene cassettes in *Salmonella enterica* serovars Oslo and Bareily from imported seafood. *J Antimicrob Chemother* **58:**1308–1310.
22. **Phan TT, Khai LT, Ogasawara N, Tam NT, Okatani AT, Akiba M, Hayashidani H.** 2005. Contamination of *Salmonella* in retail meats and shrimps in the Mekong Delta, Vietnam. *J Food Prot* **68:**1077–1080.
23. **Iyer TSG, Shrivastava KP.** 1989. Incidence and low temperature survival of *Salmonella* in fishery products. *Fish Technol.* **26:**39–42.
24. **Sanath Kumar H, Sunil R, Venugopal MN, Karunasagar I, Karunasagar I.** 2003. Detection of *Salmonella* spp. in tropical seafood by polymerase chain reaction. *Int J Food Microbiol* **88:**91–95.
25. **Fonseka TSG.** 1990. Microbial flora of pond cultured prawn (*Penaeus monodon*). *FAO Fish Rep* **401:**24–31.
26. **Rattagol P, Wongchinda N, Sanghtong N.** 1990. *Salmonella* contamination in Thai shrimp. *FAO Fish Rep* **401:**18–23.
27. **Chio TD, Chen SC.** 1981. Studies on decomposition and *Salmonella* isolated from clams and marine waters of Kuwait. *Water Air Soil Pollut* **26:**59–63.
28. **Saheki K, Kabayashi S, Kawanishi T.** 1989. *Salmonella* contamination of eel culture ponds. *Nippon Suisan Gakkai Shi* **55:**675–679.
29. **Koonse B, Burkhardt W III, Chirtel S, Hoskin GP.** 2005. *Salmonella* and the sanitary quality of aquacultured shrimp. *J Food Prot* **68:**2527–2532.
30. **Dalsgaard A, Huss HH, H-Kittikun A, Larsen JL.** 1995. Prevalence of *Vibrio cholerae* and *Salmonella* in a major shrimp production area in Thailand. *Int J Food Microbiol* **28:**101–113.
31. **Iyer TSG, Varma PRG.** 1990. Sources of contamination with *Salmonella* during processing of frozen shrimp. *Fish Technol* **27:**60–63.
32. **Reilly PJA, Twiddy DR.** 1992. *Salmonella* and *Vibrio cholerae* in brackishwater cultured tropical prawns. *Int J Food Microbiol* **16:**293–301.
33. **Fraiser MB, Koburger JA.** 1984. Incidence of *Salmonella* in clams, oysters, crabs, and mullet. *J Food Prot* **47:**343–345.
34. **Centers for Disease Control and Prevention.** 2003. Outbreaks of *Salmonella* serotype Enteritidis infection associated with eating shell eggs—United States, 1999–2001. *MMWR Morb Mortal Wkly Rep* **51:**1149–1176.
35. **Norhana MN, Poole SE, Deeth HC, Dykes GA.** 2010. Prevalence, persistence and control of *Salmonella* and *Listeria* in shrimp and shrimp products: a review. *Food Contr* **21:**343–361.
36. **Elliot EL, Kvenberg JE.** 2000. Risk assessment used to evaluate the US position on *Listeria monocytogenes* in seafood. *Int J Food Microbiol* **62:**253–260.
37. **Farber JM.** 2000. Present situation in Canada regarding *Listeria monocytogenes* and ready-to-eat seafood products. *Int J Food Microbiol* **62:**247–251.
38. **Harrison MA, Huang Y-W.** 1990. Thermal death times for *Listeria monocytogenes* (Scott A) in crabmeat. *J Food Prot* **53:**878–880.
39. **Chen YP, Andrews LS, Grodner RM.** 1996. Sensory and microbial quality of irradiated crab meat products. *J Food Sci* **61:**1239–1242.
40. **Farber JM, Peterkin PI.** 1991. *Listeria monocytogenes,* a food-borne pathogen. *Microbiol Rev* **55:**476–511.
41. **Price RJ, Tom PD.** 1992. *Environmental conditions for pathogenic bacterial growth.* Seafood Network Information Center, Davis, CA. http://seafood.oregonstate.edu/.pdf%20Links/Environmental-Conditions-for-Pathogenic-Bacterial-Growth.pdf.
42. **Centers for Disease Control and Prevention.** 1992. Cholera associated with international travel. *MMWR Morb Mortal Wkly Rep* **41:**36.
43. **Blake PA, Allegra DT, Snyder JD, Barrett TJ, McFarland L, Caraway CT, Feeley JC, Craig JP, Lee JV, Puhr ND, Feldman RA.** 1980. Cholera: a possible endemic focus in the United States. *N Engl J Med* **302:**305–309.
44. **Weber JT, Levine WC, Hopkins DP, Tauxe RV.** 1994. Cholera in the United States, 1965–1991. Risks at home and abroad. *Arch Intern Med* **154:**551–556.

45. **Centers for Disease Control and Prevention.** 2006. Two cases of toxigenic *Vibrio cholerae* O1 infection after Hurricanes Katrina and Rita: Louisiana, October 2005. *MMWR Morb Mortal Wkly Rep* **55:**31–32.
46. **Rippen TE, Hackney CR.** 1992. Pasteurization of seafood: potential for shelf-life extension and pathogen control. *Food Technol* **46:**88–94.
47. **Erdogdu F, Luzuriaga DA, Balaban MO, Chau KV.** 2001. Yield loss and moisture content changes of small tiger shrimp (*Penaeus monodon*) treated with different phosphate concentrations during thermal processing. *J Aquat Food Prod Technol* **10:**31–45.
48. **Colwell RR, Seidler RJ, Kaper J, Joseph SW, Garges S, Lockman H, Maneval D, Bradford H, Roberts N, Remmers E, Huq I, Huq A.** 1981. Occurrence of *Vibrio cholerae* serotype O1 in Maryland and Louisiana estuaries. *Appl Environ Microbiol* **41:**555–558.
49. **Centers for Disease Control and Prevention.** 1991. Epidemologic notes and reports cholera: New York. *MMWR Morb Mortal Wkly Rep* **40:** 516–518.
50. **Rasmussen T, Jensen RB, Skovgaard O.** 2007. The two chromosomes of *Vibrio cholerae* are initiated at different time points in the cell cycle. *EMBO J* **26:**3124–3131.
51. **Castro-Rosas J, Escartín EF.** 2002. Adhesion and colonization of *Vibrio cholerae* O1 on shrimp and crab carapaces. *J Food Prot* **65:** 492–498.
52. **Tamplin ML, Gauzens AL, Huq A, Sack DA, Colwell RR.** 1990. Attachment of *Vibrio cholerae* serogroup O1 to zooplankton and phytoplankton of Bangladesh waters. *Appl Environ Microbiol* **56:**1977–1980.
53. **Makukutu CA, Guthrie RK.** 1986. Behavior of *Vibrio cholerae* in hot foods. *Appl Environ Microbiol* **52:**824–831.
54. **Dadisman TA Jr, Nelson R, Molenda JR, Garber HJ.** 1972. *Vibrio parahaemolyticus* gastroenteritis in Maryland. I. Clinical and epidemiologic aspects. *Am J Epidemiol* **96:**414–426.
55. **Kaneko T, Colwell RR.** 1973. Ecology of *Vibrio parahaemolyticus* in Chesapeake Bay. *J Bacteriol* **113:**24–32.
56. **DePaola A, Hopkins LH, Peeler JT, Wentz B, McPhearson RM.** 1990. Incidence of *Vibrio parahaemolyticus* in U.S. coastal waters and oysters. *Appl Environ Microbiol* **56:**2299–2302.
57. **Cook DW, Oleary P, Hunsucker JC, Sloan EM, Bowers JC, Blodgett RJ, Depaola A.** 2002. *Vibrio vulnificus* and *Vibrio parahaemolyticus* in U.S. retail shell oysters: a national survey from June 1998 to July 1999. *J Food Prot* **65:**79–87.
58. **Daniels NA, MacKinnon L, Bishop R, Altekruse S, Ray B, Hammond RM, Thompson S, Wilson S, Bean NH, Griffin PM, Slutsker L.** 2000. *Vibrio parahaemolyticus* infections in the United States, *1973–1998*. *J Infect Dis* **181:**1661–1666.
59. **Rose JB, Epstein PR, Lipp EK, Sherman BH, Bernard SM, Patz JA.** 2001. Climate variability and change in the United States: potential impacts on water- and foodborne diseases caused by microbiologic agents. *Environ Health Perspect* **109**(Suppl 2):211–221.
60. **Faghri MA, Pennington CL, Cronholm LS, Atlas RM.** 1984. *Bacteria* associated with crabs from cold waters with emphasis on the occurrence of potential human pathogens. *Appl Environ Microbiol* **47:**1054–1061.
61. **D'Aoust J-Y.** 2007. Salmonella species, p 187–236. *In* Doyle MP, Beuchat LR (ed), *Food Microbiology: Fundamentals and Frontiers*, 3rd ed. ASM Press, Washington, DC.
62. **Centers for Disease Control and Prevention.** 1999. Outbreak of *Vibrio parahaemolyticus* infection associated with eating raw oysters and clams harvested from Long Island Sound: Connecticut, New Jersey, and New York, 1998. *MMWR Morb Mortal Wkly Rep* **48:**48–51.
63. **Centers for Disease Control and Prevention.** 2006. Foodborne outbreak tracking and reporting, p 1–31. *In OutbreakNet, Annual Listing of Foodborne Disease Outbreaks, United States, D.o.H.a.* Atlanta, GA. https://www.cdc.gov/foodborneoutbreaks/.
64. **Centers for Disease Control and Prevention.** 2008. Summary of human *Vibrio* isolates reported to CDC, 2007. http://www.cdc.gov/nationalsurveillance/PDFs/CSTEVibrio2007.pdf.
65. **Nishibuchi M, Kaper JB.** 1995. Thermostable direct hemolysin gene of *Vibrio parahaemolyticus*: a virulence gene acquired by a marine bacterium. *Infect Immun* **63:**2093–2099.
66. **Jackson JK, Murphree RL, Tamplin ML.** 1997. Evidence that mortality from *Vibrio vulnificus* infection results from single strains among heterogeneous populations in shellfish. *J Clin Microbiol* **35:**2098–2101.
67. **Price RJ.** 1990. *Seafood Safety*. Seafood Network Information Center, Davis, CA.
68. **Wright AC, Simpson LM, Oliver JD.** 1981. Role of iron in the pathogenesis of *Vibrio vulnificus* infections. *Infect Immun* **34:**503–507.
69. **Davis JW, Sizemore RK.** 1982. Incidence of *Vibrio* species associated with blue crabs (*Callinectes sapidus*) collected from Galveston Bay, Texas. *Appl Environ Microbiol* **43:**1092–1097.

70. **Centers for Disease Control and Prevention.** 2013. *National Botulism Surveillance. Annual Report.* U.S. Department of Health and Human Services, CDC, Atlanta, GA.
71. **FDA.** 2011. Fish and fishery products hazards and controls guidance, 4th ed. http://www.fda.gov/FoodGuidances.
72. **Grabow WOK.** 2002. Enteric hepatitis viruses, p 18–39. *In* Guidelines for Drinking Water Quality. 2nd ed. *Addendum. Microbiological agents in drinking water.* WHO, Geneva.
73. **Girones R, Puig M.** 1994. Detection of adenovirus and enterovirus by PCR amplification in polluted waters. *Water Sci Technol* **31:**5–6.
74. **Barry K, O'Kane JPJ.** 2009. Towards the development of a combined norovirus and sediment transport model for coastal waters. *Geophys Res Abstr* **11:**2009–12306.
75. **Mason JO, McLean WR.** 1962. Infectious hepatitis traced to the consumption of raw oysters. An epidemiologic study. *Am J Hyg* **75:** 90–111.
76. **Burkhardt W III, Calci KR.** 2000. Selective accumulation may account for shellfish-associated viral illness. *Appl Environ Microbiol* **66:** 1375–1378.
77. **Ueki Y, Shoji M, Suto A, Tanabe T, Okimura Y, Kikuchi Y, Saito N, Sano D, Omura T.** 2007. Persistence of caliciviruses in artificially contaminated oysters during depuration. *Appl Environ Microbiol* **73:**5698–5701.
78. **Richards GP.** 2001. Enteric virus contamination of foods through industrial practices: a primer on intervention strategies. *J Ind Microbiol Biotechnol* **27:**117–125.
79. **Rose JB, Sobsey MD.** 1993. Quantitative risk assessment for viral contamination of shellfish and coastal waters. *J Food Prot* **56:**1043–1050.
80. **Jaykus LA.** 2000. Enteric viruses as "emerging" agents of foodborne disease. *Ir J Agric Food Res* **39:**245–255.
81. **Thackray LB, Wobus CE, Chachu KA, Liu B, Alegre ER, Henderson KS, Kelley ST, Virgin HWT IV.** 2007. Murine noroviruses comprising a single genogroup exhibit biological diversity despite limited sequence divergence. *J Virol* **81:**10460–10473.
82. **Patel MM, Widdowson MA, Glass RI, Akazawa K, Vinjé J, Parashar UD.** 2008. Systematic literature review of role of noroviruses in sporadic gastroenteritis. *Emerg Infect Dis* **14:**1224–1231.
83. **Scallan E, Hoekstra RM, Angulo FJ, Tauxe RV, Widdowson MA, Roy SL, Jones JL, Griffin PM.** 2011. Foodborne illness acquired in the United States: major pathogens. *Emerg Infect Dis* **17:**7–15.
84. **Department of Health & Hospitals.** 2012. *DHH Recalls Oysters and Closes Oyster Harvesting Area.* Department of Health & Hospitals. http://new.dhh.louisiana.gov/index.cfm/newsroom/detail/2484.
85. **Montazeri N, Maite M, Liu D, Cormier J, Landry M, Shackleford J, Lampila LE, Achberger EC, Janes ME.** 2015. Surveillance of enteric viruses and microbial indicators in the eastern oysters (*Crassostrea virginica*) and harvest waters along Louisiana Gulf Coast. *J Food Sci* **80:**M1075–M1082.
86. **Bosch A.** 1998. Human enteric viruses in the water environment: a minireview. *Int Microbiol* **1:**191–196.
87. **Anonymous.** 1997. Hepatitis A outbreak traced to consumption of Wallis Lake oysters. *N S W Public Health Bull* **8:**1.
88. **Conaty S, Bird P, Bell G, Kraa E, Grohmann G, McAnulty JM.** 2000. Hepatitis A in New South Wales, Australia from consumption of oysters: the first reported outbreak. *Epidemiol Infect* **124:**121–130.
89. **Shieh YC, Khudyakov YE, Xia G, Ganova-Raeva LM, Khambaty FM, Woods JW, Veazey JE, Motes ML, Glatzer MB, Bialek SR, Fiore AE.** 2007. Molecular confirmation of oysters as the vector for hepatitis A in a 2005 multistate outbreak. *J Food Prot* **70:**145–150.
90. **FAO, Fisheries and Aquaculture Department.** 2010. *The State of World Fisheries and Aquaculture 2010.* FAO Publishing Management Service, Rome, Australia.
91. **Amagliani G, Brandi G, Schiavano GF.** 2011. Incidence and role of *Salmonella* in seafood safety. *Food Res Int* **45:**780–788.
92. **Hölmstrom K, Graslund S, Wahlstrom A, Poungshompoo S, Bengtsson B-E, Kautsky N.** 2003. Antibiotic use in shrimp farming and implications for environmental impacts and human health. *Int J Food Sci Technol* **38:**255–266.
93. **Newell DG, Koopmans M, Verhoef L, Duizer E, Aidara-Kane A, Sprong H, Opsteegh M, Langelaar M, Threfall J, Scheutz F, van der Giessen J, Kruse H.** 2010. Food-borne diseases: the challenges of 20 years ago still persist while new ones continue to emerge. *Int J Food Microbiol* **139**(Suppl 1):S3–S15.
94. **International Food Information Council Foundation.** 2011. 2011 Food & health survey: consumer attitudes toward food safety, nutrition & health. http://www.foodinsight.org/2011_Food_Health_Survey_Consumer_Attitudes_Toward_Food_Safety_Nutrition_Health#sthash.WsI804Ig.dpbs.
95. **Buzby JC, Ready RC.** 1996. Do consumers trust food safety information? *Food Rev* **19:**46–49.

96. **Anding J, Fletcher RD, Van Laanen P, Supak C.** 2001. The Food Stamp Nutrition Education Program's (FSNEP) impact on selected food and nutrition behaviors among Texans. *J Extension* **39**(6). http://www.joe.org/joe/2001december/rb4.html.
97. **Joy AB.** 2004. Diet, shopping and food-safety skills of food stamp clients improve with nutrition education. *Calif Agric* **58:**206–208.
98. **Heinitz ML, Johnson JM.** 1998. The incidence of *Listeria* spp., *Salmonella* spp., and *Clostridium botulinum* in smoked fish and shellfish. *J Food Prot* **61:**318–323.
99. **Pelroy GJ, Peterson M, Paranjpye R, Almond J, Eklund M.** 1994. Inhibition of *Listeria monocytogenes* in cold-process (smoked) salmon by sodium nitrite and packaging method. *J Food Prot* **57:**114–119.
100. **Rørvik LM, Yndestad M, Skjerve E.** 1991. Growth of *Listeria monocytogenes* in vacuum-packed, smoked salmon, during storage at 4 degrees C. *Int J Food Microbiol* **14:**111–117.
101. **Wikipedia contributors.** 2015. *Cooking,* on *Wikipedia, The Free Encyclopedia.* http://en.wikipedia.org/wiki/Cooking. Accessed 2 January 2017.
102. **Ryan MJ, Wall PG, Gilbert RJ, Griffin M, Rowe B.** 1996. Risk factors for outbreaks of infectious intestinal disease linked to domestic catering. *Commun Dis Rep CDR Rev* **6:**R179–R183.
103. **Chintagari S.** 2009. Determination of minimum safe cooking temperatures for shrimp to destroy foodborne pathogens. Master's thesis, Louisiana State University. http://etd.lsu.edu/docs/available/etd-06102009-081716/.
104. **Nalin DR, Daya V, Reid A, Levine MM, Cisneros L.** 1979. Adsorption and growth of *Vibrio cholerae* on chitin. *Infect Immun* **25:**768–770.
105. **McCarthy SS, Miller AL.** 1994. Effect of three biocides on Latin American and Gulf Coast strains of toxigenic *Vibrio cholera* O1. *J Food Prot* **57:**865–869.
106. **Amako K, Shimodori S, Imoto T, Miake S, Umeda A.** 1987. Effects of chitin and its soluble derivatives on survival of *Vibrio cholerae* O1 at low temperature. *Appl Environ Microbiol* **53:**603–605.
107. **Shultz LM, Rutledge JE, Grodner RM, Biede SL.** 1984. Determination of the thermal death time of *Vibrio cholerae* in blue crabs (*Callinectes sapidus*). *J Food Prot* **47:**4–6.
108. **Edwards G, Janes M, Lampila L, Supan J.** 2013. Consumer method to control *Salmonella* and *Listeria* species in shrimp. *J Food Prot* **76:**59–64.
109. **Hazard N.** 2010. *Cooking times and temperatures for safe consumption of Louisiana blue crabs* (Callinectes sapidus). *Master's thesis,* Louisiana State University. http://etd.lsu.edu/docs/available/etd-11182010-130411/.

Water for Agriculture: the Convergence of Sustainability and Safety

8

SARAH M. MARKLAND,[1] DAVID INGRAM,[2]
KALMIA E. KNIEL,[1] and MANAN SHARMA[3]

INTRODUCTION

Undoubtedly, water is one of the world's most precious commodities, and since the beginning of agriculture, irrigating crops and relocating water to hydrate livestock have been essential to sustain society. The first irrigation system is believed to have been a bucket (1) carried back and forth from a river to irrigate plants. Today, the world's most important use of water is for agriculture, more specifically for the production of crops and raising of livestock. In the United States 330 million acres of land are used for the production of food and other agricultural products (2). In 2010 alone over 126 billion gallons of water were used for irrigation, livestock, and aquaculture production, accounting for a total of 37% of total water use in the United States (3). Agriculture accounts for approximately 33% of total water use in Europe, and water use is more intensive in the southern parts of Europe, where 80% of total water consumption is for irrigation of crops (4). According to the Organization for Economic Cooperation and Development, there will be a 55% increase in the demand for water by the year 2050 due to increases in

[1]Department of Animal and Food Sciences, University of Delaware, Newark, DE 19716; [2]Food and Drug Administration, Center for Food Safety and Applied Nutrition, College Park, MD 20740; [3]Environmental Microbial Food Safety Laboratory, USDA-Agricultural Research Service, Beltsville, MD 20705.

Preharvest Food Safety
Edited by Siddhartha Thakur and Kalmia E. Kniel

doi:10.1128/microbiolspec.PFS-0014-2016

manufacturing, thermal power, and domestic industries that will put water availability for agriculture at risk (5).

Irrigation Water and Contamination

It is estimated that approximately 70% of the world's freshwater withdrawals are used for irrigation of crops, which is three times the amount of water used for agriculture 50 years ago (6). By the year 2050, it is expected that the amount of water used for irrigation will increase by another 19% (6). Therefore, sustainability and water conservation in agriculture have become equally important to water quality and management. Novel approaches to water management are needed to help conserve the declining water land base as well as balance production needs (7). The goals of these strategies are to improve water quality, to allow for conservation of water and energy, and to decrease growers' costs (8).

Irrigation water is derived from extensive supply systems that utilize both surface water and ground water (9). Surface water comes from rivers, streams, irrigation ditches, open canals, ponds, reservoirs, and lakes. Groundwater is collected from wells, and rainwater can be collected from cisterns and rain barrels. Municipal city and rural water can also be used for agriculture purposes but is rarely used due to high costs. Rural water includes well water as well as water from springs and from collected groundwater as mentioned above (10). Irrigation water can become contaminated through a variety of sources including polluted runoff from industrial sites, animal farms, barnyards, and feedlots (8). Application of irrigation water containing human pathogens can lead to the contamination of crops and subsequent outbreaks of human illness (11, 12). According to the U.S. Environmental Protection Agency (EPA), nonpoint source pollution comes from many diffuse sources (2). This type of pollution, unlike pollution from industrial and sewage treatment plants, is caused by rainfall or snowmelt moving over and through the ground (2). This water runoff picks up and carries away natural and human-made pollutants, ultimately depositing them into lakes, rivers, wetlands, coastal waters, and ground waters (2). The National Water Quality Inventory stated that agricultural nonpoint source pollution is the leading cause of river and stream pollution and the second leading cause of pollution in lakes, ponds, and reservoirs (13). Nonpoint source pollution, specifically from agriculture, includes poorly managed animal feeding operations, overgrazing, overworked land, and poorly managed application of pesticides and fertilizer (8). Nutrient management regulations and guidelines are constantly assessed at the state level in the United States to better address water pollution issues associated with land-applied nutrients.

Livestock Water and Contamination

Livestock water use is defined as water use associated with livestock watering, feedlots, dairy operations and other on-farm necessities. Total freshwater withdrawals for livestock rearing account for only 1% of the total water use in the United States (3). This type of water is used for raising cows, chickens, pigs, goats, horses, rabbits, fish (aquaculture), and pets as well as for the production of other food-producing animals for meats, eggs, and milk. Aside from animal watering, other livestock water uses include cooling of facilities for animals, cooling of facilities for animal products including milk and eggs, dairy sanitation and washing of facilities, animal waste disposal systems, and incidental waste losses (1). Another major concern for agricultural water pollution is animal feeding operations, which are livestock operations that keep animals confined for at least 45 days in a 12-month period with no grass or other vegetation available in the area during the normal growing period (14). According to the EPA, there are currently 450,000 animal feeding operations in the United States (15). Pollutants including organic matter, pathogens, solids, pesticides, hormones antibiotics, trace elements, and heavy metals from animal feeding operations have been found to be harmful to human and animal life populations (14).

Livestock water that is contaminated with animal or human pathogens can rapidly spread to animals, especially if they are drinking from a common water source (16). Pathogenic and zoonotic microorganisms can be released in the millions by infected animals into the soil, which can also contaminate other water sources (16).

REGULATING WATER

The Food Safety Modernization Act, signed into law in the United States in 2011, is considered the most significant piece of legislation addressing food safety in over 70 years. One of its main provisions in protecting consumer health from contaminated foods is the Produce Safety Rule (PSR) properly titled Standards for the Growing, Harvesting, Packing, and Holding of Produce for Human Consumption, which mandates the FDA to develop science-based, minimum standards for the safe growing, harvesting, packing, and holding of produce on farms to minimize contamination that could cause serious adverse health consequences or death. Much of the PSR addresses the regulation of agricultural water which is applied to produce. However, it should be noted that preharvest contamination of produce with foodborne pathogens is difficult to attribute to a single source since risks come from multiple sources, including contaminated agricultural water, soil amendments, wildlife, and workers.

According to the Food Safety Modernization Act, the proposed microbial standards for agricultural water require that water sourced from a nonpublic water system or supply be tested. If this water is likely to have contact with the harvestable or edible portion of produce (other than sprouts) or food contact surfaces, it must meet a specific set of microbial standards (17). These standards require that the geometric mean of the tested sample is not to exceed 126 CFU of generic *Escherichia coli* per 100 ml, and the estimate of the statistical threshold value of samples must not exceed 410 CFU of generic *E. coli* in 100 ml of water (17). If the water tested does not meet these standards, it may still be used if the producer is able to meet the microbial standard using a calculated die-off or removal rate. In this case the producer must ensure that the time between the last irrigation and harvest will allow for adequate die-off of bacteria using supporting scientific data. The producer may also choose to treat the water or establish an alternative that will allow for die-off at a rate of 0.5 log/day of bacteria prior to harvest (17). Water that is sourced from a nonpublic source and is intended to be used for agricultural teas, sprout irrigation water, postharvest washing, or handwashing must meet a more stringent standard: no detectable generic *E. coli* per 100 ml.

MITIGATION STRATEGIES FOR AGRICULTURAL WATER

Chemicals

Common chemicals used to treat agricultural water include bromine, chlorine gas (Cl_2), sodium hypochlorite (NaOCl), calcium hypochlorite ($Ca\ (OCl)_2$), chlorine dioxide (ClO_2), and activated peroxygen (H_2O_2 or CH_3COOOH). The majority of these chemicals come in a granulated, liquid, or gaseous form. Granules and liquids can be added directly to containers with water or eroded through a feeder. The mode of action of some of these oxidative biocides, such as chlorine and hydrogen peroxide, includes the removal of electrons from a susceptible chemical group, therefore altering the chemical structure of macromolecules (18). The specific action of the biocide can depend on its state (gas versus liquid) or formulation (mixture with other chemicals) as well as the target organism. Prokaryotic organisms have been found to be resistant to lower levels of these chemicals, in which case higher rates of exposure time may be needed (18, 19).

Ozone

The FDA approved ozone as a GRAS (generally recognized as safe) substance for use in bottled water in 1982 (20). Ozone, a triatomic oxygen (O_3) molecule, is a naturally occurring form of oxygen that exists as a bluish gas with a pungent odor. In nature, ozone is formed via ultraviolet irradiation from the sun and via lightening discharge; commercially, the corona discharge method is used to generate ozone (21). Ozone is partially soluble in water and is effective at killing a range of microorganisms through the oxidation of cellular membranes of vegetative cells (22). It has an oxidizing potential of 2.07 mV, which is considerably higher than that of other oxidants (21). Ozone decomposes in solution to produce hydroperoxyl (·HO2), hydroxyl (·OH), and superoxide (·O2) radicals (23). The reactivity of ozone inactivates bacteria, fungi and their spores, viruses, and protozoa to the generation of these free radical chemical species (24).

Ozone is a reactive molecule that has the ability to inactivate a broad range of microorganisms; it also reacts with nearly all organic and inorganic compounds (21). Therefore, the higher the amounts of organic matter present in water, the lower the effectiveness of ozone. Exposure of humans and animals to high levels of ozone can have detrimental effects on health, which causes concern for workers in processing plants.

Ultraviolet Radiation

Application of ultraviolet (UV) light was first used in France for disinfection of drinking water (25). UV light uses an electromagnetic spectrum from 100 to 400 nm and can be classified into four spectrum regions: UV-A (315 to 400 nm), UV-B (280 to 315 nm), UV-C (200 to 280 nm), and vacuum UV (100 to 200 nm) (26). The UV-C class is considered to be the most effective for microbial inactivation (26). The most common types of UV lamps used for water disinfection are low-pressure UV-C mercury vapor lamps with a wavelength of 254 nm (19). For UV radiation treatment, water is exposed to high doses of UV light within tubular chambers. The mechanism by which UV light inactivates microorganisms is by damaging the organism's DNA (27).

One of the disadvantages of UV light technology is the limited amount of transmittance through turbid water (26, 28). To help overcome this, UV treatment chambers have been designed so that water flows through in a thin layer (28). UV technology has become more attractive for water treatment because of concerns regarding chemical residues found with water chlorination (29). There are only a few large-scale UV water treatment plants in the United States, although there are over 2,000 located throughout Europe (29). More studies need to be performed for better optimization of UV technology for disinfection of water for multiple uses (irrigation, drinking) as well as education to increase consumer acceptance.

Copper and Silver Ionization

With the use of copper, and sometimes the addition of silver, as active ingredients, an electrical charge is passed between bars or plates made of copper and/or silver, releasing positively charged ions into the water (19). The composition of the copper:silver ratio of electrodes within the system can be customized, although it is typically set to a 70:30 copper:silver ratio (30). Copper-silver ionization uses positively charged cupric copper and silver ions, which have disinfection capabilities (31, 32). Within the system, an electric current is applied across the copper and silver electrodes into water. One of the disadvantages of this technology is that the electrodes are gradually consumed and eventually need to be replaced (30). It is recommended that a system be chosen in which copper concentrations do not exceed governmental drinking water standards and plant toxicity levels (19).

Sand Filtration

When a sand bed is created, the system supports a biologically active layer, known as a "biofilm crust" on the filter surface (19). Water can filter passively through the sand bed to remove pathogens, and clean water can be stored in a reservoir until use. Laboratory studies have shown that slow sand filters with a mature biolayer can achieve a 99.98% protozoan reduction, 90 to 99% bacterial reduction, and variable viral reduction in water (33). Pathogens are removed through physiological and biological means. Advantages of this technology include a high flow rate, simplicity of use and consumer acceptability, low maintenance requirements, and long-lasting equipment (33). Disadvantages include the lack of residual protection (such as with chlorination), harm done to the biolayer and decreased efficacy caused by routine cleaning, and the difficulty of transporting the equipment due to weight and expense (33). The size of the sand bed affects the volume of water that can be filtered at one time. Space limitations may cause problems during periods of peak irrigation demand because filtration may not be rapid enough to meet supply needs (19).

Zero-Valent Iron (ZVI) Filtration

Incorporation of ZVI into biosand filters was first shown to be effective at removing a broad range of chemical contaminants in groundwater (34). ZVI reacts with dissolved oxygen and protons to form amorphous iron hydroxides that are subsequently converted into more stable oxides and oxyhydroxides (35, 36). Iron hydroxides, oxides, and oxyhydroxides have a relatively high pH, enabling them to adsorb viruses as well as other negatively charged microorganisms through electrostatic interactions (37). ZVI is a new technology that is not yet widely utilized; research shows that ZVI filters could provide a sustainable and cost-effective way for farmers to treat irrigation water and reduce the number of illnesses spread through contaminated irrigation water.

WATER REUSE

The use of reclaimed water can provide a sustainable and safe option for agricultural water. Treated wastewater and gray water can be recycled for agricultural and landscape irrigation, industrial processes, toilet flushing, and replenishing ground water basins (38). In general, the use of recycled water can help to satisfy water demands, preserve resources, and save energy and money. By providing an additional source of water, water recycling can also help to decrease diversion of fresh water from sensitive ecosystems, decrease discharge of pollutants to sensitive water bodies, enhance wetlands and riparian habitats, and prevent pollution (38). In the United States, approximately 7 to 8% of all municipal wastewater is reused for other purposes (38) (Fig. 1). The top three uses of municipal reclaimed water are agriculture irrigation (29%), other (20%), and landscape/golf course irrigation (18%) (Fig. 2).

While water recycling technology is becoming increasingly popular for communities and businesses to help meet water supply demands, installation of distribution systems can be initially expensive compared to the alternatives, which include importing water, using ground water, and using residential gray water (38). Recycled water may also prove to be a great reliable and safe source of irrigation water. It is important to note that the EPA standards for recycled wastewater are recommendations and do not comprise a national standard; these standards are left up to the individual states to consider. However, no U.S. state has regulations allowing recycled water to be used for irrigation of a crop that is consumed raw.

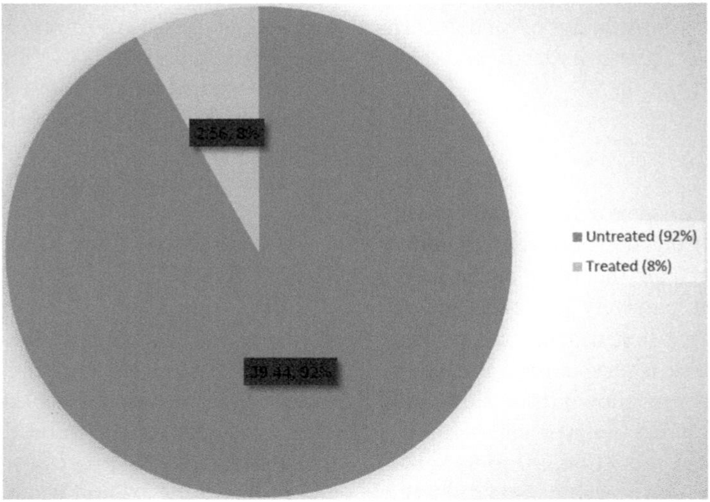

FIGURE 1 Estimated municipal effluent generated per day in the United States (billions of gallons/day). Adapted from the U.S. EPA 2012 Guidelines for Water Reuse (61).

Water Reuse Terminology

The terminology regarding water reuse can be confusing because it varies by location. Essentially, the terms "water reuse" and "water recycling" have the same meaning, although some states and countries use one, while some use the other (39). Terminology may also depend on the type and site-specific type of water reuse (Table 1).

Agricultural Water Reuse

The categories of water reuse applications according to the 2012 EPA Guidelines for Water Reuse include urban reuse, agricultural reuse, impoundments, environmental reuse, industrial reuse, groundwater recharge, groundwater recharge-nonpotable use, and potable reuse (61). The agricultural reuse category is broken down into the subcategories: water

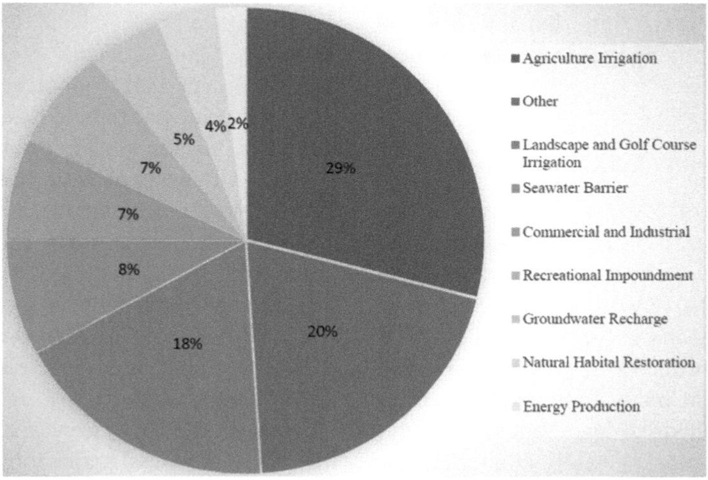

FIGURE 2 Reuse summaries of reclaimed water use in the United States. Adapted from the U.S. EPA 2012 Guidelines for Water Reuse (61).

TABLE 1 Types of water reuse and water reuse terminology[a]

Type of water reuse	Definition
De facto	A situation where reuse of water is being practiced but not officially recognized by the operation/organization*
Direct potable reuse	When reclaimed water is introduced directly into a drinking water treatment plant, either collocated or remote from the advanced wastewater treatment system*
Indirect potable reuse	Augmentation of a drinking water source (either surface or groundwater) with reclaimed water followed by an environmental buffer that precedes drinking water treatment*
Nonpotable reuse	All water reuse applications that do not involve potable reuse*
Potable reuse	Planned augmentation of a drinking water supply with reclaimed water*

*EPA 2012 Guidelines for Water Reuse. EPA/600/R-12/618. September 2012 (61).
[a]Adapted from the U.S. EPA 2012 Guidelines for Water Reuse (61).

reuse for food crops, and water reuse for processed food crops and nonfood crops. The first subcategory, agricultural reuse for food crops, is described as the use of reclaimed water to irrigate food crops that are intended for human consumption. The second, agricultural reuse for processed food crops and nonfood crops, is described as the use of reclaimed water to irrigate crops that are either processed before human consumption or are not consumed by humans.

Agricultural water reuse water quality standards differ depending on the geographical region in which the practice takes place and the governmental body that governs the region/organization (Fig. 3). Many of these regional and governmental guidelines have been adapted from the Food and Agricultural Organization's (40) water quality guidelines for irrigation. These guidelines provide measurable limits for reclaimed water, including salinity (electrical conductivity and total dissolved solids), infiltration rate, specific ion toxicity (sodium, chloride, and boron), pH, and levels of nitrate and bicarbonate.

Regulations are also different depending on the type of agricultural water reuse. In Monterey County, California, many types of crops including lettuce, broccoli, cauliflower, fennel, celery, strawberries, and artichokes have been irrigated with recycled water for over 10 years (39). This intensive 10-year pilot study was conducted to determine the safety of recycled water for use on raw agricultural commodities as well as the safety for the farmer and the environment (41). During the course of this study, researchers reported that no virus was ever detected on vegetables irrigated with reclaimed water, and levels of naturally occurring bacteria on samples from crops irrigated with recycled water were equivalent to those found on well-water-irrigated crops. In addition, there was no tendency found for metals to accumulate in soils or plant tissues. This study has led to the development of other reclaimed water use projects examining raw agricultural commodities in other states and other countries including South America, the Middle East, and Southeast Asia (39). While these studies have demonstrated the long-term safety and sustainability of recycled water for irrigating food crops, consumer acceptance remains an issue. Standards for water reuse for irrigation of processed and nonfood crops are far less complicated and much more widely accepted. While most countries use the World Health Organization (WHO) guidelines (42) for this type of water, the United States has adapted guidelines that are far more stringent than the WHO guidelines. A comparison of these guidelines by state is outlined in Table 2.

In the United States all raw wastewater is required to have a primary (mechanical) and secondary (biological) treatment (39). Primary mechanical treatment is designed to remove grass and suspended and floating solids using screening (39). Secondary biological treatment is an activated sludge process that helps to remove dissolved organic material not removed in the primary treatment process. The secondary biological treatment

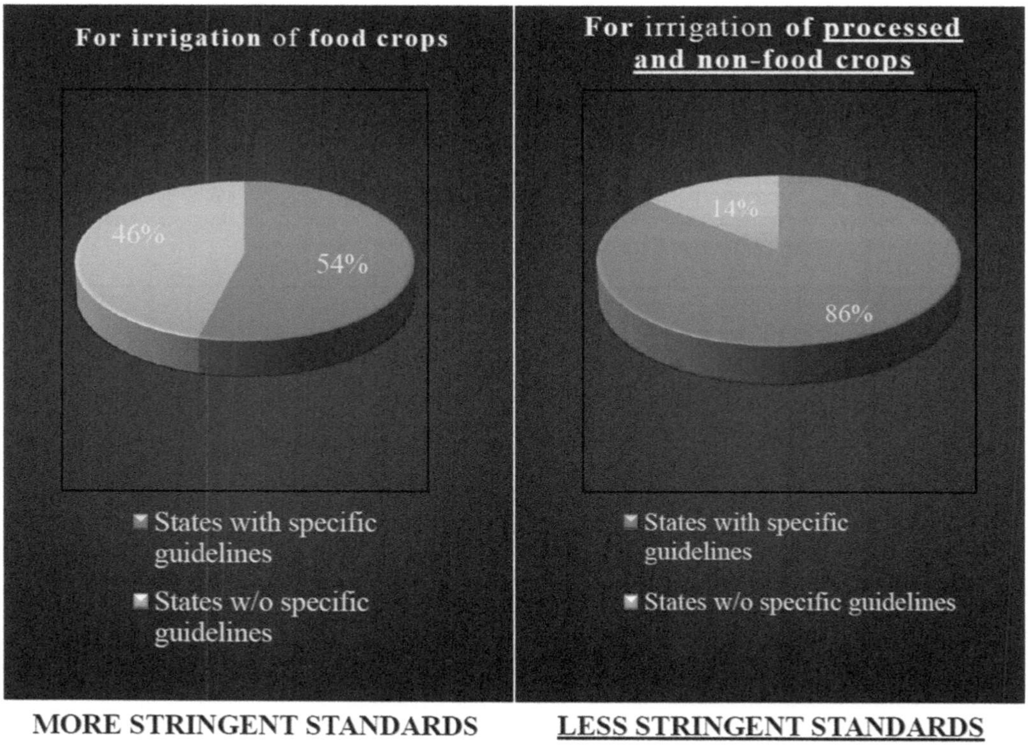

FIGURE 3 **Percentage of U.S. states with specific regulations on agricultural reuse of water. Adapted from the U.S. EPA 2012 Guidelines for Water Reuse (61).**

is achieved using bacteria and protozoa to break down the leftover organic material as well as incorporate nitrogen removal (39). This secondary treatment should also achieve removal of pathogens. Florida, Utah, and Texas use the EPA guidelines for wastewater reuse for nonfood crop irrigation (a limit of <200 fecal coliforms or generic *E. coli* per 100 ml). Arizona and New Mexico allow up to 1,000 fecal coliforms or generic *E. coli* per 100 ml. The most stringent standards belong to Puglia (southern Italy), Australia, and Germany, which require <10 fecal coliforms or generic *E. coli* per 100 ml. The least stringent standards belong to Greece and Spain, which follow the WHO standards allowing up to 10,000 fecal coliforms or generic *E. coli* per 100 ml (42).

ILLUSTRATED CASES

Foodborne Outbreaks Associated with Fruits and Vegetables Through Contact with Surface or Irrigation Water

Several outbreaks associated with the consumption of contaminated fruit and vegetable crops have been associated with irrigation or surface water containing bacterial pathogens used for irrigation or washing. While the cause cannot always be clearly defined, the cases below outline the potential risk associated with agricultural water.

In 2005, lettuce contaminated with *E. coli* O157:H7 sickened 135 individuals in Sweden (43), 11 of whom developed hemolytic uremic syndrome. Pulsed field gel electrophoresis

TABLE 2 State-by-state microbial standards for irrigation of nonfood crops with reused water[a]

State/organization	Total coliform/ 100 ml	Fecal coliform or *E. coli*/100 ml
California	≤23	
Washington	≤240	
Florida		≤200
Utah		≤200
Texas		≤200
Arizona		≤1,000
New Mexico		≤1,000
EPA		≤200
WHO		≤10,000

[a]Adapted from the U.S. EPA 2012 Guidelines for Water Reuse (61).

patterns of the outbreak isolate of *E. coli* O157:H7 from affected patients matched an isolate from cattle feces from a farm located upriver of the farm where the lettuce was grown and harvested. The farm was known to irrigate lettuce with the river water, which was of poor microbiological quality (*E. coli* populations fluctuated between 150/100 ml and 24,000/100 ml during the period of the outbreak investigation). Lettuce at this farm was irrigated with river water on the day before and the day of harvest (43). Although no specific pulsed field gel electrophoresis matches for the outbreak *E. coli* O157:H7 strain were found on the lettuce or in the water, PCR analyses indicated that several environmental samples, including one taken from the intake pipe of the irrigation hose that the farm used on the lettuce, tested positive for the same allele of the *stx2* gene that the outbreak strain contained. The combination of irrigation on the day of harvest and the potential for fecal matter containing the outbreak strain of *E. coli* O157:H7 upstream of the farm strongly implicated the river water used for irrigation in this outbreak.

In a different outbreak, in 2006, 80 *E. coli* O157:H7 infections were associated with the consumption of shredded iceberg lettuce originating from California (11). Fifty-three patients required hospitalization and eight patients developed hemolytic uremic syndrome (44). In the trace-back investigation of the outbreak, 32 of 251 samples (13%) tested positive for *E. coli* O157:H7, and 10 of these 32 (32%) samples contained the specific *E. coli* O157:H7 pulsed field gel electrophoresis pattern as the outbreak strain. Six of these ten positive samples were collected in close proximity to the suspected lettuce field where the contaminated product was harvested and included water, water mixed with sediment, soil, and equipment surface samples that were positive for the outbreak isolate of *E. coli* O157:H7. The other four samples positive for *E. coli* O157:H7 originated from dairies on the same ranch as the lettuce field (11). A blend of 20% dairy wastewater and 80% irrigation water, supplied by the local water district, was used to irrigate nonfood and forage crops grown on the ranch. However, irrigation pipes for the lettuce field and pipes for the dairy manure conveyance system appeared to be combined in several places and were not equipped with sufficient equipment to prevent backflow, which may have provided opportunities for cross-contamination between district water and dairy wastewater. The lack of backflow devices may have allowed for unplanned pressure differentials in the water lines to occur, which allowed *E. coli* O157:H7 present in the dairy manure effluent to be mixed with the irrigation water, which subsequently contaminated the growing lettuce in the field (11).

Surface water or water not containing sufficient concentrations of disinfectants has also been implicated in contaminating fruits after harvest. Mangoes and papayas contaminated with *Salmonella* spp. have been implicated in several foodborne outbreaks. The largest outbreak associated with mangoes occurred in the United States in 1999, when mangoes contaminated with *Salmonella enterica* subspecies *enterica* serovar Newport were responsible for sickening 78 people, 15 of whom required hospitalization and 2 of whom died (45). The mangoes were traced back to a Brazilian farm where unchlorinated water, used to cool the fruit after they were dipped in hot water, may have been contaminated with *S.* Newport. Hot water dips (46.1°C for 65 to 90 min) are recommended by the U.S. Department of Agriculture (USDA) Animal Plant Health Inspection Service to kill the larvae of the Mediterranean fruit fly present in mangoes (46). After this treatment, mangoes are typically cooled quickly by immersion in colder water (temperature not exceeding 22°C) to decrease quality losses. Immersing a warm piece of fruit into a colder liquid allows gases inside the fruit to contract, creating a partial vacuum inside the fruit and drawing fluid into the pores of the fruit (47). If the cooling water has not been appropriately treated with a disinfectant that can kill pathogens present in the water, pathogens such as *Salmonella* spp. may be drawn into the fruit and contaminate the internal tissues (46). The lack of sufficient chlorination in the cooling tank used for these mangos may have allowed the mangoes to internalize *S.* Newport. *S. enterica* subspecies *enterica* serovar Saintpaul was identified in an outbreak in 2001 that was associated with contaminated mangoes (48) and sickened 26 individuals in the United States. At least one contaminated mango was traced back to Peru, where again, after immersion in hot water to kill fruit fly larvae, the fruits were cooled in unchlorinated water contaminated with *S.* Saintpaul (48).

In 2006 and 2007, 27 people in the Australian states of Queensland and Western Australia became ill after eating papayas contaminated with *Salmonella enterica* subsp. *enterica* serovar Litchfield (12). Papayas tested from the grocery stores where the fruit was purchased were positive for *S.* Litchfield. In this case, river water was used to rinse the fruits in fungicide and in a 5-ppm chlorine (hypochlorite) solution. The 5-ppm level of chlorine is an insufficient concentration to eliminate the risk of pathogen transfer from water to fruit. Although the river water did not test positive for *S.* Litchfield, it contained other serovars of *Salmonella enterica* subsp. *enterica* (*S.* Chester, *S.* Eastborne, and *S.* Poona) and therefore was thought to be the source of contamination of the papayas (12). The lack of a definitive source of the pathogen from water sources illustrates the challenges of epidemiological investigations attempting to detect the presence of enteric pathogens in water sources.

Presence of Bacterial Pathogens in Surface Waters

Multiple surveys in the United States have shown that surface waters that are potentially used for irrigation contain bacterial pathogens. In a 2-year survey (2010 to 2011) on the eastern shore of Virginia, *Salmonella* spp. were prevalent in 25% (23/91) of stream water, pond water, and sediment samples examined (49). Interestingly, 0% (0/60) of irrigation water and irrigation water filter samples were positive for the presence of *Salmonella* spp. in this 2-year survey. Other investigators examining pond water, pond sediment, and irrigation ditch waters found that 14.3% (1/7) of pond water samples, 9.5% (2/21) of pond sediment samples, and 15.4% (2/13) of irrigation ditch water samples from tomato farms in the mid-Atlantic region of the United States between 2009 and 2010 were positive for *Salmonella* spp. (50). Of samples collected from on-farm or proximate (within 50 m) locations to a farm and likely used for irrigation in New York State, 12% (4/33) of samples were positive for *Salmonella* spp. (51). Lower prevalence of *Salmonella* spp. in water and sediment samples from the Central California coast was reported in a 2-year survey conducted between 2008 and 2010. These investigators determined that 6% (6/96) and 4.3% (3/67) of water and sediment samples, respectively, were positive for the presence of *Salmonella* spp. (52). A similar study conducted in Monterey County, California, showed that 7.1% of surface water samples were positive for *Salmonella* (53). These two studies from California were performed by taking grab samples (290 ml) of water. However, investigators using Moore swabs (filters made from cotton gauze) that were placed in water for up to 24 h found a higher prevalence (65%,

908/1,405) of *Salmonella* in samples taken from Central California waterways between 2011 and 2013 (54). *Salmonella* spp. were detected in 79% (57/72) of surface waters taken from the Suwanee river basin in a coastal plan in Georgia (55). These authors also determined that *Salmonella* populations fluctuate temporally and are strongly influenced by seasonal precipitation and temperature of the water. In North Carolina, 42% (5/12), 58% (20/34), 57% (16/28), and 50% (6/12) of surface waters near swine production, residential/industrial areas, forested areas, and crop agriculture, respectively, were positive for *Salmonella* spp. Overall, 55% (47/84) of water samples taken from this study contained some *Salmonella* spp. (56), with *Salmonella* isolates recovered near swine production facilities containing the most multidrug-resistant isolates.

Surveys of Shiga-toxigenic *E. coli* (STEC) and *Listeria monocytogenes* have also been conducted in surface waters. *E. coli* O157:H7 was found in 11% (110/1,386) of samples from Central California waterways, and their prevalence fluctuated by season, while non-O157 STEC was found in 11% of samples, with little seasonal variation (54). STEC was found in only 3% (1/33) of samples taken from New York state water sources (51). The presence of *L. monocytogenes* has been evaluated in several watersheds in the United States. Of the samples taken from Central California waterways, 41% (576/1,405) were positive for *L. monocytogenes*, with more positive samples detected during winter months than summer or spring months (54). In New York state, 30% (10/33) of water samples were positive for *L. monocytogenes* (51).

E. coli is commonly used as an indicator of water quality because of its utility in predicting gastrointestinal illness in fresh waters (39), but it is less reliable as an indicator of the presence of other bacterial pathogens. Surface waters examined from Georgia showed that when the *E. coli* populations exceeded 576 CFU/100 ml, salmonellae were detected 100% of the time (10/10 times) (55). However, when *E. coli* populations were less than 576 CFU/100 ml, *Salmonella* was still detected in 76% (42/55) of water samples, indicating that *E. coli* populations may not be adequate indicators for the presence of *Salmonella* in surface waters. Similarly, in a survey of agricultural waters in Central Florida, *E. coli* and *Salmonella* populations were poorly correlated using linear relationships, but logistic regression analysis provided some limited value in using *E. coli* populations in surface waters to predict *Salmonella* populations (57).

Presence of Enterohemorrhagic *E. coli* (EHEC) Virulence Factors in Surface and Harvested Rainwater

The use of specific virulence factors for STEC and EHEC have also been attempted to be correlated to *E. coli* populations in surface water and rainwater. In water samples taken from a watershed in Pennsylvania containing both forested and agricultural sites, there was no correlation between relative abundance of EHEC virulence factors and *E. coli* populations, with correlation coefficients of 0.09 and 0.02 for *eae* and *stx*, respectively (58). This work also demonstrated that there was no correlation with the high populations of *E. coli* recovered from surface waters with either *stx or eae*, potentially indicating that high *E. coli* populations may not be associated with the presence of STEC or EHEC. Roof-harvested rainwater and domestically harvested rainwater are used for various purposes (drinking, nonpotable uses). In a survey of 30 tanks containing roof-harvested rainwater in Southeast Queensland, Australia, 0% tested positive for the presence of *stx1* or *stx2*, but 30/200 (15%) were positive for the presence of the *eae* gene (59). In 16/30 (53%) tanks examined, the *E. coli* populations exceeded the standards specified for Australian drinking water (0 CFU *E. coli* /100 ml) (59). In a survey of 10 rainwater tanks from homes in Western Cape, South Africa, only 1 of 5 sampling trials revealed the presence of *stx* genes in domestically harvested rainwater, and in that one instance, 2/10 (20%) tanks were positive (60). The most commonly found *E. coli* virulence factors were from the enteroaggregative *E. coli* pathotype. Of the 92 *E. coli* isolates recovered from this study, 6 (6%) were identified as *E. coli* O157:H7 through 16S rRNA PCR identification techniques.

The presence of enteric bacterial pathogens in surface waters has led to outbreaks associated with fruit and vegetable commodities consumed raw. Numerous surveys have shown the presence of multiple bacterial pathogens in surface and harvested rainwater, and appropriate guidelines and, where necessary, mitigation techniques should be used to improve the quality of irrigation water or contact water for fruits and vegetables intended for human consumption.

SUMMARY

Agricultural water is a precious and limited resource. Increasingly more water types and sources are being explored for use in irrigation in the United States and across the globe. As outlined in this article, relatively new regulations and standards exist for application of water to fruits and vegetables consumed raw. These rules for production and use of water will continue to develop and be required as the world experiences aspects of a changing climate, including fierce storms resulting in flooding as well as drought conditions. Research continues to assess the use of agricultural water types. The increased use of reclaimed water in the United States as well as the growing population has influenced the USDA National Institute of Food and Agriculture to make funding availability for water reuse one of its key priorities. Agricultural water reuse has also been widely supported, overall, by regulatory and institutional policies. The advantages of water reuse for agricultural purposes include the generation of a highly reliable water source as the human population continues to grow, lower costs compared to use of potable water sources, accessibility of reclaimed water for irrigation of crops, and the ability of reclaimed water use to extend the availability of fresh water resources.

CITATION

Markland SM, Ingram D, Kniel KE, Sharma M. 2017. Water for agriculture: the convergence of sustainability and safety. Microbiol Spectrum 5(3):PFS-0014-2016.

REFERENCES

1. **United States Geological Survey (USGS).** 2016. Irrigation water use. http://water.usgs.gov/edu/wuir.html.
2. **EPA.** 2005. Protecting water quality from agricultural runoff. https://www.epa.gov/sites/production/files/2015-09/documents/ag_runoff_fact_sheet.pdf.
3. **Maupin MA, Kenny JF, Hutson SS, Lovelace JK, Barber NL, Linsey KS.** 2014. Estimated use of water in the United States in 2010. US Geol Surv Circ 1405. http://pubs.usgs.gov/circ/1405/.
4. **European Environment Agency.** 2012. *Towards efficient use of water resources in Europe.* EEA Report No. 1/2012. ISSN 1725-9177. Accessed at: http://www.eea.europa.eu/publications/towards-efficient-use-of-water/at_download/file.
5. **Organisation for Economic Co-operation and Development.** 2008. Environmental performance of agriculture in OECD countries since 1990. https://stats.oecd.org/Index.aspx?DataSetCode=ENVPERFINDIC_TAD_2008.
6. **Global Agriculture.** 2016. Agriculture at a crossroads: findings and recommendations for future farming: water. https://www.globalagriculture.org/report-topics/water.html.
7. **Evans RG, Sadler RG.** 2008. Methods and technologies to improve efficiency of water use. *Water Resour Res* **44:**1–15.
8. **CDC.** 2009. Agricultural water: what is agricultural water? https://www.cdc.gov/healthywater/other/agricultural/.
9. **Winter TC, Harvey JW, Franke OL, Alley WM.** 1998. Ground water and surface water: a single source. U.S. Geological Survey Circular 1139. https://pubs.usgs.gov/circ/circ1139/.
10. **CDC.** 2009. Chapter 8: rural water supplies and water-quality issues. *Healthy Housing Reference Manual.* https://www.cdc.gov/nceh/publications/books/housing/cha08.htm.
11. **California Food Emergency Response Team.** 2008. Investigation of the Taco John's *Esche-*

richia coli O157:H7 outbreak associated with iceberg lettuce. Final report. https://www.cdph.ca.gov/pubsforms/Documents/fdb%20eru%20IceLet%20TacoJohn022008.pdf.

12. **Gibbs R, Pingault N, Mazzucchelli T, O'Reilly L, MacKenzie B, Green J, Mogyorosy R, Stafford R, Bell R, Hiley L, Fullerton K, Van Buynder P.** 2009. An outbreak of *Salmonella enterica* serotype Litchfield infection in Australia linked to consumption of contaminated papaya. *J Food Prot* **72:**1094–1098.
13. **EPA.** 2002. National water quality inventory: report to Congress, 2002 reporting cyle. https://www.epa.gov/sites/production/files/2015/09/documents/2007_10_15_305b_2002report_report2002305b.pdf.
14. **CDC.** 2009. Animal feeding operations. https://www.cdc.gov/healthywater/other/agricultural/afo.html.
15. **EPA.** 2003. *Animal agriculture: concentrated animal feeding operations-livestock operation inspection.* Office of Enforcement and Compliance Assurance, Washington, DC.
16. **CDC.** 2009. Water contamination. https://www.cdc.gov/healthywater/other/agricultural/contamination.html.
17. **FDA.** 2016. FSMA final rule on produce safety: standards for the growing, harvesting, packing, and holding of produce for human consumption. http://www.fda.gov/Food/GuidanceRegulation/FSMA/ucm334114.htm.
18. **Finnegan M, Linley E, Denyer SP, McDonnell G, Simons C, Maillard JY.** 2010. Mode of action of hydrogen peroxide and other oxidizing agents: differences between liquid and gas forms. *J Antimicrob Chemother* **65:**2108–2115.
19. **Parke J, Fisher P.** 2012. *Treating irrigation water.* Oregon State University. http://c.ymcdn.com/sites/www.oan.org/resource/resmgr/imported/digger/Digger_201202_pp41-45_web.pdf.
20. **Guzel-Seydim ZB, Grene AK, Seydim AC.** 2004. Use of ozone in food industry. *LWT Food Sci Technol* **37:**453–460.
21. **Karaca H, Velioglu Y.** 2007. Ozone applications in fruit and vegetable processing. *Food Rev Int* **23:**91–106.
22. **Langlais B, Reckhow DA, Brink DR.** 1991. Practical applications of ozone: principle case study, p 133–316. *In* Langlais B, Reckhow DA, Brink DR (ed), *Ozone in Water Treatment.* Lewis Publishers, Boca Raton, FL.
23. **Hoigne J, Bader H.** 1975. Ozonation of water: role of hydroxyl radicals as oxidizing intermediates. *Science* **190:**782–784.
24. **Hirneisen KA, Black EP, Cascarino JL, Fino VR, Hoover DG, Kniel KE.** 2010. Viral inactivation in foods: a review of traditional and novel food processing technologies. *Compr Rev Food Sci Food Saf* **91:**3–20.
25. **Masschelein WJ.** 2002. *Ultraviolet Light in Water and Wastewater Sanitation* (ed for English, Rice RG). Lewis Publishers, Boca Raton, FL.
26. **Kekli NM, Krishnamurthy K, Demirco A.** 2012. Microbial decontamination of food by ultraviolet (UV) and pulsed UV light, p 344–369. *In* Demirci A, Ngadi MO (ed), *Microbial Decontamination in the Food Industry: Novel Methods and Applications.* Woodhead Publishing Series in Food Science, Technology and Nutrition. Woodhead Publishing, Cambridge, United Kingdom.
27. **Li X, Farid M.** 2016. A review on recent development in non-conventional Food sterilization technologies. *J Food Eng* **182:**33–45.
28. **Gomez-Lopez VM, Koutchma T, Linden K.** 2012. Ultraviolet and pulsed light processing of fluid foods, p 184–224. *In* Cullen PJ, Tiward BK, Valdramidis V (ed), *Novel Thermal and Non-Thermal Technologies for Fluid Foods.* Academic Press, Cambridge, MA.
29. **Oram B.** 2014. *UV disinfection drinking water: drinking water treatment with UV radiation.* Water Research Center. http://www.water-research.net/index.php/water-treatment/water-disinfection/uv-disinfection.
30. **Triantafyllidou S, Lytle D, Muhlen C, Swertfeger J.** 2016. Copper-silver ionization at a US hospital: interaction of treated drinking water with plumbing materials, aesthetics and other considerations. *Water Res* **102:**1–10.
31. **States S, Kuchta J, Young W, Conley L, Ge J, Costello M, Dowling J, Wadowsky R.** 1998. Controlling *Legionella* using copper-silver ionization. *J AWWA* **90:**563–568.
32. **Lin YS, Stout JE, Yu VL, Vidic RD.** 1998. Disinfection of water distribution systems for *Legionella. Semin Respir Infect* **13:**147–159.
33. **CDC.** 2012. Safe water system: slow sand filtration. http://www.cdc.gov/safewater/sand-filtration.html.
34. **Meggyes T, Simon G.** 2000. Removal of organic and inorganic pollutants from groundwater using permeable reactive barriers. Part 2. Engineering of permeable reactive barriers. *Land Contam Reclam* **8:**175–187.
35. **Odziemkowski MS, Schumacher TT, Gillham RW, Reardonk EJ.** 1998. Mechanism of oxide film formation on iron in simulating groundwater solutions: raman spectroscopic studies. *Corros Sci* **40:**371–389.
36. **Phillips DH, Gu B, Watson DB, Roh Y, Liang L, Lee SY.** 2000. Performance evaluation of a zero-valent iron reactive barrier: mineralogical

characteristics. *Environ Sci Technol* **34:**4169–4176.

37. **Ingram DT, Callahan MT, Ferguson S, Hoover DG, Chiu PC, Shelton DR, Millner PD, Camp MJ, Patel JR, Kniel KE, Sharma M.** 2012. Use of zero-valent iron biosand filters to reduce *Escherichia coli* O157:H12 in irrigation water applied to spinach plants in a field setting. *J Appl Microbiol* **112:**551–560.
38. **EPA.** 2016. Water recycling and reuse: the environmental benefits. https://www3.epa.gov/region9/water/recycling/.
39. **EPA.** 2012. Recreational water quality criteria. Office of Water 820-F-12-058. https://www.epa.gov/sites/production/files/2015-10/documents/rwqc2012.pdf.
40. **Food and Agriculture Organization for the United Nations.** 1985. *FAO irrigation and drainage paper. 29 Rev.1.* Food and Agriculture Organization of the United Nations, Rome, Italy.
41. **Sheikh B, Cort RP, Kirkpatrick WR, Jaques RS, Asano T.** 1990. Monterey wastewater reclamation study for agriculture. *Res J Water Pollut Control Fed* **62:**216–226.
42. **WHO.** 1989. *Health guidelines for the use of wastewater in agriculture and aquaculture.* World Health Organization Technical Report Series 778. WHO, Geneva, Switzerland.
43. **Söderström A, Osterberg P, Lindqvist A, Jönsson B, Lindberg A, Blide Ulander S, Welinder-Olsson C, Löfdahl S, Kaijser B, De Jong B, Kühlmann-Berenzon S, Boqvist S, Eriksson E, Szanto E, Andersson S, Allestam G, Hedenström I, Ledet Muller L, Andersson Y.** 2008. A large *Escherichia coli* O157 outbreak in Sweden associated with locally produced lettuce. *Foodborne Pathog Dis* **5:**339–349.
44. **CDC.** 2006. Multistate outbreak of *E. coli* O157:H7 infections linked to fresh spinach (final update). http://www.cdc.gov/ecoli/2006/spinach-10-2006.html.
45. **Sivapalasingam S, Barrett E, Kimura A, Van Duyne S, De Witt W, Ying M, Frisch A, Phan Q, Gould E, Shillam P, Reddy V, Cooper T, Hoekstra M, Higgins C, Sanders JP, Tauxe RV, Slutsker L.** 2003. A multistate outbreak of *Salmonella enterica* serotype Newport infection linked to mango consumption: impact of water-dip disinfestation technology. *Clin Infect Dis* **37:**1585–1590.
46. **Penteado AL, Eblen BS, Miller AJ.** 2004. Evidence of *Salmonella* internalization into fresh mangos during simulated postharvest insect disinfestation procedures. *J Food Prot* **67:**181–184.
47. **Solomon EB, Sharma M.** 2009. Microbial attachment and limitation of decontamination methodologies, p 21–45. *In* Sapers GM, Solomon EB, Matthews KR (ed), *The Produce Contamination Problem: Causes and Solutions.* Academic Press, New York, NY.
48. **Beatty ME, LaPorte TN, Phan Q, Van Duyne SV, Braden C.** 2004. A multistate outbreak of *Salmonella enterica* serotype Saintpaul infections linked to mango consumption: a recurrent theme. *Clin Infect Dis* **38:**1337–1338.
49. **Bell RL, Zheng J, Burrows E, Allard S, Wang CY, Keys CE, Melka DC, Strain E, Luo Y, Allard MW, Rideout S, Brown EW.** 2015. Ecological prevalence, genetic diversity, and epidemiological aspects of *Salmonella* isolated from tomato agricultural regions of the Virginia Eastern Shore. *Front Microbiol* **6:**415.
50. **Micallef SA, Rosenberg Goldstein RE, George A, Kleinfelter L, Boyer MS, McLaughlin CR, Estrin A, Ewing L, Jean-Gilles Beaubrun J, Hanes DE, Kothary MH, Tall BD, Razeq JH, Joseph SW, Sapkota AR.** 2012. Occurrence and antibiotic resistance of multiple *Salmonella* serotypes recovered from water, sediment and soil on mid-Atlantic tomato farms. *Environ Res* **114:**31–39.
51. **Weller D, Wiedmann M, Strawn LK.** 2015. Irrigation is significantly associated with an increased prevalence of *Listeria monocytogenes* in produce production environments in New York State. *J Food Prot* **78:**1132–1141.
52. **Benjamin L, Atwill ER, Jay-Russell M, Cooley M, Carychao D, Gorski L, Mandrell RE.** 2013. Occurrence of generic *Escherichia coli, E. coli* O157 and *Salmonella* spp. in water and sediment from leafy green produce farms and streams on the Central California coast. *Int J Food Microbiol* **165:**65–76.
53. **Gorski L, Parker CT, Liang A, Cooley MB, Jay-Russell MT, Gordus AG, Atwill ER, Mandrell RE.** 2011. Prevalence, distribution, and diversity of *Salmonella enterica* in a major produce region of California. *Appl Environ Microbiol* **77:**2734–2748.
54. **Cooley MB, Quiñones B, Oryang D, Mandrell RE, Gorski L.** 2014. Prevalence of Shiga toxin producing *Escherichia coli, Salmonella enterica,* and *Listeria monocytogenes* at public access watershed sites in a California Central Coast agricultural region. *Front Cell Infect Microbiol* **4:**30.
55. **Haley BJ, Cole DJ, Lipp EK.** 2009. Distribution, diversity, and seasonality of waterborne salmonellae in a rural watershed. *Appl Environ Microbiol* **75:**1248–1255.
56. **Patchanee P, Molla B, White N, Line DE, Gebreyes WA.** 2010. Tracking *Salmonella* contamination in various watersheds and

phenotypic and genotypic diversity. *Foodborne Pathog Dis* **7:**1113–1120.

57. **McEgan R, Mootian G, Goodridge LD, Schaffner DW, Danyluk MD.** 2013. Predicting *Salmonella* populations from biological, chemical, and physical indicators in Florida surface waters. *Appl Environ Microbiol* **79:**4094–4105.
58. **Shelton D, Karns JS, Coppock CR, Patel JR, Sharma M, Pachepsky YA.** 2011. Comparison of generic *E. coli* vs pathogenic *E. coli* virulence factors in an agricultural watershed: implications for irrigation water standards and leafy greens. *J Food Prot* **74:**18–23.
59. **Ahmed W, Hodgers L, Masters N, Sidhu JPS, Katouli M, Toze S.** 2011. Occurrence of intestinal and extraintestinal virulence genes in *Escherichia coli* isolates from rainwater tanks in Southeast Queensland, Australia. *Appl Environ Microbiol* **77:**7394–7400.
60. **Dobrowsky PH, van Deventer A, De Kwaadsteniet M, Ndlovu T, Khan S, Cloete TE, Khan W.** 2014. Prevalence of virulence genes associated with pathogenic *Escherichia coli* strains isolated from domestically harvested rainwater during low- and high-rainfall periods. *Appl Environ Microbiol* **80:**1633–1638.
61. **Bastian R, Murray D.** 2012. *Guidelines for Water Reuse.* U.S. EPA Office of Research and Development, Washington, DC, EPA/600/R-12/618. https://nepis.epa.gov/Adobe/PDF/P100FS7K.pdf.

Importance of Soil Amendments: Survival of Bacterial Pathogens in Manure and Compost Used as Organic Fertilizers

9

MANAN SHARMA[1] and RUSSELL REYNNELLS[2]

INTRODUCTION

Biological soil amendments (BSAs), including manure, compost, and compost teas (CTs), play an important role in conventional and organic agriculture. The use of these amendments can provide nutrients to soils, improving soil fertility and crop production. However, recent outbreaks of bacterial, viral, and parasitic infections associated with produce commodities over the past decade have focused more attention on agricultural inputs used to grow fresh fruits and vegetables. In response to these outbreaks, proposed rules from the U.S. Food and Drug Administration (FDA) have been issued, titled "Standards for the Growing, Harvesting, Packing and Holding of Produce for Human Consumption" Supplemental to the Proposed Rule (1). As part of these standards, the FDA has proposed specific rules and guidelines for how several BSAs can be applied to soils and fields intended to grow produce for human consumption.

There are many benefits to adding BSAs to soils, including improving the essential nutrient content of the soils over time, replenishing soil organic material, increasing bulk soil density, and enhancing soil water retention

[1]U.S. Department of Agriculture, Agricultural Research Service, Beltsville Area Research Center, Environmental Microbial and Food Safety Laboratory, Beltsville, MD 20705; [2]University of Maryland Eastern Shore, Department of Agriculture, Food, and Resource Science, Princess Anne, MD 21853.

Preharvest Food Safety
Edited by Siddhartha Thakur and Kalmia E. Kniel

doi:10.1128/microbiolspec.PFS-0010-2015

properties and soil structure (2, 3). The use of manure has been shown to improve the nutrient status of eroded soils, leading to higher yields of grain crops (2). Repeated compost application to soils has been shown to increase microbial biomass carbon and microbiological diversity in soils (4). While there are benefits to applying BSAs to soil, manure has been shown to be a reservoir for enteric pathogens, and there are numerous mechanisms for manure to transfer pathogens to growing fruits and vegetables to cause microbial contamination of these commodities. An outbreak of *Escherichia coli* O157:H7 infections in 2006 in the United States sickened over 70 people and was associated with lettuce thought to have been exposed to manure runoff containing the pathogen from an adjacent cattle ranch (5). In 2005, an outbreak of 135 *E. coli* O157:H7 infections was associated with the consumption of contaminated lettuce irrigated with manure-contaminated water right before harvest (6).

Manure is known to contain numerous pathogens, including *E. coli* O157:H7, *Salmonella* spp., *Listeria* spp., *Campylobacter* spp., and *Cryptosporidium parvum* (7). Therefore, the risk of potential microbial contamination to fruit and vegetable crops resulting from manure application to fields must be balanced with the benefits of applying BSAs to soils. BSAs which are applied to crops and commodities that are consumed raw and can come into contact with soil (through direct contact or soil-splash events) or that are minimally processed in a manner which does not eliminate 100% of pathogens should be of particular concern. The focus of this review is not fecal matter or scat that may be introduced to agricultural fields by feral animals in the agricultural production environment. Our work evaluates the risks solely from a food and produce safety perspective, but to be clear, other environmental perspectives (like those limiting manure runoff to watersheds from agricultural fields) also govern how BSAs are applied to agricultural fields to grow fruit and vegetable crops.

Over the past decade (2005 to 2015), several high-profile outbreaks related to leafy greens (5, 8), peppers (9), cantaloupes (10, 11), and cucumbers (12) have influenced the use of BSA in the cultivation and harvest of fruits and vegetables intended for human consumption. The burden of food-borne illness in the United States is increasingly associated with contaminated produce commodities (13). Table 1 summarizes guidelines (14), federal standards for organic production of fruits and vegetables (15), and current state laws (16) and proposed federal laws (1). While many of these requirements overlap, providing a limited amount of consistency in these guidelines and regulations related to food safety, they do not encompass all of the state-level rules that that are applicable to manure handling in agricultural environments, especially as they relate to several environmental and food safety issues.

PATHOGEN SURVIVAL IN MANURE-AMENDED SOILS

The persistence of bacterial and viral pathogens in raw animal manure may be based on the manure type, the method of application and/or incorporation into soils, the type of soil, how the manure is stored before application onto soils, and the microbial diversity present and nutrient ratios in the manure-amended soils (7, 17, 18). Persistence and survival of bacterial pathogens in manure-amended soils depends highly on geographical and environmental factors. Field conditions may be different than greenhouse or growth chamber conditions, and therefore the context of these results in the literature should be examined critically. Extensive descriptions of the benefits and limitations of each set of laboratory, growth chamber, and greenhouse setting (model systems) which could be used to evaluate the persistence of pathogens in manure are increasingly covered in the literature (19). Growth chamber or greenhouse settings which most closely model the fluctuations in

TABLE 1 Guidance and proposed rules by organizations and federal agencies in the United States on the application of biological soil amendments to agricultural fields intended to grow selected fruit and vegetable crops

Organization/program/federal agency	Manure	Compost	Compost and agricultural teas (CTs)	Reference
California Leafy Green Products Handler and Marketing Agreement	Prohibit the use of untreated manure on fields intended to grow leafy greens; must wait 1 year before planting	Active compost must achieve temperature of 131°F (55°C) for either 3 days (within vessel/aerated static pile) or 15 days/5 turnings (windrow composting); 45-day interval between application and harvest	CTs from animal manure composts not allowed on leafy green crops	14
USDA National Organic Program	Uncomposted animal manures applied at least 90 days prior to harvest of crops where edible portion does not come into contact with the soil; at 120 days prior to harvest where edible portion of plant contacts soil	Compost mixed or managed so that all feedstock heats to minimum temperature of 131°F for at least 3 days; must cure for 45 days before soil application	CT without additives can be applied without restriction if production system has been pretested to meet EPA water standards; CT with additives must adhere to 90/120-day interval before harvest (same as raw manure)	15
A Guide to Tomato Good Agricultural Practices (T-GAPS)	Untreated animal manure not allowed in tomato fields or greenhouses	Grower must have records of dates of composting, methods utilized, and application dates	None given	16
Standards for the Growing, Harvesting, Packing, and Holding of Produce for Human Consumption (FDA Supplemental Proposed Rule, 2014)	Does not take exception to the 90/120-day interval as stated in the National Organic Program	Composted soil amendment must achieve temperature of 131°F (55°C) for either 3 days (within vessel/aerated static pile) or 15 days/5 turnings (windrow composting); 0-day interval between application and harvest	Agricultural tea (CT with additive) considered as untreated manure and must be applied as such; if CT is prepared and used in <1 h, then it can be used without restriction	1

water potentials of soils, soil temperature, air temperature, relative humidity, and ultraviolet light exposure of field conditions would seem to be the most likely to predict the survival of bacterial pathogens in soils containing BSAs.

Several studies have described the prevalence of bacterial and parasitic pathogens in various types of manure. A survey of British livestock manures from dairy cattle, poultry, swine, and sheep showed that at least 30% of manures contained at least one enteric pathogen (20). Specifically, *E. coli* O157, *Salmonella* spp., and *Listeria* spp. were present in 12 to 21%, 8 to 18%, and 20 to 30%, respectively, of fresh cattle, pig, and sheep manures. Furthermore, manures in Great Britain (20) showed *C. parvum* and *Giardia intestinalis* in 5 to 29% and in 4 to 21%, respectively, of cattle, pig, poultry, and sheep wastes tested. Many factors can affect the survival of bacterial pathogens in animal manures. In a study examining the prevalence of enteric pathogens in British livestock wastes, seasonality affected the prevalence of *E. coli* O157 and *Listeria* spp. populations (7). The prevalence of *E. coli* O157 was statistically significantly higher in May and June than in other months; similarly, *Listeria* spp. in animal manures increased in the warmer months (March to June) compared to its prevalence in other months; no seasonal differences in prevalence were observed for *Salmonella* populations (7).

Effect of Soil Type on Survival of Pathogens in Manure-Amended Soils

Examining the survival of pathogens in manure and feces is important, but of greater interest is the survival of enteric pathogens in manure once it has been applied and incorporated into soil as a fertilizer or as an organic amendment. The majority of studies examining the persistence of *E. coli*, EHEC, or *Salmonella* spp. in raw (untreated) animal manure in soils have been based on the use of the manure as an organic fertilizer. *E. coli* O157:H7 B6914 containing the *gfp* (green fluorescent protein) gene inoculated at a population 7 log CFU/gdw (gram dry weight) in 1-liter pots held at a static temperature of 16°C survived in manure-amended soils that were managed both organically and conventionally for 78 and 84 days in loamy and sandy soils, respectively (18). The initial rate of *E. coli* O157:H7 population decline was significantly greater ($P < 0.05$) in sandy soils than in loamy soils, but there was no statistical difference in the survival time of *E. coli* O157:H7 in the two different soils during the course of the experiment. Survival of *E. coli* O157:H7 in sandy soils was directly and positively related to the relationship with dissolved organic carbon/biomass carbon, while dissolved organic nitrogen and microbial diversity affected the survival of *E. coli* O157:H7 in loamy soils (18). Populations of *Salmonella enterica* serovar Typhimurium declined by 1 log CFU/g in 32.1 and 32.7 days in sandy and clay soils, respectively, amended with cattle manure slurry (21). *E. coli* O157:H7 populations declined by 1 log CFU/g in 29 and 37 days, in sandy and clay soils, respectively, amended with cattle slurry manure (21). This study was performed in lysimeters (0.295 in internal diameter, 1.0 m length) which were kept outdoors and exposed to an average temperature ranging between 15 and 25°C over the first 60 days of the experiment. Swine manure containing fecal coliform bacteria and inoculated with *S. enterica* serovar Anatum was applied to a field at either the recommended agronomic rate based on phosphorus levels (37,000 liter/ha), one-half that rate (18,000 liter/ha), or twice that rate (74,000 liter/ha), and was then disked into soil (a mixture of fine loamy soils). Populations of *S.* Anatum and fecal coliforms declined by >4 log most probable number (MPN) very quickly (within 7 days) in the sandy loam soil, perhaps in part due to the prevailing dry conditions during the execution of this field study and the low soil temperature (22). No statistical correlation was observed between the persistence of S. Anatum and the rate of application of swine manure to soils, perhaps due to the short survival times of the *S.* Anatum under these field conditions.

Other field experiments have shown a shorter survival time of *E. coli* O157:H7 in manure-amended soils. Replicate experiments were conducted to assess the survival of an attenuated (*stx*-negative) strain of *E. coli* O157:H7 (ATCC 700728) in western Canada (Summerland, British Columbia at a site containing sandy chernozemic soil with a semi-arid temperate climate) and eastern Canada (Kentville, Nova Scotia, at a site containing loamy podzolic soil with a maritime climate) with plots that received liquid dairy manure at the rate of 60 kg nitrogen/acre and planted with romaine lettuce (22). No significant difference in the decay rates of the populations of *E. coli* O157:H7 was observed between the two sites or in tilled or nontilled soils within each site. *E. coli* O157:H7 populations decreased by 2 log CFU/g in the first 7 days after inoculation (22), even though air temperature and rainfall differed at the two sites. The results from this study showed that the extrinsic factors of climate, soil type, and soil treatment did not affect population declines of the attenuated *E. coli* O157:H7 strain (22). The environmental fitness (ability to withstand physiological stresses the organism may encounter in manure-amended soils) of the nalidixic acid–resistant ATCC 700728 strain used in this study cannot be discounted to have influenced the rapid decline of populations, indicating that the response of this specific strain to environmental stresses may not be as robust as other strains of *E. coli.*

Other workers evaluating the survival of *E. coli* isolates in soils using a rifampicin-resistant ATCC 700728 strain in a multistrain inoculum (including *E. coli* O157:H7 strain ATCC 43888) showed that the *E. coli* O157:H7 inoculum survived for shorter periods (7 days) than a multistrain inoculum of nonpathogenic *E. coli* strains that survived for up to 21 days (23). The nonpathogenic *E. coli* isolates TVS 353, 354, and 355 were recovered from irrigation water, on lettuce, or from soil, respectively, and may be more adapted to these nonhost environmental conditions if they survived there for extended periods of time. This potential adaptation allowed a population of these cells to develop increased environmental fitness and may have led to their relatively extended survival duration in soils compared to the attenuated *E. coli* O157:H7 inoculum. Other studies using this same inoculum of TVS 353, 354, and 355 and ATCC 700728 and 43888 have shown results similar to Gutírrez-Rodríguez et al. (24): nonpathogenic *E. coli* populations surviving for longer durations and declining less rapidly in manure-amended soils compared to the *E. coli* O157:H7 populations (24, 25). These results may indicate that survival of nonpathogenic *E. coli*, compared to the survival of attenuated *E. coli* O157:H7, in manure-amended soils may be a better indication of the risk of contamination that BSAs present to growing fruit and vegetable crops. *E. coli* isolated from soil leachates in Ireland in lysimeters have shown the ability to grow to higher populations at 10°C, 15°C, and 37°C compared to laboratory strains of *E. coli* K-12 and *E. coli* ATCC 25922 (26). Higher growth at lower temperatures may indicate adaptation to specific conditions in soil.

Survival of *E. coli* in Manure-Amended Soils in Greenhouse Trials

Whyte et al. (25) showed that *E. coli* populations (both nonpathogenic and attenuated *E. coli* O157:H7) in silt loam soils amended with different animal manures—horse manure (HM), poultry litter (PL), or unamended—responded differently after irrigation events in a greenhouse environment. Amended and unamended soils contained in larger containers (89L) were irrigated during weeks 1, 2, 4, and 8, and *E. coli* populations were determined immediately before and then 24 h after the irrigation event. Table 2 illustrates the effect of manure type on change in *E. coli* populations (preirrigation populations subtracted from postirrigation populations) in manure-amended soils. Overall, *E. coli* population

TABLE 2 Change in inoculated *E. coli* populations in manure-amended soils in large pots in a greenhouse environment after weekly irrigation events

Week	Change in populations (log CFU/g dry wt)			
	Unamended	Horse manure	Poultry litter	Overall week
1	−0.14b[a]	0.25ab	0.62a	0.24x[b]
2	0.03a	0.20a	1.98a	0.74x
4	0.60a	0.60a	0.44a	0.37xy
8	−0.13ab	−1.11b	0.47a	−0.26y
Overall manure type	0.09b	−0.15b	0.88a	

[a]Within rows (Week and Overall manure type), if letters (a.b) following mean values (changes in populations) are different, it indicates a significant difference ($P < 0.05$) between manure types within each week or for the overall study.
[b]Within the column Overall week, if letters (x.y) following mean values (changes in populations) are different, it indicates a significant difference ($P < 0.05$) in average changes in populations between weeks of the study.

increases in PL-amended soils were significantly ($P < 0.05$) higher (0.88 log CFU/gdw) than increases in unamended soils (0.09 log CFU/gdw) or HM-amended soils (–0.15 log CFU/gdw). The changes in *E. coli* populations (without regard to manure type) were significantly greater in week 1 (0.24 log CFU/gdw) than in week 8 (–0.25 log CFU/gdw).

A possible explanation for these differences is that irrigation of PL-amended soils in week 1 may alleviate the physiological (desiccation) stress placed on *E. coli* populations in soils, and *E. coli* populations are still able to take advantage of the moisture and potential nutrients (like total nitrogen) available in PL-amended soils. Poultry litter and horse manure added to soils were determined to have 3800 and 2600 mg/kg of total nitrogen, respectively, while unamended soils contained 1260 mg/kg total nitrogen. Unamended soils may not have the same nutrient levels to support resuscitation. However, *E. coli* populations in week 8 have been exposed to greater desiccation stress, along with other extrinsic physiological stresses, over time in the manure-amended soils and may not be capable of using the dwindling moisture regime to increase their populations as much as in week 1. Moisture is an additional necessity for the survival and, especially, for the regrowth of pathogens. The effects of stresses such as excess moisture or drought, temperature, and nutrient availability in amended soils need to be examined closely to understand the physiological stress placed on enteric bacterial populations in manure-amended soils; however, these studies do seem to identify the critical extrinsic and abiotic factors which affect pathogenic bacterial survival in manure-amended soils.

Role of Stress Response Genes in *E. coli* in Survival in Manure-Amended Soils

The role of stress response genes in the survival of enteric pathogens in manure or manure-amended soils has not received much attention. The *rpoS* gene in *E. coli* controls a general stress response by producing a sigma factor (σ^{38}) which allows the RNA polymerase to transcribe some of the genes involved in this stress response. *E. coli* O157:H7 isolates from a longitudinal examination of cattle manure–amended sandy soil showed that those isolates that survived for the longest duration (>200 days) had fully functional *rpoS* genes, while those isolates that survived for shorter durations (<200 days) had mutations within the *rpoS* gene, making them dysfunctional (27, 28). Mutations within the *rpoS* gene of EHEC strains can affect the phenotypic stress response displayed by these isolates (29). Other studies have shown that the *rpoS* in a single strain of *E. coli* O157:H7 was expressed at a level about 2.7-fold higher in sterilized Fox (deep, well-drained soils) silt loam soil microcosms from eastern Pennsylvania contained in pots compared to in Luria Bertani broth when held at 15°C (30). *E. coli* cells under desiccation stress in sand have shown increased synthesis of the disaccharide trehalose, which can be used as an osmoprotectant in the cell by maintaining the phospholipid bilayer, acting as a water replacement, and keeping proteins in a hydrated form (31). *E. coli* isolates recovered from Hawaiian soils showed higher internal concentrations of trehalose and a slower decline of populations in sand microcosms compared to laboratory-cultured strains, showing that *E. coli* strains exhibiting increased trehalose synthesis may survive for extended periods in soils or manure-amended soils (31). Trehalose synthesis is controlled by the *otsA/otsB* genes in *E. coli*, whereby both genes are regulated within the *rpoS*-dependent response (32). These results show that the *rpoS*-mediated stress response most likely enhances the survival of the *E. coli* isolates in manure-amended soils, and the specific roles of *rpoS*-transcribed genes in desiccation stress tolerance aiding the survival of *E. coli* and *Salmonella* spp. in manure and manure-amended soils should be more closely examined.

Effect of Application Method and Geographic Location on Survival of *E. coli* in Manure-Amended Soils

The method of manure application and incorporation into soil adds to both intrinsic edaphic factors and extrinsic environmental factors influencing the *E. coli* populations in amended soils. Graham et al. (33) showed that population dynamics of inoculated, nonpathogenic *E. coli* (23) in unamended or poultry litter- or horse manure-amended silt loam soils in southeastern Pennsylvania were different if manure was surface-applied or tilled into soils. *E. coli* populations (5.4 log CFU/m^2) were spray applied to unamended soils or soil with PL or HM. For *E. coli* populations (Fig. 1) in plots containing surface-applied manure, an increase in air temperature without a corresponding increase in soil moisture allowed an increase in *E. coli* populations by 4 to 6 log CFU/gdw from days 1 to 5. However, *E. coli* populations in plots containing tilled-in manure increased with the increase in air temperature, but the increase in population was more pronounced when accompanied by an increase in soil moisture (Fig. 2).

Pathogen Survival in Biosolids-Amended Soils

Biosolids are nutrient-rich organic materials resulting from the treatment of domestic waste at wastewater treatment facilities. Studies have shown that increasing the agronomic rate of application of biosolids used on a field may not affect the persistence of the pathogen in manure-amended soils (33).

The decay of *E. coli* and *Salmonella* populations in both loamy and sandy soils amended with dewatered biosolids at two geographic locations in Australia was significantly influenced by soil moisture, temperature, and the application of biosolids (34). T_{90} values (time to achieve a 1-log reduction) for *E. coli* and *Salmonella* populations were shorter in biosolids-amended soils (4 to 56 days) than in unamended soils (8 to 83 days), indicating that biosolids application to soils shortened the survival times of bacterial pathogens. In this study the rate of biosolids application was 6 tons/ha and 28 tons/ha at the two sites, respectively. Conversely, studies conducted in the United Kingdom showed that the application of dewatered, anaerobically digested biosolids at a rate of 10 tons/ha to sandy soils accounted for 50 to 60% of the variance in *E. coli* population levels, while time, soil moisture, and soil temperature accounted for 14, 5.5, and 5.4% of the variance, respectively (35).

The effect of biosolids hastening the decline of bacterial pathogens in amended soils (34, 35), is not in agreement with the findings from previous studies that show manure increasing the duration of bacterial survival in amended soils. The results of both studies underscore the fact that soil moisture content contributes greatly to the decline of bacterial populations regardless of the presence of biosolids. Fecal coliforms and *Salmonella* spp. were able to grow from 6.3×10^4

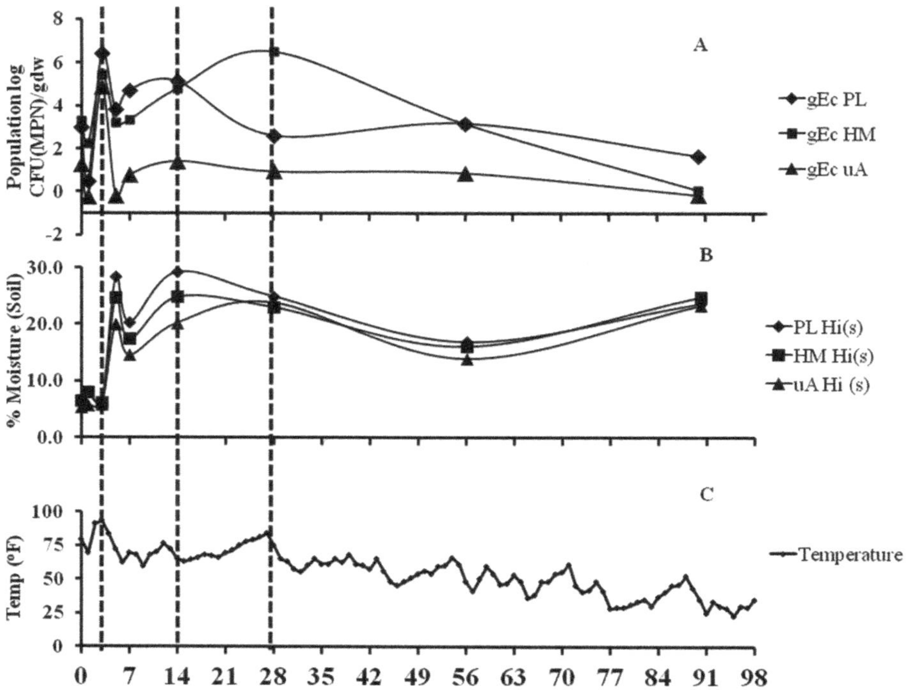

FIGURE 1 (A) The survival of a three-strain (TVS 353, 354, and 355), nonpathogenic (generic) *E. coli* inoculum in either poultry litter–amended (gEC PL), horse manure–amended (gEC HM), or unamended (gEC uA) soils over 98 days in southern Pennsylvania (Southeastern Agricultural Research and Extension Center, Pennsylvania State University, Landisville, PA). Amendments were surface applied (no till). (B) The moisture content of PL-amended soils [PL Hi(s)], HM-amended soils [HM Hi(c)], and uA soils [uA Hi (s)]. (C) The air temperature throughout the study. Values below the x-axis indicate the number of days of the study. Dashed lines indicate days when changes in *E. coli* populations occurred, potentially attributed to changes in either soil moisture content or temperature. The first dashed line indicates that an increase in air temperature (graph C) from days 0 to 5 corresponded to an increase in all *E. coli* populations in graph A; the second dashed line indicates that an increase in soil moisture content (graph B) from day 7 to day 14 corresponded to an increase in all *E. coli* populations; the third dashed line indicates that a decline in soil moisture content of PL Hi(s) from day 14 to day 28 resulted in a decline of *E. coli* populations in gEc PL. These graphs show that air temperature and soil moisture content can affect *E. coli* populations in soils amended with surface-applied manures more than *E. coli* populations in tilled in soils (Fig. 2).

and 0.09 MPN/g to 1.1 x 10^5 MPN/g and 0.7 MPN/g, respectively, in biosolids applied to soils after a rainfall event which increased the soil moisture content from 1 to 22% (36). These results show that soils amended with biosolids support the resuscitation of enteric bacterial populations after initial amendment of the soils, similar to soils amended with poultry litter (24), perhaps by improving the overall moisture-holding capacity of soils. Further, manure type and moisture content both seem to influence the resuscitation of *E. coli*, fecal coliforms, and *Salmonella* cells in amended soils. These disparate findings with regard to the effect of biosolids on the persistence of pathogens in soils support the conclusion that the specific composition of the soil amendment (type of manure or biosolid), soil type, and climate conditions specific to each geographic location affects the survival of enteric pathogens in amended soils differently.

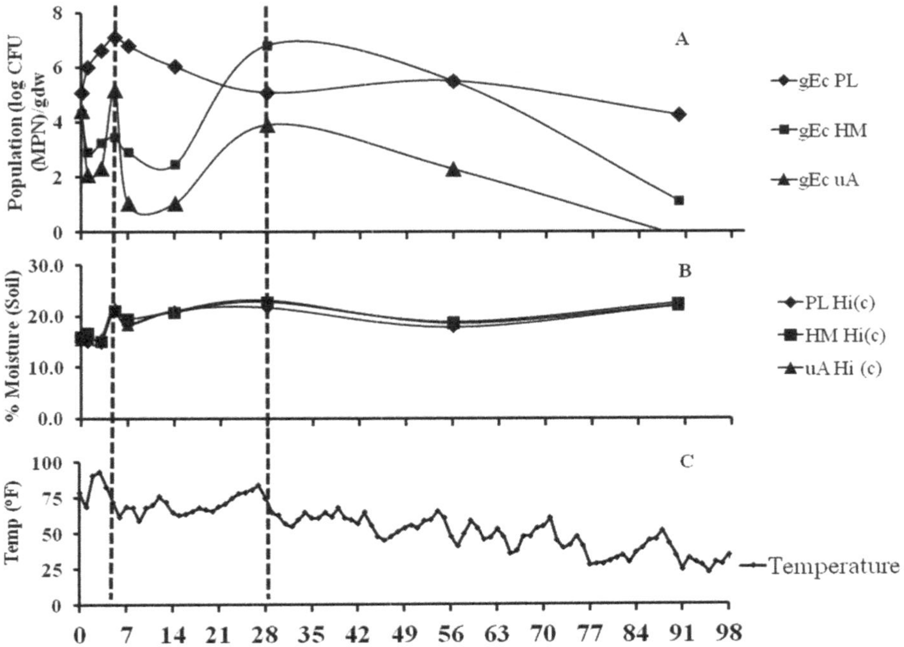

FIGURE 2 (A) The survival of a three-strain (TVS 353, 354, and 355), nonpathogenic (generic) *E. coli* inoculum in either poultry litter-amended (gEC PL), horse manure-amended (gEC HM), or unamended (gEC uA) over 98 days in soils at the Southeastern Agricultural Research and Extension Center. Amendments were tilled into soils. (B) The moisture content of tilled PL-amended [PL Hi(c)], HM-amended [HM Hi(c)], and uA soils]uA Hi(c)]. (C) The air temperature throughout the study. Values below the x-axis indicate the number of days of the study. Dashed lines indicate days where changes in *E. coli* populations occurred, potentially attributed to changes in either soil moisture content or temperature. The first dashed line indicates that all *E. coli* populations (graph A) increased on day 5 with an increase in all soil moisture contents (graph B) from day 3 to day 5. On day 7 (no dashed line), the decline in *E. coli* populations between days 5 and 7 was associated with a fall in the soil moisture content of all soils. The second dashed line indicates an increase in gEC HM and gEC uA populations from day 14 to day 28, which corresponds to an increase in HM Hi(c), and uA Hi(c) soil moisture contents (no increase in gEC PL was observed). The figure illustrates that *E. coli* populations in tilled soils amended with manure may be less responsive to changes in air temperature and more responsive to changes in soil moisture content than *E. coli* in surface manure-amended soils (Fig. 1).

E. coli and *Salmonella* Survival in Manure Dust

While contact with manure-amended soils could potentially cause contamination to fruits and vegetables, several reports suggest that manure dust transported by wind may contaminate leafy greens and other fruit or vegetable commodities. Dust containing soil and manure particles containing pathogenic *E. coli* generated from the movement of cattle herds may become airborne and subsequently contaminate growing vegetable crops. A multiseason study showed that spinach, mustard, and turnip greens planted adjacent to a cattle feedlot were contaminated with both nonpathogenic and *E. coli* O157:H7 originating from the feedlot (37). *E. coli* O157:H7 was present on 1.8% of leafy green samples planted 180 m (590 ft) away from the feedlot, which was a significantly lower prevalence than the 3.5% of leafy green samples which tested positive when grown only 60 m (197 ft) away, but not significantly different from the

2.2% of samples which tested positive at the 120 m (400 ft) distance (37).

The current California Leafy Greens Handler Marketing Agreement standards (14) state that a 400-ft distance be maintained between a concentrated animal feeding operation and a field of leafy greens to reduce the likelihood of contamination. However, this distance may be inadequate to prevent contamination of leafy greens with manure dust (37). Other work has shown that *Salmonella* spp. survival in turkey manure dust was inversely related to the moisture content and the size of the manure particles (38). In manure dust with a particle size of 125 μm, a multistrain inoculum of *Salmonella* spp. survived for 291 days at a moisture content of 5%, while at moisture contents of 10% and 15%, *Salmonella* populations survived for 88 and 68 days, respectively. This same work showed that the duration of *Salmonella* spp. survival in manure dust at 5% was inversely related to particle size. *Salmonella* spp. in manure dust on the leaves of growing spinach plants declined much less rapidly (1 to 2 log CFU) over a period of 14 days compared to *Salmonella* populations applied as a water spray (3.5 to 4.5 log CFU) to the leaves of spinach plants (38). This was observed regardless of the amount of ultraviolet light to which the plants were exposed (38). The results of these studies indicate that manure dust can not only transport pathogens for distances greater than 400 ft, but can also support the prolonged survival of bacterial pathogens and facilitate enhanced persistence on leafy green vegetable surfaces.

Prevalence of Antibiotic-Resistance Genes in Manure-Amended Soils

Bacteria in dairy manure have been known to be a reservoir of antibiotic-resistance genes associated with several enteric pathogens, and one survey found dairy manure to contain 80 distinct antibiotic-resistance genes (39). In a different study, soils which received manure as an organic fertilizer (dairy manure applied at a rate of 6,900 gal/acre, and swine manure at 3,600 gal/acre) in Ontario, Canada, contained more antibiotic-resistant bacteria than soils which did not receive manure (40). Furthermore, Marti et al. (40) reported that a greater number of antibiotic-resistant coliform bacteria were recovered from manured soils than from nonmanured soils. The presence of antibiotic-resistant bacteria on vegetables grown in manured soils was not considerably different from those grown in nonmanured soils; however, there were unique antibiotic resistance markers (macrolide-lincosamide-streptogramin type A resistance genes [*erm*] and incompatibility [*Inc*] groups) recovered from vegetables grown in manured soils which were not present in nonmanured soils. These studies suggest that manure-amended soils may increase the diversity of antibiotic-resistant coliforms in soils.

Another study has reported that the soils amended with dairy cattle manure from animals which did not receive any antibiotics may encourage the growth of bacteria expressing beta-lactamase genes originally present in the soil before the addition of the manure (41). At the outset of the study, soils amended with inorganic fertilizers contained higher populations of beta-lactamase-containing bacteria than those amended with dairy cattle manure that received no antibiotic treatments. However, from 4 days to 130 days after the application of the manure, applied at a rate of 20 kg manure/bed (unknown size), the relative abundance of specific alleles of the beta-lactamase gene (*bla*CEP-04) in manure-amended soils was higher than the abundance of *bla*CEP-04 in the soils receiving inorganic fertilizers (41). The application of manure was determined to encourage the growth of *Pseudomonas* spp., *Psychrobacter pulmonis*, and *Janthinobacterium* spp., all of which are known to contain beta-lactamases (41). The use of manure may alter the overall population of the soil microbiome by increasing the proportion of antibiotic-resistant bacteria in agricultural fields.

Detection of Non-O157 EHEC *E. coli* in Manure

Numerous studies have reported the persistence of *E. coli* O157:H7 in manure-amended soils used for growing vegetables for human consumption. However, future studies which focus on only the detection or prevalence of *E. coli* O157:H7 may underestimate the true burden of EHEC and/or Shiga-toxigenic *E. coli* (STEC) in manure-amended soils. Other non-O157 EHEC serotypes (O26, O45, O103, O111, O121, O145) have been identified as causing the majority of non-O157 EHEC outbreaks, but their prevalence remains virtually unexamined in agricultural environments including manure-amended soils (42).

Investigators examining the prevalence of the non-O157 STEC serogroups in manure, using a combination of culture and direct multiplex PCR methods, recovered isolates of serogroups O26, O103, and O111 most frequently; however, most of these recovered isolates were shown to be *stx*-negative (42). In this study a survey of 960 fecal samples from beef cattle feedlots (containing cattle from 48 background operations) revealed that the percentage of STEC serotypes O26, O45, O103, O111, O121, O145, and O157 were 23.4%, 10.8%, 11.8%, 0.01%, 16.4%, 0.03%, and 48.2%, respectively. The percentage of isolates which tested positive for *stx1*, *stx2*, and *eae* were 44.4%, 63.5%, and 77.4%, respectively (42). Other studies have shown that cattle vaccines targeting *E. coli* O157:H7 did reduce fecal shedding of STEC O157 but that the vaccination was not effective in reducing shedding of non-O157 STEC (43). Non-O157 STEC in animal manure may have received less attention than *E. coli* O157:H7 because of the difficulty of detection of the non-O157 phenotypes, which are all similar to each other. New multiplex real-time PCR methods targeting the attaching and effacing gene conserved fragment (*ecfl*) have improved the rapid detection of non-O157 STEC in manure and also accounted for potential loss of the *stx* genes upon enrichment of these samples (44).

COMPOST

Compost is organic material that has been degraded into a nutrient-stabilized humus-like substance through intense microbial activity (45). The microbial activity generating heat and high temperatures also can kill bacterial pathogens in biosolids, animal manure, and food waste, all of which can be used as feedstocks for compost (46). The application of compost to soils has been shown to improve plant nutrient availability, physical and chemical properties of soils (47), and soil microbial diversity (48, 49) and decrease the risk of transferring human pathogens to food crops (46). During the composting process, the microorganisms metabolize most of the free and fast-release simple organic compounds, by either volatilizing or incorporating the nutrients. The incorporation of nutrients into biomass, both living and dead, yields more complex matrices that stabilize the nutrients into slow-release forms. Many organic materials can be used as feedstocks including manure, plants (i.e., crop residues, yard trimmings, etc.), food waste, and various industrial products and by-products (i.e., paper, wood chips, etc.) (50). During composting, microbial activity from degrading the feedstocks can generate sufficiently high (thermophilic) temperatures (≥55°C) to kill enteric bacterial pathogens originally present in the feedstocks, assuming that proper carbon:nitrogen (C:N) ratios (recommended to be between 20:1 to 40:1), moisture (40 to 65%), and aeration levels (O_2 concentrations greater than 5%) are maintained (51). To ensure sufficient pathogen inactivation through achieving thermophilic temperatures within a compost pile, the U.S. Environmental Protection Agency (EPA) and United States Department of Agriculture (USDA) National Organic Program (NOP) recommend maintaining the pile at >55°C for at least 3 consecutive days for static and aerated piles, or at >55°C for 15 days for windrow composting. The most common composting mechanisms are static piles, windrows, and in-vessel containers (46, 50).

Survival and Resuscitation of Pathogens in Compost

Human pathogens can survive during the composting process for a variety of reasons. Detection of human pathogens in finished or "point of sale" compost is relatively rare. In an evaluation of 15 composting facilities from across the United States over several months, Ingram (52) found that only 6 out of 105 (6%) compost samples contained *Salmonella* spp., but 69% of samples contained *E. coli* populations. However, of the compost samples which contained *E. coli*, 19% exceeded the EPA Part 503 microbiological standard (1,000 MPN/g) for fecal coliforms in biosolids. Five of the seven facilities which produced compost that contained *E. coli* populations that exceeded the 1,000 MPN/g fecal coliform limit were sampled in July, indicating that the increased ambient air temperature affected the frequency of detection of *E. coli* in these samples. In a separate evaluation of 94 nonsludge composting facilities, only 1 compost sample contained *Salmonella* spp. (1%), but 28% of samples exceeded the EPA Part 503 limit for fecal coliforms, and 3% of samples were contaminated with *E. coli* O157:H7 (53). Large composting facilities and those that employed large static piles were more likely to have human pathogen contamination than smaller facilities or those that utilized turned windrows (53). Other investigators have found that compost piles using hay, straw, and cattle manure feedstocks which were turned every 2 weeks achieved higher temperatures than those which were not turned (51). Populations of *E. coli* O157:H7 were not recovered from the tops of turned piles but were recovered from the tops of static piles, along with populations of *Listeria* spp. (44).

In general, survival of the pathogen is attributed to either the entire pile or pockets within the pile not reaching 55°C (54), and areas of the pile which are not turned or are outside in cold ambient temperatures are most likely to allow the survival of enteric pathogens (55). A higher percentage (63%) of surface samples from compost piles on commercial poultry farms in South Carolina contained *E. coli* compared to 10% of internal samples during phase 1 of the composting process when temperatures within the pile had not achieved their highest level (55). Regular and complete turning of the pile ensures that all portions of the compost can be exposed to the highest temperatures within the pile and achieve temperatures sufficient to kill enteric pathogens. Covering compost piles with a 30-cm layer of finished compost has been shown to hasten reduction of *E. coli* O157:H7 and *Salmonella* spp. populations within the pile (56). The addition of the finished compost layer also increased the temperature and extended the duration of time that the pile temperatures exceeded 55°C compared to compost piles without the finished compost layer. Volatile acids generated during aerobic composting of manure feedstock may also aid in the decline of enteric bacterial populations during composting (57).

Resuscitation/Regrowth of Pathogens in Compost

Other nonmicrobial factors may also affect the regrowth and resuscitation of pathogens in finished composts. Twenty-nine commercial point of sale composts were inoculated with *E. coli* O157:H7 and *Salmonella* spp. and stored at 25°C for 3 days to determine if the feedstock (biosolids, animal manure, yard waste) or if a single physicochemical factor (C:N ratio, moisture content, pH, total organic carbon, percent volatile solids, electrical conductivity, and maturity) could predict the regrowth of *E. coli* O157:H7 or *Salmonella* spp. (58). Populations of *E. coli* O157:H7 increased or decreased by up to 2 log CFU/g in individual finished compost samples. Similarly, *Salmonella* spp. populations increased by up to 1.5 log CFU/g and decreased by up to 1.5 log CFU/g in individual finished compost samples (58). No single physicochemical factor was able to predict if a compost sample could support the regrowth of pathogens;

however, moisture content, C:N ratio, and total organic carbon levels were the three physicochemical factors which contributed the most to the regrowth potential of pathogens in finished compost. Other reports support the difficulty in identifying a single physicochemical factor which can predict resuscitation of pathogens in compost (59). Increased relative humidity (percent moisture), has been shown to prolong bacterial survival in finished composts. *S. enterica* serovar Arizonae and *E. coli* populations inoculated into finished compost were inactivated more slowly at 40% and 80% humidity than at 10% humidity when stored at room temperature over a 30-day period (60). Pathogens in compost may utilize available soluble organic matter due to the additional moisture, increasing the potential for regrowth (61). A moisture content level of 20% was the minimum determined to support *E. coli* O157:H7 survival in a compost pile (62). In effect, moisture can become a limiting factor in the growth and survival of microorganisms in compost.

Regrowth in compost teas (CTs), unheated on-farm infusions of compost used as a spray or soil drench as either a plant growth promoter or to control for phytopathogens, has been shown to be dependent on common additives (bacterial nutrient solution, kelp, humic acid, and rock dust) added to the CT (63). Populations of *E. coli* O157:H7, *S. enterica* serovar Enteritidis, and fecal coliforms increased by 1 to 4 log CFU/ml in aerated CT with additives, while no increase in populations was observed in CT without additives (64). Similarly, the addition of molasses and kelp increased populations of *E. coli* in CT formulations regardless of if the CT was aerated or not (65).

Microbial Competition Inhibiting Pathogen Growth in Compost

Competition from indigenous microorganisms in compost and amended soils has been shown to inhibit regrowth of human pathogens (59, 65, 66). Populations of *Enterococcus faecalis*, *S. enterica* serovar Infantis, and *Listeria monocytogenes* decreased by about 6 log CFU/g when inoculated into compost removed at week 8 (cooling stage) of a 14-week composting cycle and held at room temperature for 90 days (59). However, 70% of compost samples removed from between week 0 and week 6 of the 14-week composting cycle were positive for one of the three aforementioned pathogens. Compost at the week 8 stage also showed greater microbial diversity than composts aged 0 to 6 weeks, indicating that the diversity of mesophilic organisms present in the cooling stage may have inhibited pathogen growth and survival (59). Other studies have shown that *S.* Typhimurium growth was significantly greater in composted biosolids which had been sterilized than in nonsterilized composted biosolids, indicating that indigenous microorganisms play a role in inhibiting *S.* Typhimurium growth (65). Populations of actinomycetes increased by 1 log CFU/g during the composting process between 2 and 13 weeks, perhaps demonstrating that they may play a more prominent role in suppressing *Salmonella* populations than other groups of microorganisms (65). The long-term storage of compost material is likely to increase the opportunity of *Salmonella* regrowth, presumably due to a decline in *Salmonella*-suppressive microorganisms in the compost. In a separate study, coliform bacteria and Gram-positive organisms isolated from compost at a stage between 25°C and 40°C were able to suppress *S.* Typhimurium, and *S.* Typhimurium was found to be suppressed to a higher degree in compost at 55°C than compost at 70°C, where no reduction in *S.* Typhimurium populations was observed, perhaps due to the decrease in populations of *Salmonella*-suppressive bacteria present at higher temperatures (66).

Bacterial pathogens inoculated into compost at high populations (10^7 CFU/g) survived long durations in compost-amended soils (applied at a rate of 4.5 metric tons/ha) and were transferred to carrots and onions grown in these soils (67, 68). *S*. Typhimurium survived for longer durations in composted PL (231 days) than in alkaline-pH stabilized composted dairy manure (203 days) (67). *E. coli* O157:H7 populations (10^7 CFU/g) inoculated into amended soils containing composted poultry litter, composted dairy manure, and alkaline pH-stabilized composted dairy manure at 4.5 metric tons/ha survived for 196, 196, and 154 days, respectively (68). These results show that pathogen survival can be sustained for extended periods of time in compost-amended soils. These studies also raise the issue that pathogens may survive for different durations in compost-amended soils in field conditions than in the compost alone.

CONCLUSION

BSAs such as manure and compost can be applied to fields to encourage the promotion of soil health, microbial diversity, and fertility. Since manure and compost can provide valuable nutrients to soils and provide a sustainable approach to agriculture, their proper handling to minimize pathogen contamination is of the utmost importance. The current literature indicates that an appropriate interval between the application of manure to soil and the harvest of fruits and vegetables which may come into contact with soil is necessary to minimize the contamination of these plants with food-borne bacterial pathogens. The duration of that interval in the United States should balance seasonal requirements of growing fruits and vegetables, benefits to soil fertility, and plant nutrients with the need to provide safe and wholesome produce to consumers. For compost, the consistent achievement of a 55°C temperature threshold and sufficient curing times for composted feedstocks are essential to minimizing bacterial enteric pathogen populations. More attention should be paid to resuscitation and regrowth of pathogens in either manure- or compost-amended soils under field conditions to determine if these events can prolong the persistence of pathogens in these amended soils environments and increase the likelihood of transfer to fruits and vegetables. Geographic location, climate conditions, and soil and manure properties must be taken into account when determining the duration of the interval between amendment application and harvest of the fruit and vegetable commodity to minimize transfer of pathogens to growing fruits and vegetables. Overall, BSAs applied to soils used to grow fruits and vegetables should be handled in a manner that reduces the prevalence of food-borne pathogens.

ACKNOWLEDGMENTS

The authors' research was funded by the United States Department of Agriculture, Agricultural Research Service.

CITATION

Sharma M, Reynnells R. 2016. Importance of soil amendments: survival of bacterial pathogens in manure and compost used as organic fertilizers. Microbiol Spectrum 4(4):PFS-0010-2015.

REFERENCES

1. **Food and Drug Administration.** 2014. Standards for growing, harvesting, packing, and holding of produce for human consumption. FSMA Proposed Rule for Produce Safety. http://www.fda.gov/Food/GuidanceRegulation/FSMA/ucm334114.htm.
2. **Mikha M, Benjamin JG.** 2014. Remediation/restoration of degraded soil. II. Impact on crop production and nitrogen dynamics. *Agron J* **106:**261–272.
3. **Rees HW, Chow L, Zebarth B, Xing Z, Toner P, Lavoie J, Daigle J-L.** 2014. Impact of supplemental poultry manure application on potato yield and soil properties on a loam soil in north-western New Brunswick. *Can J Soil Sci* **94:**49–65.
4. **Donovan NJ, Saleh F, Chan KY, Eldridge SM, Fahey D, Muirhead L, Meszaros I, Barhcia I.** 2014. Use of garden organic compost in a long-term vegetable field trial: biological soil health. *Acta Hortic* **1018:**47–55.

5. **CalFERT (California Food Emergency Response Team).** 2008. Investigation of the Taco John's *Escherichia coli O157:H7* outbreak associated with iceberg lettuce. California Department of Public Health, Sacramento, CA.
6. **Söderström A, Osterberg P, Lindqvist A, Jönsson B, Lindberg A, Blide Ulander S, Welinder-Olsson C, Löfdahl S, Kaijser B, De Jong B, Kühlmann-Berenzon S, Boqvist S, Eriksson E, Szanto E, Andersson S, Allestam G, Hedenström I, Ledet Muller L, Andersson Y.** 2008. A large *Escherichia coli* O157 outbreak in Sweden associated with locally produced lettuce. *Foodborne Pathog Dis* **5:**339–349.
7. **Hutchison ML, Walters LD, Avery SM, Munro F, Moore A.** 2005. Analyses of livestock production, waste storage, and pathogen levels and prevalences in farm manures. *Appl Environ Microbiol* **71:**1231–1236.
8. **California Food Emergency Response Team.** 2007. Investigation of an *Escherichia coli* O157:H7 outbreak associated with Dole prepackaged spinach. https://www.cdph.ca.gov/pubsforms/Documents/fdb%20eru%20Spnch%20EC%20Dole032007wph.PDF.
9. **Centers for Disease Control and Prevention.** 2008. Multistate outbreak of *Salmonella* Saintpaul infections linked to raw produce (final update). http://www.cdc.gov/salmonella/2008/raw-produce-8-28-2008.html.
10. **Food and Drug Administration.** 2013. Environmental assessment: factors potentially contributing to the contamination of fresh whole cantaloupe implicated in a multi-state outbreak of salmonellosis. http://www.fda.gov/Food/RecallsOutbreaksEmergencies/Outbreaks/ucm341476.htm.
11. **Food and Drug Administration.** 2011. Environmental assessment: factors potentially contributing to the contamination of fresh whole cantaloupe implicated in a multi-state outbreak of listeriosis. http://www.fda.gov/food/recallsoutbreaksemergencies/outbreaks/ucm276247.htm.
12. **Angelo KM, Chu A, Anand M, Nguyen TA, Bottichio L, Wise M, Williams I, Seelman S, Bell R, Fatica M, Lance S, Baldwin D, Shannon K, Lee H, Trees E, Strain E, Gieraltowski L, Centers for Disease Control and Prevention (CDC).** 2015. Outbreak of *Salmonella* Newport infections linked to cucumbers: United States, 2014. *MMWR Morb Mortal Wkly Rep* **64:**144–147.
13. **Painter JA, Hoekstra RM, Ayers T, Tauxe RV, Braden CR, Angulo FJ, Griffin PM.** 2013. Attribution of foodborne illnesses, hospitalizations, and deaths to food commodities by using outbreak data, United States, 1998–2008. *Emerg Infect Dis* **19:**407–415.
14. **California Leafy Greens Handler Marketing Agreement.** 2013. Commodity specific food safety guidelines for the production and harvest of lettuce and leafy greens. http://www.lgma.ca.gov/wp-content/uploads/2014/09/California-LGMA-metrics-08-26-13-Final.pdf.
15. **National Organic Program.** Soil fertility and crop nutrient management practice standard. 7 C.F.R. 205.203.
16. **Tomato Best Practices Manual.** 2007. A guide to tomato good agricultural practices (T-GAP) and tomato best management practices (T-BMP) 2007. http://fvreports.freshfromflorida.com/5G_TomBPM.pdf.
17. **Avery LM, Hill P, Killham K, Jones DL.** 2004. *Escherichia coli* O157 survival following the surface and sub-surface application of human pathogen contaminated organic waste to soil. *Soil Biol Biochem* **36:**2101–2103.
18. **Franz E, Semenov AV, Termorshuizen AJ, de Vos OJ, Bokhorst JG, van Bruggen AHC.** 2008. Manure-amended soil characteristics affecting the survival of *E. coli* O157:H7 in 36 Dutch soils. *Environ Microbiol* **10:**313–327.
19. **Harris LJ, Berry ED, Blessington T, Erickson M, Jay-Russell M, Jiang X, Killinger K, Michel FC, Millner P, Schneider K, Sharma M, Suslow TV, Wang L, Worobo RW.** 2013. A framework for developing research protocols for evaluation of microbial hazards and controls during production that pertain to the application of untreated soil amendments of animal origin on land used to grow produce that may be consumed raw. *J Food Prot* **76:**1062–1084.
20. **Hutchison ML, Walters LD, Avery SM, Synge BA, Moore A.** 2004. Levels of zoonotic agents in British livestock manures. *Lett Appl Microbiol* **39:**207–214.
21. **Nyberg KA, Ottoson JR, Vinnerås B, Albihn A.** 2014. Fate and survival of *Salmonella* Typhimurium and *Escherichia coli* O157:H7 in repacked soil lysimeters after application of cattle slurry and human urine. *J Sci Food Agric* **94:**2541–2546.
22. **Gessel PD, Hansen NC, Goyal SM, Johnston LJ, Webb J.** 2004. Persistence of zoonotic pathogens in surface soil treated with different rates of liquid pig manure. *Appl Soil Ecol* **25:**237–243.
23. **Bezanson G, Delaquis P, Bach S, McKellar R, Topp E, Gill A, Blais B, Gilmour M.** 2012. Comparative examination of *Escherichia coli* O157:H7 survival on romaine lettuce and in soil at two independent experimental sites. *J Food Prot* **75:**480–487.

24. **Gutiérrez-Rodríguez E, Gundersen A, Sbodio AO, Suslow TV.** 2012. Variable agronomic practices, cultivar, strain source and initial contamination dose differentially affect survival of *Escherichia coli* on spinach. *J Appl Microbiol* **112:**109–118.
25. **Whyte C, Cotton CP, Hashem F, Sharma M, Millner P.** 2014. Irrigation, manure, and soil type influences on survival and persistence of non-pathogenic *E. coli* and *E. coli* O157:H7 in a greenhouse environment. Poster P1-129, International Association for Food Protection Annual Meeting, Indianapolis, IN, 3–6 August.
26. **Brennan FP, Abram F, Chinalia FA, Richards KG, O'Flaherty V.** 2010. Characterization of environmentally persistent *Escherichia coli* isolates leached from an Irish soil. *Appl Environ Microbiol* **76:**2175–2180.
27. **van Hoek AH, Aarts HJM, Bouw E, van Overbeek WM, Franz E.** 2013. The role of rpoS in *Escherichia coli* O157 manure-amended soil survival and distribution of allelic variations among bovine, food and clinical isolates. *FEMS Microbiol Lett* **338:**18–23.
28. **Franz E, van Hoek AHAM, Bouw E, Aarts HJM.** 2011. Variability of *Escherichia coli* O157 strain survival in manure-amended soil in relation to strain origin, virulence profile, and carbon nutrition profile. *Appl Environ Microbiol* **77:**8088–8096.
29. **Bhagwat AA, Tan J, Sharma M, Kothary M, Low S, Tall BD, Bhagwat M.** 2006. Functional heterogeneity of RpoS in stress tolerance of enterohemorrhagic *Escherichia coli* strains. *Appl Environ Microbiol* **72:**4978–4986.
30. **Duffitt AD, Reber RT, Whipple A, Chauret C.** 2011. Gene expression during survival of *Escherichia coli* O157:H7 in soil and water. *Int J Microbiol* **2011:**1–12.
31. **Kandror O, DeLeon A, Goldberg AL.** 2002. Trehalose synthesis is induced upon exposure of *Escherichia coli* to cold and is essential for viability at low temperatures. *Proc Natl Acad Sci USA* **99:**9727–9732.
32. **Zhang Q, Yan T.** 2012. Correlation of intracellular trehalose concentration with desiccation resistance of soil *Escherichia coli* populations. *Appl Environ Microbiol* **78:**7407–7413.
33. **Graham L, Wright D, Hashem F, Cotton C, Collins A, White K, Stonebraker R, Sharma M, Millner P.** 2014. The effect of manure application method on the persistence of *Escherichia coli* in manure-amended soils in southeastern Pennsylvania. Poster P1-131, International Association for Food Protection Annual Meeting, Indianapolis, IN, 3–6 August.
34. **Schwarz KR, Sidhu JPS, Pritchard DL, Li Y, Toze S.** 2014. Decay of enteric microorganisms in biosolids-amended soil under wheat (*Triticum aestivum*) cultivation. *Water Res* **59:**185–197.
35. **Lang NL, Bellett-Travers MD, Smith SR.** 2007. Field investigations on the survival of *Escherichia coli* and presence of other enteric micro-organisms in biosolids-amended agricultural soil. *J Appl Microbiol* **103:**1868–1882.
36. **Gibbs RA, Hu CJ, Ho GE, Unkovich I.** 1997. Regrowth of faecal coliforms and salmonellae in stored biosolids and soil amended with biosolids. *Water Sci Technol* **35:**269–275.
37. **Berry ED, Wells JE, Bono JL, Woodbury BL, Kalchayanand N, Norman KN, Suslow TV, López-Velasco G, Millner PD.** 2015. Effect of proximity to a cattle feedlot on *Escherichia coli* O157:H7 contamination of leafy greens and evaluation of the potential for airborne transmission. *Appl Environ Microbiol* **81:**1101–1110.
38. **Oni RA, Sharma M, Buchanan RL.** 2015. Survival of *Salmonella* spp. in dried turkey manure and persistence on spinach leaves. *J Food Prot* **78:**1791–1799.
39. **Wichmann F, Udikovic-Kolic N, Andrew S, Handelsman J.** 2014. Diverse antibiotic resistance genes in dairy cow manure. *MBio* **5:**e01017-13.
40. **Marti R, Scott A, Tien Y-C, Murray R, Sabourin L, Zhang Y, Topp E.** 2013. Impact of manure fertilization on the abundance of antibiotic-resistant bacteria and frequency of detection of antibiotic resistance genes in soil and on vegetables at harvest. *Appl Environ Microbiol* **79:**5701–5709.
41. **Udikovic-Kolic N, Wichmann F, Broderick NA, Handelsman J.** 2014. Bloom of resident antibiotic-resistant bacteria in soil following manure fertilization. *Proc Natl Acad Sci USA* **111:**15202–15207.
42. **Cernicchiaro N, Cull CA, Paddock ZD, Shi X, Bai J, Nagaraja TG, Renter DG.** 2013. Prevalence of Shiga toxin-producing *Escherichia coli* and associated virulence genes in feces of commercial feedlot cattle. *Foodborne Pathog Dis* **10:**835–841.
43. **Cernicchiaro N, Renter DG, Cull CA, Paddock ZD, Shi X, Nagaraja TG.** 2014. Fecal shedding of non-O157 serogroups of Shiga toxin-producing *Escherichia coli* in feedlot cattle vaccinated with an *Escherichia coli* O157:H7 SRP vaccine or fed a *Lactobacillus*-based direct-fed microbial. *J Food Prot* **77:**732–737.
44. **Luedtke BE, Bono JL, Bosilevac M.** 2014. Evaluation of a real time PCR assay for the detection and enumeration of enterohemorrhagic *Escherichia coli* directly from cattle feces. *J Microbiol Methods* **105:**72–79.

45. **Millner PD.** 2009. Manure management, p 79–104. *In* Sapers GM, Solomon EB, Matthews KR (ed), *The Produce Contamination Problem*. Academic Press, New York.
46. **Millner PD.** 2014. Pathogen disinfection technologies, metrics, and regulations for recycled organics used in horticulture. *Acta Hortic* **1018:** 621–630.
47. **Mylavarapu RS, Zinati GM.** 2009. Improvement of soil properties using compost for optimum parsley production in sandy soils. *Sci Hortic (Amsterdam)* **120:**426–430.
48. **Gandolfi I, Sicolo M, Franzetti A, Fontanarosa E, Santagostino A, Bestetti G.** 2010. Influence of compost amendment on microbial community and ecotoxicity of hydrocarbon-contaminated soils. *Bioresour Technol* **101:**568–575.
49. **Farrell M, Griffith GW, Hobbs PJ, Perkins WT, Jones DL.** 2009. Microbial diversity and activity are increased by compost amendment of metal-contaminated soils. *FEMS Microbiol Lett* **71:**94–105.
50. **Reynnells RE.** 2013. Comparison of pathogen detection methods in compost and compost characteristics as potential predictors of pathogen regrowth. Master's thesis, University of Maryland, College Park, MD. http://drum.lib.umd.edu/handle/1903/14819.
51. **Berry ED, Millner PD, Wells JE, Kalchayanand N, Guerni MN.** 2013. Fate of naturally occurring *Escherichia coli* O157:H7 and other zoonotic pathogens during minimally managed bovine feedlot manure composting processes. *J Food Prot* **76:**1308–1321.
52. **Ingram DT.** 2009. Assessment of foodborne pathogen survival during production and pre-harvest application of compost and compost tea. Ph.D. dissertation, University of Maryland, College Park, MD. http://drum.lib.umd.edu/bitstream/1903/9137/1/Ingram_umd_0117E_10231.pdf.
53. **Brinton WF Jr, Storms P, Blewett TC.** 2009. Occurrence and levels of fecal indicators and pathogenic bacteria in market-ready recycled organic matter composts. *J Food Prot* **72:**332–339.
54. **Wichuk KM, McCartney D.** 2007. A review of the effectiveness of current time-temperature regulations on pathogen inactivation during composting. *J Environ Eng Sci* **6:**572–586.
55. **Shepherd MW Jr, Liang P, Jiang X, Doyle MP, Erickson MC.** 2010. Microbiological analysis of composts produced on South Carolina poultry farms. *J Appl Microbiol* **108:**2067–2076.
56. **Patel JR, Yossa I, Macarisin D, Millner P.** 2015. Physical covering for control of *Escherichia coli* O157:H7 and *Salmonella* spp. in static and windrow composting processes. *Appl Environ Microbiol* **81:**2063–2074.
57. **Erickson MC, Liao J, Jiang X, Doyle MP.** 2014. Inactivation of pathogens during aerobic composting of fresh and aged dairy manure and different carbon amendments. *J Food Prot* **77:**1911–1918.
58. **Reynnells R, Ingram DT, Roberts C, Stonebraker R, Handy ET, Felton G, Vinyard BT, Millner PD, Sharma M.** 2014. Comparison of U.S. Environmental Protection Agency and U.S. Composting Council microbial detection methods in finished compost and regrowth potential of *Salmonella* spp. and *Escherichia coli* O157:H7 in finished compost. *Foodborne Pathog Dis* **11:**555–567.
59. **Paniel N, Rousseaux S, Gourland P, Poitrenaud M, Guzzo J.** 2010. Assessment of survival of *Listeria monocytogenes, Salmonella* Infantis and *Enterococcus faecalis* artificially inoculated into experimental waste or compost. *J Appl Microbiol* **108:**1797–1809.
60. **Pietronave S, Fracchia L, Rinaldi M, Martinotti MG.** 2004. Influence of biotic and abiotic factors on human pathogens in a finished compost. *Water Res* **38:**1963–1970.
61. **Soares HM, Cardenas B, Weir D, Switzenbaum MS.** 1995. Evaluating pathogen regrowth in biosolids compost. *Biocycle* **36:**70–76.
62. **Kim J, Shepherd MW Jr, Jiang X.** 2009. Evaluating the effect of environmental factors on pathogen regrowth in compost extract. *Microb Ecol* **58:**498–508.
63. **Ingram DT, Millner PD.** 2007. Factors affecting compost tea as a potential source of *Escherichia coli* and *Salmonella* on fresh produce. *J Food Prot* **70:**828–834.
64. **Kannangara T, Forge T, Dang B.** 2006. Effects of aeration molasses, kelp, compost type, and carrot juice on the growth of *Escherichia coli* in compost teas. *Compost Sci Util* **14:**40–47.
65. **Sidhu J, Gibbs RA, Ho GE, Unkovich I.** 2001. The role of indigenous microorganisms in suppression of *Salmonella* regrowth in composted biosolids. *Water Res* **35:**913–920.
66. **Millner PD, Powers KE, Enkiri NK, Burge WD.** 1987. Microbially mediated growth suppression and death of *Salmonella* in composted sewage sludge. *Microb Ecol* **14:**255–265.
67. **Islam M, Doyle MP, Phatak SC, Millner P, Jiang X.** 2005. Survival of *Escherichia coli* O157:H7 in soil and on carrots and onions grown in fields treated with contaminated manure composts or irrigation water. *Food Microbiol* **22:**63–70.
68. **Islam M, Morgan J, Doyle MP, Phatak SC, Millner P, Jiang X.** 2004. Fate of *Salmonella enterica* serovar Typhimurium on carrots and radishes grown in fields treated with contaminated manure composts or irrigation water. *Appl Environ Microbiol* **70:**2497–2502.

EMERGING ISSUES IN PREHARVEST FOOD SAFETY

Reducing Foodborne Pathogen Persistence and Transmission in Animal Production Environments: Challenges and Opportunities

10

ELAINE D. BERRY[1] and JAMES E. WELLS[1]

BACKGROUND

Preharvest measures to reduce zoonotic pathogens in food animals are critical components in farm-to-table food safety approaches, which recognize that food production and safety occurs along a continuum. The encompassing goal of an integrated food safety program is to improve public health by reducing the risk of human foodborne illness, while the more specific goal of preharvest food safety strategies is to reduce the pathogen load of animals and/or animal products (such as milk or eggs) that are brought to harvest, in order to enhance the efficacy of postharvest interventions and reduce pathogens in the final product. As an example, the presence of pathogens in cattle feces and on cattle hides has been associated with beef carcass contamination (1–4). Cattle with high levels of *Salmonella* on their hides on entry into commercial processing were often coincident with preevisceration carcasses that were contaminated with the pathogen (4). Correspondingly, studies conducted in commercial beef processing plants have demonstrated that reducing *Escherichia coli* O157:H7 prevalence on cattle hides reduces its prevalence on resultant carcasses (5–7). As another example, the risk of broiler carcass contamination is greater when there is a higher degree of *Campylobacter* intestinal colonization of birds entering slaughter (8–10).

[1]USDA, Agricultural Research Service, U.S. Meat Animal Research Center, Clay Center, NE 68933.

Preharvest Food Safety
Edited by Siddhartha Thakur and Kalmia E. Kniel

doi:10.1128/microbiolspec.PFS-0006-2014

The Meat Inspection Act of 1906 (11) heralded the modern food safety era as an act of the U.S. Congress to prevent adulterated or misbranded meat and meat products (those "which are unsound, unhealthful, unwholesome, or otherwise unfit for human food") from being sold as food and to ensure that meat animals and their products are slaughtered and processed under sanitary conditions. Provisions of the act included mandatory inspection of the live animal before slaughter, as well as mandatory postmortem inspection of each carcass. Smith et al. (12) and Oliver et al. (13) recently reviewed other historical aspects of food animal agriculture in the context of preharvest food safety, including the public health successes of the advent of widespread milk pasteurization and the reduction of human infection with *Mycobacterium bovis* as a result of the U.S. program to eradicate tuberculosis in cattle.

These examples illustrate that preharvest food safety in terms of animal health and hygiene in food production are not new concepts. However, regulatory developments in the 1990s shifted attention to the live animal and to farm-level efforts to improve the safety of food (13, 14). The enforcement of zero tolerance for *E. coli* O157:H7, the declaration of *E. coli* O157:H7 as an adulterant in raw ground beef and beef trim, and the implementation of the Pathogen Reduction/Hazard Analysis and Critical Control Point Systems regulations by the USDA Food Safety and Inspection Service brought changes to the meat industry, raising the bar on meat and poultry safety by setting pathogen performance standards on raw product (14, 15). More recently, six additional non-O157 Shiga toxin–producing *E. coli* (STEC; serogroups O26, O45, O103, O111, O121, and O145) strains have been declared to be adulterants in raw, nonintact beef products (16). The Food Safety and Inspection Service recently announced changes to poultry product inspection regulations that include the requirement for facilities to perform microbiological testing at two points in the production process to demonstrate control of *Salmonella* and *Campylobacter* (17). Current activities and discussion suggest that regulation of certain *Salmonella* serotypes as adulterants may be imminent (18).

Advancements in Meat and Poultry Safety Research

In response to these regulatory developments, as well as to the heightened awareness of foodborne illness outbreaks, there has been increased emphasis on food safety research initiatives and funding aimed at improving many aspects of the food animal production chain from farm to table (19). The National Food Safety Initiative (20) and the National Academy of Sciences' report entitled "Ensuring Safe Food from Production to Consumption" (21) further reinforced the need for research and outlined goals for improving food safety and public health. A variety of tools for slaughter sanitation and intervention have been developed or improved, validated, and in many cases, widely adopted by the meat and poultry processing industries for use at animal harvest. For example, a wide variety of antimicrobial compounds have been demonstrated to reduce pathogens on carcasses of meat and poultry species, including a variety of organic acids (lactic, citric, and acetic acids) and other antimicrobials such as peracetic acid, acidified sodium chlorite, hypobromous acid, ozone, and electrolyzed oxidized water (22). Depending upon the species or process, these antimicrobial compounds may be applied as sprays or immersion dips. Hide-on carcass washing of cattle can remove bacterial contamination, thereby reducing transfer to the carcass when the hide is removed (6, 7). Heat treatments applied to animal carcasses to reduce bacteria include steam pasteurization (18, 23), hot water washing (24, 25), and scalding and singeing (26, 27). These and other postharvest interventions used for meat and poultry carcass decontamination have been discussed in more detail in recent reviews (18, 24, 28–30).

In part through these increased research investments, food safety research has also identified many preharvest interventions with the potential to reduce pathogens in the live animal. Some of these preharvest interventions currently are in use in animal production, while others are still in development and/or have not been extensively evaluated or scientifically validated as effective at reducing pathogens in animals. For others, cost and regulatory approvals are additional barriers to implementation. Probiotics, direct-fed microbials, or competitive exclusion cultures are commonly fed to livestock to increase production efficiency but may also be useful for reducing pathogen shedding by competition for attachment sites or for nutrients, by production of antimicrobial compounds, or by promotion of immune function. Although results are inconsistent, studies have shown potential for select probiotic or competitive exclusion cultures to reduce *Salmonella* and *Campylobacter jejuni* in poultry (31, 32), *Salmonella* in pigs (33, 34), *E. coli* O157:H7 in lambs (35), and *E. coli* O157:H7 and *Salmonella* in cattle (36, 37).

In addition to their function to reduce disease in livestock, vaccines have been examined for their ability to reduce pathogens that cause foodborne illness. This approach has been explored for the reduction of *Salmonella* and *Campylobacter* in poultry (38–41), *Salmonella* in swine (42, 43), and *Salmonella* and *E. coli* O157:H7 in cattle (44–47). Vaccination to reduce *E. coli* O157:H7 in cattle has been intensively studied because of the status of this pathogen as an adulterant in ground beef and beef trim. The Epitopix *E. coli* bacterial extract vaccine with SRP technology (siderophore receptor and porin protein vaccine; Epitopix LLC, Willmar, MN) has conditional approval in the United States for use to reduce *E. coli* O157:H7 shedding in cattle (48). The Econiche vaccine (Bioniche Life Sciences, Inc., Belleville, Ontario, Canada), which targets type III secreted proteins of *E. coli* O157:H7, is licensed for use in Canada but currently is not approved for use in the United States (48). Considerable research with both vaccines demonstrates their potential to reduce *E. coli* O157:H7 in cattle, and three-dose regimens have been shown to be most effective (49, 50). However, neither vaccine has been widely used in either the United States or Canada (51, 52). A 2013 report from the USDA Animal and Plant Health Inspection Service estimated that 2.4% of large feedlots (those with a capacity of 1,000 or more head) gave cattle vaccines against *E. coli* (52). Carriage of *E. coli* O157:H7 does not affect the beef production efficiency, so the cost of the vaccinations in combination with the lack of economic incentives to cattle producers has limited their adoption to date (13, 18, 48, 53).

The USDA Food Safety and Inspection Service has approved the use of bacteriophages to reduce bacterial pathogens for limited applications, including application as a spray or wash to reduce *Salmonella*, *E. coli* O157:H7, and non-O157 STEC on the hides of live animals before slaughter and *Salmonella* on the feathers of live poultry before slaughter (22, 48). Oral dosing of bacteriophages to reduce pathogens in the gastrointestinal tract of live animals also has been explored to control *E. coli* O157:H7 in cattle and sheep (54–56), *Salmonella* in swine (57, 58), and *Salmonella* and *Campylobacter* in poultry (59–61). The addition of sodium chlorate to the water or feed of food animals before shipping and harvest has been proposed for reduction of *E. coli* O157:H7 and *Salmonella* in the intestinal tract. These pathogens contain the intracellular enzyme nitrate reductase, which reduces chlorate to chlorite, which then accumulates to lethal levels in the bacterial cells (62). Chlorate treatment has been demonstrated to reduce *Salmonella* in swine (63) and poultry (64, 65) and *E. coli* O157:H7 in swine (66), sheep (67), and cattle (68, 69). The use of chlorate to reduce pathogens in livestock currently is not approved, but the application is under review by the U.S. Food and Drug Administration. These preharvest interventions and other potential approaches for pathogen reduction in food animals (such as dietary supplementation with organic acids/fatty acids, essential oils and other phenolic compounds, prebiotic sugars) are further discussed in other chapters in this book and have also been topics of many recent reviews (13, 18, 62, 70–78).

Good Animal Management Practices Are the Foundation

Hygiene, disinfection, and biosecurity measures are at the heart of an effective preharvest food safety program and protect animal health in addition to limiting animal exposure to those zoonotic pathogens that can cause human foodborne disease. The core elements of a preharvest animal management program include (i) preventing the introduction of infection, (ii) preventing the survival and spread of infection within the herd or flock, and (iii) reducing or eliminating an established infection (79). Best management practices include provision of clean feed and water, maintenance of a clean, well-drained environment, and biosecurity procedures to isolate sick or infected animals and, to the extent possible, to exclude wildlife and such pests as rodents and insects (48, 74, 75, 80). Wild animals, birds, and insect pests as potential pathogen vehicles may be difficult or impossible to control for animals raised outside in pastures or lots, compared to animals in enclosed confinement buildings. Various treatments and additives to reduce pathogen contamination in animal feed and drinking water have been examined and are discussed in recent reviews (74, 75, 81, 82). Direct or indirect fecal–oral exposure is a significant route of pathogen transmission among animals in the production environment, so regular removal and treatment of manure will reduce an important source of pathogens. Reducing pathogens in manure will also reduce the risk for transmission of these pathogens to human food crops and water, and this is discussed further below.

Preharvest food safety research to date suggests that these good production and management practices alone will not eliminate foodborne pathogens (18, 62, 80). However, there are success stories, and notable examples are Denmark's *Salmonella* control programs for various food animal species (83–85). These programs are integrated "feed-to-food" systems that employ intensive surveillance and rigorous biosecurity measures. Preharvest controls differ depending upon species (chickens, pigs, cattle) or segment (e.g., broilers versus laying hens), but elements include elimination of infected breeding poultry flocks (83), classification of cattle and swine herds according to *Salmonella* risk (83, 86), animal movement and trade restrictions (83, 85, 86), feeding strategies (83, 87), cleaning and disinfection of houses between flocks (83), and various management recommendations (e.g., all in–all out pig flow scheme, calving management; 83, 85, 87, 88). Animals that are suspected or confirmed to be *Salmonella*-positive may be slaughtered separately using special hygienic precautions (83, 86). Using the *Salmonella* control program for swine as an example, identified risk factors for high *Salmonella* prevalence in pig herds include feed, management, and hygiene (87). The success of these programs in reducing *Salmonella* in meat and poultry and reducing the incidence of human salmonellosis illustrate the potential for preharvest management approaches to reduce foodborne pathogens (83).

PERSISTENCE AND TRANSMISSION IN PREHARVEST ENVIRONMENTS: CHALLENGES AND OPPORTUNITIES

Oliver et al. (13) discuss the on-farm contamination cycle of zoonotic foodborne pathogens, which begins with the infection of animals by ingestion of contaminated feeds and water or other oral exposure (e.g., by grooming [89]) and proceeds by shedding of pathogens in feces, which in turn contaminates additional feed, drinking water, and/or the environment, thereby causing new infections or reinfection of animals. They further illustrate a scenario in which pathogen amplification in the animal, shedding, and distribution of pathogens in the farm environment leads to the persistence of pathogens on the farm, for which the outcome is a maintained reservoir of foodborne pathogens. Current experience suggests that this scenario may be the case for many pathogens and that interrupting the infection–reinfection cycle will be key for progress in preharvest food safety. Even though complete elimination of pathogens in a preharvest animal production environment is unlikely, measures aimed at reducing infection levels, numbers, and persistence of pathogens should have a positive impact on food safety.

While recent preharvest food safety research efforts have advanced the science, they have also revealed the complexity of the problem; there are multiple pathogens of concern, multiple animal species under different production and management systems, and a variety of sources of pathogens in a farm environment, such as other livestock and domestic animals, wild animals and birds, insects, water, and feed (as reviewed in 70, 74, 75, 90). The broad problem is further complicated by the fact that some of the pathogens are not pathogenic to the animal or may result in asymptomatic infection. These research efforts have led to greater understanding of the transmission of pathogens and their ecology on the farm and have identified numerous factors that can impact pathogens on the farm, including seasonality, production systems, diet, and dietary additives. Moreover, this work has identified both challenges and potential opportunities for reducing pathogen persistence and transmission in food animals and their production environments.

Can Targeting Super-Shedders Reduce Food Safety Risk?

Enumeration of select pathogens in complex samples with high background microflora is technically difficult, so most studies determine the prevalence of the target pathogen (presence/absence). However, use of quantitative techniques to determine pathogen levels can provide information for risk assessments or for determining the relative impact of various factors on transmission and persistence of pathogens. Furthermore, the use of these techniques has led to important discoveries regarding animals that shed high concentrations of pathogens, often referred to as super-shedders. Cattle that are super-shedders of *E. coli* O157:H7 (those animals that excrete $\geq 10^4$ CFU/g in their feces) typically are a small proportion of the total animals in a herd but have a large impact on the prevalence of this pathogen in the remaining cattle and their production environment (91–97). As an example, Chase-Topping et al. (92) found that high prevalence of *E. coli* O157 among cattle on a farm was associated with the presence of a cohort animal shedding high levels of the pathogen ($>10^3$ or $>10^4$ CFU/g of feces) on that farm. Several studies have shown that the presence of super-shedders in feedlot pens is associated with higher *E. coli* O157:H7 prevalence for those pens (91, 94, 97). Super-shedding of *E. coli* O157:H7 by cattle has also been associated with higher levels of hide contamination, which can increase the risk of carcass contamination at harvest (97–100). Correspondingly, the probability of preintervention beef carcass contamination with *E. coli* O157:H7 was strongly correlated with the presence of a high-shedding animal in the same truckload of cattle brought to harvest (101). Given that preharvest interventions may not completely eliminate foodborne pathogens in animal production, these data do illustrate that reducing the load of these pathogens can have an impact on food safety. Accordingly, identifying super-shedding animals for removal or targeted interventions has been suggested for reducing *E. coli* O157:H7 prevalence in cattle (96, 102).

The host, pathogen, and/or other environmental factors or mechanisms that result in super-shedding of *E. coli* O157:H7 by cattle have remained elusive, but their discovery may suggest strategies to reduce this occurrence. Colonization at the recto-anal junction is linked both to high levels of shedding and to longer duration of shedding (93–95, 103–105), although some cattle may shed *E. coli* O157:H7 for only a short time (98, 106). In a Scottish study to identify risk factors for the presence of cattle shedding high levels of the pathogen, *E. coli* O157 from high-level shedders was more likely to be phage type 21/28, leading these researchers to hypothesize that this phage type may be a marker for a genotype or altered gene expression that may result in the tendency for high-level shedding or the ability to persist outside of the host (92, 93). Arthur et al. (107) characterized a diverse set of *E. coli* O157:H7 from super-shedding cattle using pulse-field gel electrophoresis (PFGE), phage typing, Stx-associated bacteriophage insertion site determination, lineage-specific polymorphism assay, and variant analysis of Shiga toxin, *tir*, and antiterminator *Q* genes. They found no genotype that was common to all super-shedder isolates, but the super-shedder isolates tended to have higher frequencies of traits associated with human disease isolates with regard to lineage and *tir* allele. Interestingly, this U.S. study found 19 phage types among the 102 super-shedder isolates, of which 30% were phage type 4 and none were phage type 21/28, indicating regional or global differences (92, 93, 107). Xu et al. (108) used 16S rRNA gene pyrosequencing to examine differences in bacterial communities in the feces of *E. coli* O157:H7 super-shedding and nonshedding cattle. These researchers found distinct differences in fecal microbial communities between these two groups of cattle, with a more diverse microflora associated with super-shedding animals, although the mechanism(s) for these differences is as yet uncertain. Hallewell et al. (106) found a higher prevalence of endemic bacteriophages, including T4-like phages of *Myoviridae*, in feces of cattle that were low shedders of *E. coli* O157:H7, compared to super-shedders ($<10^4$ versus $\geq 10^4$ CFU/g of feces). The T4-like phages exhibited broader host range and stronger lytic capability for *E. coli* O157:H7, further suggesting that bacteriophages may be involved in differences in shedding levels of this pathogen by cattle.

The super-shedding of *E. coli* O157:H7 by cattle is receiving intensive research attention, given the status of this pathogen as an adulterant in beef products; however, super-shedding and its impact on pathogen transmission and environmental contamination has been reported for other zoonotic pathogens and animal species, including *Mycobacterium avium* species *paratuberculosis* (MAP) in dairy cattle (109, 110), *Salmonella enterica* serovar Typhimurium in mice (111), and *Clostridium difficile* in mice (112).

What Drives Pathogen Persistence in Animal Production?

Studies that examined horizontal transmission of zoonotic pathogens among food animals, such as *E. coli* O157:H7 and MAP in cattle (89, 113, 114), *Salmonella* in cattle, pigs, chickens, and turkeys (115–118), and *Campylobacter* in chickens (116), point to the importance of environmental contamination in the pathogen transmission process. This is further suggested by research on the impact of super-shedding animals on the transmission of zoonotic pathogens to other animals in a feedlot or on a farm (91, 94, 97, 98, 109; as reviewed by Chase-Topping et al. [93]). While colonization of the gastrointestinal tract and subsequent amplification and shedding are key steps in the infection cycle, environmental persistence is also involved in the maintenance of zoonotic pathogens in animal production. As an example, *E. coli* O157:H7 and some non-O157 STEC have been demonstrated to persist for long periods in soils, manure, or feedlot surface soils (119–123; as reviewed by Berry and Wells [70]), and the presence of the pathogen in these materials has been observed to be important to the spread of *E. coli* O157:H7 to other animals in the pen or herd (124, 125). The greater persistence of *E. coli* O157:H7 in manure from cattle fed 20 and 40% corn wet distillers grains with solubles compared to corn may contribute to the higher prevalence and levels of this pathogen in feces and on hides of cattle fed wet distillers grains with solubles (126, 127). The abilities of other foodborne pathogens, including *Salmonella, Campylobacter* species, and *Listeria monocytogenes*, to survive in animal manures and/or soils have been described (128–131). As recently reviewed by Vivant et al. (132), *L. monocytogenes* is a natural inhabitant of soils, and this pathogen is commonly isolated from animal manures and the environment (133).

The significance of persistence to maintain zoonotic pathogens in preharvest production environments is further suggested by reports that have found that most isolates on a farm or feedlot are of one or a few predominant genetic subtypes of the target organism, which may persist for months or even years. Baloda et al. (128) repeatedly isolated a single PFGE subtype of *S*. Typhimurium for over a 3-year period in the animals and environment of a Danish pig farm. The persistence of predominant *Salmonella* subtypes during broiler production (134) and in beef cattle at the feedlot (135) has also been described. Petersen and Wedderkopp (136) typed fecal *C. jejuni* isolated from chickens from 12 broiler houses on 10 different farms, and found farm-specific PFGE subtypes of *C. jejuni* that persisted in houses through multiple rotations of broiler flocks. In addition, the long-term persistence of specific clones of *Campylobacter* species (137, 138) and MAP (109) in cattle has been reported. Numerous studies describe the occurrence of persistent, predominant genotypes of *E. coli* O157:H7 in cattle in feedlots and on farms (98, 124, 139–143). A recent study provided evidence of the persistence of *E. coli* O157:H7 of indistinguishable PFGE types on cattle farms for 3 to 4 years (143). The cattle production environment may be more important as a source of *E. coli* O157:H7 than are incoming cattle (139, 141).

Pathogen persistence in animals and the production environment may also be the result of persistent latent infection in animal hosts. Infection can often be asymptomatic, but in some cases the asymptomatic host can become chronically infected and the pathogen can persist in the host for long periods of time. A classic example of this is Mary Mallon (Typhoid Mary), who was a cook that inadvertently transmitted *S. enterica* serovar Typhi to numerous households in New York City a century ago. Many foodborne pathogens are asymptomatic to the food animal host, and some animals can become chronic carriers of the pathogen. In cattle, Johne's disease is a slowly progressing gastrointestinal disease caused by MAP. At slaughter, nearly all cull cattle carried MAP on their hides and more than a third of the ileocecal lymph nodes of cull cattle tested positive for MAP, whereas less than 1% of fed cattle were positive (144).

When infected with different *Salmonella* species, the bovine calf has enumerable levels for all species in the intestinal tissue, intestinal lymph nodes, spleen, liver, and peripheral lymph nodes by seven days (145). At slaughter nearly 72% of cattle tested positive for *Salmonella* in the mesenteric lymph nodes (146). Peripheral nodes can end up in ground beef, and recent research has indicated that bovine peripheral nodes were *Salmonella*-positive by culture (147). In cattle, the subiliac lymph node may be most contaminated with *Salmonella* species, with nearly 12% of these lymph nodes from fed cattle being culture positive (148). In swine the cecum and the ileocecal lymph nodes have the highest prevalence for *Salmonella* (149), and infected piglets can shed *Salmonella* for months (150). However, the subiliac (151) and the prescapular (152) lymph nodes of pigs were typically culture negative for *Salmonella* at slaughter. Poultry can harbor *Salmonella* in the ceca, liver, spleen, and reproductive organs for long durations (153, 154).

Piglets experimentally infected with *Campylobacter coli* had detectable levels in multiple tissues for a few animals, but all piglets had detectable levels in the intestines, with the highest levels in cecum and ascending colon tissues (155). Poultry are reservoirs for *C. jejuni*, which can persist at high levels in the intestinal tract (156). *L. monocytogenes* can be an intracellular pathogen (157), and infected sheep shed this pathogen from their rumen for several weeks (158). *E. coli* O157:H7 has not been shown to be a chronic colonizer of cattle.

The above examples demonstrate that greater understanding of the mechanisms that contribute to pathogen persistence is needed to develop improved strategies for breaking the infection–reinfection cycle. The ability to genotype and identify unique persistent strains by molecular or other means (e.g., by PFGE, lineage-specific polymorphism assay, multilocus sequence typing, or genome sequencing) presents opportunities to learn what these strains can do that other less persistent strains cannot, which allows for greater capacity to survive in the external environment or superior ability to colonize animals.

E. coli O157:H7 isolates of a persistent PFGE type that was shed predominantly during the finishing of feedlot cattle also had greater ability to adhere to Caco-2 human intestinal epithelial cells compared to the less persistent types, which indicated that the more prevalent strains may be better adapted to colonize and persist in the gastrointestinal tract (140). Jeong et al. (159) characterized and compared bovine *E. coli* O157:H7 strains from a dairy farm that were either persistent and predominant or less commonly isolated over a 2-year sampling period. Compared to the less persistent *E. coli* O157:H7 strains, the dominant strain utilized the most carbon sources and was the only strain to oxidize five of the carbon sources, which indicated that more flexibility to use different carbon sources may be advantageous for colonization of cattle or survival in the environment. *E. coli* O157:H7 persistence in soils and resistance to predation by protozoa was associated with the curli-negative phenotype (160). A proteomics approach was used to compare *E. coli* strains associated with persistent or transient bovine mastitis; proteins associated with swimming and swarming motility were more highly expressed in *E. coli* from persistent mastitis cases (161). While mastitis-causing *E. coli* strains are not associated with human foodborne disease, a similar approach may be useful for determining protein expression differences in other persistent versus nonpersistent pathogens from animal production environments. The persistence of *L. monocytogenes* in food processing environments is a critical food safety issue because of the increased risk of cross-contamination of finished products. Fox et al. (162) used a phenotype microarray and transcriptome sequencing to characterize persistent and nonpersistent *L. monocytogenes* and identified gene clusters they hypothesized to be important to persistence of this pathogen in environments outside the human host. With the advances in sequencing technology, and the increased application of functional genomic and comparative genomics/phenomics approaches, progress can be anticipated for determining variations among pathogen species that lead to their persistence in animals and the production environment (163–165; as reviewed by Bronowski et al. [166]). Recent reviews further discuss aspects of survival and/or persistence of *E. coli*, *Salmonella*, and/or *Campylobacter* species in preharvest animal production and other environments (78, 167).

Can We Select for Resistant Animals?

Animal selection by mankind has occurred for centuries. Humans have intentionally selected agricultural animals for meat, milk, and fiber, and in the process have selected for traits such as coat color, size, muscling, milk production, and a variety of other economically important traits. Selection for these traits has been based on obvious and easily measured phenotypes. However, most zoonotic pathogens provide no obvious phenotype for trait evaluation, and as a consequence very little research has been directed into these important avenues.

Heritability (h^2) is an estimate of a trait's variation that can be due to genetic differences in a population. Heritability is important for selective breeding, and measures of heritability have been estimated for a few pathogen-host phenotypes where the phenotype measure can be easily collected. However, heritability of a phenotype does not attribute a specific genetic cause, and the genetic contribution for a phenotype can be polygenic. With recent advances in genotyping, associations between pathogens and specific host genotypes can be better determined. Single nucleotide polymorphisms (SNPs) represent biallelic differences in the genomic sequence and are widely distributed in mammalian genomes (168). These genotypes, or markers, can be used to associate a phenotype with functional genomics and to assist with selection for resistant populations or against susceptible animals.

The poultry industry produces eggs and meat for human consumption, and poultry products have been a major source of foodborne outbreaks of *Salmonella* (169; as reviewed by Doyle and Erickson [74]). Breeds of poultry have been selected for either eggs (layers) or meat products (broilers), and both types of poultry can be chronically infected with *Salmonella*. *Salmonella* infection often occurs early in a bird's life, typically from contaminated eggs (170), and can persist into adulthood (154). *Salmonella* infection is a polygenic disease in poultry, and a variety of heritability estimates for disease susceptibility/resistance have been determined in poultry using pathogen challenge models with different breeding lines and different sites of pathogen localization (171). The resistance to becoming a *Salmonella* carrier (i.e., the carrier state) has been evaluated, and depending on the specific pathogen infection site sampled, age of the bird at infection, and breed type and line, the h^2 for *S. enterica* serovar Enteritidis infection can be as high as 0.29 (172). Similarly, candidate genes have been evaluated and localized single nucleotide polymorphisms have been associated with *Salmonella* infection. The *Slc11a1* (*Nramp1*) gene encodes for a solute carrier protein, and polymorphisms in this region have been associated with *S.* Enteritidis carriage in layers (172, 173). The *Cd28* gene encodes a transmembrane protein found on T-cells, and the *Md2* gene encodes a protein required for toll-like receptor signal recognition; polymorphisms in both genes have been associated with *S.* Enteritidis response and carriage in broilers (174). In contrast, other researchers have evaluated *S.* Typhimurium–resistant and –susceptible lines and located a large multigene region associated with infection resistance (175). Swaggerty et al. (176) have evaluated poultry for cytokine/chemokine responses, and based on selection of extreme phenotypes, they were able to generate broiler lines that differed in pathogen colonization when challenged with *S.* Enteritidis, *Enterococcus gallinarum*, or *C. jejuni*.

In swine, markers associated with *S. enterica* serovar Choleraesuis susceptibility were located throughout the genome (177), but none of these regions were mapped to candidate genes. Using inoculated piglets, Uthe et al. (178) identified a single nucleotide polymorphism in the CCT7 gene that encodes a chaperonin protein subunit (T-complex protein 1 subunit eta) that was associated with colonization and shedding of *S.* Typhimurium. As in poultry, much research in swine has been directed at the immune response and its control (179). Heritability estimates for immune response can range from 0.2 to 0.8 (180), but relationships with specific pathogens have yet to be reported. However, selection for stronger general immune function could benefit the host against multiple pathogens.

In large animals, the rationale for animal selection for pathogen traits is supported in part by studies of disease-causing organisms in production animals, such as with mastitis. Mastitis in dairy cattle is the inflammation of the udder, often a result of bacterial invasion through the teat. This is a costly disease to the dairy industry and is routinely monitored to prevent the sale of milk from diseased animals. The heritability for mastitis is low (0.01 to 0.15), but clinical records are not uniformly collected (see Pighetti and Elliott [181] for a review). The somatic cell count and its normalized transformant, the somatic cell score, are attributes correlated with mastitis, and these parameters offer greater heritability estimates. Host genetic loci associated with mastitis traits appear to be widely distributed across the bovine genome. Mastitis is a complex disease in cattle, and both Gram-positive and Gram-negative bacteria can cause the disease. Heritability estimates for any specific pathogen was low, but collectively, infections caused by Gram-positive organisms exhibited greater heritability than infections attributed to Gram-negative organisms (182). This latter observation is likely a reflection of differing immune responses for the different types of bacteria. Genetic correlations between type of bacterial infection and mastitis are high (0.7 or better), and polymorphisms in a chemokine receptor gene (*CXCR1*) have been associated with the bacterial type infecting mammary glands of dairy heifers (183).

Can the Bacteria in the Gastrointestinal Ecosystem Affect Pathogen Persistence?

Digestive systems are nutrient-rich, and animals have evolved a number of different digestive systems to digest food and absorb nutrients (184). Microbes have evolved in symbiosis with the digestive tracts, and the bacterial levels in the lumen of the gastrointestinal tract can be variable. Bacterial concentrations typically are highest in the colonic digesta and feces.

Swine have a simple digestive system and are considered omnivores. Young swine are typically susceptible to pathogens, and diet can modulate the gastrointestinal environment (185). Diets low in host nondigestible but microbial-fermentable protein can result in production of ammonia and toxic amines, whereas diets high in fermentable carbohydrates can result in beneficial short-chain fatty acids. In young piglets inoculated with *S.* Typhimurium, the shedding of the pathogen is variable (186). On day 0, animals that were later characterized as low shedders had higher levels of *Ruminococcaceae* family, whereas feces of animals that were to become high shedders had higher levels of two bacterial genera: *Phascolarctobacterium* and *Coprobacillus*. The microbial ecology was significantly altered in the high-shedders and was likely associated with immune response, similar to a study reported for *S.* Typhimurium–challenged mice (187, 188). Interestingly, when swine were challenged with *Salmonella*, the fecal microbiome of the challenged pigs was different 21 days postinfection compared to nonchallenged controls (186). Lysozyme has been demonstrated to improve performance and reduce pathogen shedding in piglets (189, 190), and this compound can reduce enterotoxigenic *E. coli* in challenged piglets (191). Maga et al. (192) noted a change in fecal microflora for piglets supplemented with lysozyme, particularly increases in *Prevotella* species, as well as increases in beneficial *Firmicutes*, such as bifidobacteria and lactobacilli.

Cattle are ruminant animals with a complex gastrointestinal tract dominated by pregastric fermentation in the reticulo-rumen. Nonetheless, like simple stomach animals, the fecal microbiota of the bovine is mainly composed of the *Firmicutes* and *Bacteroidetes* (193). In recent years, diet has been shown to influence *E. coli* O157:H7 shedding, and diets with high levels of distillers grains can increase *E. coli* O157:H7 in feces (127, 194). In addition, high levels of distillers grains in the diet increased generic *E. coli* concentrations and altered the habitat (127). The composition and diversity of the fecal microbiome can be greatly influenced by diet (195). The fecal microbiome composition of cattle fed high levels of distillers grains differed from that of cattle fed a corn diet, but no association with

E. coli O157:H7 was reported (196). Recent research with cattle characterized as supershedders of *E. coli* O157:H7 revealed more than 70 bacterial groups that differed in abundance compared to cattle negative for *E. coli* O157:H7 shedding (108). It appears possible that the microbiome composition may increase or decrease the shedding of *E. coli* O157:H7; however, most of these bacterial groups were classified as *Ruminococcaceae* family members, and there may be interactions among the *Ruminococcaceae* that associate with *E. coli* O157:H7 colonization. In contrast, *Salmonella* shedding in cattle feces may not be greatly influenced by gastrointestinal microbiota (197), but animal-to-animal variation in the microbiota may have been too great to observe significant differences for individual bacterial groups in this small study.

Poultry also have a complex digestive system with pregastric and postgastric compartmentalization. The ceca (dual cecum) in poultry are primary microbial habitats and sites for pathogen colonization, and healthy chickens can harbor high levels of pathogens asymptomatically (198). *Firmicutes* is the predominant bacterial phylum present in the broiler ceca, and there is much bird-to-bird variation (199). The microbiome between ceca has some variation within each bird but much less than variation between birds (200). Based on these previous reports, the microbiota composition of the ceca can be highly variable and, unlike cattle and swine, predominated by a few genera. Early research by Rantala and Nurmi (201) indicated that when 1-day-old chicks were pretreated with enriched gastrointestinal digesta from a *Salmonella*-free adult chicken, the chicks were resistant to *Salmonella* Infantis infection when challenged. As a consequence, a number of competitive exclusion products have been developed for poultry (202). Recently, *Bdellovibrio bacteriovorus*, a predator of Gram-negative bacteria, was effective at reducing *S.* Enteritidis infection in chicks (203).

ENVIRONMENTAL BENEFITS OF PREHARVEST PATHOGEN REDUCTION: REDUCING WATER AND PRODUCE CONTAMINATION

Animal feces and manures are significant vehicles of zoonotic pathogens for the direct or indirect contamination of produce, water, and the environment, so the preharvest reduction of pathogens in food animals may further reduce the risk for pathogen dissemination from production and the risk of human illness linked to these sources. Fresh produce has increased in prominence as a source of pathogens contributing to human foodborne illnesses and outbreaks (204, 205). In addition, water contamination and waterborne disease outbreaks have been linked to livestock production as a result of runoff from farms or manure-amended fields (206–209). The risks associated with the use of raw manures for soil amendments are well understood, and procedures for treating manures to reduce pathogens have been reviewed (70, 75, 210, 211). The recent Food Safety Modernization Act Final Rule on Produce Safety is anticipated to improve produce safety by providing guidelines for the safe use of animal manures as soil amendments for fields used to grow produce crops (212).

Recent research has examined pathogen carriage and shedding by insects and wildlife and the risks for these creatures to disseminate pathogens from livestock to contaminate produce or irrigation water (213–215). *E. coli* O157:H7 and *Campylobacter* species were isolated from feral swine living near beef cattle in a major leafy green–growing region on the central California coast; this same swine population was implicated as a potential source of *E. coli* O157:H7 for the contamination of baby spinach linked to the large 2006 foodborne illness outbreak (213, 214). Talley et al. (215) detected STEC genes in pest flies captured in a leafy green crop that was located near a cattle production area and further demonstrated in laboratory studies that house flies can transfer *E. coli* O157:

H7 to spinach leaves. Considerable work has examined zoonotic pathogen carriage by bird species that are often closely associated with cattle, including starlings (216–218), pigeons (219), brown-headed cowbirds (220), common grackles (220), and cattle egrets (220). Pathogen species found in these birds include *E. coli* O157:H7 (216, 220), *Salmonella* (217, 219, 220), *Campylobacter* (218), and MAP (217). In addition to disseminating pathogens to animals on different farms, these birds may play a role in transporting pathogens to produce crops (221–223). Additional research has been concerned with the bioaerosol dissemination of pathogens and the risks for contamination of produce grown near livestock production facilities (224, 225). Riparian buffers, windbreaks, or hedgerows can protect produce crops from airborne pathogens from animal production but may also provide habitat or harborage for wildlife that may carry pathogens and potentially contaminate crops. Comanagement of food safety objectives and conservation of water, soil, and other natural resources is a high-priority area (226).

ADDITIONAL CONSIDERATIONS FOR PREHARVEST RESEARCH AND DEVELOPMENT

The bacterial pathogen species that are discussed in this review are important in terms of their public health impact and regulatory significance but are by no means the only pathogens of concern. Preharvest control of a number of other zoonotic bacteria, as well as many species of parasitic protozoa and viruses, will be vital for improving food safety. Indeed, the long list of zoonotic pathogenic microorganisms reveals the complexity of the problem. Intensive efforts in preharvest food safety research have identified a number of interventions and management strategies that are in use and/or under investigation, with the potential to reduce zoonotic foodborne pathogens in live animals. However, preharvest reduction of foodborne pathogens in food animals and their production environments remains a significant challenge. Another important product of these on-farm food safety research efforts has been the development of greater knowledge regarding the sources, transmission, and ecology of zoonotic pathogens in animal production (Fig. 1). This work has identified both particular challenges and potential opportunities for preharvest food safety. Animals that super-shed high numbers of pathogens and predominant pathogen strains that exhibit long-term persistence appear to play large roles in maintaining the prevalence of pathogens in animals and their production environment. Continued study and advances in sequencing and other technologies should divulge the mechanisms that result in super-shedding and persistence, in addition to enhancing prospects for the selection of pathogen-resistant food animals and the understanding of the microbial ecology of the gastrointestinal tract as it relates to colonization by zoonotic pathogens. It is likely that additional research will reveal other as-yet-unrecognized challenges, which may also point to potential targets or critical control points to exploit to reduce pathogens in animal production.

Research efforts to date indicate that longitudinal studies are valuable for clarifying the nature of pathogen sources, occurrence, and transmission, as well as for identifying the factors that affect pathogens in preharvest environments and determining the efficacy of preharvest intervention treatments. For example, the fecal shedding of *E. coli* O157:H7 by cattle can fluctuate widely in terms of both prevalence (positive/negative) and the concentrations shed (98, 127, 194). Multiple samplings over time and sufficient replication (versus sampling at a single point in time) clarified the role of *E. coli* O157:H7 super-shedder cattle for hide and environmental contamination and further suggested *E. coli* O157:H7 shedding reduction targets to reduce hide contamination (98). Likewise, longitudinal studies were needed to demon-

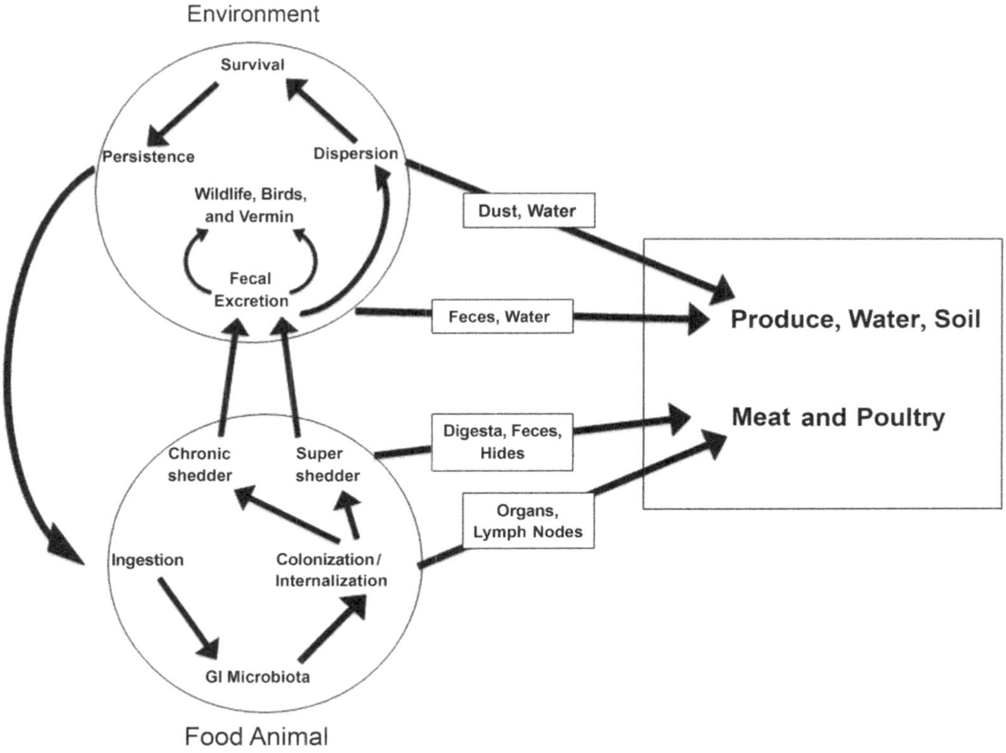

FIGURE 1 Persistence and transmission of zoonotic pathogens in food animals and the production environment.

strate the impact of feeding corn wet distillers grains on *E. coli* O157:H7 prevalence in cattle (127, 194).

Demonstration of the ability of preharvest procedures to contribute to the improved microbial safety of foods at the endpoint of the farm-to-table continuum will be essential for broad animal producer adoption of new technologies. Key to the success of these preharvest procedures will be the development and use of systems-based approaches or integrated food safety systems that link preharvest and postharvest food safety along the animal production and processing chain and avoid breaks in the chain that negate or neutralize the previous food safety efforts. Perhaps the best examples of the need for these systems approaches are demonstrated by studies showing that carriage of pathogens such as *E. coli* O157:H7 and *Salmonella* can increase in cattle or swine during transportation to and lairage at processing plants as a result of contact with one another or with contaminated feces, transport trailers, or holding pens at lairage (101, 227, 228). These exposures may nullify any benefits of previous control efforts and indicate that preserving any pathogen reduction beyond these stages would require wide adoption of preharvest practices by animal producers, and/or the application of additional antipathogen interventions during transportation and lairage. Adoption of preharvest control practices will also be favored if the procedures are economical and easy to implement. Furthermore, preharvest food safety approaches should contribute to the sustainability of farming, ranching, and feeding enterprises. Finally, successful preharvest food safety strategies directed at food animals can provide additional benefits to public health beyond the production of safer animal-based products, by reducing the risk for dissemination of zoonotic pathogens to produce, water, and the environment.

CITATION

Berry ED, Wells JE. 2016. Reducing foodborne pathogen persistence and transmission in animal production environments: challenges and opportunities. Microbiol Spectrum 4(4):PFS-0006-2014.

REFERENCES

1. **Arthur TM, Bosilevac JM, Nou X, Shackelford SD, Wheeler TL, Kent MP, Jaroni D, Pauling B, Allen DM, Koohmaraie M.** 2004. *Escherichia coli* O157 prevalence and enumeration of aerobic bacteria, *Enterobacteriaceae*, and *Escherichia coli* O157 at various steps in commercial beef processing plants. *J Food Prot* **67:**658–665.
2. **Aslam M, Nattress F, Greer G, Yost C, Gill C, McMullen L.** 2003. Origin of contamination and genetic diversity of *Escherichia coli* in beef cattle. *Appl Environ Microbiol* **69:**2794–2799.
3. **Barkocy-Gallagher GA, Arthur TM, Siragusa GR, Keen JE, Elder RO, Laegreid WW, Koohmaraie M.** 2001. Genotypic analyses of *Escherichia coli* O157:H7 and O157 nonmotile isolates recovered from beef cattle and carcasses at processing plants in the Midwestern states of the United States. *Appl Environ Microbiol* **67:**3810–3818.
4. **Brichta-Harhay DM, Guerini MN, Arthur TM, Bosilevac JM, Kalchayanand N, Shackelford SD, Wheeler TL, Koohmaraie M.** 2008. *Salmonella* and *Escherichia coli* O157:H7 contamination on hides and carcasses of cull cattle presented for slaughter in the United States: an evaluation of prevalence and bacterial loads by immunomagnetic separation and direct plating methods. *Appl Environ Microbiol* **74:**6289–6297.
5. **Nou X, Rivera-Betancourt M, Bosilevac JM, Wheeler TL, Shackelford SD, Gwartney BL, Reagan JO, Koohmaraie M.** 2003. Effect of chemical dehairing on the prevalence of *Escherichia coli* O157:H7 and the levels of aerobic bacteria and *Enterobacteriaceae* on carcasses in a commercial beef processing plant. *J Food Prot* **66:**2005–2009.
6. **Bosilevac JM, Arthur TM, Wheeler TL, Shackelford SD, Rossman M, Reagan JO, Koohmaraie M.** 2004. Prevalence of *Escherichia coli* O157 and levels of aerobic bacteria and *Enterobacteriaceae* are reduced when hides are washed and treated with cetylpyridinium chloride at a commercial beef processing plant. *J Food Prot* **67:**646–650.
7. **Bosilevac JM, Nou X, Osborn MS, Allen DM, Koohmaraie M.** 2005. Development and evaluation of an on-line hide decontamination procedure for use in a commercial beef processing plant. *J Food Prot* **68:**265–272.
8. **Hansson I, Forshell LP, Gustafsson P, Boqvist S, Lindblad J, Engvall EO, Andersson Y, Vågsholm I.** 2007. Summary of the Swedish *Campylobacter* program in broilers, 2001 through 2005. *J Food Prot* **70:**2008–2014.
9. **Lindblad M, Hansson I, Vågsholm I, Lindqvist R.** 2006. Postchill *Campylobacter* prevalence on broiler carcasses in relation to slaughter group colonization level and chilling system. *J Food Prot* **69:**495–499.
10. **Reich F, Atanassova V, Haunhorst E, Klein G.** 2008. The effects of *Campylobacter* numbers in caeca on the contamination of broiler carcasses with *Campylobacter*. *Int J Food Microbiol* **127:**116–120.
11. **Anonymous.** 1906. Meat Inspection Act of June 30, 1906, 34 Stat. 764, ch. 3913, pp. 674–679.
12. **Smith DR, Novotnaj K, Smith G.** 2010. Preharvest food safety: what do the past and the present tell us about the future? *J Agromed* **15:**275–280.
13. **Oliver SP, Patel DA, Callaway TR, Torrence ME.** 2009. ASAS centennial paper: developments and future outlook for preharvest food safety. *J Anim Sci* **87:**419–437.
14. **U.S. Department of Agriculture, Food Safety and Inspection Service (USDA-FSIS).** 1996. Pathogen reduction; hazard analysis and critical control point (HACCP) systems; final rule. *Fed Regist* **61:**38806–38989.
15. **Koohmaraie M, Arthur TM, Bosilevac JM, Brichta-Harhay DM, Kalchayanand N, Shackelford SD, Wheeler TL.** 2007. Interventions to reduce/eliminate *Escherichia coli* O157:H7 in ground beef. *Meat Sci* **77:**90–96.
16. **U.S. Department of Agriculture, Food Safety and Inspection Service (USDA-FSIS).** 2012. Shiga toxin-producing *Escherichia coli* in certain raw beef products. *Fed Regist* **77:**31975–31981.
17. **U.S. Department of Agriculture, Food Safety and Inspection Service (USDA-FSIS).** 2014. Modernization of poultry slaughter inspection; final rule. *Fed Regist* **79:**49566–49637.

18. **Wheeler TL, Kalchayanand N, Bosilevac JM.** 2014. Pre- and post-harvest interventions to reduce pathogen contamination in the U.S. beef industry. *Meat Sci* **98:**372–382.
19. **Torrence ME.** 2003. U.S. federal activities, initiatives, and research in food safety, p 3–9. *In* Torrence ME, Isaacson RE (ed), *Microbial Food Safety in Animal Agriculture: Current Topics.* Iowa State Press, Ames, IA.
20. **Binder S, Khabbaz R, Swaminathan B, Tauxe R, Potter M.** 1998. The National Food Safety Initiative. *Emerg Infect Dis* **4:**347–349.
21. **Institute of Medicine (IOM) and National Research Council (NRC) Committee to Ensure Safety Food from Production to Consumption.** 1998. *Ensuring safe food from production to consumption.* National Academy of Sciences, National Academies Press, Washington, DC.
22. **U.S. Department of Agriculture, Food Safety and Inspection Service (USDA-FSIS).** 2015. FSIS Directive 7120.1: safe and suitable ingredients used in the production of meat, poultry, and egg products. http://www.fsis.usda.gov/wps/portal/fsis/topics/regulations/directives/7000-series/safe-suitable-ingredients-related-document.
23. **Nutsch AL, Phebus RK, Riemann MJ, Kotrola JS, Wilson RC, Boyer JE Jr, Brown TL.** 1998. Steam pasteurization of commercially slaughtered beef carcasses: evaluation of bacterial populations at five anatomical locations. *J Food Prot* **61:**571–577.
24. **Huffman RD.** 2002. Current and future technologies for the decontamination of carcasses and fresh meat. *Meat Sci* **62:**285–294.
25. **Kalchayanand N, Arthur TM, Bosilevac JM, Schmidt JW, Wang R, Shackelford SD, Wheeler TL.** 2012. Evaluation of commonly used antimicrobial interventions for fresh beef inoculated with Shiga toxin-producing *Escherichia coli* serotypes O26, O45, O103, O111, O121, O145, and O157:H7. *J Food Prot* **75:**1207–1212.
26. **Pearce RA, Bolton DJ, Sheridan JJ, McDowell DA, Blair IS, Harrington D.** 2004. Studies to determine the critical control points in pork slaughter hazard analysis and critical control point systems. *Int J Food Microbiol* **90:**331–339.
27. **Wheatley P, Giotis ES, McKevitt AI.** 2014. Effects of slaughtering operations on carcass contamination in an Irish pork production plant. *Ir Vet J* **67:**1–6.
28. **Cox JM, Pavic A.** 2010. Advances in enteropathogen control in poultry production. *J Appl Microbiol* **108:**745–755.
29. **Hugas M, Tsigarida E.** 2008. Pros and cons of carcass decontamination: the role of the European Food Safety Authority. *Meat Sci* **78:** 43–52.
30. **Sofos JN, Geornaras I.** 2010. Overview of current meat hygiene and safety risks and summary of recent studies on biofilms, and control of *Escherichia coli* O157:H7 in nonintact, and *Listeria monocytogenes* in ready-to-eat, meat products. *Meat Sci* **86:**2–14.
31. **Santini C, Baffoni L, Gaggia F, Granata M, Gasbarri R, Di Gioia D, Biavati B.** 2010. Characterization of probiotic strains: an application as feed additives in poultry against *Campylobacter jejuni. Int J Food Microbiol* **141**(Suppl 1)**:** S98–S108.
32. **Zhang G, Ma L, Doyle MP.** 2007. *Salmonellae* reduction in poultry by competitive exclusion bacteria *Lactobacillus salivarius* and *Streptococcus cristatus. J Food Prot* **70:**874–878.
33. **Casey PG, Gardiner GE, Casey G, Bradshaw B, Lawlor PG, Lynch PB, Leonard FC, Stanton C, Ross RP, Fitzgerald GF, Hill C.** 2007. A five-strain probiotic combination reduces pathogen shedding and alleviates disease signs in pigs challenged with *Salmonella enterica* serovar Typhimurium. *Appl Environ Microbiol* **73:**1858–1863.
34. **Scharek-Tedin L, Pieper R, Vahjen W, Tedin K, Neumann K, Zentek J.** 2013. *Bacillus cereus* var. Toyoi modulates the immune reaction and reduces the occurrence of diarrhea in piglets challenged with *Salmonella* Typhimurium DT104. *J Anim Sci* **91:**5696–5704.
35. **Lema M, Williams L, Rao DR.** 2001. Reduction of fecal shedding of enterohemorrhagic *Escherichia coli* O157:H7 in lambs by feeding microbial feed supplement. *Small Rumin Res* **39:**31–39.
36. **Stephens TP, Loneragan GH, Karunasena E, Brashears MM.** 2007. Reduction of *Escherichia coli* O157 and *Salmonella* in feces and on hides of feedlot cattle using various doses of a direct-fed microbial. *J Food Prot* **70:**2386–2391.
37. **Tabe ES, Oloya J, Doetkott DK, Bauer ML, Gibbs PS, Khaitsa ML.** 2008. Comparative effect of direct-fed microbials on fecal shedding of *Escherichia coli* O157:H7 and *Salmonella* in naturally infected feedlot cattle. *J Food Prot* **71:**539–544.
38. **Pei Y, Parreira VR, Roland KL, Curtiss R III, Prescott JF.** 2014. Assessment of attenuated *Salmonella* vaccine strains in controlling experimental *Salmonella* Typhimurium infection in chickens. *Can J Vet Res* **78:**23–30.
39. **Penha Filho RAC, de Paiva JB, da Silva MD, de Almeida AM, Berchieri A Jr.** 2010. Control of *Salmonella* Enteritidis and *Salmonella* Gallinarum in birds by using live vaccine

candidate containing attenuated *Salmonella* Gallinarum mutant strain. *Vaccine* **28:**2853–2859.
40. **Wyszyńska A, Raczko A, Lis M, Jagusztyn-Krynicka EK.** 2004. Oral immunization of chickens with avirulent *Salmonella* vaccine strain carrying *C. jejuni* 72Dz/92 cjaA gene elicits specific humoral immune response associated with protection against challenge with wild-type *Campylobacter*. *Vaccine* **22:**1379–1389.
41. **Yeh H-Y, Hiett KL, Line JE, Seal BS.** 2014. Characterization and reactivity of broiler chicken sera to selected recombinant *Campylobacter jejuni* chemotactic proteins. *Arch Microbiol* **196:**375–383.
42. **Husa JA, Edler RA, Walter DH, Holck JT, Saltzman RJ.** 2009. A comparison of the safety, cross-protection, and serologic response associated with two commercial oral *Salmonella* vaccines in swine. *J Swine Health Prod* **17:**10–21.
43. **Schwarz P, Kich JD, Kolb J, Cardoso M.** 2011. Use of an avirulent live *Salmonella* Choleraesuis vaccine to reduce the prevalence of *Salmonella* carrier pigs at slaughter. *Vet Rec* **169:**553.
44. **Fox JT, Thomson DU, Drouillard JS, Thornton AB, Burkhardt DT, Emery DA, Nagaraja TG.** 2009. Efficacy of *Escherichia coli* O157:H7 siderophore receptor/porin proteins-based vaccine in feedlot cattle naturally shedding *E. coli* O157. *Foodborne Pathog Dis* **6:**893–899.
45. **Mizuno T, McLennan M, Trott D.** 2008. Intramuscular vaccination of young calves with a *Salmonella* Dublin metabolic-drift mutant provides superior protection to oral delivery. *Vet Res* **39:**26.
46. **Moxley RA, Smith DR, Luebbe M, Erickson GE, Klopfenstein TJ, Rogan D.** 2009. *Escherichia coli* O157:H7 vaccine dose-effect in feedlot cattle. *Foodborne Pathog Dis* **6:**879–884.
47. **Smith DR, Moxley RA, Klopfenstein TJ, Erickson GE.** 2009. A randomized longitudinal trial to test the effect of regional vaccination within a cattle feedyard on *Escherichia coli* O157:H7 rectal colonization, fecal shedding, and hide contamination. *Foodborne Pathog Dis* **6:**885–892.
48. **U.S. Department of Agriculture, Food Safety and Inspection Service (USDA-FSIS).** 2014. Pre-harvest management controls and intervention options for reducing Shiga toxin-producing *Escherichia coli* shedding in cattle: an overview of current research. http://www.fsis.usda.gov/wps/wcm/connect/d5314cc7-1ef7-4586-bca2-f2ed86d9532f/Reducing-Ecoli-Shedding-in-Cattle.pdf?MOD=AJPERES.
49. **Thomson DU, Loneragan GH, Thornton AB, Lechtenberg KF, Emery DA, Burkhardt DT, Nagaraja TG.** 2009. Use of a siderophore receptor and porin proteins-based vaccine to control the burden of *Escherichia coli* O157:H7 in feedlot cattle. *Foodborne Pathog Dis* **6:**871–877.
50. **Vogstad AR, Moxley RA, Erickson GE, Klopfenstein TJ, Smith DR.** 2013. Assessment of heterogeneity of efficacy of a three-dose regimen of a type III secreted protein vaccine for reducing STEC O157 in feces of feedlot cattle. *Foodborne Pathog Dis* **10:**678–683.
51. **Zuraw L.** 2013. Study: *E. coli* cattle vaccination could prevent 83 percent of human cases. *Food Safety News*. http://www.foodsafetynews.com/2013/09/e-coli-cattle-vaccination-could-prevent-83-percent-of-human-cases/#.VctuSPnSzJF.
52. **U.S. Department of Agriculture, Animal and Plant Health Inspection Service (USDA-APHIS).** 2013. Vaccine usage in U.S. feedlots. http://www.aphis.usda.gov/animal_health/nahms/feedlot/downloads/feedlot2011/Feed11_is_VaccineUsage.pdf.
53. **Tonsor GT, Schroeder TC.** 2015. Market impacts of *E. coli* vaccination in U.S. feedlot cattle. *Agric Food Econ* **3:**7.
54. **Callaway TR, Edrington TS, Brabban AD, Anderson RC, Rossman ML, Engler MJ, Carr MA, Genovese KJ, Keen JE, Looper ML, Kutter EM, Nisbet DJ.** 2008. Bacteriophage isolated from feedlot cattle can reduce *Escherichia coli* O157:H7 populations in ruminant gastrointestinal tracts. *Foodborne Pathog Dis* **5:**183–191.
55. **Raya RR, Oot RA, Moore-Maley B, Wieland S, Callaway TR, Kutter EM, Brabban AD.** 2011. Naturally resident and exogenously applied T4-like and T5-like bacteriophages can reduce *Escherichia coli* O157:H7 levels in sheep guts. *Bacteriophage* **1:**15–24.
56. **Rozema EA, Stephens TP, Bach SJ, Okine EK, Johnson RP, Stanford K, McAllister TA.** 2009. Oral and rectal administration of bacteriophages for control of *Escherichia coli* O157:H7 in feedlot cattle. *J Food Prot* **72:**241–250.
57. **Albino LAA, Rostagno MH, Húngaro HM, Mendonça RCS.** 2014. Isolation, characterization, and application of bacteriophages for *Salmonella* spp. biocontrol in pigs. *Foodborne Pathog Dis* **11:**602–609.
58. **Callaway TR, Edrington TS, Brabban A, Kutter B, Karriker L, Stahl C, Wagstrom E, Anderson R, Poole TL, Genovese K, Krueger N, Harvey R, Nisbet DJ.** 2011. Evaluation of phage treatment as a strategy to reduce *Salmonella* populations in growing swine. *Foodborne Pathog Dis* **8:**261–266.
59. **Atterbury RJ, Van Bergen MA, Ortiz F, Lovell MA, Harris JA, De Boer A, Wagenaar**

JA, Allen VM, Barrow PA. 2007. Bacteriophage therapy to reduce *Salmonella* colonization of broiler chickens. *Appl Environ Microbiol* **73:** 4543–4549.

60. **Fischer S, Kittler S, Klein G, Glünder G.** 2013. Impact of a single phage and a phage cocktail application in broilers on reduction of *Campylobacter jejuni* and development of resistance. *PLoS One* **8:**e78543.
61. **Loc Carrillo C, Atterbury RJ, el-Shibiny A, Connerton PL, Dillon E, Scott A, Connerton IF.** 2005. Bacteriophage therapy to reduce *Campylobacter jejuni* colonization of broiler chickens. *Appl Environ Microbiol* **71:**6554–6563.
62. **Callaway TR, Edrington TS, Nisbet DJ.** 2014. Meat Science and Muscle Biology Symposium: ecological and dietary impactors of foodborne pathogens and methods to reduce fecal shedding in cattle. *J Anim Sci* **92:**1356–1365.
63. **Anderson RC, Hume ME, Genovese KJ, Callaway TR, Jung YS, Edrington TS, Poole TL, Harvey RB, Bischoff KM, Nisbet DJ.** 2004. Effect of drinking-water administration of experimental chlorate ion preparations on *Salmonella enterica* serovar Typhimurium colonization in weaned and finished pigs. *Vet Res Commun* **28:**179–189.
64. **Byrd JA, Burnham MR, McReynolds JL, Anderson RC, Genovese KJ, Callaway TR, Kubena LF, Nisbet DJ.** 2008. Evaluation of an experimental chlorate product as a preslaughter feed supplement to reduce *salmonella* in meat-producing birds. *Poult Sci* **87:**1883–1888.
65. **Moore RW, Byrd JA, Knape KD, Anderson RC, Callaway TR, Edrington T, Kubena LF, Nisbet DJ.** 2006. The effect of an experimental chlorate product on *Salmonella* recovery of turkeys when administered prior to feed and water withdrawal. *Poult Sci* **85:**2101–2105.
66. **Anderson RC, Callaway TR, Buckley SA, Anderson TJ, Genovese KJ, Sheffield CL, Nisbet DJ.** 2001. Effect of oral sodium chlorate administration on *Escherichia coli* O157:H7 in the gut of experimentally infected pigs. *Int J Food Microbiol* **71:**125–130.
67. **Callaway TR, Edrington TS, Anderson RC, Genovese KJ, Poole TL, Elder RO, Byrd JA, Bischoff KM, Nisbet DJ.** 2003. *Escherichia coli* O157:H7 populations in sheep can be reduced by chlorate supplementation. *J Food Prot* **66:**194–199.
68. **Anderson RC, Carr MA, Miller RK, King DA, Carstens GE, Genovese KJ, Callaway TR, Edrington TS, Jung YS, McReynolds JL, Hume ME, Beier RC, Elder RO, Nisbet RJ.** 2005. Effects of experimental chlorate preparations as feed and water supplements on *Escherichia coli* colonization and contamination of beef cattle and carcasses. *Food Microbiol* **22:**439–447.
69. **Callaway TR, Anderson RC, Genovese KJ, Poole TL, Anderson TJ, Byrd JA, Kubena LF, Nisbet DJ.** 2002. Sodium chlorate supplementation reduces *E. coli* O157:H7 populations in cattle. *J Anim Sci* **80:**1683–1689.
70. **Berry ED, Wells JE.** 2010. *Escherichia coli* O157:H7: recent advances in research on occurrence, transmission, and control in cattle and the production environment. *Adv Food Nutr Res* **60:**67–117.
71. **Callaway TR, Edrington TS, Anderson RC, Harvey RB, Genovese KJ, Kennedy CN, Venn DW, Nisbet DJ.** 2008. Probiotics, prebiotics and competitive exclusion for prophylaxis against bacterial disease. *Anim Health Res Rev* **9:**217–225.
72. **Callaway TR, Carr MA, Edrington TS, Anderson RC, Nisbet DJ.** 2009. Diet, *Escherichia coli* O157:H7, and cattle: a review after 10 years. *Curr Issues Mol Biol* **11:**67–79.
73. **Callaway TR, Anderson RC, Edrington TS, Genovese KJ, Harvey RB, Poole TL, Nisbet DJ.** 2013. Novel methods for pathogen control in livestock preharvest: an update, p 275–304. *In* Sofos J (ed), *Advances in Microbial Food Safety*, 1st ed. Woodhead Publishing Limited, Sawston, Cambridge, UK.
74. **Doyle MP, Erickson MC.** 2006. Reducing the carriage of foodborne pathogens in livestock and poultry. *Poult Sci* **85:**960–973.
75. **Doyle MP, Erickson MC.** 2012. Opportunities for mitigating pathogen contamination during on-farm food production. *Int J Food Microbiol* **152:**54–74.
76. **Jacob ME, Callaway TR, Nagaraja TG.** 2009. Dietary interactions and interventions affecting *Escherichia coli* O157 colonization and shedding in cattle. *Foodborne Pathog Dis* **6:**785–792.
77. **Johnson RP, Gyles CL, Huff WE, Ojha S, Huff GR, Rath NC, Donoghue AM.** 2008. Bacteriophages for prophylaxis and therapy in cattle, poultry and pigs. *Anim Health Res Rev* **9:**201–215.
78. **Larsen MH, Dalmasso M, Ingmer H, Langsrud S, Malakauskas M, Mader A, Møretrø T, Možina SS, Rychli K, Wagner M, Wallace RJ, Zentek J, Jordan K.** 2014. Persistence of foodborne pathogens and their control in primary and secondary food production chains. *Food Control* **44:**92–109.
79. **World Health Organization, Department of Communicable Disease.** 2001. Pre-harvest safety: report of a WHO consultation with the

participation of the Food and Agriculture Organization of the United Nations and the Office of International des Epizooties. http://apps.who.int/iris/handle/10665/68889.
80. **Beef Industry Food Safety Council (BIFSCO).** 2013. Production best practices (PBP) to aid in the control of foodborne pathogens in groups of cattle. http://www.bifsco.org/CMDocs/BIFSCO/Best%20Practices/Production%20Best%20Practices.pdf.
81. **Sparks NHC.** 2009. The role of the water supply system in the infection and control of *Campylobacter* in chicken. *Worlds Poult Sci J* **65:**459–473.
82. **Wales AD, Allen VM, Davies RH.** 2010. Chemical treatment of animal feed and water for the control of *Salmonella*. *Foodborne Pathog Dis* **7:**3–15.
83. **Wegener HC, Hald T, Lo Fo Wong D, Madsen M, Korsgaard H, Bager F, Gerner-Smidt P, Mølbak K.** 2003. *Salmonella* control programs in Denmark. *Emerg Infect Dis* **9:**774–780.
84. **Hurd HS, Enøe C, Sørensen L, Wachmann H, Corns SM, Bryden KM, Greiner M.** 2008. Risk-based analysis of the Danish pork *Salmonella* program: past and future. *Risk Anal* **28:**341–351.
85. **Nielsen B, Alban L, Stege H, Sørensen LL, Møgelmose V, Bagger J, Dahl J, Baggesen DL.** 2001. A new *Salmonella* surveillance and control programme in Danish pig herds and slaughterhouses. *Berl Munch Tierarztl Wochenschr* **114:**323–326.
86. **Nielsen LR.** 2009. Current *Salmonella*-control in cattle in Denmark. http://www.fodevarestyrelsen.dk/NR/rdonlyres/24AEB101-ABAC-405C-A0EF-693B8E4B36F2/15348/MicrosoftWordBilag7FINALCurrentSalmonellacontrolin.pdf.
87. **Dahl J.** 2012. The Danish *Salmonella* control program: lessons learnt. http://www.pig333.com/salmonella/the-danish-salmonella-control-program-lessons-learnt_5977/.
88. **Nielsen TD, Vesterbæk IL, Kudahl AB, Borup KJ, Nielsen LR.** 2012. Effect of management on prevention of *Salmonella* Dublin exposure of calves during a one-year control programme in 84 Danish dairy herds. *Prev Vet Med* **105:**101–109.
89. **McGee P, Scott L, Sheridan JJ, Earley B, Leonard N.** 2004. Horizontal transmission of *Escherichia coli* O157:H7 during cattle housing. *J Food Prot* **67:**2651–2656.
90. **LeJeune JT, Wetzel AN.** 2007. Preharvest control of *Escherichia coli* O157 in cattle. *J Anim Sci* **85**(Suppl)**:**E73–E80.
91. **Bach SJ, Selinger LJ, Stanford K, McAllister TA.** 2005. Effect of supplementing corn- or barley-based feedlot diets with canola oil on faecal shedding of *Escherichia coli* O157:H7 by steers. *J Appl Microbiol* **98:**464–475.
92. **Chase-Topping ME, McKendrick IJ, Pearce MC, MacDonald P, Matthews L, Halliday J, Allison L, Fenlon D, Low JC, Gunn G, Woolhouse MEJ.** 2007. Risk factors for the presence of high-level shedders of *Escherichia coli* O157 on Scottish farms. *J Clin Microbiol* **45:**1594–1603.
93. **Chase-Topping M, Gally D, Low C, Matthews L, Woolhouse M.** 2008. Super-shedding and the link between human infection and livestock carriage of *Escherichia coli* O157. *Nat Rev Microbiol* **6:**904–912.
94. **Cobbold RN, Hancock DD, Rice DH, Berg J, Stilborn R, Hovde CJ, Besser TE.** 2007. Rectoanal junction colonization of feedlot cattle by *Escherichia coli* O157:H7 and its association with supershedders and excretion dynamics. *Appl Environ Microbiol* **73:**1563–1568.
95. **Low JC, McKendrick IJ, McKechnie C, Fenlon D, Naylor SW, Currie C, Smith DGE, Allison L, Gally DL.** 2005. Rectal carriage of enterohemorrhagic *Escherichia coli* O157 in slaughtered cattle. *Appl Environ Microbiol* **71:**93–97.
96. **Matthews L, Low JC, Gally DL, Pearce MC, Mellor DJ, Heesterbeek JAP, Chase-Topping M, Naylor SW, Shaw DJ, Reid SW, Gunn GJ, Woolhouse ME.** 2006. Heterogeneous shedding of *Escherichia coli* O157 in cattle and its implications for control. *Proc Natl Acad Sci USA* **103:**547–552.
97. **Stephens TP, McAllister TA, Stanford K.** 2009. Perineal swabs reveal effect of super shedders on the transmission of *Escherichia coli* O157:H7 in commercial feedlots. *J Anim Sci* **87:**4151–4160.
98. **Arthur TM, Keen JE, Bosilevac JM, Brichta-Harhay DM, Kalchayanand N, Shackelford SD, Wheeler TL, Nou X, Koohmaraie M.** 2009. Longitudinal study of *Escherichia coli* O157:H7 in a beef cattle feedlot and role of high-level shedders in hide contamination. *Appl Environ Microbiol* **75:**6515–6523.
99. **Arthur TM, Brichta-Harhay DM, Bosilevac JM, Kalchayanand N, Shackelford SD, Wheeler TL, Koohmaraie M.** 2010. Super shedding of *Escherichia coli* O157:H7 by cattle and the impact on beef carcass contamination. *Meat Sci* **86:**32–37.
100. **Stephens TP, McAllister TA, Stanford K.** 2008. Development of an experimental model to assess the ability of *Escherichia coli* O157:H7-inoculated fecal pats to mimic a super shedder within a feedlot environment. *J Food Prot* **71:**648–652.

101. **Fox JT, Renter DG, Sanderson MW, Nutsch AL, Shi X, Nagaraja TG.** 2008. Associations between the presence and magnitude of *Escherichia coli* O157 in feces at harvest and contamination of preintervention beef carcasses. *J Food Prot* **71:**1761–1767.
102. **Rice DH, Sheng HQ, Wynia SA, Hovde CJ.** 2003. Rectoanal mucosal swab culture is more sensitive than fecal culture and distinguishes *Escherichia coli* O157:H7-colonized cattle and those transiently shedding the same organism. *J Clin Microbiol* **41:**4924–4929.
103. **Lim JY, Li J, Sheng H, Besser TE, Potter K, Hovde CJ.** 2007. *Escherichia coli* O157:H7 colonization at the rectoanal junction of long-duration culture-positive cattle. *Appl Environ Microbiol* **73:**1380–1382.
104. **Naylor SW, Low JC, Besser TE, Mahajan A, Gunn GJ, Pearce MC, McKendrick IJ, Smith DGE, Gally DL.** 2003. Lymphoid follicle-dense mucosa at the terminal rectum is the principal site of colonization of enterohemorrhagic *Escherichia coli* O157:H7 in the bovine host. *Infect Immun* **71:**1505–1512.
105. **Davis MA, Rice DH, Sheng H, Hancock DD, Besser TE, Cobbold R, Hovde CJ.** 2006. Comparison of cultures from rectoanal-junction mucosal swabs and feces for detection of *Escherichia coli* O157 in dairy heifers. *Appl Environ Microbiol* **72:**3766–3770.
106. **Hallewell J, Niu YD, Munns K, McAllister TA, Johnson RP, Ackermann H-W, Thomas JE, Stanford K.** 2014. Differing populations of endemic bacteriophages in cattle shedding high and low numbers of *Escherichia coli* O157: H7 bacteria in feces. *Appl Environ Microbiol* **80:**3819–3825.
107. **Arthur TM, Ahmed R, Chase-Topping M, Kalchayanand N, Schmidt JW, Bono JL.** 2013. Characterization of *Escherichia coli* O157:H7 strains isolated from supershedding cattle. *Appl Environ Microbiol* **79:**4294–4303.
108. **Xu Y, Dugat-Bony E, Zaheer R, Selinger L, Barbieri R, Munns K, McAllister TA, Selinger LB.** 2014. *Escherichia coli* O157:H7 super-shedder and non-shedder feedlot steers harbour distinct fecal bacterial communities. *PLoS One* **9:**e98115.
109. **Pradhan AK, Mitchell RM, Kramer AJ, Zurakowski MJ, Fyock TL, Whitlock RH, Smith JM, Hovingh E, Van Kessel JA, Karns JS, Schukken YH.** 2011. Molecular epidemiology of *Mycobacterium avium* subsp. *paratuberculosis* in a longitudinal study of three dairy herds. *J Clin Microbiol* **49:**893–901.
110. **Mitchell RM, Whitlock RH, Stehman SM, Benedictus A, Chapagain PP, Grohn YT, Schukken YH.** 2008. Simulation modeling to evaluate the persistence of *Mycobacterium avium* subsp. *paratuberculosis* (MAP) on commercial dairy farms in the United States. *Prev Vet Med* **83:**360–380.
111. **Lawley TD, Bouley DM, Hoy YE, Gerke C, Relman DA, Monack DM.** 2008. Host transmission of *Salmonella enterica* serovar Typhimurium is controlled by virulence factors and indigenous intestinal microbiota. *Infect Immun* **76:**403–416.
112. **Lawley TD, Clare S, Walker AW, Goulding D, Stabler RA, Croucher N, Mastroeni P, Scott P, Raisen C, Mottram L, Fairweather NF, Wren BW, Parkhill J, Dougan G.** 2009. Antibiotic treatment of *Clostridium difficile* carrier mice triggers a supershedder state, spore-mediated transmission, and severe disease in immunocompromised hosts. *Infect Immun* **77:**3661–3669.
113. **Gautam R, Kulow M, Park D, Gonzales TK, Dahm J, Shiroda M, Stasic AJ, Döpfer D, Kaspar CW, Ivanek R.** 2014. Transmission of *Escherichia coli* O157:H7 in cattle is influenced by the level of environmental contamination. *Epidemiol Infect* **143:**274–287.
114. **Fecteau M-E, Whitlock RH, Buergelt CD, Sweeney RW.** 2010. Exposure of young dairy cattle to *Mycobacterium avium* subsp. *paratuberculosis* (MAP) through intensive grazing of contaminated pastures in a herd positive for Johne's disease. *Can Vet J* **51:**198–200.
115. **Cobbold RN, Rice DH, Davis MA, Besser TE, Hancock DD.** 2006. Long-term persistence of multi-drug-resistant *Salmonella enterica* serovar Newport in two dairy herds. *J Am Vet Med Assoc* **228:**585–591.
116. **Hannah JF, Wilson JL, Cox NA, Richardson LJ, Cason JA, Bourassa DV, Buhr RJ.** 2011. Horizontal transmission of *Salmonella* and *Campylobacter* among caged and cage-free laying hens. *Avian Dis* **55:**580–587.
117. **Nayak R, Stewart-King T.** 2008. Molecular epidemiological analysis and microbial source tracking of *Salmonella enterica* serovars in a preharvest turkey production environment. *Foodborne Pathog Dis* **5:**115–126.
118. **Nollet N, Houf K, Dewulf J, Duchateau L, De Zutter L, De Kruif A, Maes D.** 2005. Distribution of *Salmonella* strains in farrow-to-finish pig herds: a longitudinal study. *J Food Prot* **68:**2012–2021.
119. **Berry ED, Woodbury BL, Nienaber JA, Eigenberg RA, Thurston JA, Wells JE.** 2007. Incidence and persistence of zoonotic bacterial and protozoan pathogens in a beef cattle feedlot runoff control vegetative treatment system. *J Environ Qual* **36:**1873–1882.

120. **Wang G, Zhao T, Doyle MP.** 1996. Fate of enterohemorrhagic *Escherichia coli* O157:H7 in bovine feces. *Appl Environ Microbiol* **62:** 2567–2570.
121. **Kudva IT, Blanch K, Hovde CJ.** 1998. Analysis of *Escherichia coli* O157:H7 survival in ovine or bovine manure and manure slurry. *Appl Environ Microbiol* **64:**3166–3174.
122. **Berry ED, Miller DN.** 2005. Cattle feedlot soil moisture and manure content: II. Impact on *Escherichia coli* O157. *J Environ Qual* **34:**656–663.
123. **Ma J, Mark Ibekwe A, Crowley DE, Yang C-H.** 2014. Persistence of *Escherichia coli* O157 and non-O157 strains in agricultural soils. *Sci Total Environ* **490:**822–829.
124. **Faith NG, Shere JA, Brosch R, Arnold KW, Ansay SE, Lee M-S, Luchansky JB, Kaspar CW.** 1996. Prevalence and clonal nature of *Escherichia coli* O157:H7 on dairy farms in Wisconsin. *Appl Environ Microbiol* **62:**1519–1525.
125. **Smith D, Blackford M, Younts S, Moxley R, Gray J, Hungerford L, Milton T, Klopfenstein T.** 2001. Ecological relationships between the prevalence of cattle shedding *Escherichia coli* O157:H7 and characteristics of the cattle or conditions of the feedlot pen. *J Food Prot* **64:**1899–1903.
126. **Varel VH, Wells JE, Berry ED, Spiehs MJ, Miller DN, Ferrell CL, Shackelford SD, Koohmaraie M.** 2008. Odorant production and persistence of *Escherichia coli* in manure slurries from cattle fed zero, twenty, forty, or sixty percent wet distillers grains with solubles. *J Anim Sci* **86:**3617–3627.
127. **Wells JE, Shackelford SD, Berry ED, Kalchayanand N, Guerini MN, Varel VH, Arthur TM, Bosilevac JM, Freetly HC, Wheeler TL, Ferrell CL, Koohmaraie M.** 2009. Prevalence and level of *Escherichia coli* O157:H7 in feces and on hides of feedlot steers fed diets with or without wet distillers grains with solubles. *J Food Prot* **72:**1624–1633.
128. **Baloda SB, Christensen L, Trajcevska S.** 2001. Persistence of a *Salmonella enterica* serovar typhimurium DT12 clone in a piggery and in agricultural soil amended with *Salmonella*-contaminated slurry. *Appl Environ Microbiol* **67:**2859–2862.
129. **Inglis GD, McAllister TA, Larney FJ, Topp E.** 2010. Prolonged survival of *Campylobacter* species in bovine manure compost. *Appl Environ Microbiol* **76:**1110–1119.
130. **Ross CM, Donnison AM.** 2006. *Campylobacter jejuni* inactivation in New Zealand soils. *J Appl Microbiol* **101:**1188–1197.
131. **Hutchison ML, Walters LD, Moore A, Crookes KM, Avery SM.** 2004. Effect of length of time before incorporation on survival of pathogenic bacteria present in livestock wastes applied to agricultural soil. *Appl Environ Microbiol* **70:**5111–5118.
132. **Vivant A-L, Garmyn D, Piveteau P.** 2013. *Listeria monocytogenes*, a down-to-earth pathogen. *Front Cell Infect Microbiol* **3:**87.
133. **Hutchison ML, Walters LD, Avery SM, Synge BA, Moore A.** 2004. Levels of zoonotic agents in British livestock manures. *Lett Appl Microbiol* **39:**207–214.
134. **Pedersen TB, Olsen JE, Bisgaard M.** 2008. Persistence of *Salmonella* Senftenberg in poultry production environments and investigation of its resistance to desiccation. *Avian Pathol* **37:**421–427.
135. **Dodd CC, Renter DG, Shi X, Alam MJ, Nagaraja TG, Sanderson MW.** 2011. Prevalence and persistence of *Salmonella* in cohorts of feedlot cattle. *Foodborne Pathog Dis* **8:**781–789.
136. **Petersen L, Wedderkopp A.** 2001. Evidence that certain clones of *Campylobacter jejuni* persist during successive broiler flock rotations. *Appl Environ Microbiol* **67:**2739–2745.
137. **Hakkinen M, Hänninen M-L.** 2009. Shedding of *Campylobacter* spp. in Finnish cattle on dairy farms. *J Appl Microbiol* **107:**898–905.
138. **On SLW, Atabay HI, Corry JEL.** 1999. Clonality of *Campylobacter sputorum* bv. paraureolyticus determined by macrorestriction profiling and biotyping, and evidence for long-term persistent infection in cattle. *Epidemiol Infect* **122:**175–182.
139. **LeJeune JT, Besser TE, Rice DH, Berg JL, Stilborn RP, Hancock DD.** 2004. Longitudinal study of fecal shedding of *Escherichia coli* O157:H7 in feedlot cattle: predominance and persistence of specific clonal types despite massive cattle population turnover. *Appl Environ Microbiol* **70:**377–384.
140. **Carlson BA, Nightingale KK, Mason GL, Ruby JR, Choat WT, Loneragan GH, Smith GC, Sofos JN, Belk KE.** 2009. *Escherichia coli* O157:H7 strains that persist in feedlot cattle are genetically related and demonstrate an enhanced ability to adhere to intestinal epithelial cells. *Appl Environ Microbiol* **75:**5927–5937.
141. **Lahti E, Ruoho O, Rantala L, Hänninen M-L, Honkanen-Buzalski T.** 2003. Longitudinal study of *Escherichia coli* O157 in a cattle finishing unit. *Appl Environ Microbiol* **69:**554–561.
142. **Shere JA, Bartlett KJ, Kaspar CW.** 1998. Longitudinal study of *Escherichia coli* O157:H7 dissemination on four dairy farms in Wisconsin. *Appl Environ Microbiol* **64:**1390–1399.
143. **Herbert LJ, Vali L, Hoyle DV, Innocent G, McKendrick IJ, Pearce MC, Mellor D,**

Porphyre T, Locking M, Allison L, Hanson M, Matthews L, Gunn GJ, Woolhouse MEJ, Chase-Topping ME. 2014. *E. coli* O157 on Scottish cattle farms: evidence of local spread and persistence using repeat cross-sectional data. *BMC Vet Res* **10:**95.

144. **Wells JE, Bosilevac JM, Kalchayanand N, Arthur TM, Shackelford SD, Wheeler TL, Koohmaraie M.** 2009. Prevalence of *Mycobacterium avium* subsp. *paratuberculosis* in ileocecal lymph nodes and on hides and carcasses from cull cows and fed cattle at commercial beef processing plants in the United States. *J Food Prot* **72:**1457–1462.

145. **Smith HW, Halls S.** 1968. The simultaneous oral administration of *Salmonella dublin, S. typhimurium* and *S. choleraesuis* to calves and other animals. *J Med Microbiol* **1:**203–209.

146. **Samuel JL, O'Boyle DA, Mathers WJ, Frost AJ.** 1980. Distribution of *Salmonella* in the carcases of normal cattle at slaughter. *Res Vet Sci* **28:**368–372.

147. **Arthur TM, Brichta-Harhay DM, Bosilevac JM, Guerini MN, Kalchayanand N, Wells JE, Shackelford SD, Wheeler TL, Koohmaraie M.** 2008. Prevalence and characterization of *Salmonella* in bovine lymph nodes potentially destined for use in ground beef. *J Food Prot* **71:**1685–1688.

148. **Gragg SE, Loneragan GH, Brashears MM, Arthur TM, Bosilevac JM, Kalchayanand N, Wang R, Schmidt JW, Brooks JC, Shackelford SD, Wheeler TL, Brown TR, Edrington TS, Brichta-Harhay DM.** 2013. Cross-sectional study examining *Salmonella enterica* carriage in subiliac lymph nodes of cull and feedlot cattle at harvest. *Foodborne Pathog Dis* **10:**368–374.

149. **Bahnson PB, Fedorka-Cray PJ, Ladely SR, Mateus-Pinilla NE.** 2006. Herd-level risk factors for *Salmonella enterica* subsp. *enterica* in U.S. market pigs. *Prev Vet Med* **76:**249–262.

150. **Anderson RC, Genovese KJ, Harvey RB, Stanker LH, DeLoach JR, Nisbet DJ.** 2000. Assessment of the long-term shedding pattern of *Salmonella* serovar *choleraesuis* following experimental infection of neonatal piglets. *J Vet Diagn Invest* **12:**257–260.

151. **Wang B, Wesley IV, McKean JD, O'Connor AM.** 2010. Sub-iliac lymph nodes at slaughter lack ability to predict *Salmonella enterica* prevalence for swine farms. *Foodborne Pathog Dis* **7:**795–800.

152. **Bahnson PB, Snyder C, Omran LM.** 2006. *Salmonella enterica* in superficial cervical (prescapular) and ileocecal lymph nodes of slaughtered pigs. *J Food Prot* **69:**925–927.

153. **Janss LL, Bolder NM.** 2000. Heritabilities of and genetic relationships between *Salmonella* resistance traits in broilers. *J Anim Sci* **78:**2287–2291.

154. **Gast RK, Holt PS.** 1998. Persistence of *Salmonella enteritidis* from one day of age until maturity in experimentally infected layer chickens. *Poult Sci* **77:**1759–1762.

155. **Bratz K, Bücker R, Gölz G, Zakrzewski SS, Janczyk P, Nöckler K, Alter T.** 2013. Experimental infection of weaned piglets with *Campylobacter coli*: excretion and translocation in a pig colonisation trial. *Vet Microbiol* **162:**136–143.

156. **Lin J.** 2009. Novel approaches for *Campylobacter* control in poultry. *Foodborne Pathog Dis* **6:**755–765.

157. **Freitag NE, Port GC, Miner MD.** 2009. *Listeria monocytogenes*: from saprophyte to intracellular pathogen. *Nat Rev Microbiol* **7:**623–628.

158. **Zundel E, Bernard S.** 2006. *Listeria monocytogenes* translocates throughout the digestive tract in asymptomatic sheep. *J Med Microbiol* **55:**1717–1723.

159. **Jeong KC, Hiki O, Kang MY, Park D, Kaspar CW.** 2013. Prevalent and persistent *Escherichia coli* O157:H7 strains on farms are selected by bovine passage. *Vet Microbiol* **162:**912–920.

160. **Ravva SV, Sarreal CZ, Mandrell RE.** 2014. Strain differences in fitness of *Escherichia coli* O157:H7 to resist protozoan predation and survival in soil. *PLoS One* **9:**e102412.

161. **Lippolis JD, Brunelle BW, Reinhardt TA, Sacco RE, Nonnecke BJ, Dogan B, Simpson K, Schukken YH.** 2014. Proteomic analysis reveals protein expression differences in *Escherichia coli* strains associated with persistent versus transient mastitis. *J Proteomics* **108:**373–381.

162. **Fox EM, Leonard N, Jordan K.** 2011. Physiological and transcriptional characterization of persistent and nonpersistent *Listeria monocytogenes* isolates. *Appl Environ Microbiol* **77:**6559–6569.

163. **Tao H, Bausch C, Richmond C, Blattner FR, Conway T.** 1999. Functional genomics: expression analysis of *Escherichia coli* growing on minimal and rich media. *J Bacteriol* **181:**6425–6440.

164. **Hepworth PJ, Ashelford KE, Hinds J, Gould KA, Witney AA, Williams NJ, Leatherbarrow H, French NP, Birtles RJ, Mendonca C, Dorrell N, Wren BW, Wigley P, Hall N, Winstanley C.** 2011. Genomic variations define divergence of water/wildlife-associated *Campylobacter jejuni*

niche specialists from common clonal complexes. *Environ Microbiol* **13:**1549–1560.
165. Ihssen J, Grasselli E, Bassin C, François P, Piffaretti J-C, Köster W, Schrenzel J, Egli T. 2007. Comparative genomic hybridization and physiological characterization of environmental isolates indicate that significant (eco-) physiological properties are highly conserved in the species *Escherichia coli*. *Microbiology* **153:**2052–2066.
166. Bronowski C, James CE, Winstanley C. 2014. Role of environmental survival in transmission of *Campylobacter jejuni*. *FEMS Microbiol Lett* **356:**8–19.
167. van Elsas JD, Semenov AV, Costa R, Trevors JT. 2011. Survival of *Escherichia coli* in the environment: fundamental and public health aspects. *ISME J* **5:**173–183.
168. Fan B, Du Z-Q, Gorbach DM, Rothschild MF. 2010. Development and application of high-density SNP arrays in genomic studies of domestic animals. *Asian-Aust. J Anim Sci* **23:** 833–847.
169. Jackson BR, Griffin PM, Cole D, Walsh KA, Chai SJ. 2013. Outbreak-associated *Salmonella enterica* serotypes and food commodities, United States, 1998–2008. *Emerg Infect Dis* **19:** 1239–1244.
170. Hafez HM. 1999. Poultry meat and food safety: pre- and post-harvest approaches to reduce foodborne pathogens. *Worlds Poult Sci J* **55:** 269–280.
171. Lamont SJ. 1998. Impact of genetics on disease resistance. *Poult Sci* **77:**1111–1118.
172. Girard-Santosuosso O, Lantier F, Lantier I, Bumstead N, Elsen J-M, Beaumont C. 2002. Heritability of susceptibility to *Salmonella enteritidis* infection in fowls and test of the role of the chromosome carrying the *NRAMP1* gene. *Genet Select Evol* **34:**211–219.
173. Calenge F, Lecerf F, Demars J, Feve K, Vignoles F, Pitel F, Vignal A, Velge P, Sellier N, Beaumont C. 2009. QTL for resistance to *Salmonella* carrier state confirmed in both experimental and commercial chicken lines. *Anim Genet* **40:**590–597.
174. Malek M, Hasenstein JR, Lamont SJ. 2004. Analysis of chicken *TLR4, CD28, MIF, MD-2,* and *LITAF* genes in a *Salmonella enteritidis* resource population. *Poult Sci* **83:**544–549.
175. Fife MS, Salmon N, Hocking PM, Kaiser P. 2009. Fine mapping of the chicken salmonellosis resistance locus (*SAL1*). *Anim Genet* **40:**871–877.
176. Swaggerty CL, Pevzner IY, He H, Genovese KJ, Nisbet DJ, Kaiser P, Kogut MH. 2009. Selection of broilers with improved innate immune responsiveness to reduce on-farm infection by foodborne pathogens. *Foodborne Pathog Dis* **6:**777–783.
177. Galina-Pantoja L, Siggens K, van Schriek MGM, Heuven HCM. 2009. Mapping markers linked to porcine salmonellosis susceptibility. *Anim Genet* **40:**795–803.
178. Uthe JJ, Wang Y, Qu L, Nettleton D, Tuggle CK, Bearson SMD. 2009. Correlating blood immune parameters and a CCT7 genetic variant with the shedding of *Salmonella enterica* serovar Typhimurium in swine. *Vet Microbiol* **135:**384–388.
179. Uenishi H, Shinkai H, Morozumi T, Muneta Y. 2012. Genomic survey of polymorphisms in pattern recognition receptors and their possible relationship to infections in pigs. *Vet Immunol Immunopathol* **148:**69–73.
180. Flori L, Gao Y, Oswald IP, Lefevre F, Bouffaud M, Mercat M-J, Bidanel J-P, Rogel-Gaillard C. 2011. Deciphering the genetic control of innate and adaptive immune responses in pig: a combined genetic and genomic study. *BMC Proc* **5** (Suppl 4)**:**S32.
181. Pighetti GM, Elliott AA. 2011. Gene polymorphisms: the keys for marker assisted selection and unraveling core regulatory pathways for mastitis resistance. *J Mammary Gland Biol Neoplasia* **16:**421–432.
182. Sørensen LP, Madsen P, Mark T, Lund MS. 2009. Genetic parameters for pathogen-specific mastitis resistance in Danish Holstein Cattle. *Animal* **3:**647–656.
183. Verbeke J, Piepers S, Peelman L, Van Poucke M, De Vliegher S. 2012. Pathogen-group specific association between *CXCR1* polymorphisms and subclinical mastitis in dairy heifers. *J Dairy Res* **79:**341–351.
184. Ley RE, Lozupone CA, Hamady M, Knight R, Gordon JI. 2008. Worlds within worlds: evolution of the vertebrate gut microbiota. *Nat Rev Microbiol* **6:**776–788.
185. Rist VTS, Weiss E, Eklund M, Mosenthin R. 2013. Impact of dietary protein on microbiota composition and activity in the gastrointestinal tract of piglets in relation to gut health: a review. *Animal* **7:**1067–1078.
186. Bearson SMD, Allen HK, Bearson BL, Looft T, Brunelle BW, Kich JD, Tuggle CK, Bayles DO, Alt D, Levine UY, Stanton TB. 2013. Profiling the gastrointestinal microbiota in response to *Salmonella*: low versus high *Salmonella* shedding in the natural porcine host. *Infect Genet Evol* **16:**330–340.
187. Stecher B, Robbiani R, Walker AW, Westendorf AM, Barthel M, Kremer M, Chaffron S, Macpherson AJ, Buer J, Parkhill J, Dougan G,

von Mering C, Hardt WD. 2007. *Salmonella enterica* serovar Typhimurium exploits inflammation to compete with the intestinal microbiota. *PLoS Biol* **5:**2177–2189.

188. **Endt K, Stecher B, Chaffron S, Slack E, Tchitchek N, Benecke A, Van Maele L, Sirard JC, Mueller AJ, Heikenwalder M, Macpherson AJ, Strugnell R, von Mering C, Hardt WD.** 2010. The microbiota mediates pathogen clearance from the gut lumen after non-typhoidal *Salmonella* diarrhea. *PLoS Pathog* **6:**e1001097.
189. **May KD, Wells JE, Maxwell CV, Oliver WT.** 2012. Granulated lysozyme as an alternative to antibiotics improves growth performance and small intestinal morphology of 10-day-old pigs. *J Anim Sci* **90:**1118–1125.
190. **Oliver WT, Wells JE, Maxwell CV.** 2014. Lysozyme as an alternative to antibiotics improves performance in nursery pigs during an indirect immune challenge. *J Anim Sci* **92:** 4927–4934.
191. **Nyachoti CM, Kiarie E, Bhandari SK, Zhang G, Krause DO.** 2012. Weaned pig responses to *Escherichia coli* K88 oral challenge when receiving a lysozyme supplement. *J Anim Sci* **90:**252–260.
192. **Maga EA, Desai PT, Weimer BC, Dao N, Kültz D, Murray JD.** 2012. Consumption of lysozyme-rich milk can alter microbial fecal populations. *Appl Environ Microbiol* **78:**6153–6160.
193. **Durso LM, Harhay GP, Smith TP, Bono JL, Desantis TZ, Harhay DM, Andersen GL, Keen JE, Laegreid WW, Clawson ML.** 2010. Animal-to-animal variation in fecal microbial diversity among beef cattle. *Appl Environ Microbiol* **76:**4858–4862.
194. **Wells JE, Shackelford SD, Berry ED, Kalchayanand N, Bosilevac JM, Wheeler TL.** 2011. Impact of reducing the level of wet distillers grains fed to cattle prior to harvest on prevalence and levels of *Escherichia coli* O157:H7 in feces and on hides. *J Food Prot* **74:**1611–1617.
195. **Kim M, Kim J, Kuehn LA, Bono JL, Berry ED, Kalchayanand N, Freetly HC, Benson AK, Wells JE.** 2014. Investigation of bacterial diversity in the feces of cattle fed different diets. *J Anim Sci* **92:**683–694.
196. **Durso LM, Wells JE, Harhay GP, Rice WC, Kuehn L, Bono JL, Shackelford S, Wheeler T, Smith TP.** 2012. Comparison of bacterial communities in faeces of beef cattle fed diets containing corn and wet distillers' grain with solubles. *Lett Appl Microbiol* **55:**109–114.
197. **Patton TG, Scupham AJ, Bearson SM, Carlson SA.** 2009. Characterization of fecal microbiota from a *Salmonella* endemic cattle herd as determined by oligonucleotide fingerprinting of rDNA genes. *Vet Microbiol* **136:** 285–292.
198. **Stanley D, Hughes RJ, Moore RJ.** 2014. Microbiota of the chicken gastrointestinal tract: influence on health, productivity and disease. *Appl Microbiol Biotechnol* **98:**4301–4310.
199. **Danzeisen JL, Kim HB, Isaacson RE, Tu ZJ, Johnson TJ.** 2011. Modulations of the chicken cecal microbiome and metagenome in response to anticoccidial and growth promoter treatment. *PLoS One* **6:**e27949.
200. **Sergeant MJ, Constantinidou C, Cogan TA, Bedford MR, Penn CW, Pallen MJ.** 2014. Extensive microbial and functional diversity within the chicken cecal microbiome. *PLoS One* **9:**e91941.
201. **Rantala M, Nurmi E.** 1973. Prevention of the growth of *Salmonella infantis* in chicks by the flora of the alimentary tract of chickens. *Br Poult Sci* **14:**627–630.
202. **Schneitz C.** 2005. Competitive exclusion in poultry: 30 years of research. *Food Control* **16:** 657–667.
203. **Atterbury RJ, Hobley L, Till R, Lambert C, Capeness MJ, Lerner TR, Fenton AK, Barrow P, Sockett RE.** 2011. Effects of orally administered *Bdellovibrio bacteriovorus* on the well-being and *Salmonella* colonization of young chicks. *Appl Environ Microbiol* **77:**5794–5803.
204. **Gould LH, Walsh KA, Vieira AR, Herman K, Williams IT, Hall AJ, Cole D, Centers for Disease Control and Prevention.** 2013. Surveillance for foodborne disease outbreaks: United States, 1998–2008. *MMWR Surveill Summ* **62:**1–34.
205. **Painter JA, Hoekstra RM, Ayers T, Tauxe RV, Braden CR, Angulo FJ, Griffin PM.** 2013. Attribution of foodborne illnesses, hospitalizations, and deaths to food commodities by using outbreak data, United States, 1998–2008. *Emerg Infect Dis* **19:**407–415.
206. **O'Connor DR.** 2002. *A Summary. Report of the Walkerton Inquiry: The Events of May 2000 and Related Issues,* Part 1, p 1–35. Ontario Ministry of the Attorney General, Toronto, Ontario.
207. **Olson ME, O'Handley RM, Ralston BJ, McAllister TA, Thompson RCA.** 2004. Update on *Cryptosporidium* and *Giardia* infections in cattle. *Trends Parasitol* **20:**185–191.
208. **Jackson SG, Goodbrand RB, Johnson RP, Odorico VG, Alves D, Rahn K, Wilson JB, Welch MK, Khakhria R.** 1998. *Escherichia coli* O157:H7 diarrhoea associated with well water

and infected cattle on an Ontario farm. *Epidemiol Infect* **120:**17–20.

209. Johnson JYM, Thomas JE, Graham TA, Townshend I, Byrne J, Selinger LB, Gannon VPJ. 2003. Prevalence of *Escherichia coli* O157:H7 and *Salmonella* spp. in surface waters of southern Alberta and its relation to manure sources. *Can J Microbiol* **49:**326–335.

210. Warriner K, Huber A, Namvar A, Fan W, Dunfield K. 2009. Recent advances in the microbial safety of fresh fruits and vegetables. *Adv Food Nutr Res* **57:**155–208.

211. Millner PD, Karns J. 2005. Animal manure: bacterial pathogens and disinfection technologies, p 61–83. *In* Smith JE Jr, Millner PD, Jakubowski W, Goldstein N, Rynk R (ed), *Contemporary Perspectives on Infectious Disease Agents in Sewage Sludge and Manure.* Compost Science & Utilization/JG Press, Inc, Emmaus, PA.

212. U.S. Food and Drug Administration (USFDA). 2015. *Food Safety Modernization Act (FSMA) final rule on produce safety.* http://www.fda.gov/Food/GuidanceRegulation/FSMA/ucm334114.htm.

213. Jay MT, Cooley M, Carychao D, Wiscomb GW, Sweitzer RA, Crawford-Miksza L, Farrar JA, Lau DK, O'Connell J, Millington A, Asmundson RV, Atwill ER, Mandrell RE. 2007. *Escherichia coli* O157:H7 in feral swine near spinach fields and cattle, central California coast. *Emerg Infect Dis* **13:**1908–1911.

214. Jay-Russell MT, Bates A, Harden L, Miller WG, Mandrell RE. 2012. Isolation of *Campylobacter* from feral swine (*Sus scrofa*) on the ranch associated with the 2006 *Escherichia coli* O157:H7 spinach outbreak investigation in California. *Zoonoses Public Health* **59:**314–319.

215. Talley JL, Wayadande AC, Wasala LP, Gerry AC, Fletcher J, DeSilva U, Gilliland SE. 2009. Association of *Escherichia coli* O157:H7 with filth flies (*Muscidae* and *Calliphoridae*) captured in leafy greens fields and experimental transmission of *E. coli* O157:H7 to spinach leaves by house flies (*Diptera*: *Muscidae*). *J Food Prot* **72:**1547–1552.

216. Swirski AL, Pearl DL, Williams ML, Homan HJ, Linz GM, Cernicchiaro N, LeJeune JT. 2014. Spatial epidemiology of *Escherichia coli* O157:H7 in dairy cattle in relation to night roosts of *Sturnus vulgaris* (European starling) in Ohio, USA (2007–2009). *Zoonoses Public Health* **61:**427–435.

217. Gaukler SM, Linz GM, Sherwood JS, Dyer NW, Bleier WJ, Wannemuehler YM, Nolan LK, Logue CM. 2009. *Escherichia coli, Salmonella,* and *Mycobacterium avium* subsp. *paratuberculosis* in wild European starlings at a Kansas cattle feedlot. *Avian Dis* **53:**544–551.

218. Sanad YM, Closs G Jr, Kumar A, LeJeune JT, Rajashekara G. 2013. Molecular epidemiology and public health relevance of *Campylobacter* isolated from dairy cattle and European starlings in Ohio, USA. *Foodborne Pathog Dis* **10:**229–236.

219. Pedersen K, Clark L, Andelt WF, Salman MD. 2006. Prevalence of shiga toxin-producing *Escherichia coli* and *Salmonella enterica* in rock pigeons captured in Fort Collins, Colorado. *J Wildl Dis* **42:**46–55.

220. Callaway TR, Edrington TS, Nisbet DJ. 2014. Isolation of *Escherichia coli* O157:H7 and *Salmonella* from migratory brown-headed cowbirds (*Molothrus ater*), common grackles (*Quiscalus quiscula*), and cattle egrets (*Bubulcus ibis*). *Foodborne Pathog Dis* **11:**791–794.

221. Cooley MB, Jay-Russell M, Atwill ER, Carychao D, Nguyen K, Quiñones B, Patel R, Walker S, Swimley M, Pierre-Jerome E, Gordus AG, Mandrell RE. 2013. Development of a robust method for isolation of shiga toxin-positive *Escherichia coli* (STEC) from fecal, plant, soil and water samples from a leafy greens production region in California. *PLoS One* **8:**e65716.

222. Gorski L, Parker CT, Liang A, Cooley MB, Jay-Russell MT, Gordus AG, Atwill ER, Mandrell RE. 2011. Prevalence, distribution, and diversity of *Salmonella enterica* in a major produce region of California. *Appl Environ Microbiol* **77:**2734–2748.

223. Langholz JA, Jay-Russell MT. 2013. Potential role of wildlife in pathogenic contamination of fresh produce. *Hum Wildlife Interact* **7:**140–157.

224. Berry ED, Wells JE, Bono JL, Woodbury BL, Kalchayanand N, Norman KN, Suslow TV, López-Velasco G, Millner PD. 2015. Effect of proximity to a cattle feedlot on *Escherichia coli* O157:H7 contamination of leafy greens and evaluation of the potential for airborne transmission. *Appl Environ Microbiol* **81:**1101–1110.

225. Millner P, Suslow T. 2008. California Lettuce Research Board 2007–08 Interim Research Report Summary: concentration and deposition of viable *E. coli* in airborne particulates from composting and livestock operations. http://calgreens.org/control/uploads/Millner_and_Suslow_-_Concentration_and_deposition_of_viable_E._coli_in_airborne_particulates_from_composting_and_livestock_operations_.pdf.

226. Crohn DM, Bianchi ML. 2008. Research priorities for coordinating management of food safety and water quality. *J Environ Qual* **37:**1411–1418.

227. **Arthur TM, Bosilevac JM, Brichta-Harhay DM, Guerini MN, Kalchayanand N, Shackelford SD, Wheeler TL, Koohmaraie M.** 2007. Transportation and lairage environment effects on prevalence, numbers, and diversity of *Escherichia coli* O157:H7 on hides and carcasses of beef cattle at processing. *J Food Prot* **70:**280–286.

228. **Arthur TM, Bosilevac JM, Brichta-Harhay DM, Kalchayanand N, King DA, Shackelford SD, Wheeler TL, Koohmaraie M.** 2008. Source tracking of *Escherichia coli* O157:H7 and *Salmonella* contamination in the lairage environment at commercial U.S. beef processing plants and identification of an effective intervention. *J Food Prot* **71:**1752–1760.

Emerging Foodborne and Agriculture-Related Viruses

11

DAVID H. KINGSLEY[1]

An emerging pathogen such as a virus can be defined as a previously unknown agent causing disease or an agent that was once an infrequent cause of illness that has become more common. It can also be defined as an infectious agent introduced into a new geographic area or one that infects a new species (1). Emergence of a viral pathogen is essentially a two-step process which requires conditions suitable for host introduction and then subsequent conditions that provide the opportunity to disseminate within a new host population (2). This is evidenced by the fact that emerging viruses and pathogens often arise in specific geographic locations (2). Numerous factors can be involved in the emergence of viral pathogens. These can include human demographic changes and behavior, travel and commerce, microbial adaptation, development of new technologies and industries, environmental perturbations, and breakdown of public health measures (2). The latter may be responsible for the re-emergence of poliovirus in parts of sub-Saharan Africa, where vaccine campaigns were suspended as a result of unfounded fears or active warfare. Historically, as human civilization developed as a result of agricultural activities, it is thought that environmental changes and increased human and domestic animal population sizes and densities permitted the introduction and spread of pathogens from wildlife reservoirs (3). Retrospective evidence

[1]U.S. Department of Agriculture, Agricultural Research Service, Food Safety and Interventions Research Unit, Delaware State University, Dover, DE 19901.

Preharvest Food Safety
Edited by Siddhartha Thakur and Kalmia E. Kniel

doi:10.1128/microbiolspec.PFS-0007-2014

suggests that viruses such as measles may have evolved from canine distemper (i.e., dogs) and/or rinderpest virus, a disease of cattle and wild even-toed ungulates such as deer, antelope, and wildebeest. It has been suggested that the variola virus, more commonly known as the smallpox virus, evolved from cowpox or, alternatively, may have evolved from camelpox, a virus of camels which is the closest known relative to the smallpox virus (3).

Viruses can infect hosts by a number of different routes: respiratory (airborne), via blood and sexual contact, via biting arthropods (ticks and mosquitoes), and via oral ingestion, typically by the fecal–oral route. The latter route, via ingestion of contaminated food and water, is of paramount concern to food producers and consumers alike (4). Currently, foodborne viruses of significance include human norovirus (HuNoV) strains, which are now the most frequent cause of foodborne illness in the United States (5, 6) and are thought to cause >9 million foodborne illnesses per annum in the United States. Another foodborne virus, hepatitis A virus (HAV) is medically more serious but is becoming increasingly rare in the developed world owing to an effective vaccine which induces long-term immunity (7). HuNoVs and HAV probably represent just the tip of the proverbial iceberg regarding human-to-human virus transmission by the fecal–oral route. It has now become clear that a multitude of pathogens could be transmitted in foods or may establish themselves in the human population via food and agricultural activities.

Rather than focusing exclusively on HAV and HuNoV, which have been relatively well described elsewhere, this chapter will focus on a broader set of questions such as:

1. What are the other known and/or potential zoonotic foodborne viruses that are, or may become, a threat to food consumers?
2. What is the potential of exotic viruses to emerge as humans pathogens or enter the human population via food consumption by humans, livestock production, or companion animals?
3. What mechanisms might lead to the emergence of a new pathogenic foodborne virus?
4. From where might a new foodborne virus come?
5. What microbiological niches might be created by food and livestock production processes?
6. What are recent metagenomics studies telling us about previously unknown or uncultivable viruses that replicate in the intestinal tract of humans and other animals?

A virus is essentially a set of genetic instructions with a means of cellular entry. Viruses are ubiquitous in the environment. As an example, viruses that infect bacteria (phages) are known to outnumber bacteria by about a factor of 10, making viruses the most prevalent biologic entity on the planet (8). Thus, viruses are omnipresent and probably play a larger biological role than they receive credit for. In the case of HuNoV, a single infected person can result in the release of perhaps a trillion or more virus particles during the course of an illness. Eukaryotic viruses can be broadly classified in two categories: enveloped and nonenveloped. Enveloped viruses contain an exterior lipid envelope surrounding the virus capsid, while nonenveloped viruses lack this lipid coating. Generally speaking, the nonenveloped viruses are more resilient and persist in the outside environment more efficiently. For this reason, most viruses that are typically thought of as foodborne are nonenveloped, but as will be discussed, food and agricultural activities are the means by which both enveloped and nonenveloped viruses alike may enter the human population.

There are several evolutionary mechanisms by which viruses evolve to infect new hosts and alter their pathogenicity. Replication errors, such as substitution of a nucleotide within a codon, which as a result often encodes a new amino acid, are a common means by which viruses evolve. This is especially true for RNA viruses which have more error-prone polymerases. The net result is that multiple sequence mutations can be selected for as the virus moves from one individual host to another. These changes are so frequent that multiple mutations can even be present in a single infected organism, and therefore virus strains are sometimes referred to as quasi-species. Recombination is another common means by which viruses evolve, essentially acquiring new genes or traits. Coinfection of the same cell by two different virus strains can result in exchange of genes by homologous recombination in which viruses exchange genetic material. Segmented viruses contain numerous sections which encode individual or clusters of genes. Examples of these are the reoviruses and influenza viruses, which contain 10 and 8 segments, respectively. Coinfection of a single cell by two different virus strains can result in reassortment of these segments, producing a factorial number of new progeny strains which often have unique characteristics and novel antigenic properties. Viruses can even steal genes from host cells, which may then be utilized as-is or be altered mutagenically in subsequent replication cycles, to facilitate malevolent viral objectives. Surprisingly, the loss of gene function or deletion can result in a more virulent virus. Examples of this are the SARS coronavirus, which purportedly resulted from a genetic deletion which facilitated transmission from civets to humans (9) and virulent monkeypox (10) transmitted to humans by rodents.

Dynamic genetic change can facilitate host-shifting, which can dramatically alter the virulence of viruses. In some cases, this results in a less virulent virus that may cause an unapparent infection. However, a host shift can also result in greatly enhanced virulence, a rather cogent example of which is simian herpes B virus. This herpesvirus commonly infects primates such as macaques, readily establishing latency, and is of little health consequence to the animal. However, if a human handler becomes infected, as a result of, say, being bitten by a monkey with an active (or reactivating) herpes B infection, a severe, highly lethal, neurotropic infection often results. Although herpes B may be common in simian saliva, fortunately it is not known to be transmitted by the foodborne (fecal–oral) route.

It is estimated that two-thirds of emerging infectious diseases that affect humans are originally zoonotic, or of animal origin (11, 12). Usually, the animals originally harboring these viruses are wildlife (13). While the classic definition of a zoonosis is an animal disease that may be directly transmitted to humans (i.e., rabies virus), it is increasingly clear that emerging pathogens often have an intermediate host, which is frequently in close association with humans (1, 3). As mentioned earlier, disease emergence can be thought of as a two-step process by which an initial virus introduction occurs within a single animal, or person, followed by establishment and dissemination within the host population (2, 14). Domestic animals frequently function as a bridge that facilitates both the introduction of viruses into humans and/or facilitates its spread among a human population. Some examples of wild and domestic animals and viruses thought to have been transmitted to humans are shown in Fig 1.

Genetic relatedness between new host species and the original host species, sometimes termed the "distance effect," can influence a virus' propensity to jump to a new species, since most new host species are frequently related to the original host. Modern advances in nucleic acid sequencing now provide researchers with the routine ability to do genomic alignments and construct phylogenetic trees that infer evolutionary relationships. Identification of sequence differences can provide estimates of divergence times based on nucleotide substitution rates. While a close evolutionary relationship between hosts generally enhances the probability of infection

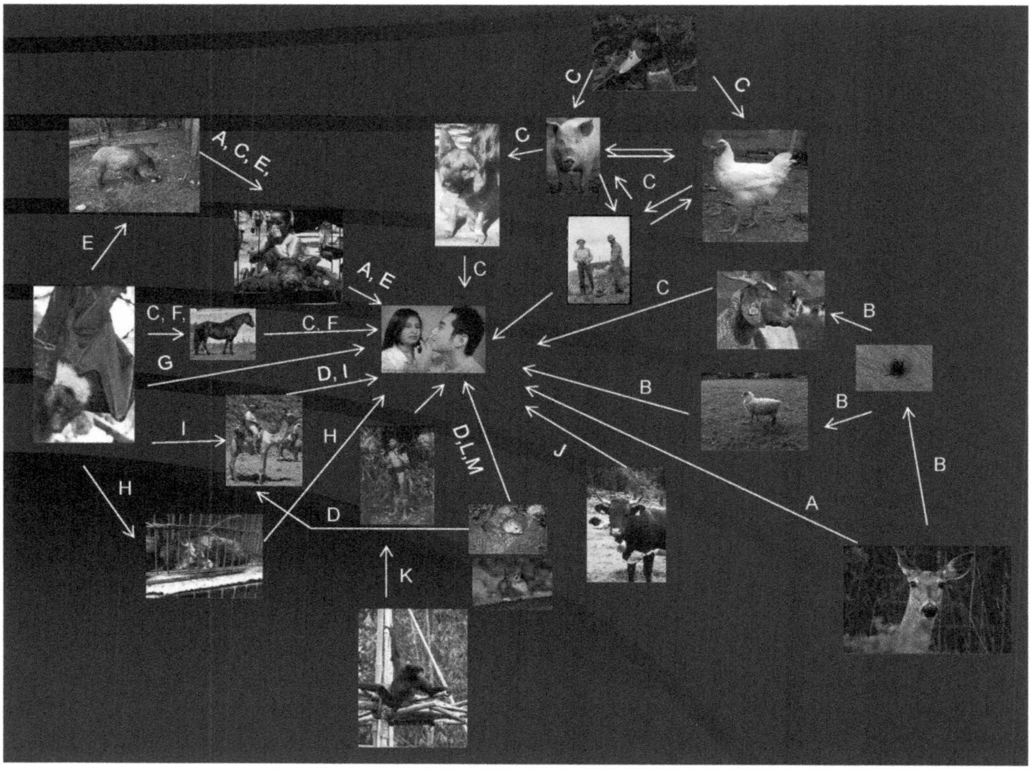

FIGURE 1 **Examples of viruses and animal hosts linked to emergence of pathogenic human viruses. Letters indicate different viruses believed to have been transmitted to other animals or humans. (A) Hepatitis E virus, (B) tick-borne encephalitis virus, (C) influenza, (D) poxvirus, (E) Nipah virus, (F) Hendra virus, (G) Ebola, (H) SARS, (I) MERS, (J) bovine leukemia virus, (K) HIV-like and retroviruses, (L) Hantaviruses, (M) Arenaviruses. All photos taken from Wiki Commons.**

(15, 16), this is not always the case since some viruses, in particular RNA viruses, sometimes use phylogenetically conserved host receptors, permitting infection of phylogenetically distant species (17). Also, jumping to a more distantly related host may have the additional advantage of allowing the virus to bypass innate host resistance mechanisms since closely related species may have similar broad resistance mechanisms to viruses (16).

Given the opportunity, viruses will evolve, allowing infection of novel hosts and new ecological niches or environmental conditions. Creation of these niches is often facilitated by, or a direct result of, anthropogenic activities, agricultural endeavors and practices in particular. Food production in the modern world increasingly exploits economies of scale, which while minimizing production costs, concentrate livestock, presumably providing optimal incubation conditions for the spread of emerging pathogens. Live markets of both traditional and exotic food animals, as well as the pet trade, can result in the mixing of animals from different geographic localities, providing unique interactions between species and pathogens. Our world is increasingly connected via global trade. Furthermore, airline travel makes it possible for animals, people, and pathogens to travel from one location on the planet to just about any other location in less than 24 hours (1). Longer human life spans with increasingly aged populations and increased proportions of immunocompromised people could conceivably increase susceptibility to infections and may provide extended opportunities for pathogen evolution (18). Beyond the virtual certainty that we will face new virus threats in the future, what is not clear is what those threats may be, what mechanisms will result in their emergence, or what novel niches will be exploited to facilitate their emergence.

What follows is an illustration of past virus emergence, as well as discussion of candidates for future foodborne threats or entrance into the human population via foods, food production, and agricultural activities.

VIRUSES POTENTIALLY SPREAD BY CONTAMINATED FOOD

Viruses that are found in human or animal feces have the potential to infect the gastrointestinal tract of other individuals or other species and therefore could be transmitted by contaminated food. New metagenomics and subsequent seroprevalence studies indicate that many previously unknown viruses are in fact quite common. This list of potential bad actors is increasing rapidly due to the accelerated pace of identification of novel viruses by high-throughput nucleic acid sequence analysis. The list below is by no means complete, and for many of these newly discovered viruses, information is relatively sparse. Some of the viruses described below are implicated in gastrointestinal illnesses, while others are implicated in respiratory disease, despite being shed in stool in both circumstances. Thus, clear evidence of enteric transmission is sometimes lacking for different viruses. In some cases, the described viruses are newly discovered and their propensity to cause morbidity is incompletely defined or unknown. In other cases, fecally shed viruses are implicated as the causative agent in systemic infections causing serious medical consequences. Lastly, while animal tissues and fluids (i.e., milk, blood) consumed for food are not ordinarily thought of as sources of foodborne virus infection, there are exceptions, as described below.

Hepatitis E Virus (HEV)

HEV is thought to have foodborne potential. It is a small nonenveloped virus that is presently classified in four genotypes (19). Types 1 and 2 are transmitted person to person, primarily endemic to the developing world, and can be medically serious, perhaps even more serious than HAV. Incubation periods for HEV range from 15 days to 60 days after exposure. It is thought that overt infection, as opposed to unapparent infection, involves a larger dose of the virus (20). In certain regions of the Indian subcontinent, there has been high maternal mortality associated with type 1 and type 2 HEV, but it has been suggested that lack of critical dietary minerals or vitamins may be influencing the maternal mortality observed (21). Previous exposure does induce some immunity against HEV, and vaccines are now becoming available (20).

HEV types 3 and 4 are currently perceived as less virulent for human hosts but are common in wild and domestic swine, as well as other animals such as deer and rodents (22). These strains are sometimes referred to as zoonotic HEV. People in close proximity to swine, such as swine farmers, often have antibodies against HEV, indicating that humans are being infected. In fact, approximately 20% of the U.S. population has been exposed to HEV type 3 (23). Foodborne transmission of type 3 is documented in cases of raw meat consumption (24, 25). Currently, a significant percentage of U.S. pork organ meat is believed to be contaminated with HEV. One U.S. market estimate indicated that 11% of pork livers tested positive for the presence of HEV (26). Similarly, HEV has been detected in swine meats in Europe (27). Bioinformatics analysis of HEV strains 3 and 4 does not suggest a clear ancestral host origin association but does infer that a cross-species infection is the origin of theses strains (28). To date, human illnesses caused by types 3 and 4 have been principally associated with immune-compromised and non-immunocompetent individuals. Why HEV types 1 and 2 do not readily infect swine, or whether they could be transmitted by swine in the future, is currently unknown.

Tick-Borne Encephalitis

Ecto-parasites such as ticks can readily transmit viruses to their hosts. In the case of tick-borne encephalitis virus, which is primarily found in Central and Eastern Europe and parts of Asia, infected ticks transmit the virus to milk-producing livestock, principally goats. Livestock can then shed this virus in milk, which when drunk unpasteurized or consumed as cheese derived from unpasteurized milk can result in very serious neurologic illness (29, 30).

Enterovirus

Enteroviruses are members of the *Picornaviridae* family, which includes viruses such as HAV, poliovirus, the coxsackieviruses, the echoviruses, the parechoviruses, and enteroviruses 68 and 71. The public health impact of these viruses can vary, with some strains being relatively benign and often causing unapparent infections, while others can be medically serious. These are shed in feces and can be transmitted human to human by the fecal–oral route and in some cases via aerosols. These viruses have been tenuously implicated in a number of chronic syndromes such as myocarditis, endocarditis, meningitis, encephalomyelitis, diabetes, amyotrophic lateral sclerosis (Lou Gehrig's disease), and dilated cardiomyopathy (31–39). In patients younger than 10, enterovirus strains echovirus 30, echovirus 6, and coxsackievirus B5 are sometimes implicated in aseptic meningitis (40, 41). Coxsackievirus A16 is known to cause hand, foot, and mouth disease (40, 41).

Enterovirus 71 is now often implicated in hand, foot, and mouth disease, sometimes with neurological complications (40, 41). Enterovirus 71 seems to be increasingly implicated in illness worldwide, and in particular, in the Asia-Pacific region. The respiratory route may be an important means of spread for enterovirus 71, in addition to its potential to spread by the fecal–oral route (40, 42, 43). Interestingly, enterovirus 68 also appears to be emerging as a significant pathogen, although its primary route of transmission appears to be respiratory (44). Another novel enterovirus is the coxsackie A2 strain identified in Hong Kong in 2012. It was reported to have been associated with respiratory symptoms that caused the deaths of two children (45). It has recently been suggested that echovirus 13, a relatively rare virus, may be responsible for a disproportionate number of cases of acute flaccid paralysis in humans on the Indian subcontinent (46). Parechoviruses are known to be quite common, normally causing mild acute respiratory and/or gastrointestinal symptoms in young children but occasionally are implicated as causing flaccid paralysis, myocarditis, and encephalitis (47). Recently, a parecho-like virus strain that was isolated in France and China from healthy piglets is now thought to be common in humans, based on the presence of antibodies to this virus (48).

Saffold Virus

The Saffold virus, a member for the *Cardiovirus* genus of the *Picornaviridae* family, was first isolated from a child with a fever of unknown etiology (49). Previously, cardioviruses had only been described in rodents, in which they were associated with myocarditis, type 1 diabetes, encephalitis, and multiple sclerosis–like symptoms (50). There are now multiple reports of its detection from patients in association with gastroenteritis, suggesting that the virus is not uncommon. For example, one report detected this virus in 7 of 454 patient stools tested, 4 of which were coinfections with other intestinal viruses and 3 of which were detected alone (51). Saffold virus has been occasionally detected in patients with acute flaccid paralysis but has not been reported in cerebrospinal fluid, so association with paralysis is tenuous at this point (50). Current thinking is that Saffold virus may infrequently cause gastroenteritis, respiratory symptoms, and neurological issues in infants, but it generally causes unapparent infections in the majority of patients (50).

Aichi Virus

Human Aichi virus was first isolated in Japan after an oyster-associated gastroenteritis outbreak in 1989. It has since been demonstrated worldwide and is classified as a member of the *Picornaviridae* family, genus *Kobuvirus* (52). Serologic evidence suggests that 90% of people over the age of 40 have been exposed to Aichi virus at some point in their life (53). While this virus has a relatively low profile from an outbreak perspective, it is clear that human Aichi virus is highly prevalent, because it is commonly identified in sewage worldwide (54) and can be foodborne.

Saliviruses and Klasseviruses

These are newly discovered genera of *Picornaviridae* which have been found to be shed during bouts of diarrhea, suggesting that they may be responsible for some unexplained gastrointestinal illness (55). A second genus of *Picornaviridae*, the Klasseviruses was isolated from gastroenteritis human stool and is now known to be common worldwide (56, 57).

Cosavirus

Cosavirus, which stands for common-stool-associated virus, is a new virus type discovered in 2008. This virus has been detected in fecal samples from children in southern Asia and has been associated with nonpolio acute flaccid paralysis (58). It is related to the *Picornaviridae* but because it is divergent from other picornaviruses, it has been proposed as a separate genus. Already, multiple strains have been detected that are highly divergent (59), and multiple reports from different continents suggest that this virus has a worldwide distribution.

Bocavirus

Although first reported in 2005, human bocavirus (HuBoV), from the *Parvoviridae*, is now known to be common worldwide, with almost 100% adult seropositivity (60–62). Multiple distinct strains have been identified in human stool (63). The virus can cause systemic infection resulting in respiratory and gastrointestinal symptoms. Recent screening of 191 patients with encephalitis in Sri Lanka demonstrated that 5 of these patients had human bocavirus in their cerebrospinal fluid (64). This finding appears to confirm an earlier report suggesting that human bocavirus was the cause of life-threatening encephalitis illnesses in Bangladesh (65).

Astroviruses

Astroviruses are small positive-stranded RNA viruses with a star-like appearance that were originally discovered in 1975 in a child suffering from diarrhea. Different strains of these viruses have subsequently been shown to infect a wide variety of animals (66). Their prevalence in the human population is high, and many astrovirus infections are asymptomatic. However, astroviruses are associated with approximately 10% of sporadic gastroenteritis cases and approximately 25% of all gastroenteritis hospitalizations. These viruses are an especially significant cause of clinical illness in toddlers and neonates. Astrovirus strains have been found that infect animals across the phylogenetic spectrum including cattle, pigs, dogs, cats, sea lions, whales, chickens, turkeys, and bats (66). Sequence analysis of multiple astroviruses indicates a high level of diversity and recombination, suggesting a propensity for cross-species infection and subsequent adaptation to the new host and/or coinfection of the same host with different astroviruses (66). In addition to diarrhea, astroviruses are known to be associated with nephritis, hepatitis, and even encephalitis, depending on the host (67).

Adenoviruses

Human adenoviruses are a common cause of childhood gastroenteritis (68). These nonenveloped viruses are composed of double-stranded DNA, making them very resilient in the environment (69, 70). They can be spread person to person and via the fecal–oral route from contaminated drinking water, food, or recreational water exposure. There are at least 57 different strains of human adenovirus (71). All strains are shed in human feces, but only enteric adenovirus strains 40 and 41 are associated with enteric infections. Adenoviruses are associated with a variety of illnesses including pneumonia, eye infections, and gastroenteritis.

Picobirnaviridae

Originally described in 1988, picobirnavirus, is now known to be common (72). It has been implicated in causing diarrhea in children and immunocompromised people. Different strains have been detected in numerous animals including rats, birds, guinea pigs, swine, rabbits, and cattle (73). Whether animals can act as reservoirs for human picobirnavirus infections is unknown at present (73), but there is some evidence of cross-species transmission (15, 74, 75). Because these viruses are bisegmented and sometimes trisegmented double RNA viruses (76), reassortment during coinfection is possible, potentially leading to increased virulence.

Rotavirus and Reovirus

These viruses, (named "REO" viruses for "respiratory enteric orphan" viruses) were originally thought to be associated with unapparent respiratory and gastrointestinal infections. However, rotaviruses, which are members of the reovirus family, are significant human pathogens. They spread by the fecal–oral route and they are principally associated with dysentery and dehydration in infants. Historically, rotaviruses have been responsible for approximately 0.5 million deaths per annum of infants and children, primarily in the developing world. However, a recently developed efficacious vaccine is now reducing rotavirus-associated infant mortality.

Anellovirus

These are small circular DNA viruses with an undefined role in disease pathogenesis. The prototype Torque Teno virus is present in 80% of human plasma samples (77, 78), indicating widespread infection of the worldwide population. Infection with multiple strains is not uncommon (79). Anelloviruses have also been detected in nonhuman primates, cats, dogs, pigs, and rodents (80, 81). Torque Teno virus is found in stools of gastroenteritis patients (82); swine Torque Teno virus strains have been shown to be present in pork and have been found in 25% of human fecal samples (83).

Cyclovirus

Found in feces, cycloviruses are single-stranded DNA viruses encoded by a 2-kb genome that encodes only two open reading frames. As a virus class, these viruses are highly diverse and are identified in the feces of farm animals, humans, and chimpanzees (84, 85). Recently, metagenomics investigation of cerebrospinal fluid samples from African patients with unexplained paraplegia found a novel cyclovirus in 15% and 10% of serum and cerebrospinal fluid samples, respectively (86).

VIRUS THREATS POTENTIALLY ASSOCIATED WITH AGRICULTURE

Agriculture can be associated with the emergence of viral pathogens. Animal husbandry and meat production bring humans into close contact with large numbers of animals and consequently with the viral pathogens that infect them. The following are examples of viruses that are less likely to be transmitted by the fecal–oral route (foodborne routes) but more likely to be transmitted from animals to humans by direct contact, blood-borne, or respiratory routes.

Influenza Viruses

Influenza viruses are negative-stranded RNA viruses comprised of eight genetic segments that are major causes of morbidity and mortality in mammals, including humans, and birds worldwide. Due to their segmented genome, influenza viruses can undergo reassortment if two different influenza viruses infect the same cell. This reassortment, termed an "antigenic shift," can result in new unique progeny viruses that may no longer be recognized by host antibodies and other adaptive immune responses. Thus, populations that were once largely immune to the previous influenza strain, either by previous exposure or via vaccination, can suddenly have no immunity to this new virus strain. As a result, large-scale pandemics can occur due to a high proportion of immunologically naive hosts. Bad animal husbandry practices can contribute to the genesis and spread of influenza strains. These practices include raising swine in close proximity to chickens, as well as waterfowl and other fowl species, feeding dead chickens to swine, and the practice of taking live chickens to open bird markets. Open bird markets are of particular concern since birds are exposed to birds from other flocks of different geographic origin, to humans, and possibly to other animals. Often, unsold live birds are then returned to their farms of origin, or sold live birds are taken to other farms and localities. In recent years, avian flu has caused angst among public health workers.

With the advent of modern nucleic acid sequencing techniques, it was recognized that the 1918 "Spanish flu" pandemic, which killed approximately 100 million people worldwide, was caused by a virus that had many gene segments which were commonly associated with avian flu strains (87). The recent appearance of H5N1 is a major concern, given that a high human mortality often occurs once this virus is contracted from chickens. While the virus has shown itself to be capable of infecting humans, fortunately to date, it has not shown a high propensity to spread from human to human, and those that have contacted this deadly virus were known to be in regular close contact with chickens. Given influenza's ability to adapt, there may be some cause for pessimism regarding H5N1's potential to remain incapable of human-to-human transmission.

A number of other influenza avian flu strains may also be a threat to human health from poultry since the H7N9 (88), H9N2, and H7N3 (89) strains have been implicated in human outbreaks (90). Curiously, H7N9 is considered a relatively less pathogenic virus for chickens, but a 29% fatality rate has been reported for infected poultry workers (91). Waterfowl are also thought to be an important reservoir of influenza virus strains. Although influenza virus is typically not considered to be disseminated via the foodborne route since it is an enveloped virus, it should be noted that people have acquired H5N1 influenza by drinking duck blood (92).

Poxviruses

Poxviruses are large double-stranded DNA viruses that can infect a large number of animals. The orthopoxvirus genus contains several species that have broad host ranges including rabbitpox, monkeypox, buffalopox, and cowpox. The deadly disease smallpox, caused by the variola virus, was eradicated via worldwide vaccination several decades ago. However, other poxviruses have relatively broad host ranges and can be transmitted and cause illness in humans. A large outbreak of buffalopox among domestic buffalo in India was accompanied by a large zoonotic infection of milk workers in India (93). Monkeypox infection is associated with human mortality in about 10% of cases (94), and camelpox is also transmissible to humans, with rodents being implicated as probable reservoirs for these two viruses (93).

Coronaviruses

Coronaviruses are the second most common cause of the "common cold" in humans. They are also associated with gastroenteritis and can be transmitted by the fecal–oral route (95). Although historically considered relatively benign for humans, the emergence of severe acute respiratory syndrome (SARS) virus and Middle East respiratory syndrome (MERS) virus highlights the high disease potential associated with coronaviruses. Indications are that live exotic food markets may have played a role in transmission of the SARS virus from civets to humans (96). The MERS virus is thought to have jumped from camels to humans (96, 97). As described below, current thinking is that civets and camels may have been just the intermediate hosts for this virus since evidence suggests that bats may have been the ultimate source of these viruses (14, 98). Also, it is worth noting that a third coronavirus, porcine epidemic diarrhea virus, is currently infecting swine worldwide, causing high levels of piglet mortality. However, to date there is no indication that porcine epidemic diarrhea virus can infect humans.

Nipah and Hendra Viruses

These viruses are members of the *Paramyxoviridae* family. In the 1990s Nipah virus emerged in Malaysia and was associated with swine farm encroachment into forested areas when fruit bats were roosting (14). Movement of pigs for slaughter then resulted in the spread of the virus to Singapore, resulting in approximately 100 human mortalities in Southeast Asia. The Hendra virus was found in association with horses exposed to fruit bat guano and subsequent transmission to a limited number of humans in Australia (99).

Bunyavirus

This virus has been associated with 10,000 cases of severe fever with thrombocytopenia syndrome in Asia (100). Surveys of farm animals in Jiangsu Province, China, showed that 47% had severe fever thrombocytopenia syndrome virus (SFTS) virus (101), and subsequent investigation in the United States has indicated that 10 to 18% of sheep, cattle, goats, deer, and elk samples in the state of Minnesota had antibodies against bunyaviruses, demonstrating that the virus is widespread in livestock and wildlife and suggesting the potential for human transmission (102). Indications are that it is spread by ticks and amplified by livestock. A novel bunyavirus causing human illness in Missouri, named the heartland virus, has also been shown to be transmitted by ticks (103). Crimean-Congo hemorrhagic fever virus is also tick-borne. It can be contracted from ticks in contact with infected people or from contact with bodies and tissue of infected livestock (104). It is endemic to Asia, Africa, and Europe. While there is no evidence that the virus causes disease in livestock, most cases are associated with agriculture or animal husbandry workers (104).

ANIMAL RESERVOIRS OF VIRUSES

As noted earlier, domestic animals generally serve as the intermediate virus hosts that facilitate the introduction of viruses to humans. However, the ultimate source of viruses, often referred to as the virus reservoir, frequently turns out to be wildlife. Among these are the usual suspects, such as mice and rats, but migratory birds, nonhuman primates, and especially, bats are increasingly implicated as sources of emerging and zoonotic viruses (105, 106).

Bats

Bats are a zoonotic reservoir for viruses that can cause a number of serious human diseases (107). Currently, the DBatVir database (http://www.mgc.ac.cn/DBatVir/) contains virus sequences of more than 4,100 bat-associated animal viruses of 23 families detected in 196 different bat species from 69 countries (108). It is increasingly clear that bats can pose a grave health threat to humans or human-associated animals, particularly when anthropogenic activities create habitat loss and encroachment and perhaps provide novel man-made habitats, such as mines (109, 110). Bats are unique animals in that they are highly diverse, comprising >1,000 species. Many bats species are highly social, living in dense groups, and many are also migratory, providing a unique caldron for virus infection and evolution (111). Bats are thought to have more viruses per species than rodents (112). Why bats are able to tolerate certain viruses that are highly pathogenic to other species is not well understood. It has been suggested that the innate immune systems of bats may have evolved to be so hyper-vigilant that they can tolerate many highly pathogenic viruses without severe illness (113). An example of how a bat virus could enter the human population can be found in the outbreaks of Nipah virus. Fruit bats (*Pteropus* sp.; i.e., flying foxes) were implicated as the original source of the outbreak. Noting that these bats were eating figs, roosting in trees above swine pens, and dropping partially eaten fruits, as well as bat feces, researchers concluded that swine probably contracted this virus via eating fruit contaminated with Nipah virus from bats. Recent and current outbreaks have implicated bats as the ultimate reservoir of Ebola and coronaviruses associated with SARS, MERS, and porcine epidemic diarrhea virus (114). Another key consideration is that many larger species of bats are hunted for food, especially in the developing world (115).

Wild Bush Meat

There have been concerns regarding the transmission of human pathogens from wild meat hunted in different parts of the world (116). Primate meat is of particular concern (116, 117). In central Africa, it has been estimated that up to 3.4 million tons of bush meat are harvested annually (1). It has been suggested that human immunodeficiency virus (HIV) entered the human population via consumption of chimpanzee and/or sooty mangabey meat (118, 119). The consensus opinion among infectious disease researchers at this point seems to be that it is unlikely that these primate viruses directly infected humans via bush meat consumption but more likely infected humans as part of the field dressing or butchering processes, which may have brought humans into contact with infected blood or tissues (118). An HIV-like gorilla strain has also been recently identified in humans not known to have directly consumed gorilla meat (120). For zoonotic HEV strains, wild swine and deer have been shown to be sources of human infection. It is also said that large fruit bats are considered quite tasty in some regions due to the fruity flavor of their meat. Hunting, butchering, and consumption of these bats could pose a potential threat of exposure to a number of different viruses including Ebola.

Rodents

Zoonotic HEV has been identified in rats and rat feces, suggesting that they could contaminate food with fecally shed viruses. It is now understood that hantaviruses, which can rapidly cause fatal respiratory infections, are common in mouse populations in the United States (121). Worldwide, other members of the *Hantavirus* genus transmitted by rodents include Hantaan, Dobrava-Belgrade, Seoul, and Puumala viruses, which cause hemorrhagic fever with renal syndrome worldwide (122). Voles have been implicated in the spread of the picornavirus Ljungan virus, which is known to be associated with the onset of diabetes, prenatal central nervous malformation, and abortion in humans. Bolivian hemorrhagic fever virus, a member of the *Arenaviridae* associated with >25% human mortality, is thought to be acquired via breathing aerosolized excreta from rodents, consumption of contaminated food, or direct mucous membrane contact with infectious particles (123, 124). The primary hosts for the highly pathogenic Lassa fever virus are mice.

Swine

A novel porcine parecho-like virus was isolated in China and found to be in high prevalence in sampled piglets. Testing of humans was negative for the virus, but a high seroprevalence of 64% was reported against a recombinant virus polypeptide antigen (48). As described earlier, swine are also a reservoir for zoonotic HEV. Recent metagenomics analysis of swine feces reveals the presence of kobuviruses, astroviruses, enteroviruses, sapoviruses, sapeloviruses, coronaviruses, bocaviruses, and teschoviruses, as well as circoviruses and parvovirus (125), indicating that swine may be a rich source of viruses.

Bovines

The delta retrovirus bovine leukemia (BLV) is common in cattle worldwide, causes bovine leucosis, and is related to the human T-cell leukemia virus. In the United States, virtually all large-scale dairy herds are infected with this virus. A decade ago, 39% of human volunteers were found to be seropositive against the BLV capsid protein, but it was unclear if this represented human exposure to inactivated virus or infection with BLV (126). Since retroviruses integrate themselves into cellular DNA, it was possible to examine human breast tissue for the integrated presence of viral DNA. Buehring et al. (127) identified BLV DNA within secretory human mammary epithelium of the milk ducts and lobules, suggesting that this virus is infecting humans and may potentially be disseminated through human breast milk. Metagenomics studies of aerosols in meat packing environments have identified the presence of potentially oncogenic bovine papilloma viruses and polyomaviruses. While the connection is certainly tenuous, it has been suggested that these viruses may account for an elevated level of lung cancer in butchers and meat packers, as well as a higher incidence of colorectal cancer among human populations that consume high quantities of beef (128, 129).

Poultry

Modern poultry production methods make full use of the economy of scale principle, making chicken one of the most economical meat sources. The downside of this is that poultry production houses hold thousands of birds together at high density, providing a highly concentrated environment for viruses to spread and evolve. For broiler production, these birds are harvested only a couple of months after hatching, meaning that chickens are immunologically immature, which enhances their susceptibility to numerous pathogens. The major zoonotic virus issue for poultry is influenza virus since there are multiple strains (i.e., H5N1, H7N9, H9N2) which can now infect humans with significant morbidity and mortality. To date, human-to-human spread of avian influenza strains has been limited. Given that influenza viruses dynamically mutate and have a propensity to cause pandemics, optimism that this will remain the case must be tempered by the knowledge that a relatively minor mutation may dramatically enhance avian influenza's ability to infect humans (130).

A plethora of other viruses infect poultry but are not known to infect humans. These include herpesviruses (i.e., Marek's disease virus) and retroviruses (i.e., Rous sarcoma virus) that cause cancer in poultry, several respiratory viruses including a coronavirus (i.e., infectious bronchitis virus) that is distantly related to SARS/MERS, and paramyxoviruses (i.e., Newcastle's disease) that are distantly related to Nipah and Hendra viruses. Reoviruses are significant pathogens for poultry. Originally isolated from mammals, these strains were viewed as relatively insignificant viruses that were detected but not frequently associated with substantial pathologies in mammalian hosts. Subsequently, it was discovered that avian reoviruses, which infect poultry, can be fairly pernicious, in part because they encode a virus fusion protein not normally found in mammalian reoviruses. This fusion protein can facilitate fusion of the membrane of infected cells with its neighbor, facilitating cell-to-cell spread of this virus as a means of avoiding extracellular neutralization by antibodies (131). Should this cell-to-cell fusion gene make its way into human reoviruses, these viruses could become a real problem.

Migratory Waterfowl

Being social, migratory, and associated with water, ducks, geese, and other migratory shorebirds are thought to be key wild reservoirs for influenza viruses since these birds essentially feed and defecate in the same waters (132). It is thought that many influenza infections of waterfowl are often unapparent. It was expected that H5N1 influenza would spread to the western hemisphere via waterfowl, but to date that has not happened, perhaps because H5N1 causes unusually high mortality in waterfowl, inhibiting migration of infected birds. Also, domestic ducks and geese are often raised together with chickens by hobbyists and small-scale farmers, providing a potential link between wild and domestic waterfowl and domestic poultry.

Bivalve Shellfish

Human pathogenic viruses do not replicate within or infect bivalve shellfish (i.e., mussels, clams, oysters). Rather, they are bioconcentrated from water by filter feeding. Conceivably, virtually any nonenveloped viruses capable of persisting in water can become bioconcentrated and sequestered within shellfish meat. The degree to which water-borne viruses persist within shellfish is a function of the shellfish's metabolic rate and may be also be a function of the virus' ability to resist the acidic digestive processes of shellfish (133). Shellfish have been directly implicated as the vector for foodborne acquisition of HuNoV, HAV, HEV, and Aichi virus as a result of exposure to human waste in growing waters. Adenovirus, enteroviruses, and astrovirus can be detected in shellfish (134, 135). As such, raw shellfish trade and consumption could conceivably provide a means of introducing waterborne virus strains to new human populations.

Exotic and Common Pets

It is clear that dogs and cats could be reservoirs of viruses that infect humans, given their close association with humans (i.e., drinking from toilet bowls, sleeping in beds, etc.). The degree to which these animals could serve as intermediates by which they might acquire a cross-species infection that could subsequently to be transmitted to humans is undetermined. Canine exposure to human viruses such as HEV, HAV, and HuNoV have been demonstrated (136, 137), in some cases among coprophagic dogs in regions where human open defecation is common. It is more likely that these are cases of zooanthroponosis rather than zoonosis, however. Exotic pets and their trade could bring humans into contact with a number of serious pathogens. Some years ago there was an outbreak of monkeypox virus associated with African rodents being kept as pets in the United States. Fruit bats are kept as pets in certain areas (i.e., Australia), and potbellied pigs are now popular companion animals. Rodents are also common pets (i.e., guinea pigs, hamsters, gerbils, mice). Although not given much thought in the developed world, canine-associated rabies kills 55,000 persons per year due to limited vaccination in some areas such as Africa and Asia (138).

Influenza has a high potential to infect pets and subsequently humans. Examples include the death of domestic cats and large zoo cats (i.e., tigers) that ate chickens infected with H5N1. H1N1 influenza has been transmitted from owners to pets (ferrets, cats, dogs), and equine H3N8 influenza has been transmitted to dogs (138). As noted earlier, some H7 and H9 influenza strains can infect humans with some propensity for fatalities (88, 89, 139). Pet birds have been shown to be susceptible to H5N1 influenza (140). A recent metagenomics study has identified a previously unknown feline astrovirus that is closely related to human astrovirus, suggesting past cross-species infection (141). Strains of feline bocavirus, feline rotavirus, feline picobirnaviruses, as well as a feline picornavirus have been identified. The later has been tentatively named as a new genus based on its diversity from other picornaviruses (142).

METAGENOMICS

Metagenomics can be defined as direct characterization (and sequence determination) of genetic material from a sample. Much of traditional virology focuses on viruses that are readily grown in tissue culture or are clearly human pathogens. Based on the ability to generate massive amounts of viral sequence data, then sort and assemble these sequences, different virus sources such as human stool (142, 143), bat guano (144), and rodent feces have been studied. These metagenomics studies are revealing a treasure trove of new information. It is now understood that the fraction of known viruses, as compared to unknown viruses, is miniscule. In addition, many virus sequences generated in virus metagenomics studies are not homologous to known viruses (145). Thus, science is beginning to identify and define the emerging and zoonotic pool of viruses that may become the source of future pathogens. Also, metagenomics has documented an increase in the enteric human virome when humans are immunosuppressed, both in terms of the total amount of virus shed and different types of viruses observed (146).

While many of these viruses appear to be commensal, their potential to emerge as foodborne threats is unknown. At the minimum, these studies provide key basic information about the virome of different species, perhaps facilitating the identification of possible future zoonotic infection sources and reservoirs (13). The identification of potential viral disease agents ahead of their implication of disease processes could facilitate future development of therapeutics and vaccines. One clear metagenomics theme is that once a novel type of virus is identified, close genetic counterparts are often identified that infect other species. This suggests that these virus strains originated from an ancient cross-over or zoonotic event followed by divergence and host adaptation. For example, human bocavirus was first identified in 2005; bocaviruses infecting sea lions, chimps, gorillas, and swine have been subsequently identified (63). An astrovirus that was recently detected in humans is closely related to animal astroviruses, suggesting zoonotic transmission (141). Mouse, bovine, canine, and human kobuviruses (Aichi) are all closely related, suggesting previous zoonotic transmission of kobuviruses, perhaps followed by independent evolution (85, 122). It is known that animal slaughterhouse workers have a higher incidence of lung cancer and hematologic neoplasm than the general population (147, 148). Metagenomics evaluation of slaughterhouse samples has revealed a high level of papilloma viruses in aerosols, as discussed earlier. One key caveat for metagenomics studies is that these studies simply reveal the presence of a virus and cannot directly indicate the degree to which the identified virus is involved in disease processes (i.e., fulfillment of Koch's disease postulates). Thus, by themselves, metagenomics studies cannot necessarily distinguish viruses that cause unapparent infections from those involved in acute or chronic disease processes.

ACKNOWLEDGMENTS

The USDA is an equal opportunity provider and employer. Mention of trade names or commercial products is solely for the purpose of providing specific information and does not imply recommendation or endorsement by the U.S. Department of Agriculture.

CITATION

Kingsley DH. 2016. Emerging foodborne and agriculture-related viruses. Microbiol Spectrum 4(4):PFS-0007-2014.

REFERENCES

1. **Brown C.** 2004. Emerging zoonoses and pathogens of public health significance: an overview. *Rev Sci Tech* **23:**435–442.
2. **Morse SS.** 2004. Factors and determinants of disease emergence. *Rev Sci Tech* **23:**443–451.
3. **Pearce-Duvet JMC.** 2006. The origin of human pathogens: evaluating the role of agriculture and domestic animals in the evolution of human disease. *Biol Rev Camb Philos Soc* **81:**369–382.
4. **Newell DG, Koopmans M, Verhoef L, Duizer E, Aidara-Kane A, Sprong H, Opsteegh M, Langelaar M, Threfall J, Scheutz F, van der Giessen J, Kruse H.** 2010. Foodborne diseases: the challenges of 20 years ago still persist while new ones continue to emerge. *Int J Food Microbiol* **139**(Suppl 1)**:**S3–S15.
5. **Mead PS, Slutsker L, Dietz V, McCaig LF, Bresee JS, Shapiro C, Griffin PM, Tauxe RV.** 1999. Food-related illness and death in the United States. *Emerg Infect Dis* **5:**607–625.
6. **Scallan E, Hoekstra RM, Angulo FJ, Tauxe RV, Widdowson MA, Roy SL, Jones JL, Griffin PM.** 2011. Foodborne illness acquired in the United States: major pathogens. *Emerg Infect Dis* **17:**7–15.
7. **Trepo C.** 2014. A brief history of hepatitis milestones. *Liver Int* **34**(Suppl 1)**:**29–37.
8. **Bergh O, Børsheim KY, Bratbak G, Heldal M.** 1989. High abundance of viruses found in aquatic environments. *Nature* **340:**467–468.
9. **Oostra M, de Haan CA, Rottier PJ.** 2007. The 29-nucleotide deletion present in human but not in animal severe acute respiratory syndrome coronaviruses disrupts the functional expression of open reading frame 8. *J Virol* **81:** 13876–13888.
10. **Kugelman JR, Johnston SC, Mulembakani PM, Kisalu N, Lee MS, Koroleva G, McCarthy SE, Gestole MC, Wolfe ND, Fair JN, Schneider BS, Wright LL, Huggins J, Whitehouse CA, Wemakoy EO, Muyembe-Tamfum JJ, Hensley LE, Palacios GF, Rimoin AW.** 2014. Genomic variability of monkeypox virus among humans, Democratic Republic of the Congo. *Emerg Infect Dis* **20:**232–239.
11. **Bengis RG, Leighton FA, Fischer JR, Artois M, Mörner T, Tate CM.** 2004. The role of wildlife in emerging and re-emerging zoonoses. *Rev Sci Tech* **23:**497–511.
12. **Woolhouse ME.** 2002. Population biology of emerging and re-emerging pathogens. *Trends Microbiol* **10**(Suppl)**:**S3–S7.
13. **Levinson J, Bogich TL, Olival KJ, Epstein JH, Johnson CK, Karesh W, Daszak P.** 2013. Targeting surveillance for zoonotic virus discovery. *Emerg Infect Dis* **19:**743–747.
14. **Field HE.** 2009. Bats and emerging zoonoses: henipaviruses and SARS. *Zoonoses Public Health* **56:**278–284.
15. **Parrish CR, Holmes EC, Morens DM, Park EC, Burke DS, Calisher CH, Laughlin CA, Saif LJ, Daszak P.** 2008. Cross-species virus transmission and the emergence of new epidemic diseases. *Microbiol Mol Biol Rev* **72:**457–470.
16. **Longdon B, Hadfield JD, Webster CL, Obbard DJ, Jiggins FM.** 2011. Host phylogeny determines viral persistence and replication in novel hosts. *PLoS Pathog* **7:**e1002260.
17. **Bergelson JM, Cunningham JA, Droguett G, Kurt-Jones EA, Krithivas A, Hong JS, Horwitz MS, Crowell RL, Finberg RW.** 1997. Isolation of a common receptor for coxsackie B viruses and adenoviruses 2 and 5. *Science* **275:**1320–1323.
18. **Cutler SJ, Fooks AR, van der Poel WHM.** 2010. Public health threat of new, reemerging, and neglected zoonoses in the industrialized world. *Emerg Infect Dis* **16:**1–7.
19. **Purcell RH, Engle RE, Rood MP, Kabrane-Lazizi Y, Nguyen HT, Govindarajan S, St Claire M, Emerson SU.** 2011. Hepatitis E virus in rats, Los Angeles, California, USA. *Emerg Infect Dis* **17:**2216–2222.
20. **Huang S-J, Liu X-H, Zhang J, Ng M-H.** 2014. Protective immunity against HEV. *Curr Opin Virol* **5:**1–6.
21. **Labrique AB, Kuniholm MH, Nelson KE.** 2010. The global impact of hepatitis E virus: new horizons for an emerging virus, p 53–93. *In* Scheld WM, Murray BE, Hughes JM

(ed), *Emerging Infections*, 9th ed. ASM press, Washington, DC.

22. **Arankalle VA, Joshi MV, Kulkarni AM, Gandhe SS, Chobe LP, Rautmare SS, Mishra AC, Padbidri VS.** 2001. Prevalence of anti-hepatitis E virus antibodies in different Indian animal species. *J Viral Hepat* **8:**223–227.
23. **Kuniholm MH, Purcell RH, McQuillan GM, Engle RE, Wasley A, Nelson KE.** 2009. Epidemiology of hepatitis E virus in the United States: results from the Third National Health and Nutrition Examination Survey, 1988–1994. *J Infect Dis* **200:**48–56.
24. **Tei S, Kitajima N, Takahashi K, Mishiro S.** 2003. Zoonotic transmission of hepatitis E virus from deer to human beings. *Lancet* **362:** 371–373.
25. **Tamada Y, Yano K, Yatsuhashi H, Inoue O, Mawatari F, Ishibashi H.** 2004. Consumption of wild boar linked to cases of hepatitis E. *J Hepatol* **40:**869–870.
26. **Feagins AR, Opriessnig T, Guenette DK, Halbur PG, Meng XJ.** 2007. Detection and characterization of infectious hepatitis E virus from commercial pig livers sold in local grocery stores in the USA. *J Gen Virol* **88:**912–917.
27. **Berto A, Grierson S, Hakze-van der Honing R, Martelli F, Johne R, Reetz J, Ulrich RG, Pavio N, Van der Poel WHM, Banks M.** 2013. Hepatitis E virus in pork liver sausage, France. *Emerg Infect Dis* **19:**264–266.
28. **Lara J, Purdy MA, Khudyakov YE.** 2014. Genetic host specificity of hepatitis E virus. *Infect Genet Evol* **24:**127–139.
29. **Hudopisk N, Korva M, Janet E, Simetinger M, Grgič-Vitek M, Gubenšek J, Natek V, Kraigher A, Strle F, Avšič-Županc T.** 2013. Tick-borne encephalitis associated with consumption of raw goat milk, Slovenia, 2012. *Emerg Infect Dis* **19:**806–808.
30. **Holzmann H, Aberle SW, Stiasny K, Werner P, Mischak A, Zainer B, Netzer M, Koppi S, Bechter E, Heinz FX.** 2009. Tick-borne encephalitis from eating goat cheese in a mountain region of Austria. *Emerg Infect Dis* **15:**1671–1673.
31. **Bendig JWA, O'Brien PS, Muir P, Porter HJ, Caul EO.** 2001. Enterovirus sequences resembling coxsackievirus A2 detected in stool and spleen from a girl with fatal myocarditis. *J Med Virol* **64:**482–486.
32. **Kim K-S, Hufnagel G, Chapman NM, Tracy S.** 2001. The group B coxsackieviruses and myocarditis. *Rev Med Virol* **11:**355–368.
33. **Legay V, Chomel JJ, Fernandez E, Lina B, Aymard M, Khalfan S.** 2002. Encephalomyelitis due to human parechovirus type 1. *J Clin Virol* **25:**193–195.
34. **Ward C.** 1978. Severe arrhythmias in coxsackievirus B3 myopericarditis. *Arch Dis Child* **53:**174–176.
35. **Andréoletti L, Bourlet T, Moukassa D, Rey L, Hot D, Li Y, Lambert V, Gosselin B, Mosnier JF, Stankowiak C, Wattré P.** 2000. Enteroviruses can persist with or without active viral replication in cardiac tissue of patients with end-stage ischemic or dilated cardiomyopathy. *J Infect Dis* **182:**1222–1227.
36. **Berger MM, Kopp N, Vital C, Redl B, Aymard M, Lina B.** 2000. Detection and cellular localization of enterovirus RNA sequences in spinal cord of patients with ALS. *Neurology* **54:**20–25.
37. **Riecanský I, Schreinerová Z, Egnerová A, Petrovicová A, Bzduchová O.** 1989. Incidence of coxsackie virus infection in patients with dilated cardiomyopathy. *Cor Vasa* **31:**225–230.
38. **Roivainen M, Ylipaasto P, Savolainen C, Galama J, Hovi T, Otonkoski T.** 2002. Functional impairment and killing of human beta cells by enteroviruses: the capacity is shared by a wide range of serotypes, but the extent is a characteristic of individual virus strains. *Diabetologia* **45:**693–702.
39. **Yin H, Berg AK, Tuvemo T, Frisk G.** 2002. Enterovirus RNA is found in peripheral blood mononuclear cells in a majority of type 1 diabetic children at onset. *Diabetes* **51:**1964–1971.
40. **Hyeon JY, Hwang S, Kim H, Song J, Ahn J, Kang B, Kim K, Choi W, Chung JK, Kim CH, Cho K, Jee Y, Kim J, Kim K, Kim SH, Kim MJ, Cheon DS.** 2013. Accuracy of diagnostic methods and surveillance sensitivity for human enterovirus, South Korea, 1999–2011. *Emerg Infect Dis* **19:**1268–1275.
41. **Trallero G, Avellon A, Otero A, De Miguel T, Pérez C, Rabella N, Rubio G, Echevarria JE, Cabrerizo M.** 2010. Enteroviruses in Spain over the decade 1998–2007: virological and epidemiological studies. *J Clin Virol* **47:**170–176.
42. **McMinn PC.** 2002. An overview of the evolution of enterovirus 71 and its clinical and public health significance. *FEMS Microbiol Rev* **26:**91–107.
43. **Liu J, Xiao H, Wu Y, Liu D, Qi X, Shi Y, Gao GF.** 2014. H7N9: a low pathogenic avian influenza A virus infecting humans. *Curr Opin Virol* **5:**91–97.
44. **Centers for Disease Control and Prevention (CDC).** 2011. Clusters of acute respiratory illness associated with human enterovirus 68: Asia, Europe, and United States, 2008–2010. *MMWR Morb Mortal Wkly Rep* **60:**1301–1304.

45. **Yip CCY, Lau SKP, Woo PC, Wong SSY, Tsang THF, Lo JYC, Lam W-K, Tsang C-C, Chan K-H, Yuen K-Y.** 2013. Recombinant coxsackievirus A2 and deaths of children, Hong Kong, 2012. *Emerg Infect Dis* **19:**1285–1288.
46. **Maan HS, Chowdhary R, Shakya AK, Dhole TN.** 2013. Genetic variants of echovirus 13, northern India, 2010. *Emerg Infect Dis* **19:**293–296.
47. **Williams CH, Panayiotou M, Girling GD, Peard CI, Oikarinen S, Hyöty H, Stanway G.** 2009. Evolution and conservation in human parechovirus genomes. *J Gen Virol* **90:**1702–1712.
48. **Yu J-M, Li X-Y, Ao Y-Y, Li LL, Liu N, Li JS, Duan ZJ.** 2013. Identification of a novel picornavirus in healthy piglets and seroepidemiological evidence of its presence in humans. *PLoS One* **8:**e70137.
49. **Jones MS, Lukashov VV, Ganac RD, Schnurr DP.** 2007. Discovery of a novel human picornavirus in a stool sample from a pediatric patient presenting with fever of unknown origin. *J Clin Microbiol* **45:**2144–2150.
50. **Tapparel C, Siegrist F, Petty TJ, Kaiser L.** 2013. Picornavirus and enterovirus diversity with associated human diseases. *Infect Genet Evol* **14:**282–293.
51. **Khamrin P, Thongprachum A, Kikuta H, Yamamoto A, Nishimura S, Sugita K, Baba T, Kobayashi M, Okitsu S, Hayakawa S, Shimizu H, Maneekarn N, Ushijima H.** 2013. Three clusters of Saffold viruses circulating in children with diarrhea in Japan. *Infect Genet Evol* **13:**339–343.
52. **Yamashita T, Sakae K, Tsuzuki H, Suzuki Y, Ishikawa N, Takeda N, Miyamura T, Yamazaki S.** 1998. Complete nucleotide sequence and genetic organization of Aichi virus, a distinct member of the *Picornaviridae* associated with acute gastroenteritis in humans. *J Virol* **72:**8408–8412.
53. **Reuter G, Boros A, Pankovics P.** 2011. Kobuviruses: a comprehensive review. *Rev Med Virol* **21:**32–41.
54. **Lodder WJ, Rutjes SA, Takumi K, de Roda Husman AM.** 2013. Aichi virus in sewage and surface water, The Netherlands. *Emerg Infect Dis* **19:**1222–1230.
55. **Li L, Victoria J, Kapoor A, Blinkova O, Wang C, Babrzadeh F, Mason CJ, Pandey P, Triki H, Bahri O, Oderinde BS, Baba MM, Bukbuk DN, Besser JM, Bartkus JM, Delwart EL.** 2009. A novel picornavirus associated with gastroenteritis. *J Virol* **83:** 12002–12006.
56. **Greninger AL, Runckel C, Chiu CY, Haggerty T, Parsonnet J, Ganem D, DeRisi JL.** 2009. The complete genome of klassevirus: a novel picornavirus in pediatric stool. *Virol J* **6:**82.
57. **Holtz LR, Finkbeiner SR, Zhao G, Kirkwood CD, Girones R, Pipas JM, Wang D.** 2009. Klassevirus 1, a previously undescribed member of the family *Picornaviridae,* is globally widespread. *Virol J* **6:**86.
58. **Kapoor A, Victoria J, Simmonds P, Slikas E, Chieochansin T, Naeem A, Shaukat S, Sharif S, Alam MM, Angez M, Wang C, Shafer RW, Zaidi S, Delwart E.** 2008. A highly prevalent and genetically diversified *Picornaviridae* genus in South Asian children. *Proc Natl Acad Sci USA* **105:**20482–20487.
59. **Kapusinszky B, Phan TG, Kapoor A, Delwart E.** 2012. Genetic diversity of the genus *Cosavirus* in the family *Picornaviridae*: a new species, recombination, and 26 new genotypes. *PLoS One* **7:**e36685.
60. **Allander T, Tammi MT, Eriksson M, Bjerkner A, Tiveljung-Lindell A, Andersson B.** 2005. Cloning of a human parvovirus by molecular screening of respiratory tract samples. *Proc Natl Acad Sci USA* **102:**12891–12896.
61. **Kantola K, Hedman L, Arthur J, Alibeto A, Delwart E, Jartti T, Ruuskanen O, Hedman K, Söderlund-Venermo M.** 2011. Seroepidemiology of human bocaviruses 1-4. *J Infect Dis* **204:**1403–1412.
62. **Vicente D, Cilla G, Montes M, Pérez-Yarza EG, Pérez-Trallero E.** 2007. Human bocavirus, a respiratory and enteric virus. *Emerg Infect Dis* **13:**636–637.
63. **Schildgen O, Qiu J, Söderlund-Venermo M.** 2012. Genomic features of the human bocaviruses. *Future Virol* **7:**31–39.
64. **Mori D, Ranawaka U, Yamada K, Rajindrajith S, Miya K, Perera HKK, Matsumoto T, Dassanayake M, Mitui MT, Mori H, Nishizono A, Söderlund-Venermo M, Ahmed K.** 2013. Human bocavirus in patients with encephalitis, Sri Lanka, 2009–2010. *Emerg Infect Dis* **19:**1859–1862.
65. **Mitui MT, Tabib SM, Matsumoto T, Khanam W, Ahmed S, Mori D, Akhter N, Yamada K, Kabir L, Nishizono A, Söderlund-Venermo M, Ahmed K.** 2012. Detection of human bocavirus in the cerebrospinal fluid of children with encephalitis. *Clin Infect Dis* **54:**964–967.
66. **De Benedictis P, Schultz-Cherry S, Burnham A, Cattoli G.** 2011. Astrovirus infections in humans and animals: molecular biology, genetic diversity, and interspecies transmissions. *Infect Genet Evol* **11:**1529–1544.

67. **Li L, Diab S, McGraw S, Barr B, Traslavina R, Higgins R, Talbot T, Blanchard P, Rimoldi G, Fahsbender E, Page B, Phan TG, Wang C, Deng X, Pesavento P, Delwart E.** 2013. Divergent astrovirus associated with neurologic disease in cattle. *Emerg Infect Dis* **19:**1385–1392.
68. **Mena KD, Gerba CP.** Waterborne adenovirus. *Rev Environ Contam Toxicol* **198:**133–167.
69. **Bosshard F, Armand F, Hamelin R, Kohn T.** 2013. Mechanisms of human adenovirus inactivation by sunlight and UVC light as examined by quantitative PCR and quantitative proteomics. *Appl Environ Microbiol* **79:**1325–1332.
70. **Fongaro G, Nascimento MA, Rigotto C, Ritterbusch G, da Silva AD, Esteves PA, Barardi CR.** 2013. Evaluation and molecular characterization of human adenovirus in drinking water supplies: viral integrity and viability assays. *Virol J* **10:**166.
71. **Buckwalter SP, Teo R, Espy MJ, Sloan LM, Smith TF, Pritt BS.** 2012. Real-time qualitative PCR for 57 human adenovirus types from multiple specimen sources. *J Clin Microbiol* **50:**766–771.
72. **Pereira HG, Flewett TH, Candeias JA, Barth OM.** 1988. A virus with a bisegmented double-stranded RNA genome in rat (*Oryzomys nigripes*) intestines. *J Gen Virol* **69:**2749–2754.
73. **Mondal ASM, Majee S.** 2014. Novel bisegmented virus (picobirnavirus) of animals, birds, and humans. *Asian Pac J Trop Dis* **4:**154–158.
74. **Bányai K, Jakab F, Reuter G, Bene J, Uj M, Melegh B, Szücs G.** 2003. Sequence heterogeneity among human picobirnaviruses detected in a gastroenteritis outbreak. *Arch Virol* **148:** 2281–2291.
75. **Bányai K, Martella V, Bogdán A, Forgách P, Jakab F, Meleg E, Bíró H, Melegh B, Szucs G.** 2008. Genogroup I picobirnaviruses in pigs: evidence for genetic diversity and relatedness to human strains. *J Gen Virol* **89:**534–539.
76. **Chandra R.** 1997. Picobirnavirus, a novel group of undescribed viruses of mammals and birds: a minireview. *Acta Virol* **41:**59–62.
77. **Takahashi K, Hoshino H, Ohta Y, Yoshida N, Mishiro S.** 1998. Very high prevalence of TT virus (TTV) infection in general population of Japan revealed by a new set of PCR primers. *Hepatol Res* **12:**233–239.
78. **Prescott LE, Simmonds P.** 1998. Global distribution of transfusion-transmitted virus. *N Engl J Med* **339:**776–777.
79. **Takayama S, Yamazaki S, Matsuo S, Sugii S.** 1999. Multiple infection of TT virus (TTV) with different genotypes in Japanese hemophiliacs. *Biochem Biophys Res Commun* **256:**208–211.
80. **Niel C, Diniz-Mendes L, Devalle S.** 2005. Rolling-circle amplification of Torque teno virus (TTV) complete genomes from human and swine sera and identification of a novel swine TTV genogroup. *J Gen Virol* **86:**1343–1347.
81. **Nishiyama S, Dutia BM, Stewart JP, Meredith AL, Shaw DJ, Simmonds P, Sharp CP.** 2014. Identification of novel anelloviruses with broad diversity in UK rodents. *J Gen Virol* **95:**1544–1553.
82. **Pinho-Nascimento CA, Leite JP, Niel C, Diniz-Mendes L.** 2011. Torque teno virus in fecal samples of patients with gastroenteritis: prevalence, genogroups distribution, and viral load. *J Med Virol* **83:**1107–1111.
83. **Jiménez-Melsió A, Parés S, Segalés J, Kekarainen T.** 2013. Detection of porcine anelloviruses in pork meat and human faeces. *Virus Res* **178:**522–524.
84. **Li L, Victoria JG, Wang C, Jones M, Fellers GM, Kunz TH, Delwart E.** 2010. Bat guano virome: predominance of dietary viruses from insects and plants plus novel mammalian viruses. *J Virol* **84:**6955–6965.
85. **Li L, Shan T, Soji OB, Alam MM, Kunz TH, Zaidi SZ, Delwart E.** 2011. Possible cross-species transmission of circoviruses and cycloviruses among farm animals. *J Gen Virol* **92:** 768–772.
86. **Smits SL, Zijlstra EE, van Hellemond JJ, Schapendonk CM, Bodewes R, Schürch AC, Haagmans BL, Osterhaus AD.** 2013. Novel cyclovirus in human cerebrospinal fluid, Malawi, 2010–2011. *Emerg Infect Dis* **19:**1511–1513.
87. **Taubenberger JK, Reid AH, Lourens RM, Wang R, Jin G, Fanning TG.** 2005. Characterization of the 1918 influenza virus polymerase genes. *Nature* **437:**889–893.
88. **Lu S, Zheng Y, Li T, Hu Y, Liu X, Xi X, Chen Q, Wang Q, Cao Y, Wang Y, Zhou L, Lowrie D, Bao J.** 2013. Clinical findings for early human cases of influenza A(H7N9) virus infection, Shanghai, China. *Emerg Infect Dis* **19:**1142–1146.
89. **Lopez-Martinez I, Balish A, Barrera-Badillo G, Jones J, Nuñez-García TE, Jang Y, Aparicio-Antonio R, Azziz-Baumgartner E, Belser JA, Ramirez-Gonzalez JE, Pedersen JC, Ortiz-Alcantara J, Gonzalez-Duran E, Shu B, Emery SL, Poh MK, Reyes-Teran G, Vazquez-Perez JA, Avila-Rios S, Uyeki T, Lindstrom S, Villanueva J, Tokars J, Ruiz-Matus C, Gonzalez-Roldan JF, Schmitt B, Klimov A, Cox N, Kuri-Morales P, Davis CT, Diaz-Quiñonez JA.** 2013. Highly pathogenic avian influenza A(H7N3) virus in poultry workers, Mexico, 2012. *Emerg Infect Dis* **19:**1531–1534.

90. **Belser JA, Davis CT, Balish A, Edwards LE, Zeng H, Maines TR, Gustin KM, Martínez IL, Fasce R, Cox NJ, Katz JM, Tumpey TM.** 2013. Pathogenesis, transmissibility, and ocular tropism of a highly pathogenic avian influenza A (H7N3) virus associated with human conjunctivitis. *J Virol* **87:**5746–5754.
91. **Pascua PN, Choi YK.** 2014. Zoonotic infections with avian influenza A viruses and vaccine preparedness: a game of "mix and match". *Clin Exp Vaccine Res* **3:**140–148.
92. **Shao D, Shi Z, Wei J, Ma Z.** 2011. A brief review of foodborne zoonoses in China. *Epidemiol Infect* **139:**1497–1504.
93. **Shchelkunov SN.** 2013. An increasing danger of zoonotic orthopoxvirus infections. *PLoS Pathog* **9:**e1003756.
94. **Jezek Z, Grab B, Szczeniowski M, Paluku KM, Mutombo M.** 1988. Clinico-epidemiological features of monkeypox patients with an animal or human source of infection. *Bull World Health Organ* **66:**459–464.
95. **La Rosa G, Fratini M, Della Libera S, Iaconelli M, Muscillo M.** 2013. Viral infections acquired indoors through airborne, droplet or contact transmission. *Ann Ist Super Sanita* **49:**124–132.
96. **Guan Y, Zheng BJ, He YQ, Liu XL, Zhuang ZX, Cheung CL, Luo SW, Li PH, Zhang LJ, Guan YJ, Butt KM, Wong KL, Chan KW, Lim W, Shortridge KF, Yuen KY, Peiris JS, Poon LL.** 2003. Isolation and characterization of viruses related to the SARS coronavirus from animals in southern China. *Science* **302:**276–278.
97. **Memish ZA, Mishra N, Olival KJ, Fagbo SF, Kapoor V, Epstein JH, Alhakeem R, Durosinloun A, Al Asmari M, Islam A, Kapoor A, Briese T, Daszak P, Al Rabeeah AA, Lipkin WI.** 2013. Middle East respiratory syndrome coronavirus in bats, Saudi Arabia. *Emerg Infect Dis* **19:**1819–1823.
98. **Raj VS, Osterhaus ADME, Fouchier RAM, Haagmans BL.** 2014. MERS: emergence of a novel human coronavirus. *Curr Opin Virol* **5:**58–62.
99. **Halpin K, Young PL, Field HE, Mackenzie JS.** 2000. Isolation of Hendra virus from pteropid bats: a natural reservoir of Hendra virus. *J Gen Virol* **81:**1927–1932.
100. **Yu XJ, Liang MF, Zhang SY, Liu Y, Li JD, Sun YL, Zhang L, Zhang Q-F, Popov VL, Li C, Qu J, Li Q, Zhang Y-P, Hai R, Wu W, Wang Q, Zhan F-X, Wang X-J, Kan B, Wang S-W, Wan K-L, Jing H-Q, Lu J-X, Yin W-W, Zhou H, Guan X-H, Liu J-F, Bi Z-Q, Liu G-H, Ren J, Wang H, Zhao Z, Song J-D, He J-R, Wan T, Zhang J-S, Fu X-P, Sun L-N, Dong X-P, Feng Z-J, Yang W-Z, Hong T, Zhang Y, Walker DH, Wang Y, Li D-X.** 2011. Fever with thrombocytopenia associated with a novel bunyavirus in China. *N Engl J Med* **364:**1523–1532.
101. **McMullan LK, Folk SM, Kelly AJ, MacNeil A, Goldsmith CS, Metcalfe MG, Batten BC, Albariño CG, Zaki SR, Rollin PE, Nicholson WL, Nichol ST.** 2012. A new phlebovirus associated with severe febrile illness in Missouri. *N Engl J Med* **367:**834–841.
102. **Xing Z, Schefers J, Schwabenlander M, Jiao Y, Liang M, Qi X, Li C, Goyal S, Cardona CJ, Wu X, Zhang Z, Li D, Collins J, Murtaugh MP.** 2013. Novel bunyavirus in domestic and captive farmed animals, Minnesota, USA. *Emerg Infect Dis* **19:**1487–1489.
103. **Savage HM, Godsey MS Jr, Lambert A, Panella NA, Burkhalter KL, Harmon JR, Lash RR, Ashley DC, Nicholson WL.** 2013. First detection of heartland virus (*Bunyaviridae*: *Phlebovirus*) from field collected arthropods. *Am J Trop Med Hyg* **89:**445–452.
104. **Ergonul O.** 2012. Crimean-Congo hemorrhagic fever virus: new outbreaks, new discoveries. *Curr Opin Virol* **2:**215–220.
105. **Chan JF-W, To KK-W, Tse H, Jin D-Y, Yuen K-Y.** 2013. Interspecies transmission and emergence of novel viruses: lessons from bats and birds. *Trends Microbiol* **21:**544–555.
106. **Hayman DTS, Bowen RA, Cryan PM, McCracken GF, O'Shea TJ, Peel AJ, Gilbert A, Webb CT, Wood JLN.** 2013. Ecology of zoonotic infectious diseases in bats: current knowledge and future directions. *Zoonoses Public Health* **60:**2–21.
107. **van der Poel WH, Lina PH, Kramps JA.** 2006. Public health awareness of emerging zoonotic viruses of bats: a European perspective. *Vector Borne Zoonotic Dis* **6:**315–324.
108. **Chen L, Liu B, Yang J, Jin Q.** 2014. DBatVir: the database of bat-associated viruses. *Database (Oxford)* **2014:**bau021.
109. **Smith I, Wang LF.** 2013. Bats and their virome: an important source of emerging viruses capable of infecting humans. *Curr Opin Virol* **3:**84–91.
110. **Wynne JW, Wang L-F.** 2013. Bats and viruses: friend or foe? *PLoS Pathog* **9:**e1003651.
111. **O'Shea TJ, Cryan PM, Cunningham AA, Fooks AR, Hayman DTS, Luis AD, Peel AJ, Plowright RK, Wood JLN.** 2014. Bat flight and zoonotic viruses. *Emerg Infect Dis* **20:**741–745.
112. **Luis AD, Hayman DT, O'Shea TJ, Cryan PM, Gilbert AT, Pulliam JR, Mills JN, Timonin ME, Willis CK, Cunningham AA, Fooks AR, Rupprecht CE, Wood JL, Webb CT.** 2013. A comparison of bats and rodents as reservoirs

of zoonotic viruses: are bats special? *Proc Biol Sci* **280:**20122753.

113. **Baker ML, Schountz T, Wang L-F.** 2013. Antiviral immune responses of bats: a review. *Zoonoses Public Health* **60:**104–116.
114. **Ithete NL, Stoffberg S, Corman VM, Cottontail VM, Richards LR, Schoeman MC, Drosten C, Drexler JF, Preiser W.** 2013. Close relative of human Middle East respiratory syndrome coronavirus in bat, South Africa. *Emerg Infect Dis* **19:**1697–1699.
115. **Jenkins RKB, Racey PA.** 2008. Bats as bushmeat in Madagascar. *Madag Conserv Dev* **3:**22–390.
116. **Wolfe ND, Daszak P, Kilpatrick AM, Burke DS.** 2005. Bushmeat hunting, deforestation, and prediction of zoonoses emergence. *Emerg Infect Dis* **11:**1822–1827.
117. **Wolfe ND, Heneine W, Carr JK, Garcia AD, Shanmugam V, Tamoufe U, Torimiro JN, Prosser AT, Lebreton M, Mpoudi-Ngole E, McCutchan FE, Birx DL, Folks TM, Burke DS, Switzer WM.** 2005. Emergence of unique primate T-lymphotropic viruses among central African bushmeat hunters. *Proc Natl Acad Sci USA* **102:**7994–7999.
118. **Aghokeng AF, Ayouba A, Mpoudi-Ngole E, Loul S, Liegeois F, Delaporte E, Peeters M.** 2010. Extensive survey on the prevalence and genetic diversity of SIVs in primate bushmeat provides insights into risks for potential new cross-species transmissions. *Infect Genet Evol* **10:**386–396.
119. **Peeters M, Courgnaud V, Abela B, Auzel P, Pourrut X, Bibollet-Ruche F, Loul S, Liegeois F, Butel C, Koulagna D, Mpoudi-Ngole E, Shaw GM, Hahn BH, Delaporte E.** 2002. Risk to human health from a plethora of simian immunodeficiency viruses in primate bushmeat. *Emerg Infect Dis* **8:**451–457.
120. **Plantier J-C, Leoz M, Dickerson JE, De Oliveira F, Cordonnier F, Lemée V, Damond F, Robertson DL, Simon F.** 2009. A new human immunodeficiency virus derived from gorillas. *Nat Med* **15:**871–872.
121. **Knust B, Rollin PE.** 2013. Twenty-year summary of surveillance for human hantavirus infections, United States. *Emerg Infect Dis* **19:**1934–1937.
122. **Phan TG, Kapusinszky B, Wang C, Rose RK, Lipton HL, Delwart EL.** 2011. The fecal viral flora of wild rodents. *PLoS Pathog* **7:**e1002218.
123. **Charrel RN, de Lamballerie X.** 2010. Zoonotic aspects of arenavirus infections. *Vet Microbiol* **140:**213–220.
124. **Patterson M, Grant A, Paessler S.** 2014. Epidemiology and pathogenesis of Bolivian hemorrhagic fever. *Curr Opin Virol* **5:**82–90.
125. **Shan T, Li L, Simmonds P, Wang C, Moeser A, Delwart E.** 2011. The fecal virome of pigs on a high-density farm. *J Virol* **85:**11697–11708.
126. **Buehring GC, Philpott SM, Choi KY.** 2003. Humans have antibodies reactive with bovine leukemia virus. *AIDS Res Hum Retroviruses* **19:**1105–1113.
127. **Buehring GC, Shen HM, Jensen HM, Choi KY, Sun D, Nuovo G.** 2014. Bovine leukemia virus DNA in human breast tissue. *Emerg Infect Dis* **20:**772–782.
128. **zur Hausen H.** 2012. Red meat consumption and cancer: reasons to suspect involvement of bovine infectious factors in colorectal cancer. *Int J Cancer* **130:**2475–2483.
129. **Giovannucci E, Goldin B.** 1997. The role of fat, fatty acids, and total energy intake in the etiology of human colon cancer. *Am J Clin Nutr* **66**(Suppl)**:**1564S–1571S.
130. **Yamada S, Suzuki Y, Suzuki T, Le MQ, Nidom CA, Sakai-Tagawa Y, Muramoto Y, Ito M, Kiso M, Horimoto T, Shinya K, Sawada T, Kiso M, Usui T, Murata T, Lin Y, Hay A, Haire LF, Stevens DJ, Russell RJ, Gamblin SJ, Skehel JJ, Kawaoka Y.** 2006. Haemagglutinin mutations responsible for the binding of H5N1 influenza A viruses to human-type receptors. *Nature* **444:**378–382.
131. **Benavente J, Martínez-Costas J.** 2007. Avian reovirus: structure and biology. *Virus Res* **123:**105–119.
132. **Fuller T, Bensch S, Müller I, Novembre J, Pérez-Tris J, Ricklefs RE, Smith TB, Waldenström J.** 2012. The ecology of emerging infectious diseases in migratory birds: an assessment of the role of climate change and priorities for future research. *EcoHealth* **9:**80–88.
133. **Provost K, Dancho BA, Ozbay G, Anderson R, Richards G, Kingsley DH.** 2011. Hemocytes are sites of persistence for enteric viruses within oysters. *Appl Environ Microbiol* **77:**8360–8369.
134. **Crossan C, Baker PJ, Craft J, Takeuchi Y, Dalton HR, Scobie L.** 2012. Hepatitis E virus genotype 3 in shellfish, United Kingdom. *Emerg Infect Dis* **18:**2085–2087.
135. **Iritani N, Kaida A, Abe N, Kubo H, Sekiguchi JI, Yamamoto SP, Goto K, Tanaka T, Noda M.** 2014. Detection and genetic characterization of human enteric viruses in oyster-associated gastroenteritis outbreaks between 2001 and 2012 in Osaka City, Japan. *J Med Virol* **86:**2019–2025.
136. **Balayan MS.** 1992. Natural hosts of hepatitis A virus. *Vaccine* **10**(Suppl 1)**:**S27–S31.
137. **Summa M, von Bonsdorff C-H, Maunula L.** 2012. Pet dogs: a transmission route for human noroviruses? *J Clin Virol* **53:**244–247.

138. **Day MJ, Breitschwerdt E, Cleaveland S, Karkare U, Khanna C, Kirpensteijn J, Kuiken T, Lappin MR, McQuiston J, Mumford E, Myers T, Palatnik-de-Souza CB, Rubin C, Takashima G, Thiermann A.** 2012. Surveillance of zoonotic infectious disease transmitted by small companion animals. *Emerg Infect Dis* http://wwwnc.cdc.gov/eid/article/18/12/12-0664_article.
139. **Liu W, Wu S, Xiong Y, Li T, Wen Z, Yan M, Qin K, Liu Y, Wu J.** 2014. Co-circulation and genomic recombination of coxsackievirus A16 and enterovirus 71 during a large outbreak of hand, foot, and mouth disease in central China. *PLoS One* **9:**e96051.
140. **Perkins LE, Swayne DE.** 2003. Varied pathogenicity of a Hong Kong-origin H5N1 avian influenza virus in four passerine species and budgerigars. *Vet Pathol* **40:**14–24.
141. **Kapoor A, Li L, Victoria J, Oderinde B, Mason C, Pandey P, Zaidi SZ, Delwart E.** 2009. Multiple novel astrovirus species in human stool. *J Gen Virol* **90:**2965–2972.
142. **Ng TFF, Marine R, Wang C, Simmonds P, Kapusinszky B, Bodhidatta L, Oderinde BS, Wommack KE, Delwart E.** 2012. High variety of known and new RNA and DNA viruses of diverse origins in untreated sewage. *J Virol* **86:** 12161–12175.
143. **Xie G, Yu J, Duan Z.** 2013. New strategy for virus discovery: viruses identified in human feces in the last decade. *Sci China Life Sci* **56:** 688–696.
144. **Donaldson EF, Haskew AN, Gates JE, Huynh J, Moore CJ, Frieman MB.** 2010. Metagenomic analysis of the viromes of three North American bat species: viral diversity among different bat species that share a common habitat. *J Virol* **84:** 13004–13018.
145. **Mokili JL, Rohwer F, Dutilh BE.** 2012. Metagenomics and future perspectives in virus discovery. *Curr Opin Virol* **2:**63–77.
146. **Lecuit M, Eloit M.** 2013. The human virome: new tools and concepts. *Trends Microbiol* **21:** 510–515.
147. **Metayer C, Johnson ES, Rice JC.** 1998. Nested case-control study of tumors of the hemopoietic and lymphatic systems among workers in the meat industry. *Am J Epidemiol* **147:**727–738.
148. **McLean D, Pearce N.** 2004. Cancer among meat industry workers. *Scand J Work Environ Health* **30:**425–437.

Toxoplasma gondii as a Parasite in Food: Analysis and Control

12

DOLORES E. HILL[1] and JITENDER P. DUBEY[1]

Common parasitic protozoal infections are frequently transmitted by food containing fecally contaminated soil or water, which may carry the environmentally resistant oocyst stage of the parasites *Cyclospora cayetanensis, Cryptosporidium* spp., *Giardia* spp., *Toxoplasma gondii*, or *Sarcocystis* spp. However, both *T. gondii* and *Sarcocystis* can also be transmitted by consumption of a cyst stage of the parasite which is present in the meat of infected animals; currently, the extent of *Toxoplasma* and *Sarcocystis* infection due to foodborne transmission is unknown (1–5). Differences in the definitive and intermediate hosts exist between these pathogens which impact their abundance and geographical distribution in the environment (Table 1). As an example, because *Giardia* and *Cryptosporidium* spp. oocysts are excreted in large numbers by cattle, other ruminants, and a wide variety of other species (more than 10^9 per day), oocysts are very commonly found in the environment, while oocysts of *Toxoplasma*, which are exclusively excreted by felids, are restricted to areas inhabited by wild and domestic cats (6, 7).

This review will focus on *T. gondii*, since it encompasses both meat-borne transmission and environmental contamination (fresh produce/soil/surface water); however, significant differences between *Toxoplasma* and the other protozoans of food safety importance will be highlighted.

[1]U.S. Department of Agriculture, Agricultural Research Service, Northeast Area, Animal Parasitic Diseases Laboratory, Beltsville Agricultural Research Center-East, Beltsville, MD 20705.

Preharvest Food Safety
Edited by Siddhartha Thakur and Kalmia E. Kniel

doi:10.1128/microbiolspec.PFS-0011-2015

TABLE 1 Protozoans of food safety importance and their hosts

Protozoan	Intermediate hosts	Definitive host
Cyclospora cayetanensis	N/A	Humans, other primates
Cryptosporidium spp.	N/A	Canids, birds, rodents, humans, ruminants, pigs, reptiles
Giardia duodenalis/Giardia lamblia	N/A	Humans, other mammals
Toxoplasma gondii	Warm-blooded animals, birds	Felids
Sarcocystis spp.	Mammals, birds, poikilothermic	Canids, birds, raccoons, primates, felids, reptiles

T. gondii is a coccidian parasite with an unusually wide range of intermediate hosts. Felids serve as definitive hosts and produce the environmentally resistant oocyst stage. *Toxoplasma* is one of the most common parasitic infections of humans, though its prevalence varies widely from place to place. Toxoplasmosis continues to be a significant public health problem in the United States, where 8 to 22% of people are infected; a similar prevalence is seen in the United Kingdom (8–13). In Central America, South America, and continental Europe, estimates of infection range from 30 to 90% (9, 14, 15). Most infections in humans are asymptomatic, but at times the parasite can produce devastating disease. Infection may be congenitally or postnatally acquired. In the United States, nationwide serological surveys demonstrated that seroprevalence in people remained stable at 23% from 1990 until 1998 (10), while recent surveys have demonstrated a significant decrease in age-adjusted seroprevalence over the past 15 years to 10.8% (12) and 12.4% (13). Similar decreases in seroprevalence have been observed in many European countries (14).

It is estimated that 1,075,242 people are infected with *T. gondii* each year in the United States, and approximately 2,839 people develop symptomatic ocular disease annually (16). The cost of illness caused by *Toxoplasma* in the United States has been estimated to be nearly $3 billion, with an $11,000 quality-adjusted life year loss annually (17, 18). Recent publications have linked suicide and schizophrenia to *Toxoplasma* infection (19, 20).

T. gondii also infects food animals, including sheep, goats, pigs, chickens, and many game animal species. Infected animals harbor tissue cysts, and human consumers can be infected by ingestion of these cysts in raw or undercooked meat. Virtually all edible portions of an animal can harbor viable *T. gondii* tissue cysts (21), and tissue cysts can survive in food animals for years.

Unlike *T. gondii*, which is the only species in the genus, and whose only definitive hosts are members of the *Felidae*, there are perhaps hundreds of species of *Sarcocystis*. *Sarcocystis* spp. infect all vertebrate animals; these animals can serve as both intermediate or definitive hosts (22). Humans act as intermediate hosts for an unknown number of *Sarcocystis* spp. and act as definitive hosts for at least two species of *Sarcocystis*: *Sarcocystis hominis* and *Sarcocystis suihominis*, which are acquired by eating infectious sarcocysts in undercooked beef and pork muscle, respectively (23).

Cyclospora, *Giardia*, and *Cryptosporidium* spp. require only one host to complete their life cycles, and infections are transmitted to humans strictly through inadvertent consumption of infectious oocysts in contaminated food, water, or soil.

MORPHOLOGY, STRUCTURE, AND LIFE CYCLE

T. gondii belongs to phylum *Apicomplexa* (Levine, 1970), class *Sporozoasida* (Leukart, 1879), subclass *Coccidiasina* (Leukart, 1879), order *Eimeriorina* (Leger, 1911), and family *Toxoplasmatidae* (Biocca, 1956). There is only one species of *Toxoplasma*, *T. gondii*. Coccidia

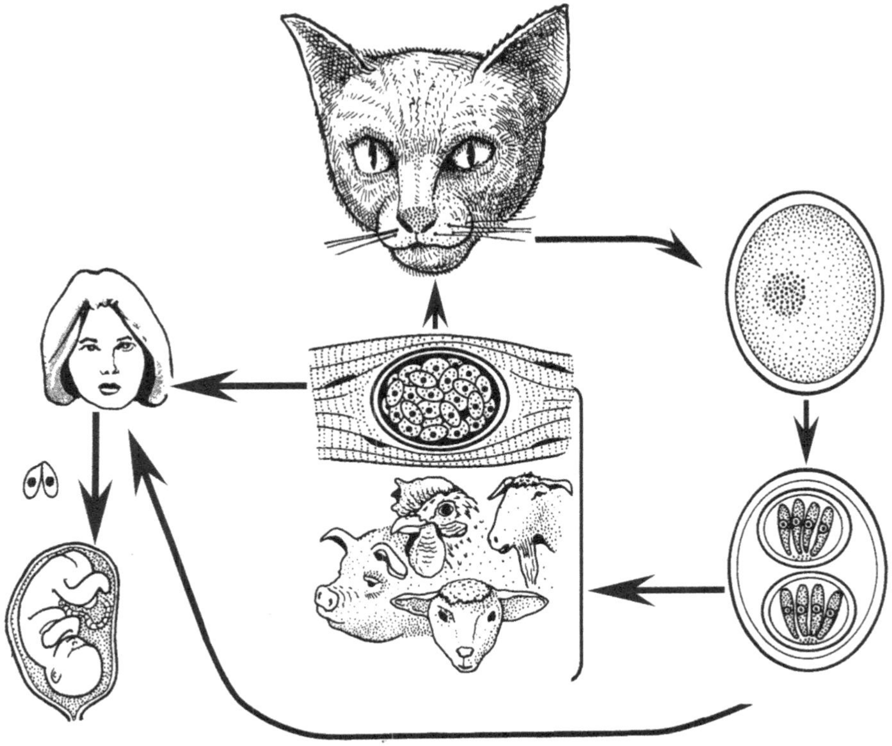

FIGURE 1 **Life cycle of *Toxoplasma gondii*.**

in general have complicated life cycles. Most coccidia are host-specific and are transmitted via a fecal–oral route. Transmission of *T. gondii* occurs via the fecal–oral route (Fig. 1), as well as through consumption of infected meat and by transplacental transfer from mother to fetus (8, 24).

The name *Toxoplasma* (toxon = arc, plasma = form) is derived from the crescent shape of the tachyzoite stage (Fig. 2). There are three infectious stages of *T. gondii*: the tachyzoites (in groups) (Fig. 3A), the bradyzoites (in tissue cysts) (Fig. 3B, C), and the sporozoites (in oocysts) (Fig. 3G). The tachyzoite is often crescent-shaped and is approximately the size (2 × 6 μm) of a red blood cell (Fig. 4). The anterior end of the tachyzoite is pointed, and the posterior end is round. It has a pellicle (outer covering), several organelles including subpellicular microtubules, mitochondrium, smooth and rough endoplasmic reticulum, a Golgi apparatus, an apicoplast, ribosomes, a micropore, and a well-defined nucleus. The nucleus is usually situated toward the posterior end or in the central area of the cell. The tachyzoite enters the host cell by active penetration of the host-cell membrane and can tilt, extend, and retract as it searches for a host cell. After entering the host cell, the tachyzoite becomes ovoid in shape and becomes surrounded by a parasitophorous vacuole (Pv in Fig. 4). *T. gondii* in a parasitophorous vacuole is protected from host defense mechanisms. The tachyzoite multiplies asexually within the host cell by repeated divisions in which two progenies form within the parent parasite, consuming it (Fig. 5A–D). Tachyzoites continue to divide until the host cell is filled with parasites.

After a few divisions, *T. gondii* forms tissue cysts that vary in size from 5 to 70 μm and

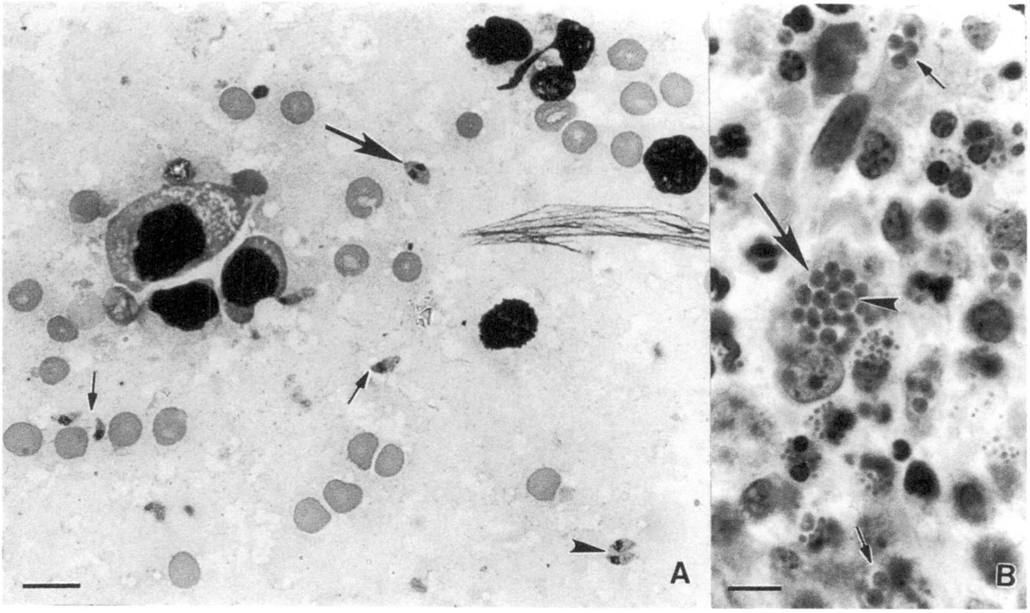

FIGURE 2 Tachyzoites of *Toxoplasma gondii*. Bars = 10 μm. (A) Individual (small arrows), binucleate (large arrow), and divided (arrowhead) tachyzoites. Impression smear of lung. Compare size with red blood cells and leukocytes. Giemsa stain. (B) Tachyzoites in a group (large arrow) and in pairs (small arrows) in section of a mesenteric lymph node. Note that organisms are located in parasitophorous vacuoles and some are dividing (arrowhead). Hematoxylin and eosin stain.

remain intracellular (Fig. 6A–F). The tissue cyst wall is elastic, thin (<0.5 μm), and may enclose hundreds of the crescent-shaped, slender *T. gondii* stage known as bradyzoites (Fig. 7). The bradyzoites are approximately 7 × 1.5 μm in size and differ structurally only slightly from tachyzoites. They have a nucleus situated toward the posterior end, whereas the nucleus in tachyzoites is more centrally located. Bradyzoites are more slender and less susceptible to destruction by proteolytic enzymes than tachyzoites. Although tissue cysts containing bradyzoites may develop in visceral organs, including lungs, liver, and kidneys, they are more prevalent in muscular and neural tissues, including the brain (Fig. 6A–F), eye, and skeletal and cardiac muscle. Intact tissue cysts probably do not cause any harm and can persist for the life of the host (Table 2).

After the ingestion of tissue cysts by cats, the tissue cyst wall is dissolved by proteolytic enzymes in the stomach and small intestine. The released bradyzoites penetrate the epithelial cells of the small intestine and initiate development of numerous generations of asexual and sexual cycles of *T. gondii* (25). Bradyzoites penetrate the lamina propria of the feline intestine and multiply as tachyzoites. Within a few hours after infection of cats, *T. gondii* may disseminate to extraintestinal tissues. *T. gondii* persists in the intestinal and extraintestinal tissues of cats for at least several months and possibly for the life of the cat.

As the enteroepithelial cycle progresses, *T. gondii* multiplies profusely in the intestinal epithelial cells of cats (enteroepithelial cycle), and these stages, represented by five distinct morphological types (types A to E), are known as schizonts (Fig. 3D). Several generations of each type are produced, and daughter organisms known as merozoites are

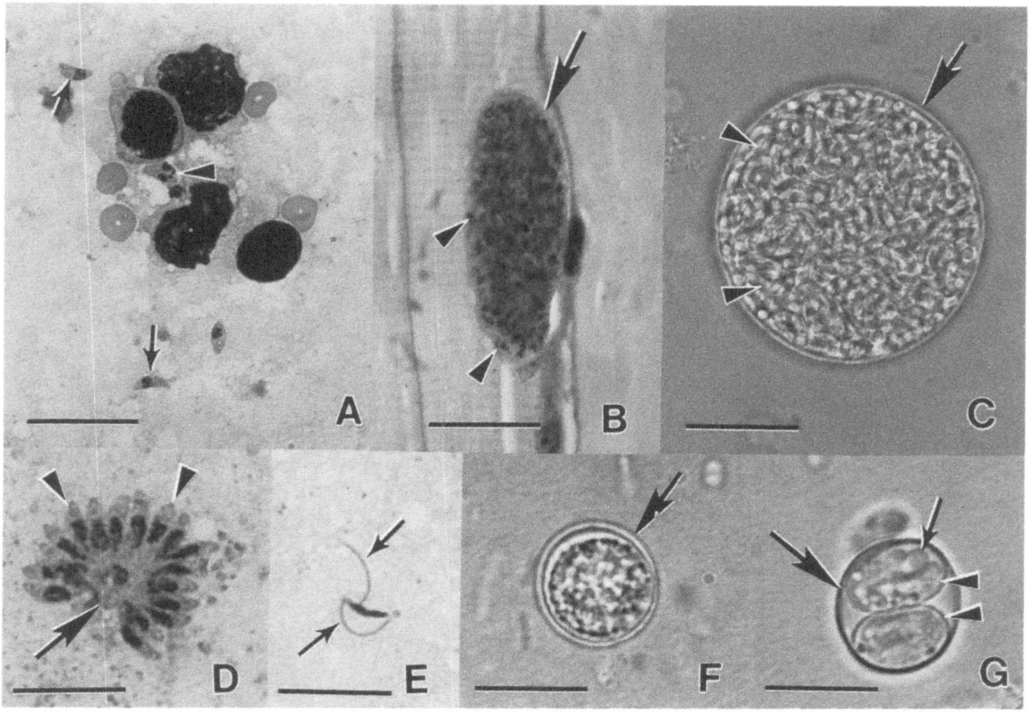

FIGURE 3 Stages of *Toxoplasma gondii*. Bars in A–D = 20 μm, in E–G = 10 μm. (A) Tachyzoites in impression smear of lung. Note crescent-shaped individual tachyzoites (arrows) and dividing tachyzoites (arrowheads) compared with size of host red blood cells and leukocytes. Giemsa stain. (B) Tissue cysts in section of muscle. The tissue cyst wall is very thin (arrow) and encloses many tiny bradyzoites (arrowheads). Hematoxylin and eosin stain. (C) Tissue cyst separated from host tissue by homogenization of infected brain. Note tissue cyst wall (arrow) and hundreds of bradyzoites (arrowheads). Unstained. (D) Schizont (arrow) with several merozoites (arrowheads) separating from the main mass. Impression smear of infected cat intestine. Giemsa stain. (E) A male gamete with two flagella (arrows). Impression smear of infected cat intestine. Giemsa stain. (F) Unsporulated oocyst in fecal float of cat feces. Unstained. Note double-layered oocyst wall (arrow) enclosing a central undivided mass. (G) Sporulated oocyst with a thin oocyst wall (large arrow) and two sporocysts (arrowheads). Each sporocyst has four sporozoites (small arrow), which are not in complete focus. Unstained.

formed coincident with the last nuclear division. Merozoites give rise to gametes, the sexual stages of the organism. The microgamont (male gamont) has two flagella (Fig. 3E), and it swims to, enters, and fertilizes the macrogamont (female gamont), forming a zygote. After fertilization, oocyst wall formation begins around the zygote. Oocysts are discharged into the intestinal lumen by the rupture of intestinal epithelial cells.

Oocysts are environmentally resistant and are formed only in felids, probably in all members of the *Felidae* (Fig. 3G). Cats shed oocysts after ingesting any of the three infectious stages of *T. gondii*, i.e., tachyzoites, bradyzoites, and sporozoites (25–28). Prepatent periods (time to the shedding of oocysts after initial infection) and frequency of oocyst shedding vary according to the stage of *T. gondii* ingested (Table 3). Prepatent periods occur 3 to 10 days after ingesting tissue cysts and 19 days or more after ingesting tachyzoites or oocysts (25–27). Less than 50% of cats shed oocysts after ingesting tachyzoites or oocysts,

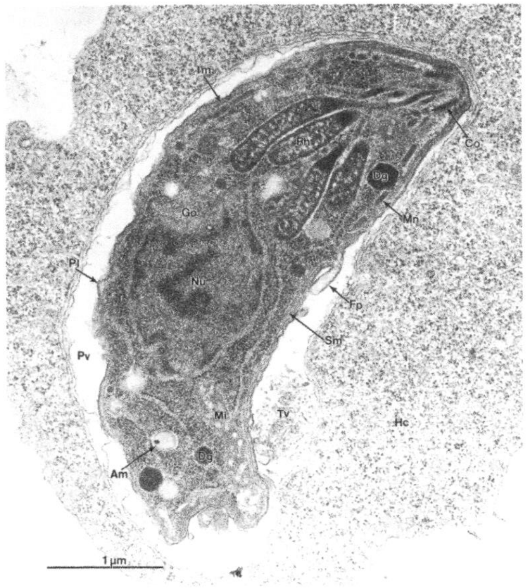

FIGURE 4 Transmission electron micrograph of a tachyzoite of *Toxoplasma gondii* in a mouse peritoneal exudate cell. Am, amylopectin granule; Co, conoid; Dg, electron-dense granule; Fp, finger-like projection of tachyzoite plasmalemma; Go, Golgi complex; Hc, host cell cytoplasm; Im, inner membrane complex; Mi, mitochondrion; Mn, microneme; Nu, nucleus; Pl, plasmalemma; Pv, parasitophorous vacuole; Rh, rhoptry; Sm, subpellicular microtubule; Tv, tubulovesicular membranes. Bar = 1 µm.

whereas nearly all cats shed oocysts after ingesting tissue cysts (26). In freshly passed feces, oocysts are unsporulated (noninfective; Fig. 3F). Unsporulated oocysts are subspherical to spherical and are 10 × 12 µm in diameter. They sporulate (become infectious) outside the cat within 1 to 5 days depending upon aeration and temperature. Sporulated oocysts contain two ellipsoidal sporocysts (Fig. 3G), and each sporocyst contains four sporozoites. The sporozoites are 2 × 6 to 8 µm in size. The nucleus of *T. gondii* is haploid except during sexual recombination in the gut of the cat. Fourteen chromosomes are present, encompassing a 65-Mb genome (14, 29).

Most *T. gondii* isolates from human and animal sources in North America, Europe, and Africa have been grouped into one of three clonal lineages (types I, II, and III) (29–35) and are biologically and genetically different from isolates from Brazil and Columbia (36–40). A number of recent studies suggest that only a few ancestral strains have given rise to the three dominant clonal lineages and the existing genetic diversity seen in various geographic regions through a process of limited, mostly asexual, recombination (31, 34, 40, 41). Recent genotyping studies of isolates from pigs, lambs, and goats demonstrate that the type II lineage predominates in food animals in the United States, followed by type III isolates and atypical genotypes; type I isolates have rarely been found in farm animals (35, 42–44).

EPIDEMIOLOGY

Toxoplasmosis may be acquired by ingestion of oocysts or by ingestion of tissue-inhabiting stages of the parasite. Contamination of the environment by oocysts is widespread, a result of fecal contamination of soil and ground water by the estimated 140 million domestic and feral cats in the United States and other members of the *Felidae*. After ingesting as few as one bradyzoite or one tissue cyst (and many tissue cysts may be present in one infected mouse), each cat can deposit hundreds of millions of oocysts in feces during infection (8, 14, 24, 45–47). Domestic cats are the major source of contamination because oocyst formation is greatest in domestic cats, and cats are extremely common. Sporulated oocysts survive for long periods under most ordinary environmental conditions and even in harsh environments for months. They can survive in moist soil, for example, for months and even years (8).

Oocysts in soil can be mechanically transmitted by invertebrates such as flies, cockroaches, dung beetles, and earthworms, which can spread oocysts into human food and animal feeds. Infection rates in cats are determined by the rate of infection in local avian and rodent populations because cats typically

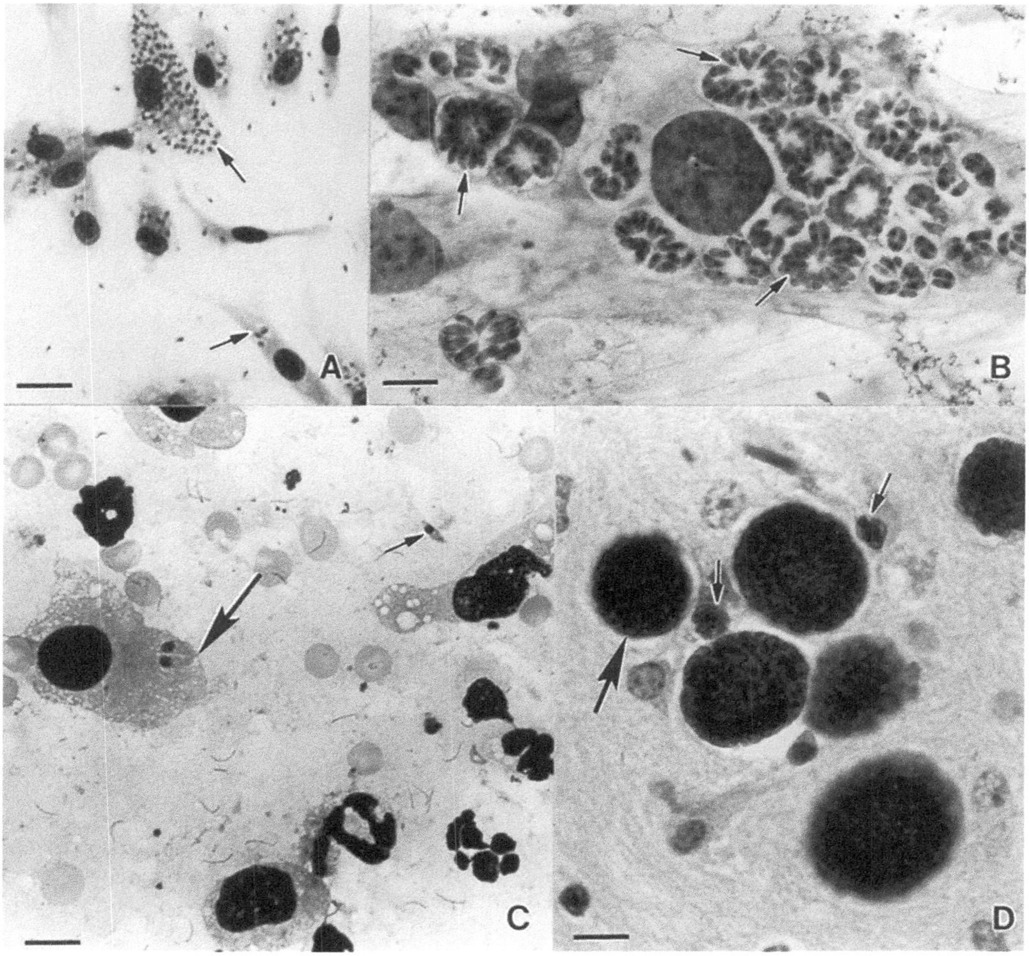

FIGURE 5 ***Toxoplasma gondii*** **stages in** ***in vitro*** **and** ***in vivo*** **preparations. (A) Tachyzoites in culture of human foreskin fibroblast cells. Giemsa stain. Bar = 25 μm. (B) Rosettes of tachyzoites in human foreskin fibroblasts. Immunohistochemical stain with antitachyzoite-specific antibody. Smear. Bar = 10 μm. (C) Tachyzoites in a cytospin smear of pleural fluid from a cat with pneumonia. Giemsa stain. Compare the size of tachyzoites (arrow) with host cells. Giemsa stain. Bar = 10 μm. (D) Tachyzoites (arrows) and tissue cysts (large arrow) in section of mouse brain. Immunohistochemical stain with** ***T. gondii*****–specific antibody. Bar = 10 μm.**

become infected by eating these animals. The more oocysts there are in the environment, the more likely it is that prey animals would be infected, and this in turn would increase the infection rate in cats. In certain areas of Brazil, up to 50% of 6- to 8-year-old children have antibodies to *T. gondii* linked to the ingestion of oocysts from an environment heavily contaminated with *T. gondii* oocysts (48, 49). The largest recorded outbreak of clinical toxoplasmosis in humans in North America was epidemiologically linked to drinking water from a municipal water reservoir in British Columbia, Canada (50, 51). This water reservoir was supposedly contaminated with *T. gondii* oocysts excreted by cougars (*Felis concolor*). Although attempts to recover *T. gondii* oocysts from water samples in the

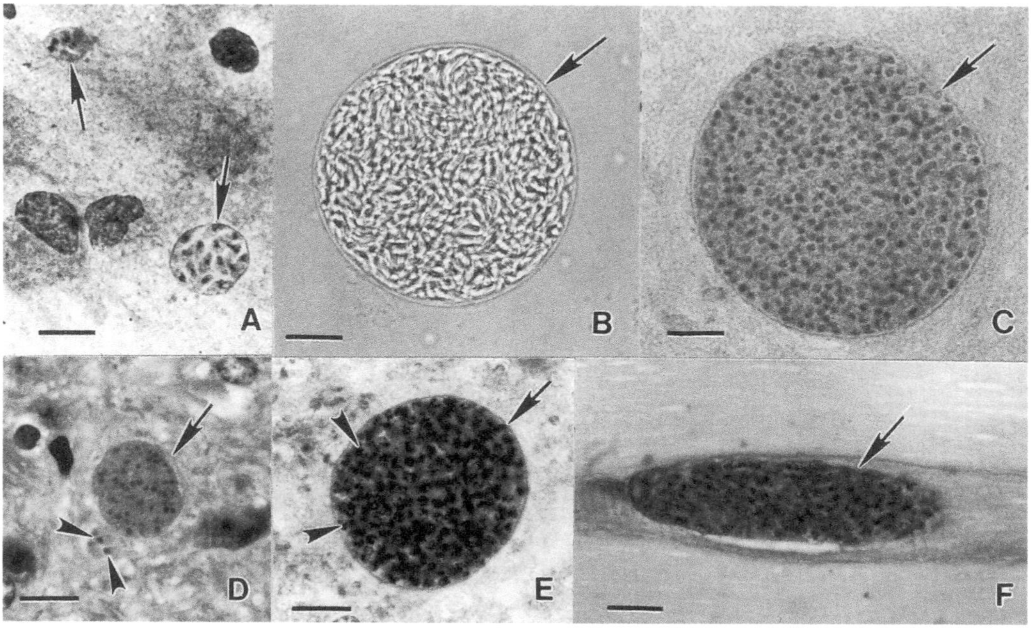

FIGURE 6 Tissue cysts of *Toxoplasma gondii*. Bar = 10 μm. (A) Two tissue cysts (arrows). Note the thin cyst wall enclosing bradyzoites. Impression smear of mouse brain. Silver impregnation and Giemsa stain. (B) A tissue cyst freed from mouse brain by homogenization in saline. Note the thin cyst wall (arrow) enclosing many bradyzoites. Unstained. (C) A large tissue cyst in a section of rat brain 14 months postinfection. Note the thin cyst wall (arrow). Hematoxylin and eosin stain. (D) A small tissue cyst with intact cyst wall (arrow) and four bradyzoites (arrowheads) with terminal nuclei adjacent to it. Section of mouse brain 8 months postinfection. Hematoxylin and eosin stain. (E) A tissue cyst in a section of mouse brain. Note Periodic acid Schiff (PAS)–negative cyst wall (arrow) enclosing many PAS-positive bradyzoites (arrowheads). The bradyzoites stain bright red with PAS but they appear black in this photograph. PAS hematoxylin stain. (F) An elongated tissue cyst (arrow) in a section of skeletal muscle of a mouse. PASH stain.

British Columbia outbreak were unsuccessful, methods to detect oocysts were reported (52). At present, no commercial reagents are available to reliably detect *T. gondii* oocysts in the environment.

Widespread infection in aquatic mammals indicates contamination and survival of oocysts in sea-water contaminated by runoff surface water entering the marine environment (53, 54). Wild populations of endangered southern sea otters off the west coast of the United States have been significantly impacted by exposure to *Toxoplasma* oocysts, presumably by eating filter-feeding mollusks in near-shore environments, resulting in devastating disease (55–57). Given the levels of oocyst contamination in surface waters, the potential for contamination of produce with oocysts is high, especially where surface waters are used for irrigation purposes. However, little is known of the contribution of oocyst-contaminated fruits and vegetables to human infection in the United States, but recent studies (4, 58) have suggested that oocyst exposure is the predominate route of *Toxoplasma* transmission to humans in the United States. Animal infections with *Toxoplasma*, especially infections in non-meat-eating ruminants, birds, wild herbivores, and pigs raised in confinement, also likely result from environmental exposure to *T. gondii* oocysts (59).

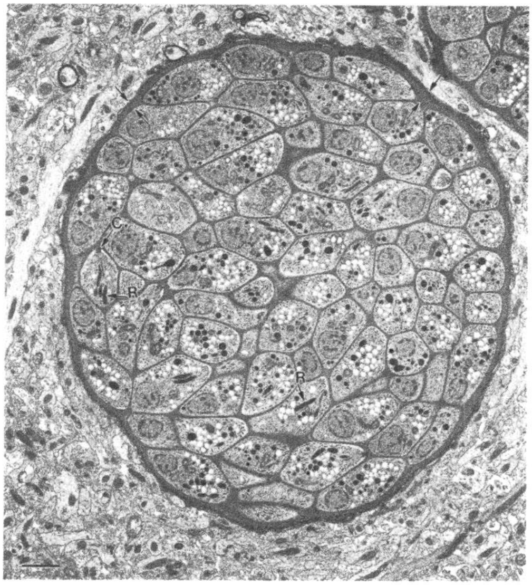

FIGURE 7 Transmission electron micrograph of a tissue cyst in the brain of a mouse 6 months postinfection. Note the thin cyst wall (opposing arrows), numerous bradyzoites each with a conoid (C), and electron-dense rhoptries (R). Bar = 3.0 μm.

FOODBORNE TRANSMISSION

The relative contribution of meat-borne sources of *Toxoplasma* infection versus oocyst transmission of *Toxoplasma* to human infection is unknown and difficult to quantify, since meat from infected animals may undergo postharvest treatments such as heating, freezing, salting, or pumping (injection of water and salt-based solutions to retard microbial growth) that can render the parasite nonviable (60, 61), and few comprehensive assessments have been completed in meat available for retail purchase. Complicating matters is the fact that the number of *T. gondii* organisms in meat from naturally infected food animals is very low, making the parasite difficult and expensive to detect by direct methods. It is estimated that as few as one tissue cyst may be present in 100 g of meat (62). In addition, there is no predilection site for *Toxoplasma* in meat animals; virtually all edible portions of an animal can harbor viable *T. gondii* tissue cysts (21), and tissue cysts can remain viable in food animals for years.

Despite the uncertainty of human infection sources, *Toxoplasma* is increasingly recognized as a foodborne risk, and various studies have suggested widely disparate estimates of foodborne transmission. Mead et al. (63) suggested that *T. gondii* is one of three pathogens (along with *Salmonella* and *Listeria*) which account for >75% of all deaths due to foodborne disease in the United States. Roghmann et al. (64) suggested that 50% of *Toxoplasma* infections in the United States could be ascribed to foodborne transmission. Scallan et al. (65) estimated that *Toxoplasma* caused 8% of hospitalizations and 24% of deaths resulting from foodborne illnesses.

Dubey et al. (43) determined the prevalence of *T. gondii* in 383 U.S. lambs (<1 year old) from Maryland, Virginia, and West Virginia. Hearts of 383 lambs were obtained from a slaughter house on the day of killing, and blood was removed from each heart and tested for antibodies to *T. gondii* using the modified agglutination test (MAT). Anti-

TABLE 2 Host tissues invaded by protozoan parasites

Species	Tissue parasitized in intermediate host	Tissue parasitized in definitive host
Cyclospora cayetanensis	N/A	Intestinal lining
Cryptosporidium spp.	N/A	Epithelial cells of the respiratory system or the intestine
Giardia duodenalis/ *Giardia lamblia*	N/A	Brush border villous epithelium surface of small intestinal mucosa
Toxoplasma gondii	Many types of tissue, including muscle and epithelial cells	Intestinal epithelial cells followed by invasion of extraintestinal tissues and organs
Sarcocystis spp.	Meronts in endothelial cells of organ blood vessels, sarcocysts in muscle	Lamina propria of the small intestine

TABLE 3 Oocyst characteristics of protozoans of food safety importance

Species	Oocyst size	Number of organisms per cyst in feces	Days to infectivity
Cyclospora cayetanensis	8–10 μm	4	~14 days
Cryptosporidium spp.	4–6 μm	4	Immediately infective
Giardia duodenalis/Giardia lamblia	9–12 μm	2	Immediately infective
Toxoplasma gondii	10–12 μm	8	1–7 days
Sarcocystis spp.	12.3–14.6 μm	8 in oocyst/4 in sporocyst	Immediately infective

bodies (MAT, 1:25 or higher) to *T. gondii* were found in 104 (27.1%) of 383 lambs. Hearts of 68 seropositive lambs were used for isolation of viable *T. gondii* by bioassay in cats, mice, or both. For bioassays in cats, the entire myocardium was chopped and fed to one cat per heart, and feces of the recipient cats were examined for shedding of *T. gondii* oocysts. For bioassays in mice, 50 g of the myocardium was digested in an acid pepsin solution, and the digest was inoculated into mice; the recipient mice were then examined for *T. gondii* infection. In total, 53 isolates of *T. gondii* were obtained from 68 seropositive lambs. A similar study was carried out using hearts from 234 goats (44). Antibodies to *T. gondii* were found in 125 (53.4%) of 234 goats. Hearts of 112 goats were used for isolation of viable *T. gondii* by bioassays in mice. *T. gondii* was isolated from 29 goats. Taken together, these results indicate high parasite prevalence of *T. gondii* in lambs and goats destined for human consumption in the United States.

Chickens are considered one of the most important hosts in the epidemiology of *T. gondii* infection because they are an efficient source of infection for cats that excrete the environmentally resistant oocysts and because humans may become infected with this parasite after eating undercooked infected chicken meat. In a recent review of *T. gondii* infection in chickens worldwide (66), a very high prevalence of the parasite was found in chickens raised in backyards (up to 100%) and free-range organic (30 to 50%) establishments.

Beef, chicken, and pork are the main meat types consumed in the United States. In a case control study of 148 recently (<6 months) infected individuals, Jones et al. (67) identified an elevated risk of infection associated with eating raw ground beef, rare lamb, locally produced cured, dried, or smoked meat, raw oysters, clams, or mussels; working with meat; and drinking unpasteurized goat's milk. The relative risk to U.S. consumers of acquiring *T. gondii* infection from undercooked meat was recently determined in a nationwide survey of retail chicken, beef, and pork (68). The survey of 698 retail outlets in 28 metropolitan statistical areas (MSAs as defined by the U.S. Census Bureau) covered 80% of the U.S. population. No viable *Toxoplasma* was found in any of 2,094 beef or 2,094 chicken samples by bioassay. Only pork was found to harbor viable *T. gondii* tissue cysts, which were isolated from 0.38% of samples (7/2,094) by cat bioassay, and 0.57% of pork samples were suspected to be infected based on positive enzyme-linked immunosorbent assay (ELISA) results. No beef samples were positive by ELISA, while 1.4% of chickens were positive by ELISA only. The northeastern United States had a higher number of positive pork samples than other regions of the country, reflecting the higher risk of pig infection due to regional management practices (outdoor versus confinement rearing).

Modern biosecure management practices on swine farms, which include restriction of human entry into pig barns, stringent rodent control, secure feed, and exclusion of cats and other wildlife, have resulted in reduced levels of *Toxoplasma* infection in confinement-raised swine over the past 20 years from 23% to 2.7% in the United States (69, 70).

While *Toxoplasma* infection in confinement-raised market pigs has decreased significantly, infection levels in pigs with access to the outdoors can be quite high, reaching 50 to 90% in recent studies (71, 72). In the

U.S. national meat survey mentioned above, two of seven *Toxoplasma*-positive pork samples were derived from "naturally raised" pigs. An upsurge in consumer demand for "organically raised," "humanely raised," and "free-range" pork products has resulted in increasing numbers of hogs being raised in nonconfinement systems (73). Swine producers have been recruited to produce animals for the organic market to fulfill a consumer demand that has increased 20% per year in sales since 1990 (74). National Organic Program standards (http://www.ams.usda.gov/nop/) require that all organically raised animals must have access to the outdoors. Though humanely raised and free-range products have standards that are less stringently defined, outdoor access is also considered a requirement for labeling.

Few studies have determined the prevalence of *Toxoplasma* infection in swine raised in organic management systems. Kijlstra et al. (75) found 0 of 621 conventionally raised pigs to be seropositive for *Toxoplasma*, while 38 of 1,295 (2.9%) pigs raised in "animal friendly" management systems were seropositive for *Toxoplasma*. In another study, 22 of 324 (6.8%) free-range pigs in North Carolina were seropositive for *Toxoplasma*, while 3 of 292 conventionally raised pigs (1.1%) tested were seropositive (76). Dubey et al. (77), determined *T. gondii* prevalence in organically raised pigs from two farms in Michigan. Serum and tissue samples from 33 pigs from the farms were available for *T. gondii* evaluation at slaughter. Serological testing was performed using both ELISA and MAT. Antibodies to *T. gondii* were detected by both ELISA and MAT in 30 of 33 animals, with MAT titers of 1:25 in 3, 1:50 in 6, 1:100 in 7, 1:200 in 13, and 1:400 in 1. The hearts of all 33 pigs were bioassayed for *T. gondii* in mice; *T. gondii* was isolated from 17 pigs including one from a seronegative (both ELISA and MAT) pig, revealing very high prevalence of *T. gondii* in organic pigs for the first time in the United States, indicating potentially increased health risk of consuming organic swine products.

Access to organic material contaminated with cat feces or to rodents or wildlife potentially infected with *Toxoplasma* during outdoor pasturage substantially increases the risk of exposure of pigs to *Toxoplasma* (42, 75, 78). These pigs enter the food chain, since there is no national system of identifying individual pigs slaughtered in the United States and no routine testing for *Toxoplasma*. Commercial meat cuts may therefore contain viable *T. gondii* tissue cysts; Dubey et al. (21, 79) demonstrated that virtually every edible portion of an infected pig carcass may contain viable tissue cysts. A single *T. gondii*–infected pig can be a source of infection for many people, since one market-weight hog (100 kg or more) can yield over 600 individual servings of meat. Prevention of exposure to infectious stages is the only way to stop infection of pigs, since there is no vaccine or treatment available, and once infected, tissue cysts persist in pig tissues for the life of the hog.

S. hominis (with its thick-walled cyst) is commonly found as infectious cysts in the muscles of slaughtered cattle worldwide (80, 81). Numerous studies of slaughtered cattle in Europe, South America, and Asia have demonstrated prevalence in cattle of over 40% in some countries (82–85). Tissue prevalence of *S. suihominis* in pigs is also high, especially in Asian countries and Europe (86, 87). Human intestinal *Sarcocystis* accompanied by the production of sporocysts has been shown to range from 1 to 23%, depending upon the country surveyed (5, 86–88).

C. cayetanensis has a direct fecal–oral transmission route, most likely from consumption of fecally contaminated food or water (89, 90). Extraintestinal phases are not present, and noninfectious oocysts are excreted in human feces; complete sporulation occurs within 1 to 2 weeks. Humans appear to be the only host of *C. cayetanensis*. Prevalence surveys of humans in South American and Asian countries demonstrated fecal oocysts in between 0.5 and 6% of the population, while fecal oocysts have not been found in surveys of domestic animals (22, 91).

Nearly two dozen species of *Cryptosporidium* occur in wild and domestic animals and humans worldwide. *Cryptosporidium hominis* and *Cryptosporidium parvum* are the most common cause of infections in humans. Like *Cyclospora*, transmission occurs through a direct fecal–oral route, and oocysts are immediately infective upon release in feces. Huge outbreaks of human infection have occurred in the United States as a result of contamination of municipal drinking water (92, 93), and consumption of fresh produce contaminated with oocysts has also been identified as a source of infection (94, 95).

Giardia duodenalis is a common cause of foodborne illness, even in developed nations. Host specificity has recently been clarified (96, 97), resulting in taxonomic separation of a number of species with more or less broad host ranges. *G. duodenalis* (group A or B) infects humans and other animals, while five other species do not appear to be zoonotic (98). The direct, one-host life cycle and immediate infectivity of the cyst upon release in feces results in millions of human infections with *Giardia* spp. each year worldwide. Transmission results from inadvertent consumption of contaminated soil or water and from direct contamination from feces of infected humans or animals.

PATHOGENESIS AND CLINICAL FEATURES

T. gondii usually parasitizes the host, definitive and intermediate, without producing clinical disease. Only rarely does it produce severe clinical manifestations (Table 4). The bradyzoites from the tissue cysts, or sporozoites from the oocyst, penetrate intestinal epithelial cells and multiply in the intestine as tachyzoites within 24 h of infection. *T. gondii* may spread first to mesenteric lymph nodes and then to distant organs by invasion of lymphatics, and blood and can multiply in virtually any cell in the body. All extracellular forms of the parasite are directly affected by antibody, but intracellular forms are not. More virulent strains of *Toxoplasma* have developed effective defensive mechanisms using ROP18, a rhoptry-associated serine/threonine kinase, to inactivate p47 GTPases, which are generated by the infected cell to rupture the vacuole containing the parasite, resulting in digestion of the organism (99). It is believed that cellular factors, including lymphocytes and lymphokines, are more important than humoral factors in immune-mediated destruction of *T. gondii* (100–102). Immunity does not eradicate infection. *T. gondii* tissue cysts persist for years after acute infection. The fate of tissue cysts is not fully known. Whether bradyzoites can form new tissue cysts directly without transforming into tachyzoites is not known. It has been proposed that tissue cysts may at times rupture during the life of the host. The released bradyzoites may be destroyed by the host's immune responses or there may be formation of new tissue cysts. In immunosuppressed patients, such as those given large doses of immunosuppressive agents in preparation for organ transplants and in those with acquired immunodeficiency syndrome (AIDS), rupture

TABLE 4 Clinical signs in humans[a]

Species	Signs
Cyclospora cayetanensis	Watery diarrhea, nausea, cramping, weight loss; symptoms can persist or relapse for weeks or months
Cryptosporidium spp.	Profuse diarrhea, nausea, cramping which lasts for 1–2 weeks; life threatening infections can occur in immunocompromised individuals
Giardia duodenalis/ Giardia lamblia	Asymptomatic to abdominal cramps, flatulence, foul smelling, greasy diarrhea, nausea, malabsorption of nutrients and failure to thrive, especially in young children (98)
Sarcocystis spp.	Tissue *Sarcocystis* may result in fever, swelling, muscle pain; intestinal *Sarcocystis* may result in abdominal pain, nausea, and diarrhea; both are frequently asymptomatic

[a]Table information (except *Giardia*) derived from reference 22.

of a tissue cyst may result in transformation of bradyzoites into tachyzoites and renewed multiplication. The immunosuppressed host may die from toxoplasmosis unless treated. It is not known how corticosteroids cause relapse, but it is unlikely that they directly cause rupture of the tissue cysts.

Pathogenicity of *T. gondii* is determined by the virulence of the strain and the susceptibility of the host species (103). *T. gondii* strains may vary in their pathogenicity in a given host. Certain strains of mice are more susceptible than others, and the severity of infection in individual mice within the same strain may vary. Mice of any age are susceptible to clinical *T. gondii* infection (14). However, adult rats do not become ill, while young rats can die of toxoplasmosis. Adult dogs, like adult rats, are resistant, whereas puppies are fully susceptible to clinical toxoplasmosis. Certain species are genetically resistant to clinical toxoplasmosis. Cattle and horses are among the hosts more resistant to clinical toxoplasmosis, whereas certain marsupials and New World monkeys are highly susceptible to *T. gondii* infection (8, 14).

Nothing is known concerning genetically determined susceptibility to clinical toxoplasmosis in higher mammals, including humans. Infection in humans may be congenitally or postnatally acquired. Congenital infection occurs when a woman becomes infected during pregnancy. Congenital infections acquired during the first trimester are more severe than those acquired in the second and third trimesters (104, 105). While the mother rarely has symptoms of infection, she does have a temporary parasitemia. Focal lesions develop in the placenta, and the fetus may become infected. At first there is generalized infection in the fetus. Later, infection is cleared from the visceral tissues and may localize in the central nervous system. A wide spectrum of clinical diseases occurs in congenitally infected children (104). Mild disease may consist of slightly diminished vision, whereas severely diseased children may have the full tetrad of lesions of the eye, hydrocephalus, convulsions, and intracerebral calcification. Of these, hydrocephalus is the least common but most significant lesion of toxoplasmosis. So far, the most common sequela of congenital toxoplasmosis is ocular disease (104, 105). The socioeconomic impact of toxoplasmosis in human suffering and the cost of care of sick children, especially those with mental retardation and blindness, are enormous (106, 107). The testing of all pregnant women for *T. gondii* infection is compulsory in some European countries, including France and Austria (108, 109). The cost benefits of such mass screening are being debated in many other countries (105, 110, 111). Recently, Stillwaggon et al. (112) provided an extensive guideline for estimating costs of preventive maternal screening for and the social costs resulting from toxoplasmosis based on studies in Europe and the United States. When estimating these costs, the value of all resources used or lost should be considered, including the cost of medical and nonmedical services, wages lost, cost of in-home care, and indirect costs of psychological impacts borne by the family for lifetime care of a substantially cognitively impaired child; the cost of a fetal death was estimated to be $5 million dollars (112).

Postnatally acquired infection may be localized or generalized. Infection may occur in any organ. Oocyst-transmitted infections may be more severe than tissue cyst-induced infections (8, 113–117). Enlarged lymph nodes are the most frequently observed clinical form of toxoplasmosis in humans (113, 114). Lymphadenopathy may be associated with fever, fatigue, muscle pain, sore throat, and headache. Although the condition may be benign, its diagnosis is vital in pregnant women because of the risk to the fetus. In a British Columbia outbreak, of the 100 people who were diagnosed with acute infection, 51 had lymphadenopathy and 20 had retinitis (50, 51). Encephalitis is the most important manifestation of toxoplasmosis in immunosuppressed patients because it causes the most severe damage to the patient (8, 118). Patients may have headache, disorientation, drowsiness, hemiparesis, reflex changes, and convulsions, and many become comatose. Encephalitis caused by *T. gondii* is now recognized with great frequency in patients treated with immunosuppressive agents.

Toxoplasmosis ranked high on the list of diseases which lead to the death of patients with AIDS, and approximately 10% of AIDS patients in the United States and up to 30% in Europe have died from toxoplasmosis (118). In AIDS patients, although any organ may be involved, including the testis, dermis, and spinal cord, infection of the brain is most frequently reported. Most AIDS patients suffering from toxoplasmosis have bilateral, severe, and persistent headaches which respond poorly to analgesics. As the disease progresses, the headaches may give way to a condition characterized by confusion, lethargy, ataxia, and coma. The predominant lesion in the brain is necrosis, especially of the thalamus (119). Since the advent of highly active antiretroviral therapy (HAART) in the mid-1990s, the number of AIDS patients suffering from toxoplasmic encephalitis has fallen dramatically, at least partially due to the impact of protease inhibitors used in HAART on *Toxoplasma* proteases (120–122).

DIAGNOSIS AND TREATMENT

Diagnosis is made by biologic, serologic, or histologic methods or by some combination of the above (Table 5). Clinical signs of toxoplasmosis are nonspecific and are not sufficiently characteristic for a definite diagnosis. Toxoplasmosis mimics several other infectious diseases.

Detection of *T. gondii* antibody in patients may aid diagnosis. Numerous serologic procedures are available for the detection of humoral antibodies; these include the Sabin-Feldman dye test, MAT, the indirect hemagglutination test, the indirect fluorescent antibody assay (IFA), the direct agglutination test, the latex agglutination test, ELISA, and the immunosorbent agglutination assay test (IAAT). The IFA, IAAT, and ELISA have been modified to detect IgM antibodies (105). The IgM antibodies appear sooner after infection than the IgG antibodies and disappear faster than IgG antibodies after recovery, though a small percentage of infected people produce IgG first (105, 123). Progress has been made in the diagnosis of human infection with *Toxoplasma* using PCR (124). Infection has been diagnosed using nested, stage-specific primers and cerebrospinal fluid from AIDS patients with suspected toxoplasmic encephalitis (125, 126), in immunocompromised patients undergoing hematopoietic stem cell transplantation (127), and in suspected cases of fetal toxoplasmosis using amniotic fluid (128). Improved sensitivity and performance standards for in-house methods and commercially available PCR kits is needed, because recent studies have shown that these

TABLE 5 Diagnosis and treatment[a]

Species	Diagnosis	Treatment
Cyclospora cayetanensis	Oocysts in stool most common; PCR assays have been designed and used successfully for diagnosis	Most commonly treated with trimethoprim-sulfamethoxazole or ciprofloxacin
Cryptosporidium spp.	Oocysts in stool	Nitazoxanide is the only FDA-approved treatment for immunocompetent patients, along with supportive therapy
Giardia duodenalis/ *Giardia lamblia*	Oocysts in stool	5-nitroimidazole and benzimidazole derivatives, quinacrin, furazolidone, paromomycin, and nitazoxanide are all approved for treatment
Sarcocystis spp.	Oocysts in stool	Anticoccidials such as sulfanilamides are thought to be effective

[a]Table information derived from reference 140.

PCR tests may not perform well using experimental or clinical samples (129–131).

Sulfadiazine and pyrimethamine (Daraprim) are two drugs widely used for the treatment of toxoplasmosis (132, 133). While these drugs have a beneficial action when given in the acute stage of the disease process when there is active multiplication of the parasite, they do not usually eradicate infection. It is believed that these drugs have little effect on subclinical infections, but the growth of tissue cysts in mice has been restrained with sulfonamides. Certain other drugs, such as diaminodiphenylsulfone, atovaquone, spiramycin, and clindamycin are also used to treat toxoplasmosis in difficult cases.

PREVENTION

To prevent infection of human beings by *Toxoplasma*, *Cryptosporidium*, *Cyclospora*, *Sarcocystis*, and *Giardia*, precautions should be taken to boil surface water before drinking to kill oocysts. General good hygiene should be observed, with frequent hand washing or wearing of gloves when in contact with soil or feces from animals or humans. Vegetables should be washed thoroughly before eating since they may have been contaminated with animal or human feces in irrigation water or soil.

Additionally, for *Toxoplasma* and *Sarcocystis*, the hands of people handling meat should be washed thoroughly with soap and water before they go to other tasks (8, 134). All cutting boards, sink tops, knives, and other materials coming in contact with uncooked meat should be washed with soap and water. *T. gondii* organisms in meat can be killed by exposure to extreme cold or heat. Tissue cysts in meat are killed by heating the meat throughout to 67°C (135) and by cooling to −13°C (136). *Toxoplasma* in tissue cysts is also killed by exposure to 0.5 kilorads of gamma irradiation (137). Meat of any animal should be cooked to at least 67°C before consumption, and tasting meat while cooking or while seasoning should be avoided. Thermal death curves for *Sarcocystis* have not yet been generated.

For *Toxoplasma*, due to the risk to the fetus, pregnant women should avoid contact with cats, soil, and raw meat. Pet cats should be fed only dry, canned, or cooked food. The cat litter box should be emptied every day, preferably not by a pregnant woman. Gloves should be worn while gardening. Expectant mothers should be made aware of the dangers of toxoplasmosis (138, 139).

At present there is no vaccine to prevent infection with *Toxoplasma*, *Cyclospora*, *Cryptosporidium*, *Giardia*, or *Sarcocystis* in humans.

CITATION

Hill DE, Dubey JP. 2016. Toxoplasma gondii as a parasite in food: analysis and control. Microbiol Spectrum 4(4):PFS-0011-2015.

REFERENCES

1. **Eckert J.** 1996. Workshop summary: food safety: meat- and fish-borne zoonoses. *Vet Parasitol* **64:**143–147.
2. **Anantaphruti MT.** 2001. Parasitic contaminants in food. *Southeast Asian J Trop Med Public Health* **32**(Suppl 2):218–228.
3. **Macpherson CN.** 2005. Human behaviour and the epidemiology of parasitic zoonoses. *Int J Parasitol* **35:**1319–1331.
4. **Hill D, Coss C, Dubey JP, Wroblewski K, Sautter M, Hosten T, Muñoz-Zanzi C, Mui E, Withers S, Boyer K, Hermes G, Coyne J, Jagdis F, Burnett A, McLeod P, Morton H, Robinson D, McLeod R.** 2011. Identification of a sporozoite-specific antigen from *Toxoplasma gondii*. *J Parasitol* **97:**328–337.
5. **Poulsen CS, Stensvold CR.** 2014. Current status of epidemiology and diagnosis of human sarcocystosis. *J Clin Microbiol* **52:**3524–3530.
6. **Dawson D.** 2005. Foodborne protozoan parasites. *Int J Food Microbiol* **103:**207–227.
7. **Clayton R.** 2011. Cryptosporidium *in Water Supplies*. 3rd ed. Foundation for Water Research, Marlow, Bucks, UK, http://www.fwr.org/cryptosp.pdf.
8. **Dubey JP, Beattie CP.** 1988. *Toxoplasmosis of Animals and Man*. CRC Press, Boca Raton, FL.
9. **Dubey JP, Jones JL.** 2008. *Toxoplasma gondii* infection in humans and animals in the United States. *Int J Parasitol* **38:**1257–1278.

10. **Jones JL, Kruszon-Moran D, Wilson M, McQuillan G, Navin T, McAuley JB.** 2001. *Toxoplasma gondii* infection in the United States: seroprevalence and risk factors. *Am J Epidemiol* **154:**357–365.
11. **Jones JL, Kruszon-Moran D, Wilson M.** 2003. *Toxoplasma gondii* infection in the United States, 1999–2000. *Emerg Infect Dis* **9:**1371–1374.
12. **Jones JL, Kruszon-Moran D, Sanders-Lewis K, Wilson M.** 2007. *Toxoplasma gondii* infection in the United States, 1999–2004, decline from the prior decade. *Am J Trop Med Hyg* **77:**405–410.
13. **Jones JL, Kruszon-Moran D, Rivera HN, Price C, Wilkins PP.** 2014. *Toxoplasma gondii* seroprevalence in the United States 2009–2010 and comparison with the past two decades. *Am J Trop Med Hyg* **90:**1135–1139.
14. **Dubey JP.** 2010. *Toxoplasmosis of Animals and Humans*, 2nd ed. CRC Press, Boca Raton, FL.
15. **Minbaeva G, Schweiger A, Bodosheva A, Kuttubaev O, Hehl AB, Tanner I, Ziadinov I, Torgerson PR, Deplazes P.** 2013. *Toxoplasma gondii* infection in Kyrgyzstan: seroprevalence, risk factor analysis, and estimate of congenital and AIDS-related toxoplasmosis. *PLoS Negl Trop Dis* **7:**e2043.
16. **Jones JL, Holland GN.** 2010. Annual burden of ocular toxoplasmosis in the US. *Am J Trop Med Hyg* **82:**464–465.
17. **Batz MB, Hoffmann S, Morris JG Jr.** 2012. Ranking the disease burden of 14 pathogens in food sources in the United States using attribution data from outbreak investigations and expert elicitation. *J Food Prot* **75:**1278–1291.
18. **Hoffmann S, Batz MB, Morris JG Jr.** 2012. Annual cost of illness and quality-adjusted life year losses in the United States due to 14 foodborne pathogens. *J Food Prot* **75:**1292–1302.
19. **Pedersen MG, Mortensen PB, Norgaard-Pedersen B, Postolache TT.** 2012. *Toxoplasma gondii* infection and self-directed violence in mothers. *Arch Gen Psychiatry* **69:**1123–1130.
20. **Torrey EF, Bartko JJ, Yolken RH.** 2012. *Toxoplasma gondii* and other risk factors for schizophrenia: an update. *Schizophr Bull* **38:** 642–647.
21. **Dubey JP, Murrell KD, Fayer R, Schad GA.** 1986. Distribution of *Toxoplasma gondii* tissue cysts in commercial cuts of pork. *J Am Vet Med Assoc* **188:**1035–1037.
22. **Gajadhar AA, Lalonde LF, Al-Adhami B, Singh BB, Lobanov V.** 2015. Foodborne apicomplexan protozoa: *Coccidia*. *In* Gajadhar AA (ed), *Foodborne Parasites in the Food Supply Web. Occurrence and Control.* Woodhead Publishing, Cambridge, UK.
23. **Dubey JP, Caleo-Bernal R, Rosenthal BM, Speer CA, Fayer R.** 2016. Sarcocystis *of Animals and Humans.* 2nd ed. CRC Press, Boca Raton, FL.
24. **Frenkel JK, Dubey JP, Miller NL.** 1970. *Toxoplasma gondii* in cats: fecal stages identified as coccidian oocysts. *Science* **167:**893–896.
25. **Dubey JP, Frenkel JK.** 1972. Cyst-induced toxoplasmosis in cats. *J Protozool* **19:**155–177.
26. **Dubey JP, Frenkel JK.** 1976. Feline toxoplasmosis from acutely infected mice and the development of *Toxoplasma* cysts. *J Protozool* **23:**537–546.
27. **Dubey JP.** 1996. Infectivity and pathogenicity of *Toxoplasma gondii* oocysts for cats. *J Parasitol* **82:**957–961.
28. **Dubey JP.** 2002. Tachyzoite-induced life cycle of *Toxoplasma gondii* in cats. *J Parasitol* **88:**713–717.
29. **Khan A, Fux B, Su C, Dubey JP, Darde ML, Ajioka JW, Rosenthal BM, Sibley LD.** 2007. Recent transcontinental sweep of *Toxoplasma gondii* driven by a single monomorphic chromosome. *Proc Natl Acad Sci USA* **104:**14872–14877.
30. **Dardé ML, Bouteille B, Perstreal M.** 1992. Isoenzyme analysis of 35 *Toxoplasma gondii* isolates and the biological and epidemiologic implications. *J Parasitol* **78:**909–912.
31. **Howe DK, Sibley LD.** 1995. *Toxoplasma gondii* comprises three clonal lineages: correlation of parasite genotype with human disease. *J Infect Dis* **172:**1561–1566.
32. **Ajzenberg D, Bañuls AL, Tibayrenc M, Dardé ML.** 2002. Microsatellite analysis of *Toxoplasma gondii* shows considerable polymorphism structured into two main clonal groups. *Int J Parasitol* **32:**27–38.
33. **Ajzenberg D, Cogné N, Paris L, Bessières MH, Thulliez P, Filisetti D, Pelloux H, Marty P, Dardé ML.** 2002. Genotype of 86 *Toxoplasma gondii* isolates associated with human congenital toxoplasmosis, and correlation with clinical findings. *J Infect Dis* **186:**684–689.
34. **Su C, Evans D, Cole RH, Kissinger JC, Ajioka JW, Sibley LD.** 2003. Recent expansion of *Toxoplasma* through enhanced oral transmission. *Science* **299:**414–416.
35. **Velmurugan GV, Su C, Dubey JP.** 2009. Isolate designation and characterization of *Toxoplasma gondii* isolates from pigs in the United States. *J Parasitol* **95:**95–99.
36. **Dubey JP, Graham DH, Blackston CR, Lehmann T, Gennari SM, Ragozo AMA,**

Nishi SM, Shen SK, Kwok OCH, Hill DE, Thulliez P. 2002. Biological and genetic characterisation of *Toxoplasma gondii* isolates from chickens (*Gallus domesticus*) from São Paulo, Brazil: unexpected findings. *Int J Parasitol* **32:**99–105.

37. **Dubey JP, Cortés-Vecino JA, Vargas-Duarte JJ, Sundar N, Velmurugan GV, Bandini LM, Polo LJ, Zambrano L, Mora LE, Kwok OCH, Smith T, Su C.** 2007. Prevalence of *Toxoplasma gondii* in dogs from Colombia, South America and genetic characterization of *T. gondii* isolates. *Vet Parasitol* **145:**45–50.
38. **Lehmann T, Marcet PL, Graham DH, Dahl ER, Dubey JP.** 2006. Globalization and the population structure of *Toxoplasma gondii*. *Proc Natl Acad Sci USA* **103:**11423–11428.
39. **Dubey JP, Su C.** 2009. Population biology of *Toxoplasma gondii*: what's out and where did they come from. *Mem Inst Oswaldo Cruz* **104:** 190–195.
40. **Su C, Khan A, Zhou P, Majumdar D, Ajzenberg D, Dardé ML, Zhu XQ, Ajioka JW, Rosenthal BM, Dubey JP, Sibley LD.** 2012. Globally diverse *Toxoplasma gondii* isolates comprise six major clades originating from a small number of distinct ancestral lineages. *Proc Natl Acad Sci USA* **109:**5844–5849.
41. **Grigg ME, Bonnefoy S, Hehl AB, Suzuki Y, Boothroyd JC.** 2001. Success and virulence in *Toxoplasma* as the result of sexual recombination between two distinct ancestries. *Science* **294:**161–165.
42. **Dubey JP, Hill DE, Sundar N, Velmurugan GV, Bandini LA, Kwok OCH, Pierce V, Kelly K, Dulin M, Thulliez P, Iwueke C, Su C.** 2008. Endemic toxoplasmosis in pigs on a farm in Maryland: isolation and genetic characterization of *Toxoplasma gondii*. *J Parasitol* **94:**36–41.
43. **Dubey JP, Sundar N, Hill D, Velmurugan GV, Bandini LA, Kwok OCH, Majumdar D, Su C.** 2008. High prevalence and abundant atypical genotypes of *Toxoplasma gondii* isolated from lambs destined for human consumption in the USA. *Int J Parasitol* **38:**999–1006.
44. **Dubey JP, Rajendran C, Ferreira LR, Martins J, Kwok OC, Hill DE, Villena I, Zhou H, Su C, Jones JL.** 2011. High prevalence and genotypes of *Toxoplasma gondii* isolated from goats, from a retail meat store, destined for human consumption in the USA. *Int J Parasitol* **41:**827–833.
45. **Dubey JP.** 2001. Oocyst shedding by cats fed isolated bradyzoites and comparison of infectivity of bradyzoites of the VEG strain *Toxoplasma gondii* to cats and mice. *J Parasitol* **87:**215–219.
46. **Dubey JP, Saville WJ, Stanek JF, Reed SM.** 2002. Prevalence of *Toxoplasma gondii* antibodies in domestic cats from rural Ohio. *J Parasitol* **88:**802–803.
47. **Levy JK, Crawford PC.** 2004. Humane strategies for controlling feral cat populations. *J Am Vet Med Assoc* **225:**1354–1360.
48. **Bahia-Oliveira LM, Jones JL, Azevedo-Silva J, Alves CC, Oréfice F, Addiss DG.** 2003. Highly endemic, waterborne toxoplasmosis in north Rio de Janeiro state, Brazil. *Emerg Infect Dis* **9:**55–62.
49. **Dubey JP, Lago EG, Gennari SM, Su C, Jones JL.** 2012. Toxoplasmosis in humans and animals in Brazil: high prevalence, high burden of disease, and epidemiology. *Parasitology* **139:**1375–1424.
50. **Aramini JJ, Stephen C, Dubey JP.** 1998. *Toxoplasma gondii* in Vancouver Island cougars (*Felis concolor vancouverensis*): serology and oocyst shedding. *J Parasitol* **84:**438–440.
51. **Aramini JJ, Stephen C, Dubey JP, Engelstoft C, Schwantje H, Ribble CS.** 1999. Potential contamination of drinking water with *Toxoplasma gondii* oocysts. *Epidemiol Infect* **122:** 305–315.
52. **Isaac-Renton J, Bowie WR, King A, Irwin GS, Ong CS, Fung CP, Shokeir MO, Dubey JP.** 1998. Detection of *Toxoplasma gondii* oocysts in drinking water. *Appl Environ Microbiol* **64:**2278–2280.
53. **Cole RA, Lindsay DS, Howe DK, Roderick CL, Dubey JP, Thomas NJ, Baeten LA.** 2000. Biological and molecular characterizations of *Toxoplasma gondii* strains obtained from southern sea otters (*Enhydra lutris nereis*). *J Parasitol* **86:**526–530.
54. **Lindsay DS, Dubey JP.** 2009. Long-term survival of *Toxoplasma gondii* sporulated oocysts in seawater. *J Parasitol* **95:**1019–1020.
55. **Miller MA, Gardner IA, Kreuder C, Paradies DM, Worcester KR, Jessup DA, Dodd E, Harris MD, Ames JA, Packham AE, Conrad PA.** 2002. Coastal freshwater runoff is a risk factor for *Toxoplasma gondii* infection of southern sea otters (*Enhydra lutris nereis*). *Int J Parasitol* **32:**997–1006.
56. **Miller MA, Miller WA, Conrad PA, James ER, Melli AC, Leutenegger CM, Dabritz HA, Packham AE, Paradies D, Harris M, Ames J, Jessup DA, Worcester K, Grigg ME.** 2008. Type X *Toxoplasma gondii* in a wild mussel and terrestrial carnivores from coastal California: new linkages between terrestrial mammals, runoff and toxoplasmosis of sea otters. *Int J Parasitol* **38:**1319–1328.
57. **Conrad PA, Miller MA, Kreuder C, James ER, Mazet J, Dabritz H, Jessup DA, Gulland F, Grigg ME.** 2005. Transmission of *Toxo-*

plasma: clues from the study of sea otters as sentinels of *Toxoplasma gondii* flow into the marine environment. *Int J Parasitol* **35**:1155–1168.

58. **Boyer K, Hill D, Mui E, Wroblewski K, Karrison T, Dubey JP, Sautter M, Noble AG, Withers S, Swisher C, Heydemann P, Hosten T, Babiarz J, Lee D, Meier P, McLeod R, Toxoplasmosis Study Group.** 2011. Unrecognized ingestion of *Toxoplasma gondii* oocysts leads to congenital toxoplasmosis and causes epidemics in North America. *Clin Infect Dis* **53**:1081–1089.
59. **Hill DE, Chirukandoth S, Dubey JP.** 2005. Biology and epidemiology of *Toxoplasma gondii* in man and animals. *Anim Health Res Rev* **6**:41–61.
60. **Hill DE, Sreekumar C, Gamble HR, Dubey JP.** 2004. Effect of commonly used enhancement solutions on the viability of *Toxoplasma gondii* tissue cysts in pork loin. *J Food Prot* **67**:2230–2233.
61. **Hill DE, Benedetto SMC, Coss C, McCrary JL, Fournet VM, Dubey JP.** 2006. Effects of time and temperature on the viability of *Toxoplasma gondii* tissue cysts in enhanced pork loin. *J Food Prot* **69**:1961–1965.
62. **Dubey JP.** 2009. Toxoplasmosis in pigs: the last 20 years. *Vet Parasitol* **164**:89–103.
63. **Mead PS, Slutsker L, Dietz V, McCaig LF, Bresee JS, Shapiro C, Griffin PM, Tauxe RV.** 1999. Food-related illness and death in the United States. *Emerg Infect Dis* **5**:607–625.
64. **Roghmann MC, Faulkner CT, Lefkowitz A, Patton S, Zimmerman J, Morris JG Jr.** 1999. Decreased seroprevalence for *Toxoplasma gondii* in Seventh Day Adventists in Maryland. *Am J Trop Med Hyg* **60**:790–792.
65. **Scallan E, Hoekstra RM, Angulo FJ, Tauxe RV, Widdowson MA, Roy SL, Jones JL, Griffin PM.** 2011. Foodborne illness acquired in the United States: major pathogens. *Emerg Infect Dis* **17**:7–15.
66. **Dubey JP.** 2010. *Toxoplasma gondii* infections in chickens (*Gallus domesticus*): prevalence, clinical disease, diagnosis and public health significance. *Zoonoses Public Health* **57**:60–73.
67. **Jones JL, Dargelas V, Roberts J, Press C, Remington JS, Montoya JG.** 2009. Risk factors for *Toxoplasma gondii* infection in the United States. *Clin Infect Dis* **49**:878–884.
68. **Dubey JP, Hill DE, Jones JL, Hightower AW, Kirkland E, Roberts JM, Marcet PL, Lehmann T, Vianna MCB, Miska K, Sreekumar C, Kwok OCH, Shen SK, Gamble HR.** 2005. Prevalence of viable *Toxoplasma gondii* in beef, chicken, and pork from retail meat stores in the United States: risk assessment to consumers. *J Parasitol* **91**:1082–1093.
69. **Dubey JP, Leighty JC, Beal VC, Anderson WR, Andrews CD, Thulliez P.** 1991. National seroprevalence of *Toxoplasma gondii* in pigs. *J Parasitol* **77**:517–521.
70. **Hill DE, Haley C, Wagner B, Gamble HR, Dubey JP.** 2010. Seroprevalence of and risk factors for *Toxoplasma gondii* in the US swine herd using sera collected during the National Animal Health Monitoring Survey (Swine 2006). *Zoonoses Public Health* **57**:53–59.
71. **Gamble HR, Brady RC, Dubey JP.** 1999. Prevalence of *Toxoplasma gondii* infection in domestic pigs in the New England states. *Vet Parasitol* **82**:129–136.
72. **Dubey JP, Gamble HR, Hill D, Sreekumar C, Romand S, Thuilliez P.** 2002. High prevalence of viable *Toxoplasma gondii* infection in market weight pigs from a farm in Massachusetts. *J Parasitol* **88**:1234–1238.
73. **Honeyman MS, Pirog RS, Huber GH, Lammers PJ, Hermann JR.** 2006. The United States pork niche market phenomenon. *J Anim Sci* **84**:2269–2275.
74. **Dimitri C, Greene C.** 2002. Recent growth patterns in the U.S. organic foods market. Agriculture Information Bulletin No. (AIB777) , 1–42; www.ers.usda.gov/publications/aib777/.
75. **Kijlstra A, Eissen OA, Cornelissen J, Munniksma K, Eijck I, Kortbeek T.** 2004. *Toxoplasma gondii* infection in animal-friendly pig production systems. *Invest Ophthalmol Vis Sci* **45**:3165–3169.
76. **Gebreyes WA, Bahnson PB, Funk JA, McKean J, Patchanee P.** 2008. Seroprevalence of *Trichinella, Toxoplasma,* and *Salmonella* in antimicrobial-free and conventional swine production systems. *Foodborne Pathog Dis* **5**:199–203.
77. **Dubey JP, Hill DE, Rozeboom DW, Rajendran C, Choudhary S, Ferreira LR, Kwok OC, Su C.** 2012. High prevalence and genotypes of *Toxoplasma gondii* isolated from organic pigs in northern USA. *Vet Parasitol* **188**:14–18.
78. **Kijlstra A, Jongert E.** 2008. Control of the risk of human toxoplasmosis transmitted by meat. *Int J Parasitol* **38**:1359–1370.
79. **Dubey JP, Murrell KD, Fayer R.** 1984. Persistence of encysted *Toxoplasma gondii* in tissues of pigs fed oocysts. *Am J Vet Res* **45**:1941–1943.
80. **Vangeel L, Houf K, Chiers K, Vercruysse J, D'Herde K, Ducatelle R.** 2007. Molecular-based identification of *Sarcocystis hominis* in Belgian minced beef. *J Food Prot* **70**:1523–1526.
81. **Moré G, Pantchev A, Skuballa J, Langenmayer MC, Maksimov P, Conraths FJ, Venturini MC, Schares G.** 2014. *Sarcocystis sinensis* is the most

prevalent thick-walled *Sarcocystis* species in beef on sale for consumers in Germany. *Parasitol Res* **113:**2223–2230.

82. **Pena HF, Ogassawara S, Sinhorini IL.** 2001. Occurrence of cattle *Sarcocystis* species in raw kibbe from Arabian food establishments in the city of São Paulo, Brazil, and experimental transmission to humans. *J Parasitol* **87:**1459–1465.
83. **Domenis L, Peletto S, Sacchi L, Clementi E, Genchi M, Felisari L, Felisari C, Mo P, Modesto P, Zuccon F, Campanella C, Maurella C, Guidetti C, Acutis PL.** 2011. Detection of a morphogenetically novel *Sarcocystis hominis*-like in the context of a prevalence study in semi-intensively bred cattle in Italy. *Parasitol Res* **109:**1677–1687.
84. **Hornok S, Mester A, Takács N, Baska F, Majoros G, Fok É, Biksi I, Német Z, Hornyák Á, Jánosi S, Farkas R.** 2015. *Sarcocystis*-infection of cattle in Hungary. *Parasit Vectors* **8:**69.
85. **Nourollahi-Fard SR, Kheirandish R, Sattari S.** 2015. Prevalence and histopathological finding of thin-walled and thick-walled *Sarcocysts* in slaughtered cattle of Karaj abattoir, Iran. *J Parasit Dis* **39:**272–275.
86. **Chhabra MB, Samantaray S.** 2013. *Sarcocystis* and sarcocystosis in India: status and emerging perspectives. *J Parasit Dis* **37:**1–10.
87. **Fayer R.** 2004. *Sarcocystis* spp. in human infections. *Clin Microbiol Rev* **17:**894–902.
88. **Fayer R, Esposito DH, Dubey JP.** 2015. Human infections with *Sarcocystis* species. *Clin Microbiol Rev* **28:**295–311.
89. **Kozak GK, MacDonald D, Landry L, Farber JM.** 2013. Foodborne outbreaks in Canada linked to produce: 2001 through 2009. *J Food Prot* **76:**173–183.
90. **Giangaspero A, Marangi M, Koehler AV, Papini R, Normanno G, Lacasella V, Lonigro A, Gasser RB.** 2015. Molecular detection of *Cyclospora* in water, soil, vegetables and humans in southern Italy signals a need for improved monitoring by health authorities. *Int J Food Microbiol* **211:**95–100.
91. **Tandukar S, Ansari S, Adhikari N, Shrestha A, Gautam J, Sharma B, Rajbhandari D, Gautam S, Nepal HP, Sherchand JB.** 2013. Intestinal parasitosis in school children of Lalitpur district of Nepal. *BMC Res Notes* **6:**449.
92. **Mac Kenzie WR, Hoxie NJ, Proctor ME, Gradus MS, Blair KA, Peterson DE, Kazmierczak JJ, Addiss DG, Fox KR, Rose JB, Davis JP.** 1994. A massive outbreak in Milwaukee of *Cryptosporidium* infection transmitted through the public water supply. *N Engl J Med* **331:**161–167.
93. **Hlavsa MC, Roberts VA, Kahler AM, Hilborn ED, Mecher TR, Beach MJ, Wade TJ, Yoder JS, Centers for Disease Control and Prevention (CDC).** 2015. Outbreaks of illness associated with recreational water: United States, 2011–2012. *MMWR Morb Mortal Wkly Rep* **64:**668–672.
94. **Dixon B, Parrington L, Cook A, Pollari F, Farber J.** 2013. Detection of *Cyclospora, Cryptosporidium*, and *Giardia* in ready-to-eat packaged leafy greens in Ontario, Canada. *J Food Prot* **76:**307–313.
95. **Dixon BR.** 2015. Transmission dynamics of foodborne parasites on fresh produce. *In* Gajadhar AA (ed), *Foodborne Parasites in the Food Supply Web. Occurrence and Control.* Woodhead Publishing, Cambridge, UK.
96. **Thompson RCA.** 2011. *Giardia* infections, p 522–535. *In* Palmer SR, Soulsby EJL, Torgerson P, Brown D (ed), *Zoonoses.* Oxford University Press, Oxford.
97. **Thompson RCA.** 2015. Fooborne, enteric, non-apicomplexan unicellular parasites. *In* Gajadhar AA (ed), *Foodborne Parasites in the Food Supply Web. Occurrence and Control.* Woodhead Publishing, Cambridge, UK.
98. **Cacció SM, Lalle M.** 2015. Giardia. *In* Xiao L, Ryan U, Feng Y (ed), *Biology of Foodborne Parasites.* CRC Press, Boca Raton, FL.
99. **Fentress SJ, Sibley LD.** 2011. The secreted kinase ROP18 defends *Toxoplasma's* border. *BioEssays* **33:**693–700.
100. **Gigley JP, Fox BA, Bzik DJ.** 2009. Cell-mediated immunity to *Toxoplasma gondii* develops primarily by local Th1 host immune responses in the absence of parasite replication. *J Immunol* **182:**1069–1078.
101. **Vouldoukis I, Mazier D, Moynet D, Thiolat D, Malvy D, Mossalayi MD.** 2011. IgE mediates killing of intracellular *Toxoplasma gondii* by human macrophages through CD23-dependent, interleukin-10 sensitive pathway. *PLoS One* **6:**e18289.
102. **Koshy AA, Dietrich HK, Christian DA, Melehani JH, Shastri AJ, Hunter CA, Boothroyd JC.** 2012. *Toxoplasma* co-opts host cells it does not invade. *PLoS Pathog* **8:**e1002825.
103. **Hunter CA, Sibley LD.** 2012. Modulation of innate immunity by *Toxoplasma gondii* virulence effectors. *Nat Rev Microbiol* **10:**766–778.
104. **Desmonts G, Couvreur J.** 1974. Congenital toxoplasmosis. A prospective study of 378 pregnancies. *N Engl J Med* **290:**1110–1116.
105. **Remington JS, McLeod R, Desmonts G.** 1995. Toxoplasmosis, p 140–243. *In* Remington JS, Klein J (ed), *Infectious Diseases of the Fetus and*

Newborn Infant. W.B. Saunders, Philadelphia, PA.

106. **Roberts T, Frenkel JK.** 1990. Estimating income losses and other preventable costs caused by congenital toxoplasmosis in people in the United States. *J Am Vet Med Assoc* **196:**249–256.
107. **Roberts T, Murrell KD, Marks S.** 1994. Economic losses caused by foodborne parasitic diseases. *Parasitol Today* **10:**419–423.
108. **Thiébaut R, Leproust S, Chêne G, Gilbert R, SYROCOT (Systematic Review on Congenital Toxoplasmosis) Study Group.** 2007. Effectiveness of prenatal treatment for congenital toxoplasmosis: a meta-analysis of individual patients' data. *Lancet* **369:**115–122.
109. **Petersen E.** 2007. Prevention and treatment of congenital toxoplasmosis. *Expert Rev Anti Infect Ther* **5:**285–293.
110. **Cortina-Borja M, Tan HK, Wallon M, Paul M, Prusa A, Buffolano W, Malm G, Salt A, Freeman K, Petersen E, Gilbert RE, European Multicentre Study on Congenital Toxoplasmosis (EMSCOT).** 2010. Prenatal treatment for serious neurological sequelae of congenital toxoplasmosis: an observational prospective cohort study. *PLoS Med* **7:**e1000351.
111. **Remington JS, McLeod R, Thulliez P, Desmonts G.** 2001. Toxoplasmosis, p 205–346. *In* Remington JS, Klein J (ed), *Infectious Diseases of the Fetus and Newborn Infant.* W.B. Saunders, Philadelphia, PA.
112. **Stillwaggon E, Carrier CS, Sautter M, McLeod R.** 2011. Maternal serologic screening to prevent congenital toxoplasmosis: a decision-analytic economic model. *PLoS Negl Trop Dis* **5:**e1333.
113. **Teutsch SM, Juranek DD, Sulzer A, Dubey JP, Sikes RK.** 1979. Epidemic toxoplasmosis associated with infected cats. *N Engl J Med* **300:**695–699.
114. **Benenson MW, Takafuji ET, Lemon SM, Greenup RL, Sulzer AJ.** 1982. Oocyst-transmitted toxoplasmosis associated with ingestion of contaminated water. *N Engl J Med* **307:**666–669.
115. **Smith JL.** 1993. Documented outbreaks of toxoplasmosis: transmission of *Toxoplasma gondii* to humans. *J Food Prot* **56:**630–639.
116. **Bowie WR, King AS, Werker DH, Isaac-Renton JL, Bell A, Eng SB, Marion SA, The BC Toxoplasma Investigation Team.** 1997. Outbreak of toxoplasmosis associated with municipal drinking water. *Lancet* **350:**173–177.
117. **Burnett AJ, Shortt SG, Isaac-Renton J, King A, Werker D, Bowie WR.** 1998. Multiple cases of acquired toxoplasmosis retinitis presenting in an outbreak. *Ophthalmology* **105:**1032–1037.
118. **Luft BJ, Remington JS.** 1992. Toxoplasmic encephalitis in AIDS. *Clin Infect Dis* **15:**211–222.
119. **Renold C, Sugar A, Chave JP, Perrin L, Delavelle J, Pizzolato G, Burkhard P, Gabriel V, Hirschel B.** 1992. *Toxoplasma* encephalitis in patients with the acquired immunodeficiency syndrome. *Medicine (Baltimore)* **71:**224–239.
120. **Palella FJ Jr, Delaney KM, Moorman AC, Loveless MO, Fuhrer J, Satten GA, Aschman DJ, Holmberg SD, HIV Outpatient Study Investigators.** 1998. Declining morbidity and mortality among patients with advanced human immunodeficiency virus infection. *N Engl J Med* **338:**853–860.
121. **Pozio E.** 2004. Highly Active AntiRetroviral Therapy and opportunistic protozoan infections. *Parassitologia* **46:**89–93. (In Italian.)
122. **Pozio E, Morales MA.** 2005. The impact of HIV-protease inhibitors on opportunistic parasites. *Trends Parasitol* **21:**58–63.
123. **Fricker-Hidalgo H, Cimon B, Chemla C, Darde ML, Delhaes L, L'ollivier C, Godineau N, Houze S, Paris L, Quinio D, Robert-Gangneux F, Villard O, Villena I, Candolfi E, Pelloux H.** 2013. Toxoplasma seroconversion with negative or transient immunoglobulin M in pregnant women: myth or reality? A French multicenter retrospective study. *J Clin Microbiol* **51:**2103–2111.
124. **Rahumatullah A, Khoo BY, Noordin R.** 2012. Triplex PCR using new primers for the detection of *Toxoplasma gondii*. *Exp Parasitol* **131:**231–238.
125. **Contini C, Cultrera R, Seraceni S, Segala D, Romani R, Fainardi E, Cinque P, Lazzarin A, Delia S.** 2002. The role of stage-specific oligonucleotide primers in providing effective laboratory support for the molecular diagnosis of reactivated *Toxoplasma gondii* encephalitis in patients with AIDS. *J Med Microbiol* **51:**879–890.
126. **Joseph P, Calderón MM, Gilman RH, Quispe ML, Cok J, Ticona E, Chavez V, Jimenez JA, Chang MC, Lopez MJ, Evans CA.** 2002. Optimization and evaluation of a PCR assay for detecting toxoplasmic encephalitis in patients with AIDS. *J Clin Microbiol* **40:**4499–4503.
127. **Lewis JS Jr, Khoury H, Storch GA, DiPersio J.** 2002. PCR for the diagnosis of toxoplasmosis after hematopoietic stem cell transplantation. *Expert Rev Mol Diagn* **2:**616–624.
128. **Antsaklis A, Daskalakis G, Papantoniou N, Mentis A, Michalas S.** 2002. Prenatal diagnosis of congenital toxoplasmosis. *Prenat Diagn* **22:**1107–1111.

129. **Hill DE, Chirukandoth S, Dubey JP, Lunney JK, Gamble HR.** 2006. Comparison of detection methods for *Toxoplasma gondii* in naturally and experimentally infected swine. *Vet Parasitol* **141:**9–17.
130. **Morelle C, Varlet-Marie E, Brenier-Pinchart MP, Cassaing S, Pelloux H, Bastien P, Sterkers Y.** 2012. Comparative assessment of a commercial kit and two laboratory-developed PCR assays for molecular diagnosis of congenital toxoplasmosis. *J Clin Microbiol* **50:**3977–3982.
131. **Mikita K, Maeda T, Ono T, Miyahira Y, Asai T, Kawana A.** 2013. The utility of cerebrospinal fluid for the molecular diagnosis of toxoplasmic encephalitis. *Diagn Microbiol Infect Dis* **75:**155–159.
132. **Guerina NG, Hsu H-W, Meissner HC, Maguire JH, Lynfield R, Stechenberg B, Abroms I, Pasternack MS, Hoff R, Eaton RB, Grady GF, The New England Regional Toxoplasma Working Group.** 1994. Neonatal serologic screening and early treatment for congenital *Toxoplasma gondii* infection. *N Engl J Med* **330:**1858–1863.
133. **Chirgwin K, Hafner R, Leport C, Remington J, Andersen J, Bosler EM, Roque C, Rajicic N, McAuliffe V, Morlat P, Jayaweera DT, Vilde JL, Luft BJ.** 2002. Randomized phase II trial of atovaquone with pyrimethamine or sulfadiazine for treatment of toxoplasmic encephalitis in patients with acquired immunodeficiency syndrome: ACTG 237/ANRS 039 Study. AIDS Clinical Trials Group 237/Agence Nationale de Recherche sur le SIDA, Essai 039. *Clin Infect Dis* **34:**1243–1250.
134. **Lopez A, Dietz VJ, Wilson M, Navin TR, Jones JL.** 2000. Preventing congenital toxoplasmosis. *MMWR Recomm Rep* **49**(RR-2)**:**59–68.
135. **Dubey JP, Kotula AW, Sharar A, Andrews CD, Lindsay DS.** 1990. Effect of high temperature on infectivity of *Toxoplasma gondii* tissue cysts in pork. *J Parasitol* **76:**201–204.
136. **Kotula AW, Dubey JP, Sharar AK, Andrew CD, Shen SK, Lindsay DS.** 1991. Effect of freezing on infectivity of *Toxoplasma gondii* tissue cysts in pork. *J Food Prot* **54:**687–690.
137. **Dubey JP, Thayer DW.** 1994. Killing of different strains of *Toxoplasma gondii* tissue cysts by irradiation under defined conditions. *J Parasitol* **80:**764–767.
138. **Foulon W, Naessens A, Derde MP.** 1994. Evaluation of the possibilities for preventing congenital toxoplasmosis. *Am J Perinatol* **11:**57–62.
139. **Foulon W, Naessens A, Ho-Yen D.** 2000. Prevention of congenital toxoplasmosis. *J Perinat Med* **28:**337–345.
140. **Xiao L, Ryan U, Feng Y (ed).** 2015. *Biology of Foodborne Parasites*. CRC Press, Boca Raton, FL.

Local Food Systems Food Safety Concerns

13

BENJAMIN CHAPMAN[1] and CHRIS GUNTER[2]

Researchers have stated that pathogens can contaminate food at any point along the food chain—at the farm, processing plant, transportation vehicle, retail store, or foodservice operation and the home. By understanding where potential problems exist, it is possible to develop strategies to reduce risks of contamination (1).

While there have been over 600 known produce-related disease outbreaks in North America since 2004, resulting in about 20,000 illnesses (2), there have probably been thousands of unknown outbreaks. The growing desire of consumers in developed countries to consume fresh fruits and vegetables coupled with the expanding global nature of produce distribution has led to an increase in the reported pathogens, outbreaks, and incidence of contamination. Laboratory studies have found that fresh produce can support the growth and survival of organisms such as *Salmonella* spp., *Shigella* spp., *Escherichia coli* O157:H7, *Listeria monocytogenes*, enteric viruses, and parasites (1–8).

Many recent outbreaks are representative of a growing trend of large, geographically dispersed outbreaks caused by sporadic or low-level contamination of widely distributed food items (9). Most recently, epidemiological and traceability investigations determined that cases of cyclosporiasis in Pennsylvania were caused by contaminated snow peas sourced from

[1]Department of Agricultural and Human Sciences, NC State University, Raleigh, NC 27695; [2]Department of Horticulture Sciences, NC State University, Raleigh, NC 27695.

Preharvest Food Safety
Edited by Siddhartha Thakur and Kalmia E. Kniel

doi:10.1128/microbiolspec.PFS-0020-2017

Guatemala (10). Analysis of the surveillance of foodborne pathogens has revealed that there is a significant increase in the reporting of outbreaks in the summer months, which may be due to greater consumption of fresh fruits and vegetables during this season (3).

Enhancing the safety of agricultural commodities that are also ready-to-eat foods, particularly fruits and vegetables, presents a major challenge to industry, regulators, and public health officials (9). Risk management strategies are difficult because the potential pathogen sources in fresh fruit and vegetable production are numerous (3). The source of many human pathogens associated with produce was once thought to be only of animal origin. Figure 1 illustrates the cycle of pathogens from humans and animals to the environment and the impact this has on produce (11). Efforts need to be made all along the farm-to-fork continuum (in production, harvest, processing, transport, retail, and home) to reduce potential food safety risks. Researchers have reported that the use of pathogen-free water for washing minimizes the risk of contamination (3, 12). The efficacy of disinfectants varies with commodities, surface characteristics, temperature, and pathogen (3). *L. monocytogenes* is more resistant to disinfectants than are *Salmonella* spp., *E. coli* O157:H7, and *Shigella* spp. (3). Little is known about parasites and viruses in terms of pathogen removal or disinfection, though there has been an increase in research in this area in recent years (3). Rafferty and colleagues (13) demonstrated that *E. coli* could spread on the farm in plant production cuttings from one contaminated source, magnifying an outbreak to a whole farm (13).

Allwood and colleagues (14) examined 40 items of fresh produce taken from a retail setting in the United States that had been preprocessed (including cut, shredded, chopped, or peeled) at or before the point of purchase. They found fecal contamination indicators (*E. coli*, F-specific coliphages, and noroviruses) in 48% of samples. Although it is uncertain where the contamination originated, this suggests that fecal matter does persist despite the employment of risk management strategies at processing sites. Parasites have also been shown to exist on produce at retail. In 2001, researchers tested 475 samples in Norway for *Cryptosporidium* oocysts and *Giardia* cysts. A total of 6% of the samples were positive; this was the first time parasites had been detected without an outbreak occurring (15).

As discussed above, pathogens can contaminate food at any point along the food chain. Raw produce can become contaminated with pathogenic and nonpathogenic microorganisms at a number of different stages by several means, from production through to consumption. Likely causes of this contamination include workers' hands, water used

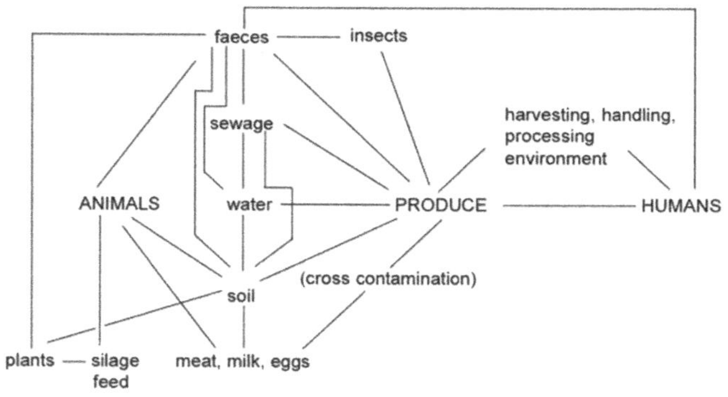

FIGURE 1 Cycle of pathogens in fresh fruit and vegetable production (source: reference 11).

during crop production and handling, and contaminated surfaces contacting produce.

FOOD SAFETY MODERNIZATION ACT

The fresh produce industry is facing major changes in production practices due to the implementation of the Food Safety Modernization Act. Within the new rules the standards associated with water quality and testing are among the most contested by industry associations. Current guidelines require water which will be in direct contact with the crop to meet specific microbiological thresholds based on the 2012 EPA recreational water standards. However, this water standard does not focus on the use of this water for agricultural uses. Alternative provisions to comply with these rules have been allowed by FDA when water cannot meet these standards, but these provisions have not been fully defined. One of these options takes into consideration a microbial die-off rate of 0.5 log per day that may occur naturally between irrigation and harvest events and/or via preharvest-postharvest intervention.

Pre- and postharvest risk factors are outlined in Table 1 (16). Consequently, methods of growing, handling, processing, packaging, and distributing fresh produce have received increased attention in terms of identifying and minimizing microbiological hazards.

TABLE 1 Sources of pathogenic microorganisms for produce[a]

Preharvest
Feces
Soil
Irrigation water
Water used to apply fungicides, insecticides
Green or inadequately composted manure
Air (dust)
Wild and domestic animals (including fowl and reptiles)
Insects
Human handling
Postharvest
Feces
Human handling (workers, consumers)
Harvesting equipment
Transport containers (field to packing shed)
Wild and domestic animals (including fowl and reptiles)
Insects
Air (dust)
Wash and rinse water
Sorting, packing, cutting, and further processing equipment
Ice
Transport vehicles
Improper storage (temperature, physical environment)
Improper packaging (including new packaging technologies)
Cross-contamination (order of foods in storage, preparation, and display areas)
Improper display temperature

[a]Source: reference 16.

FACTORS OF PATHOGEN CONTAMINATION IN PRODUCE

Soil

It is thought that the presence of many pathogens in soil is from the historical application or environmental presence of feces or untreated sewage (16). Pathogens existing in soil or water can be the source of both pre- and postharvest contamination. These organisms remain viable until consumption and can be protected by surface features of the fruit or vegetable. Flesh scarring can provide a suitable environment for pathogen growth and decreases the value of employing sanitizers, either in the packing shed or by consumers (17). *E. coli* O157 can survive in soil for over 3 months, thus making manure application within three months of harvest risky for many crops (18, 19). *E. coli* O157 has also been recovered from aerosol sources such as soil and sawdust up to 42 weeks following contamination by animals (20). Researchers in Minnesota recently conducted a small-scale comparative study of organic versus conventionally grown produce. They found that while all samples were virtually free of pathogens, *E. coli* was 19 times more prevalent on produce acquired from the organic farms (21). They estimated that this was due to the common use of manure aged for less than a year. Use of cattle manure was found to have a higher risk because *E. coli* was found 2.4 times more often on farms using it than on those using other animal manures (21).

Irrigation and Surface Runoff Waters

Production water can be a source of pathogenic microorganisms that contaminate fruits and vegetables in the field. Irrigation water containing raw sewage or improperly treated effluents from sewage treatment plants may contain hepatitis A, Norwalk viruses, or enteroviruses, in addition to bacterial pathogens such as *E. coli* O157:H7, *Salmonella* spp., and *Shigella* spp. (3). Solomon and colleagues (22) discovered that the transmission of *E. coli* O157:H7 to lettuce was possible through both spray and drip irrigation. They also found that the pathogen persisted on the plants for 20 days following application, and submerging the lettuce in a solution of 200 ppm chlorine did not eliminate all viable *E. coli* O157:H7 cells. This suggests that irrigation water of unknown microbial quality should be avoided in lettuce production (22). In a follow-up experiment, Solomon and colleagues (23) explored the transmission of *E. coli* O157:H7 from manure-contaminated soil and irrigation water to lettuce plants. The researchers recovered viable cells from the inner tissues of the lettuce plants and found that the cells migrated to internal locations in plant tissue and were thus protected from the action of topical sanitizing agents. These experiments demonstrated that *E. coli* O157:H7 can enter the lettuce plant through the root system and migrate throughout the edible portion of the plant (23). Researchers have described anywhere between 50 and 400 ppm concentration of total available chlorine as adequate to eliminate viable bacteria in water (either wash or irrigation, which is easily measurable with chlorine strips) (12). The occurrence of human pathogenic parasites in irrigation waters used for food crops traditionally eaten raw has been found in the United States and several Central American countries. The average concentrations in samples from Central America containing *Giardia* cysts and *Cryptosporidium* oocysts were a magnitude of 10 higher, suggesting that there may be a greater risk of infection to consumers who come in contact with or eat these products from Central American countries (24).

The risk of contamination of produce due to *Salmonella* spp. was found to be increased when soil and water were present, and soil and water actually act as reservoirs of the pathogen. Guo and colleagues (17) found that soil and water were factors in the infiltration of *Salmonella* into the tissues of tomato. This supports the theory that preharvest contact with contaminated soil or water increases the contamination potential by certain pathogens and can lead to problems in pathogen removal and the efficacy of sanitizers.

Birds

Birds have been identified as a risk factor, though it is difficult to quantify their impact. Pathogenic bacteria are picked up as a result of birds feeding on garbage, sewage, fish, or lands that are grazed with cattle or have had applications of fresh manure. Control of preharvest contamination of fruits and vegetables with pathogenic bacteria carried by wild birds would be difficult to implement and monitor (16). Monitoring for the presence of this potential source of contamination and logging the pattern of occurrence could allow producers to choose field locations less likely to be encountered by birds.

Washing Produce

Contaminated water used during postharvest operations can transmit diseases that decay the produce or adversely affect human health (25). Chlorine has generally been used as a disinfectant in wash waters to keep them free of detectable generic *E. coli*. It is difficult to achieve a standard for wash waters to eliminate pathogens that may be present and that remain effective over 8 hours of fruit or vegetable washing. Suslow (12) described anywhere between 50 and 400 ppm of total available chlorine as adequate to eliminate viable bacteria in water. Pathogens can be present in the water source (if ground or well water is used) and can enter wash water on the product itself or through transfer from workers with poor hygiene. Waxing of fruits may increase the adhesion potential, decreasing the effectiveness of chlorine inactivation of pathogens; this has been demonstrated with cucumbers (7).

Chlorine has been widely used as a sanitizer to maintain the microbial quality and safety of fresh-cut produce; however, chlorine can lack efficacy for pathogen reduction, especially when the fresh-cut processing water contains heavy organic loads (3). Organic matter reacts with chlorine in solution to lower its effective concentration. During sanitation studies using chlorine spraying, the World Health Organization research team (3) found that tomato disinfection was significantly more effective at 2,000 ppm than at 200 ppm but was no more effective at 3,000 ppm. Inactivation of microorganisms occurred within 1 minute of the 200 ppm chlorine application. Chlorine has been shown to be effective in reducing microbial loads in asparagus as well (26).

Hydrogen peroxide is an effective, though less practical, method for sanitation of fresh produce. McWatters et al. (27) found that hydrogen peroxide at 40°C for 15 minutes was effective in reducing foodborne bacterial pathogens on raw apples. However, the effects of this treatment on an apple's sensory characteristics were not well explored. Further, having to treat produce for 15 minutes is a limitation to the procedure.

Other sanitizing agents for wash water include acidified sodium chlorite, but like chlorine, the organic load in process water significantly decreases effectiveness (28). Acidic electrolyzed water has been demonstrated to be effective in its frozen form for the inactivation of *L. monocytogenes* and *E. coli* O157:H7 on lettuce as well as a quality preservation step (29). Combinations of lactic acid and hydrogen peroxide have also been shown to be effective sanitizers, resulting in a 4-log reduction of *E. coli* O157:H7 and *Salmonella enterica* serovar Enteritidis at a 60-second exposure time (30). The hydrogen peroxide residue is also not detectable after a cold-water rinse, demonstrating that it is a viable sanitizer alternative with no negative impact on quality (30). Venkitanarayanan and colleagues (31) demonstrated that apples, oranges, and tomatoes could be sanitized with a solution of 0.5% lactic acid plus 1.5% hydrogen peroxide for 15 minutes at 40°C, which could be used at the packing stage as an effective sanitizer. Han and colleagues (32) found that injured surfaces on green peppers protected bacteria from treatment, reducing the effectiveness of all sanitizers due to inaccessibility and a lack of contact time.

Internalization of Pathogens

Produce can also be contaminated with pathogens due to the internalization of pathogens through the root system and through flesh or stem scars. Evidence of infiltration of bacteria into vegetables is reported in several articles (33–37). Clear evidence exists to conclude that pathogens can be incorporated into fresh produce. So far, this evidence is based on laboratory experiments, not real-world situations. Past research suggests that pathogens can enter lettuce plants through their roots and end up in the edible leaves. Small gaps in growing roots through which plant pathogens infect tissue may also allow *E. coli* entry (23, 38, 39).

The uptake of *Salmonella* spp. by the roots of hydroponically grown tomato plants has also been shown. Within 1 day of exposure to a high-concentration mixture of *Salmonella* spp., pathogen cells were found in the hypocotyls, cotyledons, stems, and leaves of young plants, though whether fruit is affected is not known at this time (40).

Farmers' Markets

Farmers' markets are a growing retail channel within North Carolina and provide a significant opportunity for many traditional producers to receive a premium for a portion of their products that would otherwise have been sold wholesale. There are at least 150 active (at least weekly) state-recognized farmers' markets throughout the state, with an estimated 300 additional smaller tailgate markets operating in parking lots and community spaces. An analysis of outbreaks from 1990 to 2003 found that 12% of outbreaks and 20% of outbreak-related illnesses were associated with produce. The potential for microbiological risks to be associated with the growing farmers' market sector provides a dangerous situation for this resurging business outlet.

Local Foods and Community Supported Agriculture

Shared kitchens, community supported agriculture (CSA), and food hubs are growing phenomena in the United States and around the world. Each of these food entities may include public food preparation areas, storage areas, distribution areas, or a combination of these. Shared kitchens, also known as kitchen incubators, community kitchens, or shared time kitchens are fully equipped, commercial kitchens with storage space available to use through plans (41). Shared kitchens are popularly located in urban areas, with a few more facilities in the western region of the United States (42) than elsewhere in the country. It has been found that "most facilities include rental kitchens, dry storage, refrigerated storage, and freezer storage, as well as a variety of other facilities" and that "the large majority of [these facilities] were established in the past five years" (42). In addition, it has been stated that one of the challenges of these shared kitchen facilities is cleaning and sanitation (42).

In May of 2012 it was confirmed that Smiling Hara brand tempeh had been contaminated with *S. enterica* Paratyphi B detected during a routine inspection. The outbreak occurred in North Carolina, South Carolina, New York, Tennessee, and Georgia and infected a total of 46 people. The shared kitchen, Blue Ridge Food Ventures, where the tempeh was produced was also home to over 20 other businesses at the time of the outbreak. After the outbreak occurred, cleaning and sanitizing steps were followed at the shared kitchen, but with increased food safety knowledge prior to the incident this outbreak could have possibly been prevented (43). Prior to the *Salmonella* outbreak in the tempeh at the shared kitchen, in August of 2011 in Oregon, there was an outbreak of *E. coli* O157:H7 on strawberries at a food hub. The strawberry producer sold the berries to buyers who then distributed them to farmers' markets and roadside stands. One person died and approximately 16 were sickened with *E. coli* due to this outbreak, because of the lack of food safety knowledge surrounding food hubs. The berries were not sold with a label and could no be traced and recalled (44). An additional outbreak occurred in a shared kitchen in California near the middle of June in 2012. Canned soups manufactured by One Gun Ranch at the Santa Barbara Organic Soup Kitchen had the potential to be contaminated with the toxin that produces foodborne botulism. The soups were then sold at local farmers' markets. There were no reported illnesses linked to the soups, but this was a major crisis averted due to the serious harm of botulism (45). Most recently, in Seattle, Washington, in September 2015 there was an *E. coli* O157:H7 outbreak linked to Mexican food prepared at the Bellevue Community Kitchen. Six cases of *E. coli* were reported; three patients had to be hospitalized, and one of those was a 4-year-old girl. The community kitchen's space and equipment were being shared with approximately a dozen other food businesses, but after news of the outbreak the kitchen was told to cease operation and close (46). With these foodborne illness cases and increasing numbers of others around the United States, it is necessary that food safety and foodborne illness prevention methods be taught, prioritized, and implemented in shared kitchens, food hubs, CSAs, and other local food systems.

CSAs are where "consumers buy shares in [a] farm before planting begins and receive a portion of whatever is available each week of the growing season" (47). The U.S. Department of Agriculture (USDA) collected data about CSAs in 2012 and found that 12,617 active farms across the United States participate in this marketing strategy (48). Similar to shared kitchens, food hubs are public areas that offer "a combination of production, aggregation, distribution, and marketing services" (49) to entrepreneurs in the food industry who would otherwise have no location to start their business. The 2013 National Food Hub Survey found that "more than half of the food hubs indicated that they offer product storage," while "less than 20% of hubs offer canning, cutting, or freezing services" (50). These food hubs are pretty evenly spread out across the United States, with the majority in the Middle Atlantic and Northeast regions and the least in the south-central region; in these regions, "75% of food hubs were located in metro counties" (50). Overall, each of these food entities located across the United States is a public food area with little to no food safety precautions being taught or implemented.

GAPs and GAP Audits

The use of good agricultural practices (GAPs) is a method for assessing and managing some of the risks to the health and safety of raw agricultural commodities that result from the production, harvesting, transportation, storage, and handling of those commodities on the farm. The GAP guidelines developed by Cornell University and published in the U.S. Food and Drug Administration's Guide to Minimize Microbial Food Safety Hazards for Fresh Fruits and Vegetables are widely accepted in agriculture and are primarily concerned with pathogen contamination, along with physical hazards such as broken glass and, to some extent, with contamination by chemical poisons. Following GAPs does not ensure that food is safe.

The USDA GAP audit program is a widely accepted system for third-party assessment of a farm's practices for managing some health and safety risks in the production, harvest, transportation, storage, and handling of raw agricultural commodities. The USDA audit is based on the Cornell/FDA GAPs and is composed of multiple modules, including general questions, farm review, field harvest, storage and transportation, and packing house operation. The USDA audit applies point values to each of the elements of the Cornell GAP rubric. To receive certification under any module, the audited entity must earn 80% of the points available in that module. GAP audit certification does not ensure that food is safe.

The storage and transportation and house packing modules are applicable in nonfarm settings as well as farms and are more accurately referred to as good handling practices (GHP). Buyers vary in their audit certification requirements of sellers, with some requiring only elements of the farm modules, others requiring only GHP certification, and others requiring both GAP and GHP.

GAPs versus the National Organic Program

Many of the farms participating in this research program were already third-party certified under the USDA's National Organic Program (NOP), and others followed NOP practices but were not certified organic. For produce to be sold as "organic," it must be certified as being grown in accordance with practices and standards developed by the Agricultural Marketing Service as part of NOP. These regulations detail the practices that are accepted and prohibited in the growing, cleaning, packaging, and marketing of organically labeled products. The regulations include requirements for maintaining and improving soil health and fertility and specifically prohibit the use of commercial fertilizers, soil and product fumigants, and chemical pest control practices in organic production. In addition, sewage sludge may not be used in the production of crops, and crops may not be processed using ionizing radiation. The regulations include the National List of Allowed and Prohibited Substances to guide growers and certifying entities in the use of the label.

One concern identified by operators of organic farms regarding GAP certification is that once approved, a farm's organic certification applies to all produce, grain, and oilseed crops produced on that farm, whereas a GAP audit only certifies specific crops. Organic farmers expressed concern that this crop-by-crop approach would require multiple audits and duplicative efforts compared to organic certification. In our research, this concern proved to be less significant than expected.

A GAP inspector can certify multiple crops at once and will certify types of similar crops as opposed to individual cultivars; GAP certification is based on the way crops are handled in harvesting and packing. For instance, if requested, an auditor will certify all leafy greens from observing one particular leafy green on your farm—you do not have to get kale, chard, collards, leaf lettuces, etc., certified individually.

Moreover, under certain circumstances, an inspector may approve GAP certification for all crops grown on the farm, if requested. Table 2 lists some of the major similarities and differences between the NOP and GAPs.

Investigating Audit Barriers and Compliance

In 2011, information on GAP costs and barriers was collected from 12 small, diversified North Carolina farms, defined as operations with fewer than 30 acres under cultivation and growing at least eight commodities, as they went through the steps that could lead to USDA GAP certification. These steps include addressing risks and collecting documentation on food safety practices. Many of the participating farms were motivated to open their farms up for this investigation in hopes of creating and communicating approaches to food safety that are practical for small farms.

Two to four visits were made to each of the 12 farms over the period from May through November 2011, to observe on-farm activities

TABLE 2 Comparing NOP to USDA GAPs

	NOP	USDA GAPs
Type of audit	Audit farm inputs/outputs from documentation regarding organic growing methods	Audit farm practices from documentation and observation regarding risks of contamination
Cost	Baseline price plus price per acre	\$92/hour (includes conducting the audit, travel time, and preparatory time), \$50 website administrative fee
Duration of certification	Annual	Annual
Raw manure application	Incorporated into the soil Not less than 120 days prior to the harvest of a product whose edible portion has direct contact with soil Not less than 90 days prior to the harvest of a product whose edible portion does not have direct contact with the soil surface or soil particle	Incorporated into the soil At least 2 weeks prior to planting or a minimum of 120 days prior to harvest. Not used on commodities harvested within 120 days of planting (i.e., vegetables, especially crops like leafy greens)
Similar documentation	Field map, land history, seed stock documentation, manure application, compost monitoring, cleaning records of equipment and transportation, traceability	

and identify risks associated with on-farm processes. During an initial visit, farmers described their production activities and processes and some of the potential barriers to GAP certification and safety risk reduction. Follow-up visits tested possible solutions and tracked the implementation of risk reduction strategies.

Activities included reviewing document, surveying farmers on the projected capital and labor costs necessary to meet certification requirements, tracking the time required to document practices as required by the USDA GAP audit protocols, tracking actual capital investments such as sorting tables or product washing facilities, tracking water testing costs, and tracking costs related to changes in fertility management made in response to GAP protocols. Farmers also kept journals (video, audio, and text) documenting their food safety risk reduction activities and specifically discussing any problems they encountered, including tips on solutions for audit-related issues.

Barriers to Produce Food Safety Revealed from Data Collection

Audit expenses and crop diversity

One of the main concerns identified as a potential barrier for small farms to attain GAP certification is a perceived requirement that every individual crop that a farm might wish to sell to a buyer must go through a separate GAP audit. In other words, the common understanding is that every individual crop on a farm must have its own GAP audit, creating an untenable burden of inspection expenses, regardless of time, paperwork, and capital costs. Discussions with grading services staff have shown that this understanding is not necessarily correct. Grading services inspectors conducting an on-farm GAP audit will include in their certification all crops that they can observe being harvested at the time of the audit. Additionally, grading services advise that small farms near one another may cooperate to request inspections on the same date, so that inspectors' travel costs can be spread over multiple farms, reducing the inspection cost for each.

Variation in buyer needs

It is critical for farms to understand the specific GAP or GHP modules for which their buyers seek certification. Some buyers may be satisfied by a GHP audit of a packing or warehousing facility, whether the farm's own or a shared/copacking facility. Defining the scope of the certification required creates an opportunity for the producer to educate the buyer on the appropriate level of food safety assessment for a given farm and buyer relationship.

Documentation/time

This has proven to be a significant concern for small farms where the owner/operators are the primary farm workers/record keepers. Farms can do everything right under GAP standards in terms of risk reduction activities, but they won't pass the audit if they do not write it down every time they do it. You have to record that you conducted training with your workers on hygiene, that you scouted the field for wildlife intrusion prior to harvest, that you washed harvest containers and tools, that you took rodent-control steps in your storage facility, that your icemaker was sanitized, that your harvest crates were sanitized, that you made sure your refrigerator thermometer was working properly, and so on. There is also a challenge for small farmers in accessing sample standard operating procedures and other documentation forms that can be adapted to their operations, and for organic-certified operations it can be a challenge to determine what portions of their organic certification documentation may also be used for GAP documentation.

Use of compost and manure

Many small farms follow organic standards regarding soil fertility, including the use of compost and manure. GAP audit standards for the use of manure are more stringent than the federal organic standards when it comes to crops where the edible portion of the crop does not come into contact with the soil, potentially compromising the ability of these producers to achieve desired crop yields. There is also very little guidance available on what is sufficient "passive or active treatment" for aged manure to make it microbially safe; obviously, manure—raw, composted, and aged—is widely used as a soil amendment, and safely, in many cases.

Water source, treatment, and testing

Producers using surface waters for irrigation (at least half of study farms do so) face considerable water testing expenses because they may be required to test repeatedly during the production season. Ensuring a supply of potable water for washing produce is also expensive to provide and to document. Mastering water sampling procedures is time-consuming, and guidelines for where and how to test are inconsistent. Not washing harvested product is a step many small farms might wish to consider.

Animals

Management of domestic animals to exclude them from fields in active produce production is a common issue among the participating farms. There are also issues of preventing cross-contamination when moving between livestock and produce fields. The practice of rotating livestock through fields that are out of production may trigger burdensome soil testing requirements. There is little proven guidance on what the correct response should be when wildlife feces are found during a preharvest field evaluation.

Equipment and facilities

Small farms may need to improve their cleaning procedures for harvest tools. If CSA distribution is a major part of the farm operation, sanitizing reused subscriber boxes is difficult or impossible, and providing new boxes every week is a significant cost.

Traceability

Conducting an effective mock recall is difficult for farmers currently depending on direct marketing for a major portion of their sales; a mock recall exercise is supposed to be completed within 2 hours and account for 100% of the potentially contaminated product. Documentation of where the product is distributed and what part of the farm it came from may be difficult to maintain, although organic-certified producers already are required to keep this information.

CITATION

Chapman B, Gunter C. 2018. Local food systems food safety concerns. Microbiol Spectrum 6(2):PFS-0020-2017.

REFERENCES

1. **Tauxe R, Kruse H, Hedberg C, Potter M, Madden J, Wachsmuth K.** 1997. Microbial hazards and emerging issues associated with produce, a preliminary report to the National Advisory Committee on Microbiologic Criteria for Foods. *J Food Prot* **60:**1400–1408.
2. **Center for Science in the Public Interest.** 2015. A review of foodborne Illness in the U.S. from 2004–2013. https://cspinet.org/sites/default/files/attachment/outbreak-alert-2015.pdf
3. **World Health Organization.** 1998. Surface decontamination of fruits and vegetables eaten raw: a review. WHO/FSF/FOS/Publication 98.2. World Health Organization, Geneva, Switzerland.
4. **Beuchat LR.** 2002. Ecological factors influencing survival and growth of human pathogens on raw fruits and vegetables. *Microbes Infect* **4:**413–423.
5. **Bracket RE.** 1999. Incidence, contributing factors, and control of bacterial pathogens in produce. *Postharvest Biol Technol* **15:**305–311.
6. **Konowalchuk J, Speirs JI.** 1975. Survival of enteric viruses on fresh fruit. *J Milk Food Technol* **38:**598–600.
7. **Reina LD, Fleming HP, Breidt F Jr.** 2002. Bacterial contamination of cucumber fruit through adhesion. *J Food Prot* **65:**1881–1887.
8. **Sewell AM, Farber JM.** 2001. Foodborne outbreaks in Canada linked to produce. *J Food Prot* **64:**1863–1877. https://www.ncbi.nlm.nih.gov/pubmed/11726177.

9. **Hedberg CW, Angulo FJ, White KE, Langkop CW, Schell WL, Stobierski MG, Schuchat A, Besser JM, Dietrich S, Helsel L, Griffin PM, McFarland JW, Osterholm MT, The Investigation Team.** 1999. Outbreaks of salmonellosis associated with eating uncooked tomatoes: implications for public health. *Epidemiol Infect* **122:** 385–393.
10. **Painter JA, Hoekstra RM, Ayers T, Tauxe RV, Braden CR, Angulo FJ, Griffin PM.** 2013. Attribution of foodborne illnesses, hospitalizations, and deaths to food commodities by using outbreak data, United States, 1998–2008. *Emerg Infect Dis* **19:**407–415.
11. **Beuchat LR.** 1996. Pathogenic microorganisms associated with fresh produce. *J Food Prot* **59:** 204–216.
12. **Suslow T.** 1997. Postharvest chlorination: Basic properties and key points for effective disinfection. University of California. Publication 8003.
13. **Rafferty SM, Williams S, Falkiner FR, Cassell AC.** 2000. Persistence in *in vitro* cultures of cabbage (*Brassica oleracea* var capitata l.) of human food poisoning pathogens: *Escherichia coli* and *Serratia marcescens*. ISHS Acta Horticulturae 530: International Symposium on Methods and Markers for Quality Assurance in Micropropagation. Cork, Ireland.
14. **Allwood PB, Malik YS, Maherchandani S, Vought K, Johnson LA, Braymen C, Hedberg CW, Goyal SM.** 2004. Occurrence of *Escherichia coli*, noroviruses, and F-specific coliphages in fresh market-ready produce. *J Food Prot* **67:**2387–2390.
15. **Robertson LJ, Gjerde B.** 2001. Occurrence of parasites on fruits and vegetables in Norway. *J Food Prot* **64:**1793–1798.
16. **Beuchat LR, Ryu JH.** 1997. Produce handling and processing practices. *Emerg Infect Dis* **3:**459–465.
17. **Guo X, Chen J, Brackett RE, Beuchat LR.** 2002. Survival of *Salmonella* on tomatoes stored at high relative humidity, in soil, and on tomatoes in contact with soil. *J Food Prot* **65:**274–279.
18. **Kudva IT, Blanch K, Hovde CJ.** 1998. Analysis of *Escherichia coli* O157:H7 survival in ovine or bovine manure and manure slurry. *Appl Environ Microbiol* **64:**3166–3174.
19. **Maule A.** 2000. Survival of verocytotoxigenic *Escherichia coli* O157 in soil, water and on surfaces. *Symp Ser Soc Appl Microbiol* **29:**71S–78S.
20. **NASPHV, Centers for Disease Control and Prevention, Council of State and Territorial Epidemiologists, American Veterinary Medical Association.** 2009. Compendium of measures to prevent disease associated with animals in public settings, 2009: National Association of State Public Health Veterinarians, Inc. (NASPHV). *MMWR Recomm Rep* **58** (RR-5):1–21.
21. **Mukherjee A, Speh D, Dyck E, Diez-Gonzalez F.** 2004. Preharvest evaluation of coliforms, *Escherichia coli, Salmonella,* and *Escherichia coli* O157:H7 in organic and conventional produce grown by Minnesota farmers. *J Food Prot* **67:**894–900.
22. **Solomon EB, Yaron S, Matthews KR.** 2002. Transmission of *Escherichia coli* O157:H7 from contaminated manure and irrigation water to lettuce plant tissue and its subsequent internalization. *Appl Environ Microbiol* **68:**397–400.
23. **Solomon EB, Potenski CJ, Matthews KR.** 2002. Effect of irrigation method on transmission to and persistence of *Escherichia coli* O157:H7 on lettuce. *J Food Prot* **65:**673–676.
24. **Thurston-Enriquez JA, Watt P, Dowd SE, Enriquez R, Pepper IL, Gerba CP.** 2002. Detection of protozoan parasites and microsporidia in irrigation waters used for crop production. *J Food Prot* **65:**378–382.
25. **Sanderson PG, Spotts RA.** 1995. Postharvest decay of winter pear and apple fruit caused by species of *Penicillium*. *Phytopathology* **85:**103–110.
26. **Simon A, Gonzolez-Fandos E, Tobar V.** 2004. Influence of washing and packaging on the sensory and microbiological quality of fresh peeled white asparagus. *J Food Sci* **69:**FMS6–FMS12.
27. **McWatters KH, Doyle MP, Walker SL, Rimal AP, Venkitanarayanan K.** 2002. Consumer acceptance of raw apples treated with an antibacterial solution designed for home use. *J Food Prot* **65:**106–110.
28. **Gonzalez RJ, Luo Y, Ruiz-Cruz S, McEvoy JL.** 2004. Efficacy of sanitizers to inactivate *Escherichia coli* O157:H7 on fresh-cut carrot shreds under simulated process water conditions. *J Food Prot* **67:**2375–2380.
29. **Koseki S, Isobe S, Itoh K.** 2004. Efficacy of acidic electrolyzed water ice for pathogen control on lettuce. *J Food Prot* **67:**2544–2549.
30. **Lin CM, Moon SS, Doyle MP, McWatters KH.** 2002. Inactivation of *Escherichia coli* O157:H7, *Salmonella enterica* serotype enteritidis, and *Listeria monocytogenes* on lettuce by hydrogen peroxide and lactic acid and by hydrogen peroxide with mild heat. *J Food Prot* **65:**1215–1220.
31. **Venkitanarayanan KS, Lin CM, Bailey H, Doyle MP.** 2002. Inactivation of *Escherichia coli* O157:H7, *Salmonella enteritidis,* and

Listeria monocytogenes on apples, oranges, and tomatoes by lactic acid with hydrogen peroxide. *J Food Prot* **65:**100–105.
32. **Han Y, Sherman DM, Hinton RH, Nielson SS, Nelson PE.** 2000. The effects of washing and chlorine dioxide gas on survival and attachment of *Escherichia coli* O157:H7 to green pepper surfaces. *Food Microbiol* **17:**521–533.
33. **Bartz JA.** 1982. Infiltration of tomatoes immersed at different temperatures to different depths in suspensions of *Erwinia carotovora* subsp. carotovora. *Plant Dis* **66:**302–305.
34. **Bartz JA, Showalter RK.** 1981. Infiltration of tomatoes by aqueous bacterial suspensions. *Phytopathology* **71:**515–518.
35. **Burnett SL, Chen J, Beuchat LR.** 2000. Attachment of *Escherichia coli* O157:H7 to the surfaces and internal structures of apples as detected by confocal scanning laser microscopy. *Appl Environ Microbiol* **66:**4679–4687.
36. **Seo KH, Frank JF.** 1999. Attachment of *Escherichia coli* O157:H7 to lettuce leaf surface and bacterial viability in response to chlorine treatment as demonstrated by using confocal scanning laser microscopy. *J Food Prot* **62:**3–9.
37. **Zhuang RY, Beuchat LR, Angulo FJ.** 1995. Fate of *Salmonella montevideo* on and in raw tomatoes as affected by temperature and treatment with chlorine. *Appl Environ Microbiol* **61:** 2127–2131.
38. **Warriner K, Ibrahim F, Dickinson M, Wright C, Waites WM.** 2003. Internalization of human pathogens within growing salad vegetables. *Biotechnol Genet Eng Rev* **20:**117–134.
39. **Warriner K, Ibrahim F, Dickinson M, Wright C, Waites WM.** 2003. Interaction of *Escherichia coli* with growing salad spinach plants. *J Food Prot* **66:**1790–1797.
40. **Guo X, van Iersel MW, Chen J, Brackett RE, Beuchat LR.** 2002. Evidence of association of salmonellae with tomato plants grown hydroponically in inoculated nutrient solution. *Appl Environ Microbiol* **68:**3639–3643.
41. **Anonymous.** 2016. *Shared Kitchens.* http://www.sharedkitchens.com/index.html.
42. **Econsult Solutions.** 2013. U.S. kitchen incubators: an industry snapshot. http://www.econsultsolutions.com/wp-content/uploads/2013/08/ESI-SharedKitchenReport_2013.pdf
43. **Chapman B.** 2012. Tempeh outbreak has lots of nasty consequences: illnesses are the biggest. http://www.barfblog.com/2012/05/tempeh-outbreak-has-lots-of-nasty-consequences-illnesses-are-the-biggest/
44. **Terry L.** 2011. E. coli *in Oregon strawberries: 1 death, as many as 16 illnesses traced to Washington County farm.* http://www.oregonlive.com/washingtoncounty/index.ssf/2011/08/1_death_16_illnesses_traced_to.html
45. **News Desk.** 2012. CA recalls more farmers market soups for botulism potential. *Food Safety News* 12 June. www.foodsafetynews.com/2012/06/ca-recalls-more-farmers-market-soups-for-botulism-potential
46. **Siegner C.** 2015. Update: 9 confirmed, 1 probable case in Seattle *E. coli* outbreak. *Food Safety News* 4 September. http://www.foodsafetynews.com/2015/09/6-e-coli-cases-linked-to-mexican-food-sold-at-washington-farmers-markets/
47. **Brown C, Miller S.** 2008. The impacts of local markets: a review of research on farmers markets and community supported agriculture (CSA). *Am J Agric Econ* **90:**1296–1302.
48. **Community Supported Agriculture.** 2016. *Surveys and statistics.* United States Department of Agriculture. Available from http://afsic.nal.usda.gov/community-supported-agriculture-3. Accessed 23 May 2016.
49. **Barham J, Tropp D, Enterline K, Farbman J, Fisk J, Kiraly S.** 2012. Regional food hub resource guide. United States Department of Agriculture. https://www.ams.usda.gov/sites/default/files/media/Regional%20Food%20Hub%20Resource%20Guide.pdf. Accessed 20 December 2017.
50. **Fischer M, Hamm M, Pirog R, Fisk J, Farbman J, Kirlay S.** 2013. Findings of the 2013 National Food Hub Survey. Michigan State University. http://kresge.org/sites/default/files/2013-national-food-hub-survey.pdf. Acessed 23 May 2016.

Preharvest Food Safety Under the Influence of a Changing Climate

14

KALMIA E. KNIEL[1] and PATRICK SPANNINGER[1]

"It seems clear to me also that climate change is a problem which can no longer be left to our future generation."

Pope Francis during his visit to
Washington, DC, 23 September 2015

GLOBAL DEVELOPMENTS

Providing food for the 9.5 billion people of the world by 2050 is one of the world's greatest challenges. The change in population from 2010 to 2015 was +1.18%, with the continuing persistence of nearly 800 million chronically undernourished people (1). To meet these challenges, scientists are addressing various aspects of agriculture, including how environmental factors affect this task. Perhaps no two words have brought about more angst, frustration, and passion than "climate change," including being a focus of political satire. It is difficult to show persuasive change over billions of years given that our technological advances have occurred most recently over the past half-century or so; however, we see change even in this relatively short period. Like many areas of science, facts associated with climate change may be formatted to

[1]Department of Animal and Food Sciences, College of Agriculture and Natural Resources, University of Delaware, Newark, DE 19716.

Preharvest Food Safety
Edited by Siddhartha Thakur and Kalmia E. Kniel

doi:10.1128/microbiolspec.PFS-0015-2016

persuade the intended audience. The scientific basis of climate change is complex historical data involving analysis of a myriad of sciences including physics, chemistry, and biology. Undoubtedly, there is great interest in climate change, with an increase in publications in the area of climate change relative to the life sciences and disease by 42% and 21%, respectively, from 2012 to 2013; in 2013 alone, 21 papers were cited related to climate change and food safety (MEDLINE trend performed 16 October 2016). The controversial nature of climate change may be due in part to the fact that people experience climate change so differently across the globe (Fig. 1).

While agriculture is undoubtedly a significant contributor to climate change, advanced adaptations and changes are greatly reducing carbon emissions and increasing agricultural productivity compared to that from 50 to 100 years ago (2). Though our attempts continue to reduce the impact of agriculture on climate change, food production continues to be impacted by changes in climate and differing production systems. Food safety and security are impacted as well. Hunger and food insecurity is not only a problem in the developing world, but is a critical problem within the United States and other developed countries where significant populations are affected by insufficient food resources and undernutrition (3). Changes in climate and extreme weather events can interrupt food production, harvest, and distribution without advance warning.

The Intergovernmental Panel on Climate Change (IPCC) reported in 2014 that due to projected climate change, by the mid-21st century and beyond, productivity of fisheries will be effected by altered marine biodiversity (4). Similarly, temperature increases will affect crop growth (wheat, rice, and maize) in tropical and temperate regions. Climate change is projected to reduce renewable surface water and groundwater resources in most dry subtropical regions, intensifying competition for water, while increased precipitation may negatively affect crop production in other areas, reinforcing the fact that the irregular effects of climate change are a significant impact. The need for more sustainable agricultural practices magnifies the perplexing relationship of climate change to agriculture. According to the IPCC (4), changes in many extreme weather and climate events have been observed since about 1950; some of these changes have been linked to human influences, including a decrease in cold temperature extremes, an increase in warm temperature extremes, an increase in extreme high sea levels, and an increase in the number of heavy precipitation events in a number of regions (5). The scientists in this panel used the term "unequivocal," which carries a good deal of weight in this situation. While scientists may agree on the facts of climate change, it is with the use of the data and political interests where the majority of people start to disagree (6).

For this article, climate change will be framed by its correlation to preharvest food safety. The IPCC states that human influence on our climate system is clear and that recent anthropogenic emissions of greenhouse gases are the highest in recorded history. Factors of climate change that directly impact food safety and security include trends in stronger storms, increased frequency of events with heavy precipitation, and extended dry periods (7). It is widely conceived that most actions causing climate change originated in the developed world, yet the public health burden will be largely felt within the developing world (8). Among the published studies there is a deficit of studies in developing regions, and quantitative studies needed for making sound policies are rare; therefore, climate change research remains a global priority (9). Food safety risks may occur at many places in the preharvest arena for all commodities. There is a need to update policy related to climate change across the board at local and global levels. Interestingly, Gelband and Laxminarayan (10) suggest that antimicrobial resistance is a global issue like climate change; however, regarding climate change

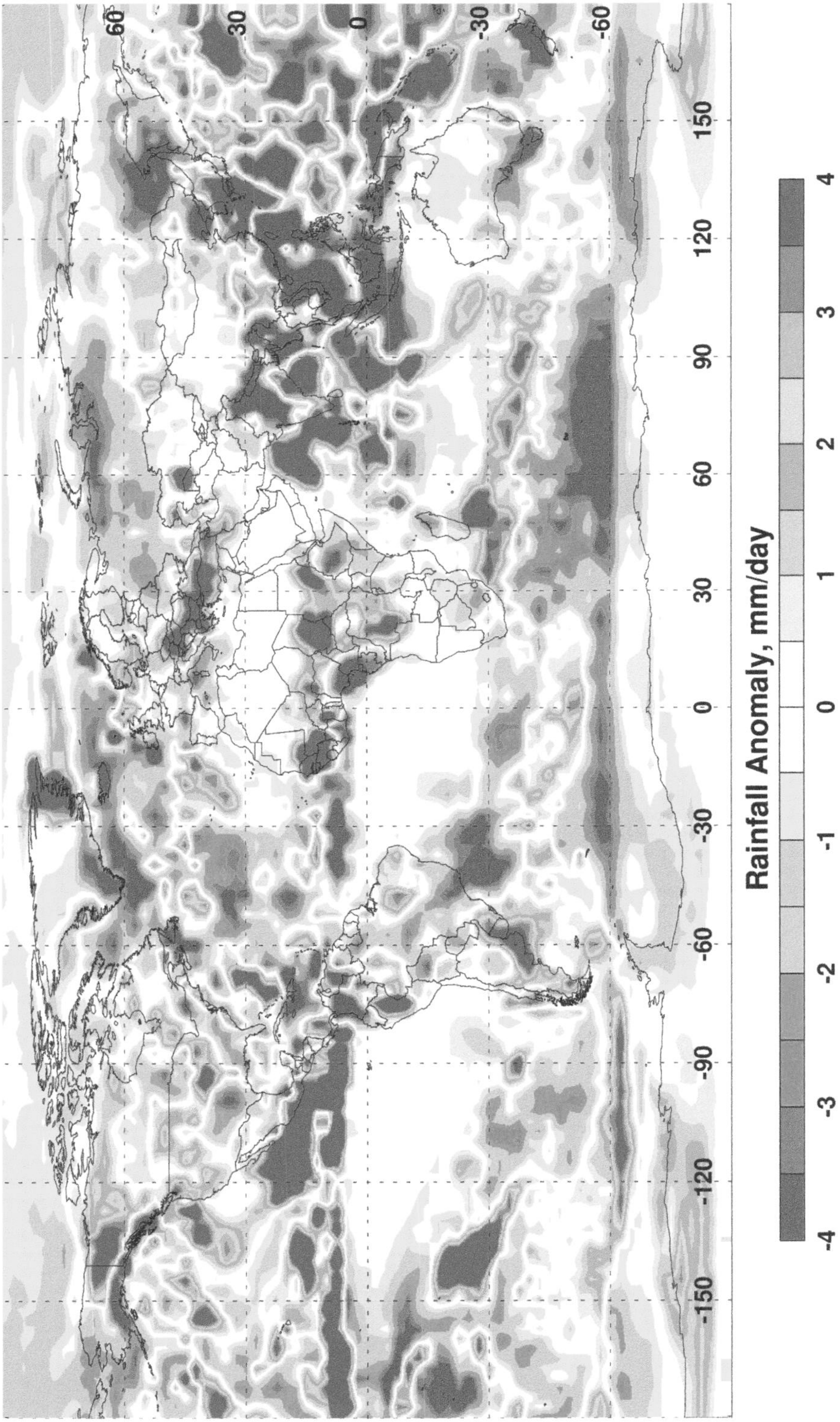

FIGURE 1 Global seasonal rainfall anomalies (August to October 2014). Shifts in climate change affect the globe differently. Reused from reference 5, with permission.

there is a lack of direct localized benefits that can improve the global situation. To date, climate change concerns in North America have not been as great as those across the globe, but this is expected to change (2). Across the United States, farmers are increasingly dealing with challenges to food safety and security, such as hurricanes, extreme rains, and likewise, extreme droughts. Perhaps this influences why climate change ignites so much controversy, because people experience it so differently across the globe and even across the United States. Here we will review several illustrated cases of climate change related to preharvest food safety risks including waterborne disease and seafood production, crop production, and growth of mycotoxins.

ILLUSTRATIONS

Waterborne Disease

Two classic case studies of waterborne disease include the outbreak of gastroenteritis caused by *Escherichia coli* O157:H7 in Walkerton, Ontario, in 2000 and the 2003 outbreak of cryptosporidiosis in Milwaukee, Wisconsin, caused by the protozoan parasite *Cryptosporidium parvum* (11, 12). Both outbreaks resulted in thousands of ill people, and both involved fecal-contaminated, overwhelmed water systems. At the conclusion of each outbreak, resolutions and mitigation efforts were identified to address civic water needs in wells and water treatment plants. From a biological perspective, both outbreaks marked a turning point for more basic studies and investigation into the survival and transmission of pathogens. Likewise, prevention and control measures developed following these outbreaks addressed better testing procedures and assistance for water-treatment facilities. Both of these outbreaks were unexpected and showcase the role and magnitude of municipal water associated with disease transmission. Climate-induced changes to municipal water systems are poorly understood, especially in regard to effects on the ecosystem and public health.

The National Institute of Environmental Health Sciences stated eloquently in their report titled "A Human Health Perspective on Climate Change" that complex human interactions with water intersect several critical issues relative to climate change, including sea-level rise, precipitation, flooding, drought, and intense storms (13). A myriad of impacts on human and public health may result from these interactions. As the report states, sound science must be the basis for future policies. While this report was written in 2010, much of it holds today, as does the consistent need for informed policy. Specific areas of research related to waterborne disease include the study of changes in water flow, sewage, surface and groundwater supplies, along with drinking water distribution systems and irrigation water sources. A better understanding of these systems and their interactions (1) in relation to climate change can predict and reduce the risk of human exposure to pathogens and biotoxins.

Outcomes of global climate change can vary from excessive rainfall leaving flooding or storm surges to extreme drought (14). The frequency and strength of precipitation events has increased since the 1970s (15). Heavy rainfall can increase run-off from fields, increasing the levels of organics in rivers and coastal waters. Movement of microorganisms in the field can impact food safety risks associated with field-grown crops. Extreme precipitation can lead to overwhelmed water-filtration plants and increase risk of human exposure to waterborne pathogens through direct exposure to water or through consumption of contaminated seafood. In a systemic review of scientific literature published up to 12 May 2010, more than 50% of the outbreaks of human illness following extreme water-related weather events reported heavy rainfall and flooding as the most common combination of events leading to contamination and illness (14). An important area of research related to climate change is the way microbial pathogens adapt to changing environmental conditions. In the reported outbreaks in the study mentioned above, changes in pathogen survival due to environmental conditions occurred in nearly 10% of the outbreak reports. Waterborne pathogens associated with environmental exposure from extreme weather events included *Leptospira* spp., *Cryptosporidium* spp., norovirus, and *Vibrio vulnificus*. In this same review, *Campylobacter* spp. and *Cryptosporidium* spp. were associated with extreme weather-related events that affected water treatment facilities. This systemic review provides important consideration for the future, given that global climate change implies an increased frequency and intensity of water-related weather events, and this review suggests that extreme water-related weather events result in increased risk of illness. Human adaptation of policy and regulatory change may vary but can increase chances of quick recovery and change depending on available information and technology.

Limited water supply is a separate concern related to climate change. Diminishing water sources have led to the exploration of wastewater reuse as a means to cope with conventional water resources (1). While wastewater reuse may have benefits as an additional means of fertilization (16), wastewater has the potential to introduce pathogens into the environment as well. Policy, regulatory developments, and risk management must balance the benefits of the use of wastewater with the potential for contamination. This is not simple given that all the risks have yet to be identified.

Seafood

Warmer air temperatures, warmer seawater temperatures, increased ocean acidity, increased sea water level, and changes in the intensity of precipitation and wind patterns are aspects of climate change that affect seafood production and harvest (15). The world's oceans are expected to take the brunt of climate change effects, including sea-level rise by 0.18 to 0.59 m by the end of the 21st century due to thermal warming and melting glaciers and ice sheets (17). These changes will alter seafood production, on which about 520 million people, 8% of the world's population, depend as a protein source. Changes in water temperature may exacerbate water pollution, including increases in sediment, nutrients, dissolved organic carbon, pathogens, pesticides, toxic metals, and salts (18). Of course, this is not a surface issue, but rather one that may be exacerbated by animal physiology, water salinity, and the concentration of metals and organic chemicals. A vivid example of these external factors is the harmful algal blooms that have already damaged some environments and threatened human and aquatic health (19).

Bacteria in the marine environment that have the capacity to be pathogenic may be transmitted to people through ingestion of contaminated shellfish or through direct contact with wounds in the skin. An epidemiological study conducted in 2010 of outbreaks associated with *Vibrio parahaemolyticus* identified a great change in the recent global spread of this bacterium associated with seafood. *V. parahaemolyticus* is a bacterium naturally found in marine environments of warm and temperate regions and is a leading cause of gastroenteritis associated with seafood. Illness from *V. parahaemolyticus* usually results from ingestion of raw or improperly cooked or recontaminated seafood. Infections in Chile and Alaska have become more frequent, while previously this pathogen was rarely reported (20). Emergence in Chile, Peru, Texas, Spain, and Madagascar were all associated with a pandemic clonal complex, a serotype with specific genetic characteristics (20, 21). While routes of transmission of these strains are still not well understood, ballast water discharges are recognized as a vehicle for worldwide invasions by marine species (22). Nutrient availability and bacterial survival in marine waters change with seawater temperature variability. Martinez-Urtaza et al. (20) state that *V. parahaemolyticus* population growth and transmission are dependent on rising temperatures, suggesting that illnesses caused by this organism may be especially influenced by climate change. This is compounded by the fact that *V. parahaemolyticus* has one of the shortest generation times of any bacterium, at <10 min, with an optimum growth temperature of 37°C (23). Even a slight warming of the environment will affect the growth rates of *V. parahaemolyticus* and reduce the restrictions that cold winters have on their life cycle (24). It is hypothesized that ballast water discharges combined with warmer sea waters may account for the emergence and proliferation of *V. parahaemolyticus* (25). Through routine surveillance of agents of foodborne disease, the Centers for Disease Control and Prevention have observed an increase in laboratory-confirmed *Vibrio* infections in 2014 compared to 2006–2008 (Fig. 2). The exact role of climate change is unclear and complex, as discussed above.

Fruit and Vegetable Production

The connection between food safety and climate conditions has been recognized (26), and scientists suggest that changes in climate may have already affected the occurrence of human pathogens in preharvest production areas. This incidence may occur on varying levels. In a multiregional study across the mid-Atlantic, generic *E. coli* levels on leafy greens and tomatoes spiked following a one-time irrigation event with manure-contaminated

2015 FOOD SAFETY REPORT

Measuring progress toward Healthy People 2020 goals

Pathogen	Healthy People 2020 Target Rate*	2015 Rate†	Change Compared with 2006-2008§	
Campylobacter	8.5	12.97	↑ 9%	☹
E. coli O157¶	0.6	0.95	↓ 30%	☺
Listeria	0.2	0.24	No change	😐
Salmonella	11.4	15.89	No change	😐
Vibrio	0.2	0.39	↑ 34%	☹
Yersinia	0.3	0.29	No change	😐

U.S. Department of Health and Human Services
Centers for Disease Control and Prevention

CS264717-A April 2016

*Per 100,000 population
†Culture-confirmed infections per 100,000 population
§2006-2008 were the baseline years used to establish Healthy People 2020 targets
¶Shiga toxin-producing *Eschericha coli* O157

For more information, visit www.cdc.gov/foodnet

FIGURE 2 Comparative long-term progress in laboratory-confirmed infections of foodborne pathogenic bacteria, as determined by CDC Foodnet (http://www.cdc.gov/foodnet/pdfs/foodnet-mmwr-progress-508-final.pdf).

water that ranged from 10,000 cells/ml to 100 cells/ml (27). Significant increases in *E. coli* levels were observed within a 10-day harvest period following rainfall events greater than 1 to 2 inches. These spikes in *E. coli* levels may be due to regrowth of bacteria to detectable levels given the moisture-rich environment rather than recontamination events. This is an especially important area needing further research in light of the bacterial metrics used in the Produce Rule of the Food Safety Modernization Act.

While air temperature variation may be the most noticeable and measurable indicator of climate change, it is likely that the number of heavy precipitation events are intensifying due to the interconnected nature of hydrological cycles (28). Precipitation is one of the indicators used from climatological data to detail climate change. The Global Precipitation Climatology Project has been collecting data on precipitation since 1979 (5). One reason for the collection of data is to identify El Nino-associated infectious disease activity (29). Heavy precipitation events may contaminate edible portions of crops by splashing fecal material or dissemination by field runoff into irrigation water sources compounding produce food safety dramatically. Similarly, following high-humidity conditions from dew, rain, or irrigation water, human pathogens may have the ability to experience significant population growth (30). According to the IPCC assessment report, since 1950 it is possible that the number of heavy precipitation events has increased in many regions more than it has decreased (4). Further, based on averages from 1981 to 2000, the frequency of hurricanes in the North Atlantic has been above normal 9 out of 11 years, with 2005 being a record-breaking season (28). Land areas such as eastern parts of North and South America, Northern Europe, and Northern and Central Asia have experienced an increase in overall precipitation amounts, with heavy precipitation events likely being a contributing factor. Similarly, heavy precipitation events are likely to increase in California, even though in 2015 the state faced the worst water shortage in recent memory. One possible explanation for the rise in heavy precipitation events is the increase in overall atmospheric moisture content. The atmosphere becomes wetter as both air temperature and land precipitation increase, which also raises soil moisture content.

Researchers have pointed out that salmonellosis cases associated with leafy greens follow seasonal correlations related to temperature and moisture in England, Wales, Poland, The Netherlands, the Czech Republic, and Switzerland (31). The effect of climate on leafy green vegetables is not simple to deduce and may be direct or indirect. After recording these observations, the researchers set out to assess variation in *E. coli* presence and concentration. They determined that temperature and management variables influence *E. coli* presence and concentration together. Management variables in this study included application of inorganic fertilizer and composted manure. The study explored several possible findings. While *E. coli* may be able to utilize the inorganic fertilizer more readily at warmer soil temperatures, the origin of the *E. coli* is unclear. The manure may increase the concentration of *E. coli* in the soil, but the nutrients are less readily available for growth of the bacteria. Other studies determined that manure application and region were determining factors (32, 33). Across these studies the effect of precipitation was difficult to determine depending on occurrence and location. Climate change may influence the warming trends in the soil, allowing increased proliferation of bacteria; however, from a microbiological perspective, *E. coli* is anticipated to have reduced survival with increasing temperature in soil, water, or manure (31). These intertwined relationships (climate variables, management variables, *E. coli* presence, *E. coli* concentration) are difficult to study independently. Meta-analysis of published studies allows opportunities to synthesize information across regions to better evaluate the effects of precipitation and temperature on bacteria persistence and concentration across the globe.

Mycotoxins

Some filamentous fungi produce toxic low-molecular-weight compounds known as mycotoxins. Regulations and end product testing are in place across the globe to reduce exposure to mycotoxins through contaminated grains and ground nuts. Multiple factors affect the presence or production of mycotoxins on crops that may be ingested by humans or animals, and several of these may be influenced by changes in climate. Climate is the most important factor, since mycotoxin production is highly susceptible to temperature and available moisture (34).

As mentioned previously, changes in weather patterns threaten food security, including increased mold production. Extreme weather conditions are predicted over the 21st century, including milder wetter winters, hotter drier summers, and overall more extreme weather events; all of that will alter the development of diseases worldwide (34). As climates warm and weather patterns change, mycotoxin production could increase in restricted areas where profitable crops are grown. Maize is a critical example of a vulnerable crop grown in warm regions throughout Africa, Asia, and the Americas; crop vulnerability to lethal aflatoxicosis in Kenya is increasing (35). However, due to the complex nature of host-pathogen interactions that lead to mycotoxin production, in some cases climate change may decrease the severity of epidemics (36). Excessively high temperatures coupled with reduced moisture could decrease mycotoxin production. Future solutions will require scientific models and forecasting that can accurately direct policy and practice to counter emerging threats. Such a situation exists for cottonseed aflatoxin where contamination was correlated temporally and spatially with rain amounts as a positive correlation over a 5-year period (37).

Assessing the factors of climate change that impact plant disease is a central aspect of successful risk management (38). One example of this is that plants grown in elevated carbon dioxide levels can become more infected with some diseases that can interact with or compete with mycotoxigenic fungi (36). Changing farming practices due to climate conditions may also impact mycotoxigenic fungi through the establishment of new microhabitats (34). Genetic studies have identified mutations as part of these adaptations (39), and such mutations could be shared with surrounding fungi. Climatic differences may influence the types of aflatoxins present on crops; where both regional differences and the abundance of different aflatoxins have been noted (40).

Fungi can be spread by insect vectors. An indirect factor of climate change may be changes in feeding habits and the presence of arthropod vectors which may increase exposure of crops to mycotoxigenic fungi. This was detailed in the warmer night time temperatures of Switzerland in 2003, when during this heat wave higher crop contamination was noted (41). Invasion by nonnative species and parasites has also been correlated with changes in climate. Changes in climate affect host susceptibility. For example, elevated temperatures can result in heat and drought stress in plants that can increase susceptibility to molds. This has been observed in pistachios, which develop hull cracking or early split, and maize kernel integrity can be compromised by increased silk cut, which all lead to a greater threat of aflatoxin production (37). Likewise, peanuts growing under conditions of increased rain during pod maturation can show increased aflatoxin. Both of the above mentioned factors can lead to an increased amount of mycotoxin production. Crops grown in warmer areas consistently have a greater chance of infection by mycotoxin-producing molds, and infection has been noted in some places only when temperatures rise in association with drought (34).

CONCLUSIONS

Environmental changes can impact infectious disease transmitted in both coastal areas and inland areas. Globally, temperature and precipitation anomalies and extreme events have increased cholera, dengue, malaria, and Rift valley fever (5), while *Cryptosporidium* and noncholera *Vibrio* species have increased in regional hot spots (42, 43). Research tools to evaluate these types of trends are more frequently being used. Schijven et al. (44) used a decision support tool to link foodborne infection and/or illness risk associated with climate change. Results suggested that increasing temperatures lead to increasing risk of *Campylobacter* associated with undercooked poultry and *Vibrio* infection from water exposure, and increasing drought leads to increasing risk of norovirus and *Cryptosporidium*. More publications aligning disease risk with climate trends are likely. The Veg-i-Trade project is an example of field studies trying to quantify the impacts of climate change on food safety in the European Union (45). While scientists have tools to assess and link data sets, political policies and funding are critical to draw attention to the climatic effects on agriculture that are less noticeable than the monsoon-like flooding events that may occur. As publications continue to mount linking disease and climate change, scientists will continue to knock on the doors of local, national, and international leaders sharing the need for acknowledgment and better understanding.

CITATION

Kniel KE, Spanninger P. 2017. Preharvest food safety under the influence of a changing climate. Microbiol Spectrum 5(2):PFS-0015-2016.

REFERENCES

1. **Nguyen-The C, Bardin M, Berard A, Berge O, Brillard J, Broussolle V, Carlin F, Renault P, Tchamitchian M, Morris CE.** 2016. Agrifood systems and the microbial safety of fresh produce: trade-offs in the wake of increased sustainability. *Sci Total Environ* **562:**751–759.
2. **Burney JA, Davis SJ, Lobell DB.** 2010. Greenhouse gas mitigation by agricultural intensification. *Proc Natl Acad Sci USA* **107:**12052–12057.
3. **Nord M.** 2009. *Household Food Security in the United States, 2008*. U.S. Deparment of Agriculture, Washington, DC.
4. **IPCC.** 2014. *Climate Change 2014. Synthesis Report*. IPCC, Geneva, Switzerland.
5. **Chretien JP, Anyamba A, Small J, Britch S, Sanchez JL, Halbach AC, Tucker C, Linthicum KJ.** 2015. Global climate anomalies and potential infectious disease risks: 2014–2015. *PLoS Curr* **7:**7.
6. **Lineman M, Do Y, Kim JY, Joo GJ.** 2015. Talking about climate change and global warming. *PLoS One* **10:**e0138996.
7. **FAO/WHO.** 2008. *Climate Change: Implications for Food Safety*. Rome, Italy.
8. **Campbell-Lendrum D, Corvalán C, Neira M.** 2007. Global climate change: implications for international public health policy. *Bull World Health Organ* **85:**235–237.
9. **Hosking J, Campbell-Lendrum D.** 2012. How well does climate change and human health research match the demands of policymakers? A scoping review. *Environ Health Perspect* **120:**1076–1082.
10. **Gelband H, Laxminarayan R.** 2015. Tackling antimicrobial resistance at global and local scales. *Trends Microbiol* **23:**524–526.
11. **Auld H, MacIver D, Klaassen J.** 2004. Heavy rainfall and waterborne disease outbreaks: the Walkerton example. *J Toxicol Environ Health A* **67:**1879–1887.
12. **MacKenzie WR, Hoxie NJ, Proctor ME, Gradus MS, Blair KA, Peterson DE, Kazmierczak JJ, Addiss DG, Fox KR, Rose JB, Davis JP.** 1994. A massive outbreak in Milwaukee of cryptosporidium infection transmitted through the public water supply. *N Engl J Med* **331:**161–167.
13. **Portier CJ, Thigpen Tart K, Carter SR, Dilworth CH, Grambsch AE, Gohlke J, Hess J, Howard SN, Luber G, Lutz JT, Maslak T, Prudent N, Radtke M, Rosenthal JP, Rowles T, Sandifer PA, Scheraga J, Schramm PJ, Strickman D, Trtanj JM, Whung P-Y.** 2010. *A Human Health Perspective On Climate Change: A Report Outlining the Research Needs on the Human Health Effects of Climate Change*. Environmental Health Perspectives/National Institute of Environmental Health Sciences, Research Triangle Park, NC.
14. **Cann KF, Thomas DR, Salmon RL, Wyn-Jones AP, Kay D.** 2013. Extreme water-related weather events and waterborne disease. *Epidemiol Infect* **141:**671–686.

15. **Solomon S, Qin D, Manning Z, Chen M, Marquis M, Averyt KB, Tignor M, Miller HL (ed).** 2007. *The Physical Science Basis. Contribution of Working Group I to the Fourth Assessment Report of the Intergovernmental Panel on Climate Change, 2007.* Cambridge University Press, Cambridge, United Kingdom.
16. **Keraita B, Jimenez B, Breschsel P.** 2008. Extent and implications of agricultural reuse of untreated, partly treated and diluted wastewater in developing countries. *CAB Rev* **3:**1–15.
17. **Marques A, Nunes ML, Moore SK, Strom MS.** 2010. Climate change and seafood safety: human health implications. *Food Res Int* **43:** 1766–1779.
18. **Boorman DB.** 2003. LOIS in-stream water quality modelling. Part 2. Results and scenarios. *Sci Total Environ* **314–316:**397–409.
19. **Erdner DL, Dyble J, Parsons ML, Stevens RC, Hubbard KA, Wrabel ML, Moore SK, Lefebvre KA, Anderson DM, Bienfang P, Bidigare RR, Parker MS, Moeller P, Brand LE, Trainer VL.** 2008. Centers for Oceans and Human Health: a unified approach to the challenge of harmful algal blooms. *Environ Health* **7**(Suppl 2)**:**S2.
20. **Martinez-Urtaza J, Huapaya B, Gavilan RG, Blanco-Abad V, Ansede-Bermejo J, Cadarso-Suarez C, Figueiras A, Trinanes J.** 2008. Emergence of Asiatic *Vibrio* diseases in South America in phase with El Niño. *Epidemiology* **19:**829–837.
21. **González-Escalona N, Cachicas V, Acevedo C, Rioseco ML, Vergara JA, Cabello F, Romero J, Espejo RT.** 2005. *Vibrio parahaemolyticus* diarrhea, Chile, 1998 and 2004. *Emerg Infect Dis* **11:** 129–131.
22. **Niimi AJ.** 2004. Role of container vessels in the introduction of exotic species. *Mar Pollut Bull* **49:**778–782.
23. **Miles DW, Ross T, Olley J, McMeekin TA.** 1997. Development and evaluation of a predictive model for the effect of temperature and water activity on the growth rate of *Vibrio parahaemolyticus*. *Int J Food Microbiol* **38:**133–142.
24. **Harvell CD, Mitchell CE, Ward JR, Altizer S, Dobson AP, Ostfeld RS, Samuel MD.** 2002. Climate warming and disease risks for terrestrial and marine biota. *Science* **296:**2158–2162.
25. **McLaughlin JB, DePaola A, Bopp CA, Martinek KA, Napolilli NP, Allison CG, Murray SL, Thompson EC, Bird MM, Middaugh JP.** 2005. Outbreak of *Vibrio parahaemolyticus* gastroenteritis associated with Alaskan oysters. *N Engl J Med* **353:**1463–1470.
26. **Jacxsens L, Uyttendaele M, Devlieghere F, Rovira J, Gomez SO, Luning PA.** 2010. Food safety performance indicators to benchmark food safety output of food safety management systems. *Int J Food Microbiol* **141**(Suppl 1)**:** S180–S187.
27. **Spanninger P, Shortlidge K, Ferelli AM, Brown R, Markland S, Kniel K.** 2013. The Effects of contaminated irrigation water on bacterial persistence and transmission of coliforms on tomatoes, abstr P3-126, *IAFP Annual Meeting*, Charlotte, NC, July 2013.
28. **IIPoC.** 2007. *Contribution of Working Group II to the Fourth Assessment Report of the Intergovernmental Panel on Climate Change, 2007.* Cambridge University Press, Cambridge, United Kingdom.
29. **Kovats RS, Bouma MJ, Hajat S, Worrall E, Haines A.** 2003. El Niño and health. *Lancet* **362:**1481–1489.
30. **Brandl MT, Mandrell RE.** 2002. Fitness of *Salmonella enterica* serovar Thompson in the cilantro phyllosphere. *Appl Environ Microbiol* **68:**3614–3621.
31. **Liu C, Hofstra N, Franz E.** 2016. Impacts of climate and management variables on the contamination of preharvest leafy greens with *Escherichia coli*. *J Food Prot* **79:**17–29.
32. **Park S, Navratil S, Gregory A, Bauer A, Srinath I, Szonyi B, Nightingale K, Anciso J, Jun M, Han D, Lawhon S, Ivanek R.** 2015. Multifactorial effects of ambient temperature, precipitation, farm management, and environmental factors determine the level of generic *Escherichia coli* contamination on preharvested spinach. *Appl Environ Microbiol* **81:**2635–2650.
33. **Park S, Navratil S, Gregory A, Bauer A, Srinath I, Szonyi B, Nightingale K, Anciso J, Jun M, Han D, Lawhon S, Ivanek R.** 2014. Farm management, environment, and weather factors jointly affect the probability of spinach contamination by generic *Escherichia coli* at the preharvest stage. *Appl Environ Microbiol* **80:**2504–2515.
34. **Paterson RRM, Lima N.** 2010. How will climate change affect mycotoxins in food? *Food Res Int* **43:**1902–1914.
35. **Lewis L, Onsongo M, Njapau H, Schurz-Rogers H, Luber G, Kieszak S, Nyamongo J, Backer L, Dahiye AM, Misore A, DeCock K, Rubin C, Kenya Aflatoxicosis Investigation Group.** 2005. Aflatoxin contamination of commercial maize products during an outbreak of acute aflatoxicosis in eastern and central Kenya. *Environ Health Perspect* **113:**1763–1767.
36. **Chakraborty S, Murray GM, Magarey PA, Yonow T, O'Brien RG, Croft BJ, Barbetti MJ, Sivasithamparam K, Old KM, Dudzinski MJ, Sutherst RW.** 1998. Potential impact of climate change on plant diseases of economic

significance to Australia. *Australas Plant Pathol* **27:**15–35.

37. **Cotty PJ, Jaime-Garcia R.** 2007. Influences of climate on aflatoxin producing fungi and aflatoxin contamination. *Int J Food Microbiol* **119:**109–115.
38. **van der Fels-Klerx HJ, de Rijk TC, Booij CJ, Goedhart PW, Boers EA, Zhao C, Waalwijk C, Mol HG, van der Lee TA.** 2012. Occurrence of *Fusarium* head blight species and *Fusarium* mycotoxins in winter wheat in the Netherlands in 2009. *Food Addit Contam Part A Chem Anal Control Expo Risk Assess* **29:**1716–1726.
39. **Brault AC, Powers AM, Ortiz D, Estrada-Franco JG, Navarro-Lopez R, Weaver SC.** 2004. Venezuelan equine encephalitis emergence: enhanced vector infection from a single amino acid substitution in the envelope glycoprotein. *Proc Natl Acad Sci USA* **101:**11344–11349.
40. **Horn BW, Dorner JW.** 1999. Regional differences in production of aflatoxin B1 and cyclopiazonic acid by soil isolates of *Aspergillus flavus* along a transect within the United States. *Appl Environ Microbiol* **65:**1444–1449.
41. **van Munster JM, Daly P, Delmas S, Pullan ST, Blythe MJ, Malla S, Kokolski M, Noltorp EC, Wennberg K, Fetherston R, Beniston R, Yu X, Dupree P, Archer DB.** 2014. The role of carbon starvation in the induction of enzymes that degrade plant-derived carbohydrates in *Aspergillus niger*. *Fungal Genet Biol* **72:**34–47.
42. **Semenza JC, Höuser C, Herbst S, Rechenburg A, Suk JE, Frechen T, Kistemann T.** 2012. Knowledge mapping for climate change and food- and waterborne diseases. *Crit Rev Environ Sci Technol* **42:**378–411.
43. **Semenza JC, Herbst S, Rechenburg A, Suk JE, Höser C, Schreiber C, Kistemann T.** 2012. Climate change impact assessment of food- and waterborne diseases. *Crit Rev Environ Sci Technol* **42:**857–890.
44. **Schijven J, Bouwknegt M, de Roda Husman AM, Rutjes S, Sudre B, Suk JE, Semenza JC.** 2013. A decision support tool to compare waterborne and foodborne infection and/or illness risks associated with climate change. *Risk Anal* **33:**2154–2167.
45. **Jacxsensa L, Luningb PA, van der Vorstc JGAJ, Devliegherea F, Leemansd R, Uyttendaelea M.** 2010. Simulation modelling and risk assessment as tools to identify the impact of climate change on microbiological food safety: the case study of fresh produce supply chain. *Food Res Int* **43:** 1925–1935.

Potential for Meta-Analysis in the Realm of Preharvest Food Safety

15

JAN M. SARGEANT[1] and ANNETTE M. O'CONNOR[2]

INTRODUCTION

Meta-analysis refers to the statistical combination (pooling) of data from multiple original research studies. The results from different studies on the same topic can vary, and meta-analysis provides a means of summarizing a parameter or effect across studies to develop a more precise estimate of the outcome of interest (1). The combination of data from multiple studies can be undertaken using two broad approaches: combining individual-level data from multiple studies or combining study-level results (effect sizes) from multiple studies (1). The former requires the meta-analyst to have access to all of the original data from each study subject for each study and is therefore not commonly seen in the preharvest food safety literature (2), although databases of microbial growth and inactivation kinetics for foodborne pathogens are available and growing (3). Therefore, this chapter will focus on meta-analysis in the context of combining effect sizes from multiple studies to calculate a summary effect size.

Meta-analyses should be conducted as the statistical component of a systematic review. The systematic review methodology aims to provide a

[1]Center for Public Health and Zoonoses and Department of Population Medicine, Ontario Veterinary College, University of Guelph, Guelph, ON, Canada N1G 2W1; [2]Department of Veterinary Diagnostic and Production Animal Medicine, Iowa State University College of Veterinary Medicine, Ames, IA 50100.

Preharvest Food Safety
Edited by Siddhartha Thakur and Kalmia E. Kniel

doi:10.1128/microbiolspec.PFS-0004-2014

transparent and comprehensive summary of the scientific evidence for a specific clinical or policy question. Nesting a meta-analysis within a systematic review helps to ensure that all available literature is considered in the summary effect size estimate. Further, the systematic review methodology explicitly includes an assessment of the risk of bias in the included studies, thereby reducing the potential for biased data to influence the results of the meta-analysis.

The advantages of meta-analysis include an increased sample size for estimating a parameter or an effect size and therefore greater precision in the estimate; the confidence intervals around a summary estimate can be calculated and displayed; sensitivity analysis can be conducted to explore the contribution of specific studies to the summary estimate; individual study results may be weighted by the precision of the effect size estimate (often partly a function of sample size); and graphical methods are available for displaying the results of the individual studies contributing to the summary estimate (4). Technically, a meta-analysis can be conducted if at least two studies with the same outcome are included in the review, although larger sample sizes increase the utility of a meta-analysis.

A meta-analysis can address a number of types of questions. These fall into four general categories: descriptive questions (e.g., prevalence or incidence), intervention questions, exposure questions, and questions about diagnostic test accuracy (4). All of these question types are of potential relevance to preharvest food safety. The results of meta-analyses could be used to inform clinical or policy decision-making (5) or could be used as inputs to quantitative risk assessment or other models (3, 6, 7).

Examples of published meta-analyses for intervention or exposure questions related to preharvest food safety include evaluations of the effect of direct-fed microbials on fecal shedding of *Escherichia coli* O157 in cattle (8), of vaccination on fecal shedding of *E. coli* O157 in cattle (9, 10), and of interventions to reduce *Salmonella* in broiler chickens (11–13) and in swine (14) and comparisons between the prevalence of enteric pathogens in organic and traditional production systems (15). These meta-analyses provide a summary of the totality of information that is available to address each review question. Therefore, a policy person, veterinarian, producer, or manufacturer can read one meta-analyses to obtain the scientific information needed to make a decision on implementing a preharvest intervention rather than having to read all of the primary literature. Examples of meta-analyses related to a descriptive question in preharvest food safety include an evaluation of factors influencing the prevalence of *Salmonella* on swine farms (16) and a study to estimate the prevalence of *E. coli* O157 in cattle (17). The latter meta-analysis also explored possible explanations for differences in the prevalence of *E. coli* O157, such as region, cattle type, and diagnostic method. For diagnostic test accuracy questions, meta-analysis has been used to evaluate differences in the diagnostic accuracy of culture versus PCR methods for detection of *Salmonella* in swine (18).

META-ANALYSIS IN THE CONTEXT OF SYSTEMATIC REVIEWS

Ideally, meta-analyses are conducted as the statistical component of a systematic review. Not all systematic reviews include a meta-analysis due to an insufficient quantity of data addressing the same question using the same outcome. However, including a systematic review approach increases the transparency and rigor of a meta-analysis.

A number of resources are available that describe the process of systematic review and meta-analysis in the context of preharvest food safety (2, 4, 19–23). The steps of a systematic review include defining the systematic review question, searching for studies, selecting relevant studies based on eligibility criteria, extracting data from relevant studies, assessing the risk of bias in the selected studies, synthesizing the results qualitatively or quantitatively (meta-analysis), and presenting and interpreting the results (4).

A systematic review question is built around key elements, which should be defined in the review question (4). The definition of the key elements, and associated eligibility criteria, set the scope of the systematic review and, subsequently, the meta-analysis. The broad example of *Salmonella* in swine is used for illustration. For a descriptive question, the key elements are the population and the outcome. Thus, a systematic review question might be "What is the prevalence of *Salmonella* (outcome) in market-weight swine (population)?" For an intervention question, the key elements are the population, intervention, comparator, and outcome (PICO). An example would be "What is the effect of vaccination (intervention) on fecal shedding of *Salmonella* (outcome) compared to no vaccination (comparator) in market-weight swine (population)?" The key elements of a diagnostic accuracy question are the population, index test, and target organism. An example would be "What is the sensitivity and specificity of PCR-based methods (index test) to identify *Salmonella* spp. (target organism) from the feces of market-weight swine (population)?" The key elements of an exposure question are the population, exposure, comparator, and outcome (PECO). Exposure questions are common with toxin exposures. An example would be "Is there a dose-response relationship between high dose exposure to toxin X (exposure) and liver toxicity (outcome) in humans (population) compared to low dose exposure to toxin X (comparator)?"

The key elements of the question can then be used to create eligibility criteria for inclusion of original research studies in the review. The type of review question also will determine which study designs are appropriate to address that question, and study design may be used as an eligibility criterion. For instance, a systematic review on the efficacy of a therapeutic intervention may restrict eligibility of original research studies to only randomized controlled trials (RCTs), because this design has the highest evidentiary value for questions about intervention efficacy (21).

In preharvest food safety, it is possible to conduct clinical trials in the species of interest (animal or plant) with an induced disease outcome ("challenge trials"). For instance, if researchers wished to evaluate a vaccine for *E. coli* O157 in cattle, they could randomly allocate cattle to receive or not receive the vaccine and then wait for the animals to be naturally exposed to *E. coli* O157 (natural exposure trial), or they could deliberately expose all of the cattle to *E. coli* O157, perhaps by inoculating the bacteria into the water or food source (challenge trial). The advantage of a deliberate disease exposure is that the researchers can be sure that exposure will take place (in the RCT example, above, it is possible that the experimental animals would not end up being exposed to *E. coli* O157 during the course of the trial). In a challenge trial, all of the animals in the trial are exposed to the pathogen of interest, meaning that a smaller sample size will be needed in a challenge trial compared to an RCT. However, while challenge trials often are an efficient way to evaluate proof-of-concept for an intervention in the species of interest, they are of lower evidentiary value compared to natural disease exposure trials (21). The disease challenge may not represent natural exposure to disease; the challenge may involve a higher dose of the infectious disease agent or exposure by a different route than is typical in a natural exposure. Additionally, with many foodborne agents, some degree of biocontainment may be needed for challenge trials, meaning that, for logistical reasons, the experimental animal population may differ in a meaningful way (i.e., smaller or younger) from the animals for which the intervention would be used in the field.

There is some empirical evidence that the results of challenge trials differ from the results of natural exposure trials for preharvest food safety interventions (24). However, for some interventions, data from challenge trials may be the only form of evidence available; as an example, 63 of 66 studies included in a meta-analysis of competitive exclusion products and *Salmonella* in broiler chickens were challenge trials (11). Thus, researchers conducting a meta-analysis will need to consider how, or whether, to include challenge trials. If there is a sufficient body of literature using natural exposure trials, then it might be appropriate to exclude challenge trials, However, if this is not the case, then the researcher may decide to include both types of trials. If challenge trials are included in a review, the researcher should evaluate whether the results differ between natural exposure trials and challenge trials as a component of the meta-analysis.

The next step of a systematic review is to conduct a comprehensive and documented search of the peer-reviewed and "gray" (non-peer-reviewed) literature to identify all possible publications providing information to address the review question. The validity of the findings of a systematic review is directly related to the comprehensiveness of the search and to the reproducibility of the search protocol (4, 19). Search terms are created based on the key elements of the question and the eligibility criteria. These search terms are entered into electronic databases or used to search other sources. Although no guidelines have been published on methods of searching the literature specifically for preharvest food safety, guidelines are available for searching the medical literature (25–27) and the veterinary literature (28, 29).

The intention is to create a highly sensitive search. Thus, the specificity of the search may be low, and it is necessary to screen the titles and abstracts identified by the search to select the original research studies that are relevant to the review question (30). This is done using a small number of questions designed to rapidly identify nonrelevant articles. For instance, if the purpose of the review is to compare the prevalence of *Campylobacter* spp. between organic and traditional swine farms, the review questions might include the following: (i) Does the title/abstract describe a primary research study? (ii) Is *Campylobacter* in swine a measured outcome? and (iii) Does the study include both traditional and organic farms? If the answer to any of these questions is "no," then the publication would be excluded from further stages of the review. If the answer to any of the questions is "yes" or "unclear," then the full publication would be acquired and the publication would be included in subsequent steps of the review. The titles and abstracts of all citations identified by the search are independently assessed by at least two reviewers using the screening questions, with any disagreements between reviewers resolved by consensus. Abstracts that are not relevant are excluded from the review at this stage.

Once relevant studies have been identified, full articles are obtained for the relevant studies and data are extracted. These data include information on study characteristics such as the population (including animal/plant characteristics and sample sizes in each intervention group), the study setting and details on the intervention, and information on the outcome and results (23). The data are extracted using structured forms, with at least two reviewers independently extracting data from each article. Any disagreements are resolved by discussion or reference to an additional reviewer.

The risk of bias also is assessed for each of the relevant studies included in the review. The important sources of bias, and the methods used to reduce them, vary between study designs and review question type. For intervention questions using RCTs, a common instrument used to assess bias is the Cochrane risk of bias tool (31), which focuses on five domains of bias: selection bias, performance bias, detection bias, attrition bias, and reporting bias. To determine the potential for bias in an original research study, the reviewer considers sources of bias, including random sequence generation, allocation concealment, blinding, incomplete outcome data, selective outcome reporting, and any other potential sources of bias specific to the context of the review question. A risk of bias graph can then be generated to demonstrate the proportion of studies with each judgment or the individual assessments for each study (31). Examples using hypothetical data are shown in Fig. 1 and Fig. 2.

Risk of bias tools developed for human intervention studies may not address all of the important issues in preharvest food safety studies (23). However, the Cochrane risk of bias tool could be modified for use in evaluating preharvest food safety trials. For instance, reviewers may want to include a consideration of whether the exposure to a pathogen of interest was due to natural exposure or whether there was a deliberate pathogen challenge. If the unit of intervention allocation was at the group level (for instance, a pen of animals), the reviewer may want to include a description of whether or not the potential for clustering of data was included in the analysis.

For questions that relate to diagnostic test evaluation, a commonly used risk of bias tool that can be modified for food safety is the QUADAS-2 tool (32) and the Cochrane website devoted to systematic reviews of diagnostic tests (http://methods.cochrane.org/sdt/). Authors of reviews about prevalence will likely need to design a risk of bias tool specific to the review topic. This tool should consider the representativeness of the sample (potential for sampling bias), the reliability of the test used to assess the outcome and the potential for information bias, and if applicable, the loss to follow-up (attrition bias) (33, 34). Risk of bias tools for nonrandomized studies used for questions about etiology are being developed. None are specific to preharvest food safety.

The risk of bias assessment should be conducted by at least two reviewers working independently, with any disagreements resolved by consensus. Information on the risk of bias in individual studies can be used in meta-analysis as a potential explanatory variable in meta-regression or as a factor in subgroup analysis (see sections below). Once the data on study characteristics, study outcomes, and risk of bias are collected, a meta-analysis can be undertaken.

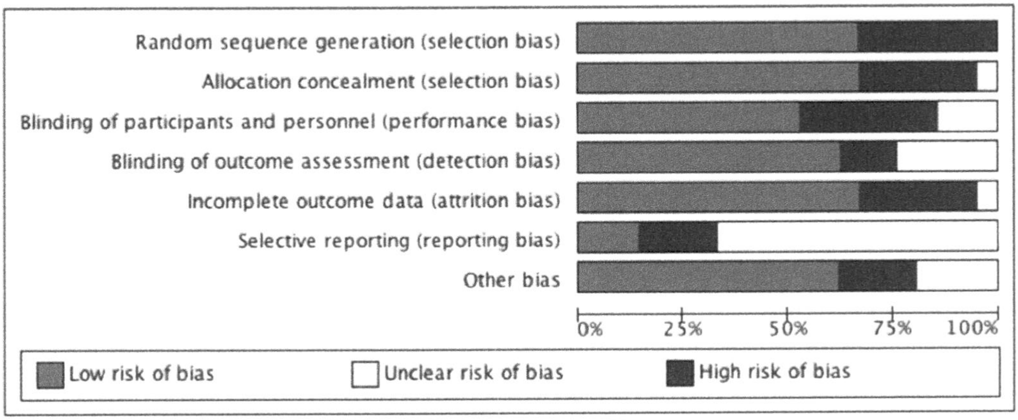

FIGURE 1 Example of a risk of bias graph using hypothetical data (created in Revman version 5.2). Each study included in the review has been evaluated for the risk of bias based on the domains shown in this figure. Each row of the figure summarizes the proportion of studies classified as low risk of bias, high risk of bias, or unclear risk of bias for that domain.

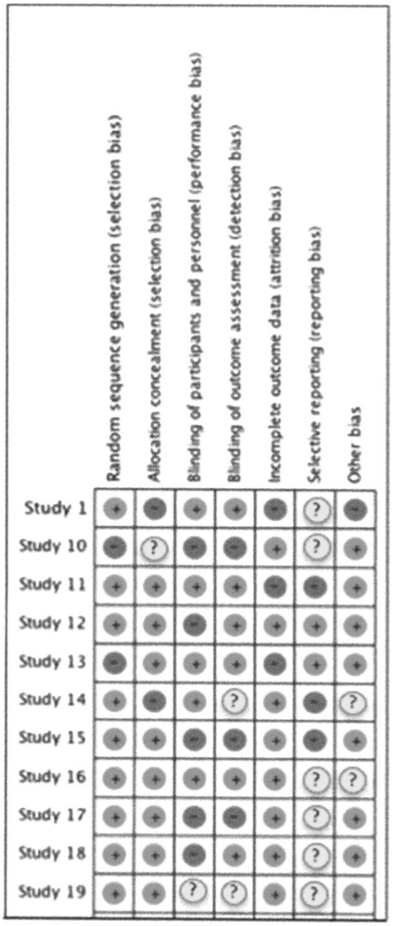

FIGURE 2 Example of a risk of bias summary using hypothetical data (created in Revman version 5.2). The results of the risk of bias assessment for each study for each risk of bias domain are shown, where "+" (green circles) corresponds to a low risk of bias in a specific study for that domain, "-" (red circles) corresponds to a high risk of bias, and "?" (yellow circles) corresponds to an unclear risk of bias.

THE PROCESS OF META-ANALYSIS

Visualizing the Results from Individual Studies

Meta-analysis involves the calculation of a summary effect size as a weighted average of the results from individual studies. Describing the terminology surrounding study results can be confusing because studies can have several layers of outcome. A study subject may have an outcome (e.g., the organism was detected: yes or no), those individual results can be compiled to create an outcome for within an intervention group (e.g., the proportion positive), and the comparison of the intervention among groups could correspond to the outcome (e.g., the ratio of the group proportions). For meta-analysis, the outcome of interest is the comparison level outcome, i.e., the one number that describes the result of the comparison.

The actual outcome that is used will vary depending on the question being addressed by the meta-analysis. For instance, if the intent is to estimate the prevalence of a foodborne pathogen, then the outcome will be the proportion positive (number positive / number at risk). If the intent of the meta-analysis is to estimate the efficacy of an intervention or to evaluate the impact of an exposure, the outcome will be a comparison of two groups, such as a ratio for categorical outcomes (risk ratio, odds ratio) or the difference in a continuous outcome (mean difference or standardized mean difference). For questions about diagnostic tests, the outcome will be sensitivity and specificity. The discussion in this section will focus on intervention questions, although the principles apply to any question type.

Data on the outcome are extracted from each of the individual studies. This may consist of data for each intervention group (e.g., proportion positive in each group) or intervention effects at the comparison level (e.g., relative risk of an event in the treated group versus the control group). A measure of variability also is extracted or calculated. For preharvest studies where there is the potential for nonindependence between intervention groups (e.g., clustering of observations due to animal grouping), the intervention effect and measure of variability corrected for nonindependence should be used.

Intervention effects from multiple studies often are displayed using a forest plot (35)

(Fig. 3). Each row in the forest plot represents an intervention comparison (e.g., intervention group versus control group). The *x* axis represents the summary outcome of interest. In Fig. 3, the outcome of interest is a relative risk (risk ratio). However, the *x* axis could be another relative measure such as odds ratio or a parameter such as prevalence, depending on the specific question being addressed by the meta-analysis. For each intervention comparison, the center of the solid square box corresponds to the estimated risk ratio obtained for that comparison, and the size of the box corresponds to the weight given to that comparison in the meta-analysis. The horizontal line on either side of the box corresponds to the precision of the estimate (generally the 95% confidence interval).

When using risk ratios or odds ratios, the effects are usually plotted on a log-scale to produce symmetrical confidence intervals.

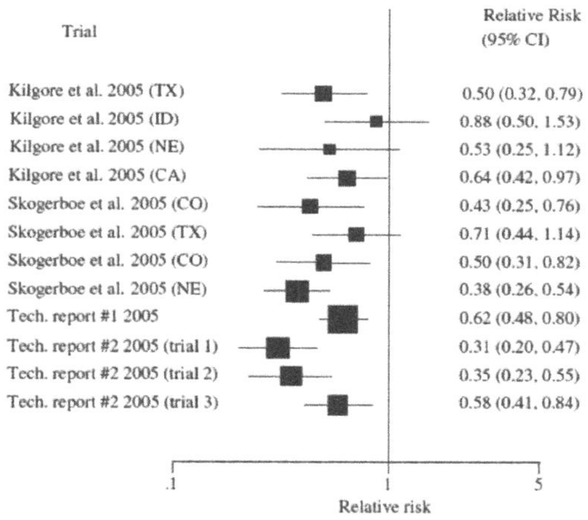

FIGURE 3 Forest plot illustrating relative risk of retreatment for bovine respiratory disease following treatment with tulathromycin compared to other available antibiotic treatments (61). Each row corresponds to treatment comparison, with the box representing the relative risk estimate for the comparison and the line corresponding to the 95% confidence interval around that estimate. The size of the box is representative of the relative amount of information contributed for that comparison (study weighting). The vertical line represents the null effect (relative risk of 1).

A vertical line drawn at a risk ratio or odds ratio value of 1.0 represents no difference in the effect between the intervention groups. For desirable outcomes, a risk ratio or odds ratio value >1 indicates that the intervention was effective in improving the outcome compared to the control group. For example, if the outcome is being "pathogen negative," and the intervention group is 80% pathogen negative and the control group is 20% pathogen negative, the risk ratio is 4 (0.8/0.2), and the desirable outcome occurs at a 4-fold higher level in the intervention group compared to the control group. However, in many food safety studies, the outcome is undesirable (e.g., shedding of a foodborne pathogen in feces). If that is the case, then relative risk or odds ratio values <1 indicate a beneficial intervention (i.e., a reduction in shedding). For example, if the outcome is being "pathogen positive," and the intervention group is 20% pathogen positive and the control group is 80% pathogen positive, the risk ratio is 0.25 (0.2/0.8), and the undesirable outcome occurs in the intervention group at one-quarter of the rate it does in the control group. The confidence interval describes our certainty about the magnitude of the effect size. A confidence interval overlapping the vertical line of "no effect" represents the lack of a statistically significant association at the 5% alpha level for that comparison.

If the outcome is continuous, such as plant yield, average daily gain, or log colony forming units of a pathogen, then the summary effect may be the difference in means between intervention groups. The vertical line in a forest plot will then present a 0 difference in means, the null value. The interpretation of the negative and positive values is based on how the mean difference is calculated. For example, if the mean difference is calculated as A – B, then a negative value implies that group B has a higher mean and a positive value means that group A has a higher mean.

While a graphic display of the results from individual studies provides a useful visual representation of intervention effectiveness, it is inappropriate to generate inferences based on a graphical display. It is necessary to conduct a formal meta-analysis to statistically combine the results from the individual studies to make inferences regarding the statistical significance of the summary effect size.

Calculating a Summary Effect

Meta-analysis involves the calculation of a weighted mean of the results from the individual studies; thus, each observation in the meta-analysis is the result from an original research study. A simple arithmetic mean of the results across all trials would give misleading results, because small studies are more subject to random error and therefore should be given less weight (36). The larger the weight given to a specific study, the more the results of that study will contribute to the weighted average summary effect measure. The weights are therefore chosen to reflect the amount of information that each study contains.

Meta-analysis is based on one of two basic statistical models, either a fixed effect or a random effects model (1, 31). The underlying assumption of a fixed effect model is that there exists a "true effect size" (intervention efficacy), and therefore any difference between studies is the result of sampling error (chance). In contrast, the random effects approach assumes that there may be different effect sizes underlying different studies (for example, the effect size may vary based on characteristics of the study populations) (1). Random effects models generally produce a point estimate of the summary effect size that is similar to that obtained from a fixed-effects model, but with wider confidence intervals. Random effects models are generally more appropriate and provide a more conservative estimate. In the absence of heterogeneity (differences among studies), a fixed effect model will produce the same results as a random effects model.

In a fixed effect meta-analysis, it is assumed that the observed intervention effect varies between studies only because of the random error inherent in each study. Therefore, the weight assigned to each study is commonly based on the inverse of the study's variance (1). With this approach, larger studies (which have smaller standard errors/variance) are given more weight than smaller studies (which have larger standard errors). There are three common approaches to estimating the weights: the inverse variance method, the Mantel-Haenszel method, and the Peto's odds ratio method (31, 37).

In the inverse variance method, the weight assigned to each study is the reciprocal of the squared standard error. The inverse variance method has wide applicability, because this approach can be used for either dichotomous or continuous data and requires that each study provide only an intervention effect estimate and a standard error. In some instances, an intervention effect such as the odds ratio (for categorical data) can be calculated from the data provided in a publication. Again, for categorical data, when zero cells (no events in one arm of a study) are present in a study, it is necessary to add a small quantity (generally 0.5) to that cell to allow calculation of the odds ratio (37, 38). When data are sparse, either because of low event rates or small trials, estimates of the standard error are poor. In these instances, the Mantel-Haenszel or Peto method may be preferred.

In a random effects meta-analysis, it is necessary to compute both the within study variance and the between study variance (the variance of the intervention effect size across the population of studies). A common method is the DerSimonian and Laird model (39), in which the study effects are assumed to follow a normal distribution, with the variance of that distribution estimated from the data.

The random effects method tends to be more conservative (wider confidence intervals on the summary effect size) compared to the fixed effect method and gives relatively more weight to smaller studies. The DerSimonian and Laird random effects method has the same wide applicability as fixed effect models, in that it can be used for studies with any type of outcome data as long as an intervention effect and standard error are provided in the individual studies or can be calculated from the data presented (37).

Evaluating Heterogeneity

While meta-analysis produces an estimate of the summary effect size, it is equally important to understand the consistency of that effect size across studies. For instance, the applicability of an intervention that consistently reduced fecal shedding of a foodborne pathogen by 50% across all studies would differ from an intervention that resulted in an average reduction in fecal shedding of 50% but ranged from 10% to 90% in the individual studies that contributed to that average.

Heterogeneity refers to the differences in effect sizes among studies (31). Variations in effect size between studies may be due to random error (chance) or "true" heterogeneity. True heterogeneity may occur due to differences between studies in characteristics of the populations, due to interventions and outcomes ("clinical heterogeneity"), and/or due to differences between studies in study design or risk of bias ("methodological heterogeneity") (40). For instance, if an intervention had a different effect in preweaned animals compared to adult animals, and studies with different ages of animals were included in the meta-analysis, then "age" would be a source of clinical heterogeneity in the results. Alternatively, if an antibiotic had a different efficacy when administered intravenously compared to intramuscularly, and if studies evaluating both methods of administration were included in the meta-analysis, then "route of administration" would be a source of clinical heterogeneity. If studies that employed blinding of the outcome assessor tended to have a different effect size compared to studies that did not use blinding, and if both types of studies were included in the meta-analysis, then "blinding of outcome assessor" would be a source of methodological heterogeneity. If substantive heterogeneity exists, it may not be meaningful or appropriate to calculate a summary effect size.

Heterogeneity should be evaluated as a component of all meta-analyses; the evaluation may be used to provide context on the degree of difference among studies in the review, to determine whether pooling of results should be undertaken, or as a first step for evaluating possible sources of heterogeneity. A variety of formal statistical tests are available to evaluate heterogeneity among a group of studies (41). One commonly used test to determine whether or not heterogeneity is present (at a predetermined P value cutpoint) is Cochran's chi-square test of homogeneity (sometimes known as Cochran's Q or simply Q). This test measures deviation of observed effect sizes from an underlying overall effect size. The null hypothesis for the Q test is that there is homogeneity of effect sizes among the studies included in the analysis; therefore, rejection of the null hypothesis implies that heterogeneity is present. However, the Q test may be a poor indicator of true heterogeneity among studies if a small number of studies are included in the calculation, resulting in low power to detect heterogeneity. It is common that a P value cut-off of 0.10 or larger be used as a significance level, although this approach carries the risk that when many large studies are available, a clinically unimportant difference could be significantly heterogeneous.

In addition to testing for the presence or absence of heterogeneity, it is possible to quantify the amount of heterogeneity. The I^2 test provides an estimate of the proportion of total variability that can be attributed to heterogeneity beyond chance. I^2 lies between 0% and 100%, with a value of 0% indicating no observed heterogeneity and larger values corresponding to increasing heterogeneity beyond that expected by chance. The I^2 is less affected by the number of studies in the analysis compared to the Q test (41), although this has been disputed (42). Given that the Q test and I^2 statistic both contribute to understanding heterogeneity in a meta-analysis, both should be calculated and reported.

One of the most difficult questions in meta-analysis is How much heterogeneity is too much for calculating a summary effect size? Heterogeneity will always exist, and the question is actually What is meaningful heterogeneity (43)? As a general guideline, I^2 values of 0 to 40% are likely unimportant, 30 to 60% represents moderate heterogeneity, 50 to 90% represents substantial heterogeneity, and 75 to 100% represents considerable heterogeneity (31). As an example of the range of heterogeneity estimates that have been reported in the food safety literature, Totton et al. (13) conducted 21 random-effects meta-analyses on different feed and water additives to evaluate their effect on *Salmonella* concentration or prevalence in the ceca of broiler chickens. The I^2 estimates for the meta-analyses ranged from 0 to over 95%; the authors did not report summary estimates for meta-analyses where heterogeneity was statistically significant.

Exploring Causes of Heterogeneity: Meta-Regression and Subgroup Analysis

Testing and reporting the magnitude of heterogeneity does not help to explain the factors associated with heterogeneity. Subgroup analysis and meta-regression are tools in meta-analysis that are used to understand factors associated with heterogeneity.

Subgroup analysis involves splitting the original studies into subgroups to make comparisons between them (for example, based on geographic location or other population characteristics or based on indicators of study quality such as randomized versus nonrandomized trials). Subgroup analysis may be conducted to explore heterogeneity or to address specific questions about populations or types of studies (31). For instance, researchers might be interested in evaluating the efficacy of vaccination for a foodborne pathogen separately for dairy cattle and for beef cattle, or in calves versus adult cattle. Snedeker et al. (9) used subgroup analysis to evaluate two vaccine types for reducing fecal shedding of *E. coli* O157 and found that reduction in fecal shedding was similar between the two vaccine types. Subgroup analysis is limited to comparisons of a single variable that is categorical. However, numerous subgroup analyses can be conducted. Subgroup analysis can be a useful tool for exploring data but should be used with caution, particularly if the subgroups were not identified *a priori*, due to the increased probability of false-positive significance tests as multiple tests are performed.

If it is of interest to explore multiple potential explanatory variables or a continuous covariate as a source of heterogeneity, then meta-regression can be used. Meta-regression is a weighted regression where the unit of concern is the study, and the outcome of interest is the effect of the intervention at the study level. The weights in meta-regression are frequently the inverse variance of each study's result. The regression coefficient describes how the outcome variable changes with a unit increase in the explanatory variable. If the summary effect is a ratio measure, such as a risk ratio or odds ratio, the log-transformed value of the intervention effect is used, and the exponent of the regression coefficient gives an estimate of the relative change in intervention effect size with a unit increase in the explanatory variable. Meta-regression allows the exploration of potential sources of heterogeneity or potential biases (such as indicators of study quality). Examples of meta-regression in the food safety literature include an evaluation of study-level predictors on the prevalence of *Salmonella* on swine farms (16) and an investigation of both study-level variables and methodological criteria as explanatory variables in a meta-regression on the effect of competitive exclusion products on *Salmonella* in broiler chickens (11). Wisener et al. (24) used meta-regression and subgroup analyses to evaluate whether intervention effect sizes differed between challenge trials and natural exposure trials for three food safety pathogen–commodity group combinations.

An excellent review of the advantages and disadvantages of meta-regression is available (41). If the number of studies is limited, factors might be investigated one at a time in univariate regressions, or if there are sufficient data, a multivariable regression model could be built. However, investigation of multiple factors is often not possible due to inadequate numbers of studies. Meta-regression should generally not be considered when there are few studies; a rough guideline is that more than 10 studies should be available before considering meta-regression as an approach to exploring heterogeneity (31).

Limitations of Meta-Analysis

A meta-analysis can only assess the literature that is available to the investigators. A large amount of the actual research conducted for many interventions may not be available in the public domain, and this may serve as a source of bias in the estimation of the summary effect measure if availability is associated with the effect size. This type of bias is called reporting bias and has several subcategories: publication bias, time lag bias, duplicate publication bias, location bias, citation bias, language bias, and outcome reporting bias (31, 39, 44).

Publication bias refers to the tendency for studies that do not show a significant effect not to be published, particularly if the sample size was small (either because they are not submitted for publication or because journals are more reluctant to publish them) (45).

Time lag bias is related to the tendency for trials showing a positive intervention effect to be published faster than trials that report nonstatistically significant results. Thus, recent nonsignificant trials may not be available in the public domain as quickly as trials showing a significant difference. Trials showing a significant difference may be more likely to be available in multiple forums (e.g., conference proceeding, report, and peer-reviewed publication), thereby resulting in duplicate publication bias. Trials showing a significant intervention effect also tend to be published in higher-profile journals, which are more likely to be indexed in electronic databases and are more likely to be cited. This results in easier identification for inclusion in a meta-analysis using standard literature search methods, resulting in location bias and citation bias. Literature published in the English language also is easier to identify in searches, and if non-English articles are identified, there may not be the resources to translate the articles for inclusion in the meta-analysis. There is some evidence in the human health care literature that published non-English trials tend to be smaller, of lower quality, and more likely to report a statistically significant difference, although the effect on the overall intervention estimate may not be large (46). Finally, there is evidence that, even within published trials, statistically significant outcomes are more likely to be reported than nonsignificant outcomes, leading to outcome reporting bias (47). A study of conference proceeding abstracts reporting on food safety intervention studies at the preharvest and abattoir level reported that less than half of the research reported in conference abstracts was published in the peer-reviewed literature within 4 years (48). The same study found that conference abstracts reporting at least one positive outcome were more likely to be subsequently published and were published faster than those reporting nonsignificant findings.

One approach to evaluating publication bias in a meta-analysis is to create a funnel plot, which can be visually examined for signs of asymmetry (47, 49). Funnel plots are simple scatter plots of the intervention effect sizes estimated from individual studies (on the x axis) against some measure of each study's precision (y axis), usually the standard error. The name "funnel" is used because precision in the estimation of the true intervention effect size should increase as the sample size of the studies included in the meta-analysis increases. Therefore, the expectation would be that effect sizes from small studies would scatter more widely at the bottom of the graph, with smaller differences in the estimated effect size among larger studies. Theoretically, in the absence of bias, the plot should resemble a symmetrical inverted funnel. Therefore an assessment of the symmetry of a funnel plot is actually an assessment of small study effects. Although funnel plots are often said to assess publication bias, they actually assess whether small studies have different effect sizes compared to larger studies. Publication bias may be one cause of small study effects. Other possible sources of asymmetry in funnel plots can include (i) selection bias, location bias, language bias, citation bias, or multiple publication bias; (ii) poor methodological quality of smaller studies; (iii) true heterogeneity where the magnitude of the intervention effect size differs according to study size (for example, due to differences in the intensity of interventions); (iv) artifactual (sampling variation can lead to an association between the intervention and its standard error); or (v) chance. There are a number of statistical tests to formally evaluate asymmetry in a funnel plot rather than relying on a visual assessment for details (see 45, 50, 51).

Reporting the Results of a Meta-Analysis

As with any type of study, it is important that meta-analyses (and systematic reviews) are reported in sufficient detail to enable the reader to determine how the meta-analysis was conducted and to evaluate the potential for biased results. The PRISMA statement (Preferred Reporting Items for Systematic Reviews and Meta-Analyses; www.prisma-statement.org) provides guidelines for reporting the results of systematic reviews, including the meta-analysis component (52). An accompanying elaboration document provides explanations for each of the recommended items and examples of good reporting (53).

CHALLENGES

The main challenges associated with meta-analysis are publication bias (as discussed in the previous section), insufficient quantity of research available to address the question of interest, lack of comprehensive reporting of the effect sizes, and high risk of bias in the primary studies. Theoretically, a meta-analysis can be done if two relevant studies are available. However, meta-analysis becomes more informative when there are more studies for inclusion. When the number of studies does not warrant a meta-analysis, a narrative summary may still be useful and can provide insights and directions for further research.

The challenge with poor-quality studies in meta-analysis is the "garbage in, garbage out" paradigm. There is evidence in the human health care literature that trials that do not report important methodological features aimed at reducing bias have exaggerated intervention effects (54–57), and there is some evidence that the same is true in preharvest food safety trials (58). Guidelines are available to describe the items that should be reported in preharvest food safety clinical trials (59, 60; www.reflect-statement.org). While it is possible that a well-reported trial was conducted poorly, it is essential that the reader of a trial know what was done in the trial in order to make an informed decision about the likelihood of bias. A major advantage of conducting meta-analysis as a component of a systematic review is that risk of bias assessment will have been conducted on each of the studies. This information can be included in a meta-analysis and in the interpretation of the results of the meta-analysis.

SUMMARY

Meta-analysis is the statistical pooling of data from multiple studies. Including meta-analysis as the analytical component of a systematic review increases the transparency and rigor of the analysis. The steps to a meta-analysis include extracting the data from the original research studies and displaying the data using a forest plot, combining the data using either a fixed effect or a random effects models, and evaluating heterogeneity. If the amount of heterogeneity suggests that a single summary effect size does not well represent the data, it may not be appropriate to report a summary effect size. Reasons for heterogeneity can be explored using meta-regression or subgroup meta-analysis. Meta-analysis may be a useful methodology for preharvest food safety research to aid in policy or clinical decision-making or to provide inputs to quantitative risk assessment or other models.

CITATION

Sargeant JM, O'Connor AM. 2016. Potential for meta-analysis in the realm of preharvest food safety. Microbiol Spectrum 4(5):PFS-0004-2014.

REFERENCES

1. **Borenstein M, Hedges LV, Higgins JPT, Rothstein HR.** 2009. *Introduction to Meta-Analysis*. John Wiley and Sons, West Sussex, United Kingdom.
2. **O'Connor AM, Sargeant JM, Wang C.** 2014. Conducting systematic reviews of intervention questions. III. Synthesizing data from intervention studies using quantitative approaches (meta-analysis). *Zoonoses Public Health* **61** (Suppl. 1)**:**52–63.
3. **den Besten HMW, Zwietering MH.** 2012. Meta-analysis for quantitative microbiological risk assessments and benchmarking data. *Trends Food Sci Technol* **25:**34–39.
4. **European Food Safety Authority.** 2010. Application of systematic review methodology to food and feed safety assessments to support decision making. *Eur Food Saf Auth J* **8:**1637–1727.
5. **Sargeant JM, Rajic A, Read S, Ohlsson A.** 2006. The process of systematic review and its application in agri-food public-health. *Prev Vet Med* **75:**141–151.
6. **Gonzales-Barron U, Butler F.** 2011. The use of meta-analytical tools in risk assessment for food safety. *Food Microbiol* **28:**823–827.
7. **Vialette M, Pinon A, Leporq B, Dervin C, Membré JM.** 2005. Meta-analysis of food safety information based on a combination of a relational database and a predictive modeling tool. *Risk Anal* **25:**75–83.
8. **Wisener LV, Sargeant JM, O'Connor AM, Faires MC, Glass-Kaastra SK.** 2015. The use of direct-fed microbials to reduce shedding of *Escherichia coli* O157 in beef cattle: a systematic review and meta-analysis. *Zoonoses Public Health* **62:**75–89.
9. **Snedeker KG, Campbell M, Sargeant JM.** 2012. A systematic review of vaccinations to reduce the shedding of *Escherichia coli* O157 in the faeces of domestic ruminants. *Zoonoses Public Health* **59:**126–138.
10. **Varela NP, Dick P, Wilson J.** 2013. Assessing the existing information on the efficacy of bovine vaccination against *Escherichia coli* O157:H7: a systematic review and meta-analysis. *Zoonoses Public Health* **60:**253–268.
11. **Kerr AK, Farrar AM, Waddell LA, Wilkins W, Wilhelm BJ, Bucher O, Wills RW, Bailey RH, Varga C, McEwen SA, Rajić A.** 2013. A systematic review-meta-analysis and meta-regression on the effect of selected competitive exclusion products on *Salmonella* spp. prevalence and concentration in broiler chickens. *Prev Vet Med* **111:**112–125.
12. **Totton SC, Farrar AM, Wilkins W, Bucher O, Waddell LA, Wilhelm BJ, McEwen SA, Rajic A.** 2012. A systematic review and meta-analysis of the effectiveness of biosecurity and vaccination in reducing *Salmonella* spp. in broiler chickens. *Food Res Int* **45:**617–627.
13. **Totton SC, Farrar AM, Wilkins W, Bucher O, Waddell LA, Wilhelm BJ, McEwen SA, Rajić A.** 2012. The effectiveness of selected feed and water additives for reducing *Salmonella* spp. of public health importance in broiler chickens: a systematic review, meta-analysis, and meta-regression approach. *Prev Vet Med* **106:**197–213.
14. **Wilhelm B, Rajić A, Parker S, Waddell L, Sanchez J, Fazil A, Wilkins W, McEwen SA.** 2012. Assessment of the efficacy and quality of evidence for five on-farm interventions for *Salmonella* reduction in grow-finish swine: a systematic review and meta-analysis. *Prev Vet Med* **107:**1–20.
15. **Young I, Rajić A, Wilhelm BJ, Waddell L, Parker S, McEwen SA.** 2009. Comparison of the prevalence of bacterial enteropathogens, potentially zoonotic bacteria and bacterial resistance to antimicrobials in organic and conventional poultry, swine and beef production: a systematic review and meta-analysis. *Epidemiol Infect* **137:**1217–1232.
16. **Sanchez J, Dohoo IR, Christensen J, Rajic A.** 2007. Factors influencing the prevalence of *Salmonella* spp. in swine farms: a meta-analysis approach. *Prev Vet Med* **81:**148–177.

17. **Islam MZ, Musekiwa A, Islam K, Ahmed S, Chowdhury S, Ahad A, Biswas PK.** 2014. Regional variation in the prevalence of *E. coli* O157 in cattle: a meta-analysis and meta-regression. *PLoS One* **9:**e93299.
18. **Wilkins W, Rajić A, Parker S, Waddell L, Sanchez J, Sargeant J, Waldner C.** 2010. Examining heterogeneity in the diagnostic accuracy of culture and PCR for *Salmonella* spp. in swine: a systematic review/meta-regression approach. *Zoonoses Public Health* **57**(Suppl 1):121–134.
19. **O'Connor AM, Anderson KM, Goodell CK, Sargeant JM.** 2014. Conducting systematic reviews of intervention questions. I. Writing the review protocol, formulating the question and searching the literature. *Zoonoses Public Health* **61**(Suppl 1):28–38.
20. **Sargeant JM, O'Connor AM.** 2014. Introduction to systematic reviews in animal agriculture and veterinary medicine. *Zoonoses Public Health* **61**(Suppl 1):3–9.
21. **Sargeant JM, Kelton DF, O'Connor AM.** 2014. Study designs and systematic reviews of interventions: building evidence across study designs. *Zoonoses Public Health* **61**(Suppl 1):10–17.
22. **Sargeant JM, Kelton DF, O'Connor AM.** 2014. Randomized controlled trials and challenge trials: design and criterion for validity. *Zoonoses Public Health* **61**(Suppl 1):18–27.
23. **Sargeant JM, O'Connor AM.** 2014. Conducting systematic reviews of intervention questions. II. Relevance screening, data extraction, assessing risk of bias, presenting the results and interpreting the findings. *Zoonoses Public Health* **61**(Suppl 1):39–51.
24. **Wisener LV, Sargeant JM, O'Connor AM, Faires MC, Glass-Kaastra SK.** 2014. The evidentiary value of challenge trials for three pre-harvest food safety topics: a systematic assessment. *Zoonoses Public Health* **61:**449–476.
25. **Conn VS, Isaramalai SA, Rath S, Jantarakupt P, Wadhawan R, Dash Y.** 2003. Beyond MEDLINE for literature searches. *J Nurs Scholarsh* **35:**177–182.
26. **Crumley ET, Wiebe N, Cramer K, Klassen TP, Hartling L.** 2005. Which resources should be used to identify RCT/CCTs for systematic reviews: a systematic review. *BMC Med Res Methodol* **5:**24.
27. **McKibbon KA, Wilczynski NL, Haynes RB, Hedges Team.** 2009. Retrieving randomized controlled trials from MEDLINE: a comparison of 38 published search filters. *Health Info Libr J* **26:**187–202.
28. **Alpi KM, Stringer E, Devoe RS, Stoskopf M.** 2009. Clinical and research searching on the wild side: exploring the veterinary literature. *J Med Libr Assoc* **97:**169–177.
29. **Grindlay DJ, Brennan ML, Dean RS.** 2012. Searching the veterinary literature: a comparison of the coverage of veterinary journals by nine bibliographic databases. *J Vet Med Educ* **39:**404–412.
30. **Meade MO, Richardson WS.** 1997. Selecting and appraising studies for a systematic review. *Ann Intern Med* **127:**531–537.
31. **Higgins JPT, Green S (ed).** 2011. Cochrane Handbook for Systematic Reviews of Interventions Version 5.1.0 [updated March 2011]. The Cochrane Collaboration. www.handbook.cochraneorg.
32. **Whiting PF, Rutjes AW, Westwood ME, Mallett S, Deeks JJ, Reitsma JB, Leeflang MM, Sterne JA, Bossuyt PM, QUADAS-2 Group.** 2011. QUADAS-2: a revised tool for the quality assessment of diagnostic accuracy studies. *Ann Intern Med* **155:**529–536.
33. **Broen MP, Braaksma MM, Patijn J, Weber WE.** 2012. Prevalence of pain in Parkinson's disease: a systematic review using the modified QUADAS tool. *Mov Disord* **27:**480–484.
34. **Shamliyan TA, Kane RL, Ansari MT, Raman G, Berkman ND, Grant M, Janes G, Maglione M, Moher D, Nasser M, Robinson KA, Segal JB, Tsouros S.** 2011. Development quality criteria to evaluate nontherapeutic studies of incidence, prevalence, or risk factors of chronic diseases: pilot study of new checklists. *J Clin Epidemiol* **64:**637–657.
35. **Lewis S, Clarke M.** 2001. Forest plots: trying to see the wood and the trees. *BMJ* **322:**1479–1480.
36. **Egger M, Smith GD, Phillips AN.** 1997. Meta-analysis: principles and procedures. *BMJ* **315:** 1533–1537.
37. **Egger M, Davey Smith G, Altman DG (ed).** 2001. *Systematic Reviews in Health Care: Meta-Analysis in Context,* 2nd ed. BMJ Publishing Group, London, United Kingdom.
38. **Friedrich JO, Adhikari NK, Beyene J.** 2007. Inclusion of zero total event trials in meta-analyses maintains analytic consistency and incorporates all available data. *BMC Med Res Methodol* **7:**5.
39. **DerSimonian R, Laird N.** 1986. Meta-analysis in clinical trials. *Control Clin Trials* **7:**177–188.
40. **Khan KS, Kunz R, Kleijnen J, Antes G.** 2005. *Systematic Reviews to Support Evidence-Based Medicine: How to Review and Apply Findings of Healthcare Research.* Royal Society of Medicine Press, London, United Kingdom.
41. **Higgins JPT, Thompson SG.** 2002. Quantifying heterogeneity in a meta-analysis. *Stat Med* **21:**1539–1558.

42. **Rücker G, Schwarzer G, Carpenter JR, Schumacher M.** 2008. Undue reliance on I(^{2}) in assessing heterogeneity may mislead. *BMC Med Res Methodol* **8:**79.
43. **Higgins JP.** 2008. Commentary: heterogeneity in meta-analysis should be expected and appropriately quantified. *Int J Epidemiol* **37:**1158–1160.
44. **Egger M, Smith GD.** 1998. Bias in location and selection of studies. *BMJ* **316:**61–66.
45. **Thornton A, Lee P.** 2000. Publication bias in meta-analysis: its causes and consequences. *J Clin Epidemiol* **53:**207–216.
46. **Jüni P, Holenstein F, Sterne J, Bartlett C, Egger M.** 2002. Direction and impact of language bias in meta-analyses of controlled trials: empirical study. *Int J Epidemiol* **31:**115–123.
47. **Dwan K, Altman DG, Arnaiz JA, Bloom J, Chan AW, Cronin E, Decullier E, Easterbrook PJ, Von Elm E, Gamble C, Ghersi D, Ioannidis JP, Simes J, Williamson PR.** 2008. Systematic review of the empirical evidence of study publication bias and outcome reporting bias. *PLoS One* **3:**e3081.
48. **Snedeker KG, Totton SC, Sargeant JM.** 2010. Analysis of trends in the full publication of papers from conference abstracts involving preharvest or abattoir-level interventions against foodborne pathogens. *Prev Vet Med* **95:**1–9.
49. **Sterne JAC, Egger M, Smith GD.** 2001. Systematic reviews in health care: investigating and dealing with publication and other biases in meta-analysis. *BMJ* **323:**101–105.
50. **Begg CB, Mazumdar M.** 1994. Operating characteristics of a rank correlation test for publication bias. *Biometrics* **50:**1088–1101.
51. **Egger M, Davey Smith G, Schneider M, Minder C.** 1997. Bias in meta-analysis detected by a simple, graphical test. *BMJ* **315:**629–634.
52. **Moher D, Liberati A, Tetzlaff J, Altman DG, Group P, PRISMA Group.** 2009. Preferred reporting items for systematic reviews and meta-analyses: the PRISMA statement. *J Clin Epidemiol* **62:**1006–1012.
53. **Liberati A, Altman DG, Tetzlaff J, Mulrow C, Gøtzsche PC, Ioannidis JP, Clarke M, Devereaux PJ, Kleijnen J, Moher D.** 2009. The PRISMA statement for reporting systematic reviews and meta-analyses of studies that evaluate health care interventions: explanation and elaboration. *PLoS Med* **6:**e1000100.
54. **Jüni P, Altman DG, Egger M.** 2001. Systematic reviews in health care: assessing the quality of controlled clinical trials. *BMJ* **323:**42–46.
55. **Kjaergard LL, Villumsen J, Gluud C.** 2001. Reported methodologic quality and discrepancies between large and small randomized trials in meta-analyses. *Ann Intern Med* **135:**982–989.
56. **Kunz R, Oxman AD.** 1998. The unpredictability paradox: review of empirical comparisons of randomised and non-randomised clinical trials. *BMJ* **317:**1185–1190.
57. **Moher D, Pham B, Jones A, Cook DJ, Jadad AR, Moher M, Tugwell P, Klassen TP.** 1998. Does quality of reports of randomised trials affect estimates of intervention efficacy reported in meta-analyses? *Lancet* **352:**609–613.
58. **Sargeant JM, Saint-Onge J, Valcour J, Thompson A, Elgie R, Snedeker K, Marcynuk P.** 2009. Quality of reporting in clinical trials of preharvest food safety interventions and associations with treatment effect. *Foodborne Pathog Dis* **6:**989–999.
59. **O'Connor AM, Sargeant JM, Gardner IA, Dickson JS, Torrence ME, Dewey CE, Dohoo IR, Evans RB, Gray JT, Greiner M, Keefe G, Lefebvre SL, Morley PS, Ramirez A, Sischo W, Smith DR, Snedeker K, Sofos J, Ward MP, Wills R, Consensus Meeting Participants.** 2010. The REFLECT statement: methods and processes of creating reporting guidelines for randomized controlled trials for livestock and food safety by modifying the CONSORT statement. *Zoonoses Public Health* **57:**95–104.
60. **Sargeant JM, O'Connor AM, Gardner IA, Dickson JS, Torrence ME, Dohoo IR, Lefebvre SL, Morley PS, Ramirez A, Snedeker K, Consensus Meeting Participants.** 2010. The REFLECT statement: reporting guidelines for randomized controlled trials in livestock and food safety: explanation and elaboration. *Zoonoses Public Health* **57:**105–136.
61. **Wellman NG, O'Connor AM.** 2007. Meta-analysis of treatment of cattle with bovine respiratory disease with tulathromycin. *J Vet Pharmacol Ther* **30:**234–241.

EMERGING SOLUTIONS TO PREHARVEST FOOD SAFETY

Phage Therapy Approaches to Reducing Pathogen Persistence and Transmission in Animal Production Environments: Opportunities and Challenges

16

ANNA COLAVECCHIO[1] and LAWRENCE D. GOODRIDGE[1]

INTRODUCTION

One of the major challenges to current global food production and food security is the presence of antibiotic-resistant bacteria in animals (ruminants, poultry, swine) from which foods of animal origin are produced. Foodborne diseases significantly impact public health globally, with the World Health Organization (WHO) estimating that 1 in 10 people, or approximately 600 million people worldwide, are sickened and 420,000 die annually from foodborne illnesses (1). There is concern that many foodborne bacterial pathogens are either resistant or increasing their resistance to antimicrobials commonly used for medical treatment. For example, the Centers for Disease Control and Prevention reported that in 2013, the percentage of human *Campylobacter jejuni* isolates with macrolide resistance increased from 1.8% in 2012 to 2.2% in 2013, and from 9.0% in 2012 to 17.6% among *Campylobacter coli* isolates (2). In addition, the percentage of human *Salmonella* ser. I 4,[5],12:i:- isolates with resistance to ampicillin, streptomycin, sulfonamide, and tetracycline continued to increase, from 17% in 2010 to 45.5% in 2013 (2). *Campylobacter* spp. (845,024 cases per year) and nontyphoidal *Salmonella* spp. (1,027,561 cases per year) are the two most prevalent causes of foodborne

[1]Department of Food Science and Agricultural Chemistry, Food Safety and Quality Program, McGill University, Ste Anne de Bellevue, Quebec, H9X 3V9, Canada.

Preharvest Food Safety
Edited by Siddhartha Thakur and Kalmia E. Kniel

doi:10.1128/microbiolspec.PFS-0017-2017

illness in the United States, accounting for 51% of annual foodborne illnesses due to known bacterial agents (3) and highlighting the fact that an increasing number of foodborne illnesses are becoming more difficult to treat with antibiotics.

The economic health care burden caused by antimicrobial resistance (AMR) is estimated to be $4 to 5 billion annually (4). Furthermore, AMR pathogens also result in economic losses in the food animal production industry as a result of treatment failures in livestock. As such, global agencies such as the WHO and the Food and Agricultural Organization of the United Nations have expressed concern regarding the effect of antibiotic resistance on global food security (5) and have conducted studies aimed at understanding the magnitude of the problem and determining solutions. Yet in spite of the widely appreciated magnitude of the AMR problem and the efforts by industry and government to curb the rise in AMR organisms, novel strains of resistant bacteria continue to emerge and disseminate.

Reduction (and elimination) of antibiotic use in agriculture has led to some success in lowering the frequency of antimicrobial-resistant bacteria in the food chain (6). In North America, the use of antibiotics in food production is a major topic, with many companies suspending the use of antibiotics during production of food animals. The decrease in antibiotic use has placed an emphasis on finding alternative approaches to treat animal diseases as well as alternative agents that would aid in growth promotion. The discovery of bacteriophages (phages) approximately 100 years ago (7) ushered in a new field of microbiology in which phages were employed as antimicrobial agents. The discovery and production of antibiotics in the 1930s to 1940s (8) led to the cessation of most of the research aimed at utilizing phages as antimicrobial agents. The emergence of multidrug-resistant bacteria in the 1980s rekindled the interest in this approach, and the use of phage therapy (the application of phages to bodies, substances, or environments to effect the biocontrol of pathogenic bacteria) was reborn in the Western world. To be effective as a treatment, phages must remain viable until they reach the site of bacterial infection (in animals) or contamination (in foods), where they must be able to attach to and infect bacteria and kill those bacteria. The era of genomics has allowed for characterization and manipulation of phages for use as antimicrobials to treat animal infections with a level of precision never before realized. As more research in this emerging area of bacterial control has been conducted, several opportunities and challenges with respect to implementation of phage therapy have been identified (Fig. 1). These opportunities and challenges can be further separated into intrinsic and extrinsic factors, as described below.

OPPORTUNITIES

The ability of phages to kill antibiotic-resistant bacteria, when combined with their ubiquitous nature, specificity, prevalence in the biosphere, and low inherent toxicity, makes them a safe, natural, and sustainable technology for control of animal diseases. These unique qualities of phages have led to several opportunities with respect to emerging trends in infectious disease treatment.

Simultaneous Control of Multiple Pathogens

One disadvantage of phage therapy is that phages typically have narrow host ranges and tend to be only capable of infecting bacteria within a given species or individual bacterial strains within a species. While this is often described as a positive aspect of phage therapy, in that the normal bacterial flora is only minimally affected (unlike broad-spectrum antibiotics, for example), difficulty arises when multiple bacterial species must be controlled. Such is the case in food animals, where antibiotic growth promoters are currently employed to maximize growth of the

	Intrinsic	Extrinsic
Opportunities	Simultaneous control of multiple pathogens	Genetically modified phages as antimicrobials
	Phage mediated modulation of the immune system	Regulatory approval of phage-based antimicrobials
Challenges	Phage clearance by the immune system	Stability of phages in the gastrointestinal tract
	Bacteriophage induced release of endotoxin	Development of Bacterial Insensitive Mutants (BIMs)

FIGURE 1 Intrinsic and extrinsic characteristics that may contribute to the success or failure of bacteriophage therapy.

animal prior to harvest. While the mechanism of action is unclear, it is thought that growth promoters suppress sensitive populations of bacteria in the intestines. Other benefits of the use of antibiotic growth promoters include control of zoonotic pathogens including *Salmonella*, *Campylobacter*, *Escherichia coli*, and *Enterococci* (9). For growth promotion, the traditional approach to phage therapy, which involves developing a cocktail of phages against the target bacteria, would be impractical, since it is likely that dozens of phages would be needed to control all of the bacterial species necessary to effect growth in the animal.

One potential solution to this problem is the application of temperate bacteriophages that reside in the bacterial chromosome as an antimicrobial. Current phage therapy dogma holds that temperate phages are not suitable candidates for use in treating bacterial diseases because they may harbor and transfer antibiotic resistance and virulence determinants to other bacteria. Furthermore, the tendency of temperate phages to lysogenize rather than lyse their host means that their ability to decrease bacterial concentrations is severely decreased. However, these statements arise from the premise that in phage therapeutic applications, the phages must be supplied exogenously. Cadieux et al. (10) proposed that instead of external addition of phages, prophages could be induced from the chromosomes of the target bacterial species. In this scenario, the inducing agent would lead to prophage excision from the chromosome and lytic growth, resulting in death of the host cell. Many prophage inducers have been described in the literature. While many of these are antibiotics (11, 12) and would therefore be unsuitable for use, several nonantibiotic compounds including hydrogen peroxide (13), EDTA, and sodium citrate (14) have also been demonstrated to induce prophages.

The use of prophage induction has several advantages compared to traditional phage therapy in which virulent phages are added exogenously. For example, the issue of host range (phages typically only infect one or a few bacterial strains, and a large number of different phages are therefore needed to treat infections because these tend to be caused by genetically different strains of bacteria [15]) and the challenge of bacteria developing resistance to the phages is solved. The majority of bacterial genome sequences deposited in the NCBI database contain prophage sequences (16). Direct evidence of this is provided by Kang (17), who extracted all bacterial genomes from the SEED database, which contains genome sequence and annotation information that is used by the prophage identification program PhiSpy. The genomes contained at least one representative of every bacterial genus currently known. At the time of analysis, the SEED database contained 11,941 bacterial genomes spanning 34 different phyla from Archaea and Bacteria. Kang (17) identified 36,465 prophages in 9,883 bacterial hosts (82.76% of the bacteria have at least one prophage), demonstrating that most bacterial genomes contain prophages.

Since the prophages already reside within the bacterial cell, there is no issue of bacterial resistance developing. Second, a single compound can lead to induction of multiple prophages within a single bacterial cell, as well as across species, meaning that multiple bacterial species can be controlled simultaneously. Third, because the inducing agent is a chemical, many of the physical impediments to using phages as drugs are eliminated (including the fact that the large size of the phage particles limits the dose that can be delivered; the large size of phages also means that they cannot be given in a sufficiently concentrated solution [15]).

Finally, the use of prophage induction has a major advantage over traditional phage therapy in that it should be capable of treating bacteria that reside intracellularly. For example, one potential source of pathogenic bacteria in ground beef is the lymph nodes of cattle. Arthur et al. (18) determined the prevalence of *Salmonella* in bovine lymph nodes associated with lean and fat trimmings that might be utilized in ground beef production. Lymph nodes from the flanks of cow and bull carcasses had the highest prevalence, at 3.86%, whereas lymph nodes from the chuck region of fed cattle carcasses had the lowest prevalence, at 0.35%. The presence of *Salmonella* in lymph nodes could be a food safety concern, and prophage induction could provide a viable means of controlling *Salmonella* organisms that reside in lymph nodes.

The major concern with the use of prophages as therapeutics is the potential for antibiotic resistance genes and/or virulence genes to be horizontally transferred by prophage-mediated transduction to other bacterial cells. Horizontal gene transfer due to prophage induction has been the subject of much discussion in the scientific literature. However, most of that discussion is either theoretical or based on *in vitro* studies or metagenomic studies. Recent studies have suggested that the role that phages play in the spread of antibiotic resistance genes may be overestimated. For example, Enault et al. (19) used exploratory bioinformatic strategies to identify antibiotic resistance genes in phage genomes and found that antibiotic resistance gene abundances in 1,181 phage genomes were vastly overestimated, due to low similarities and matches to proteins unrelated to antibiotic resistance. Reanalysis of human- and mouse-associated viromes for antibiotic resistance genes demonstrated that these genes attributed to phages in viromes were previously overestimated. The authors concluded that their findings reassert the traditional dogma that antibiotic resistance genes are rarely encoded in phages.

It should also be noted that very few studies have been conducted with respect to *in vivo* transduction of antibiotic resistance genes. Thus, it is important to analyze such studies before making general comments regarding prophage induction and horizontal gene transfer. In one study, Allen et al. (20) used metagenomics to evaluate the effect of two in-feed antibiotics (carbadox and ASP250 [chlortetracycline, sulfamethazine, and penicillin]) on swine intestinal viromes. The authors also monitored the bacterial communities using 16S rRNA gene sequencing. ASP250, but not carbadox, caused significant population shifts in both the phage and bacterial communities. The authors also observed that while the abundance of phage integrase-encoding genes was significantly increased in the viromes of medicated swine over that in the viromes of nonmedicated swine, which demonstrated the induction of prophages with antibiotic treatment, the number of antibiotic resistance genes, such as multidrug resistance efflux pumps, identified in the viromes was not significantly increased when in-feed antibiotics were applied. This means that while prophages were induced in the swine gut, this did not result in horizontal gene transfer of antibiotic resistance genes from the prophages to other bacteria in the swine gut. Additionally, the concept of superinfection immunity (21) means that any prophages released from a lysed bacterial strain will be unable to infect other (unlysed) cells of that same strain. When coupled with the aforementioned fact that phages tend to have narrow host ranges, the ability of temperate phages to transfer antibiotic resistance genes and virulence genes to other bacteria may be overstated. Finally, the selective pressure exerted by the presence of prophage inducers should drive temperate phage infection away from lysogeny toward lytic growth.

Still, potential concerns related to prophage transduction of antibiotic resistance or virulence genes could be alleviated by specific induction of prophages that do not carry any of these genes. Irbe et al. (22) demonstrated that a prophage could be induced in a specific manner with the use of oligo-deoxynucleotides. The authors showed that lysogenized prophage Φ80 virions were induced in permeabilized *E. coli* K-12 cells that were exposed to specific oligo-deoxynucleotides. Of the oligo-deoxynucleotides tested, only d(A-G), d(G-G), and d(I-G) induced Φ80. Interestingly, the oligo-deoxynucleotides were capable of inducing Φ80, but not phage λ. For example, while induction by d(G-G) and d(A-G) was observed in plasmolyzed cells prepared from phage λ lysogens, the efficiency of the induction was much lower (<5%) than the level observed for Φ80 induction. These observations raise the possibility of using oligo-deoxynucleotides to specifically induce prophages which either do not carry antibiotic resistance genes or are incapable of lysogenizing additional cells due to defects in their chromosomes.

Genetically Modified Phages as Antimicrobials

The availability of modern molecular biology tools has led to the development of genetically modified phages, based on modification of phage properties such as host range and lytic ability. One recent phage therapy approach involves the creation of recombinant phages that have been altered to encode lethal genes, which are delivered to the host bacteria. The expression of the lethal gene produces a gene product that inactivates the target cell. Alternatively, phages may be physically labeled with toxic molecules which destroy the bacterial cell when brought into close proximity (due to binding of the phage). These lethal agent delivery systems (LADS) (which kill their host cell without lysis) are helpful in reducing the presence of large amounts of endotoxin, which is released upon lysis of Gram-negative bacterial cells. Lethal agents encoded and delivered by LADS include naturally occurring genes such as *doc*, *chpBK*, and *gef*. The product of the *doc* gene, for example, has been shown to be lethal in *E. coli* and is either bacteriocidal or bacteriostatic in other Gram-negative and Gram-positive bacteria (23). Gef and ChpBK exert a bactericidal effect on *E. coli* by interfering with cell growth (24).

Several approaches based on the general principle of LADS have been developed. Westwater and colleagues (24) produced a recombinant phage that carried a postsegregational killing system and demonstrated the efficacy of the phage to kill target bacteria. Postsegregational killing systems consist of two components: a stable toxin and an unstable antitoxin. One of the best-known systems is the host killing/suppressor of killing (hok/sok) system of the plasmid R1 of *E. coli* (25). Other examples include the *pemI-pemK* genes of plasmid R100, the *phd-doc* genes of phage P1, and the *ccdA-ccdB* genes of plasmid F (26–29). All postsegregational killing systems work in a similar fashion. The system is controlled by two genes which code for a toxin with a long half-life and an antitoxin with a shorter half-life. After cell division, daughter cells that do not contain a copy of the plasmid die because the toxin from the parent cell has a longer half-life than that of the antitoxin. Therefore, only cells that carry a plasmid can produce more antitoxin and survive.

Another LADS is based on a transfer plasmid carrying the genes encoding lethal products (23). The plasmid is maintained in a phage P1 lysogen unable to package its own DNA. Induction of the lysogen by a temperature shift results in multiplication of DNA, packaging of the transfer plasmid into P1 phage heads, and lysis of the production strain. The purified virions deliver the transfer plasmid to the host cell. Following delivery to the target bacterial cell, the plasmid DNA recircularizes, and subsequent expression of the lethal agent results in rapid cell death (23). Other researchers have genetically manipulated phages to enable them to degrade biofilms. For example, Lu and Collins (30) engineered phage T7 to express a biofilm-degrading enzyme during infection, which allowed this phage to simultaneously attack the bacterial cells within the biofilm and the biofilm itself. Results indicated that the engineered phage substantially reduced bacterial biofilm cell counts by 4.5 logs, which was approximately 2 orders of magnitude better than that of the wild-type (nonenzymatic) phage.

The emergence of the clustered, regularly interspaced, short palindromic repeat (CRISPR) technology has facilitated the use of targeted genome editing using engineered nucleases such as Cas9, with customizable specificities. Genome editing mediated by these nucleases has been used to efficiently modify endogenous genes in a number of diverse cell types and in organisms that have traditionally been challenging to manipulate genetically. Several studies have shown that phage-transferable CRISPR-Cas systems are capable of killing pathogens and resensitizing them to antibiotics (31, 32). One important result from these studies was the demonstration that the transferred CRISPR-Cas system was capable of eliminating specific bacterial populations. Yosef and colleagues (33) applied this interesting application by using phages to deliver a CRISPR-associated (Cas) system to reverse antibiotic resistance and eliminate the transfer of resistance between strains. Briefly, a CRISPR array, encoding spacers that target conserved sequences of the resistance genes *ndm-1* and *CTX-M-15* (which encode β lactamases that confer resistance to carbapenems), was introduced into a phage λ lysogen immediately downstream of the cas genes. Induced λ prophages were then used to lysogenize *E. coli* bacteria, enabling the modified CRISPR-Cas system to be transferred to bacteria via lysogenization.

An important observation of this work was that lysogenized bacteria outcompeted bacteria harboring the ndm-1 and CTX-M-15 resistance plasmids, indicating that the genetic fitness cost of the transferred prophage was smaller than that of the resistance plasmids. For the system to work efficiently, it was imperative that a selective advantage be conferred upon lysogenized bacteria that harbored the modified CRISPR-Cas system. To facilitate this need, the authors engineered lytic T7 phages encoding protospacers that were identical to the ndm-1 and CTX-M-15 protospacers targeted by the transferred CRISPR-Cas system. The use of identical protospacers ensured that the lysogens could not lose the sensitizing element without also losing phage resistance. The genetically modified T7 phages would thus be targeted concomitantly with the resistance genes.

Both the modified Crisper-Cas λ prophages and the genetically modified T7 phages were combined to form an effective strategy to counteract the emerging threat of antibiotic-resistant bacteria. Instead of directly killing bacterial pathogens, as in traditional phage therapy, the bacteria are instead resensitized to the antibiotics that they are resistant to via the λ prophage, while the T7 phages ensure the enrichment of the resensitized bacteria by killing of the nonsensitized and therefore T7-susceptible bacteria.

The advantage of this approach over traditional phage therapy includes the fact that it does not require administration of phages into host tissues. In addition, it aims not to directly kill treated bacteria but rather to sensitize them to antibiotics and to kill the resistant bacteria. Therefore, there is no counterselection against the sensitization. Still, there are some drawbacks. First, while antibiotic-resistant bacteria that have not been lysogenized with the λ phage would be initially susceptible to the T7 phage, there is nothing stopping T7-resistant subpopulations (via a resistance mechanism other than the modified CRISPR-Cas system) from developing. In this scenario, treatment with the antibiotic would destroy the sensitized bacteria but leave the nonsensitized and still antibiotic-resistant bacteria intact, leading to treatment failure. Second, this approach represents a proof of concept, since it would only work on bacteria that phages λ and T7 could infect. This means that analogous systems utilizing different phage combinations would need to be developed for different bacteria, and given the number of antibiotic-resistant bacterial species that can cause infection or transfer their antibiotic resistance to other bacteria, this approach seems unrealistic. Nevertheless, the collective studies described above indicate that genetically modified phages have great potential for the control of bacterial pathogens in veterinary settings to effectively treat infected animals.

Phage-Mediated Modulation of the Immune System

While the consequences of phage interaction with the mammalian immune system and its components are still not well defined, an increasing body of evidence suggests that phages play a role in direct and indirect regulation of immunity. For example, phages are highly immunogenic foreign antigens that can interact with the innate immune system to induce specific humoral as well as cellular immune responses (34). Studies also show that phages influence mammalian immune responses, including the attenuation of specific and nonspecific immune reactions, and maintenance of local immune tolerance to gut microorganism-derived antigens, and that phages are capable of acting as powerful immunomodulators.

Phages may be used as nonspecific agents in disease treatment due to their ability to trigger anti-inflammatory responses *in vivo*. Phages impart anti-inflammatory action by at least two mechanisms: a direct mechanism, which acts directly on components of the immune system responsible for the development of inflammatory processes, and an indirect action based on the ability of phages to kill their host bacteria. For example, Górski et al. (35) showed that phages can decrease cellular infiltration of allogeneic skin transplants in mice and activation of the nuclear transcription factor NF-kappa B (which leads to expression of proinflammatory cytokines, chemokines, and adhesion molecules). Additionally, the same group demonstrated that recombinant gp12 of phage T4 (a short tail fiber protein that binds to lipopolysaccharide [LPS] and promotes adsorption of phage T4 to *E. coli*) counteracts proinflammatory effects of LPS *in vivo*, causing the reduction of serum interleukin-1 and interleukin-6 levels as well as a decrease of inflammatory infiltration in the spleen and liver (36). These studies suggest that phages or their components could be used as treatments to counter the inflammatory effects of a number of diseases or, perhaps, could be used to counter the side effects of antibiotic-induced endotoxin release.

Reactive oxygen species (ROS) interact with a variety of cellular components, including proteins, lipids, and nucleic acids, and are employed by neutrophils and mononuclear phagocytes to neutralize endocytosed antigenic material. However, overproduction of ROS molecules may cause inactivation of the body's endogenous antioxidant defense mechanisms, leading to oxidative stress and resulting in serious tissue damage (37) or apoptosis of immune cells. Thus, suppression of ROS production may be beneficial in controlling sequelae caused by ROS. Phages have been reported to reduce ROS production by polymorphonuclear leukocytes in the presence of bacteria or their LPS (38). For example, Weber-Dabrowska et al. (39) and Hoshino and colleagues (40) both showed that a purified phage T4 preparation with low-endotoxin content could significantly diminish the luminol-dependent chemiluminescence of peripheral blood polymorphonuclear leukocytes stimulated by LPS. These observations led to speculation that phage-mediated inhibition of LPS or bacteria-stimulated ROS production by polymorphonuclear leukocytes may be attributed not only to phage-polymorphonuclear leukocyte interactions but also to phage-LPS interactions and bacterial lysis (39). It should be noted that T4 phage and the related T4-like phages utilize LPS as one of two bacterial receptors for binding to and subsequent infection of the host cell (41), so the interaction of T4 phage with LPS and the consequences of those interactions vis à vis ROS production is plausible.

The interaction of phages with the immune system will likely have consequences for major organs within the body. Górski and Weber-Dąbrowska (42) proposed that in the gastrointestinal tract (GIT), phages may mediate immunomodulatory, probiotic-like functions and that translocation from the intestines might contribute to phages mediating such probiotic-like functions in other parts of the body (43). Kim et al. (38) showed that phages could behave as probiotics by promoting growth in pigs. In this work, the authors compared the feed addition of 0.5, 1.0, and 1.5 g/kg of a commercial bacteriophage product that contained a cocktail of bacteriophages active against *Salmonella* spp., *Staphylococcus aureus*, *E. coli*, and *Clostridium perfringens* types A and C to a fermented probiotic product that contained *Lactobacillus acidophilus*, *Bacillus subtilis*, and *Saccharomyces cerevisiae* (added to feed at a concentration of 3.0 g/kg). In one experiment, increasing levels of the phage product linearly improved the average daily gain, average daily feed intake, and apparent total tract digestibility of dry matter. Furthermore, after 35 days, pigs that were fed diets supplemented with increasing levels of phages had greater total fecal anaerobic bacteria, *Bifidobacterium* spp., and *Lactobacillus* spp. and fewer fecal *Clostridium* spp. and coliforms.

In a second experiment, pigs were fed a basal diet without any antimicrobials (control) and basal diets supplemented with 3.0 g/kg fermented probiotic product (P), 1.0 g/kg bacteriophage (B), and a combination of 1.0 g/kg bacteriophage and 3.0 g/kg fermented probiotic product (BP). Pigs that were fed the B and BP diets had greater average daily gain, average daily feed intake, gain:feed, and apparent total tract digestibility of dry matter, crude protein, and gross energy than that of pigs that were fed the control and P diets. Pigs fed the P diet had greater average daily gain, average daily feed intake and apparent total tract digestibility of crude protein than that of pigs fed the control diet. At day 35, pigs fed the BP diet had more total fecal anaerobic bacteria, *Bifidobacterium* spp., and *Lactobacillus* spp. and fewer *Clostridium* spp. and coliforms than pigs fed the control diet. Also, pigs fed the P and B diets had more *Lactobacillus* spp. and fewer coliforms at day 35 than did pigs fed the control diet. In both experiments, the dietary treatments had no effect ($P > 0.05$) on serum immunoglobulin concentrations. These results show that while phages and probiotics both improve performance aspects in grower pigs, performance was enhanced to a greater effect when phages were employed.

In addition to the consequences of phage-immune system interactions mentioned above, the nature of phage interactions with the immune system and the various local and systemic responses to those interactions may be further summarized, as by Kaur et al. (44) as follows: (i) the physicochemical properties of the phage, including the size and number of different surface epitopes, determines the extent of immune response to the phage; (ii) the route of immunization plays a critical role in immune response, with some routes leading to active immunity and others conferring tolerance; (iii) the dose of phage administered is also important, since low doses of antigen have been shown to stimulate cell-mediated Th1 cytokine profiles, while higher doses activate the Th2 pathway; (iv) the primary antibody response to phages in naive individuals is much less efficient in clearing phage antigens than secondary exposure in primed individuals due to B-cell maturation, isotype switching, and affinity maturation; and (v) the antibody and cytokine profiles of different isotypes impart different functions (44). A note of caution should be extended when interpreting these results. It is estimated that there are 10^{31} phages in the biosphere (45), with many of these belonging to phage families that have yet to be characterized. Therefore, it is likely that the results of the above studies are relevant only to the specific phages that were studied and should not be considered relevant for phages in general. Given the large number of unstudied phage strains that are associated with the human microbiome and in light of the described studies, many further studies related to phage interactions with animals are needed to advance the use of phages as immune modulators for priming the immune system to withstand pathogen invasion or as adjuncts to enhance traditional phage therapeutic approaches.

Regulatory Approval of Phage-Based Antimicrobials

While the use of phages as natural biocontrol agents to control pathogenic bacteria in food has gained governmental acceptance with the regulatory approval of several diverse phage products (Table 1), there are only a few approved phage products for use in food animal production. These include Finalyse, a phage preparation targeting *E. coli* O157:H7, which received U.S. Department of Agriculture (USDA) Food Safety and Inspection Service approval for commercialization and application as a spray mist or wash on live animals prior to slaughter to decrease pathogen transfer to meat, and Armament, a phage preparation targeting *Salmonella*, which received USDA Food Safety and Inspection Service approval for commercialization and application as a spray mist or wash on the feathers of live poultry prior to slaughter to decrease pathogen transfer to meat (46). Both of these products are used to treat food production animals to reduce foodborne pathogens on their surfaces. As yet, there are currently no products approved for reduction of bacteria within the living animal. Such approval may become easier to achieve since the European Food Safety Authority's Biohazards Panel has endorsed the use of phages as a treatment for foods of animal origin including carcasses, meat, and dairy products. The panel concluded that some phages, under specific conditions, have been demonstrated to be very effective in the targeted elimination of specific pathogens from meat, milk, and products thereof (47). A logical next step would be to use phages to target zoonotic pathogens in the food animals themselves.

A major hurdle to routine phage-based treatment of infections in animals is the absence of well-defined guidelines for regulatory approval of phage-based products. Currently, phage products are regulated according to existing guidelines developed for antibacterials (48). The U.S. Food and Drug Administration (FDA) defines a new animal drug as any drug intended for use in animals other than humans, including any drug intended for use in animal feed but not including the animal feed, the composition of which is such that the drug is not generally regarded as safe and effective for use under the conditions prescribed, recommended, or suggested in the labeling of the drug (49). In the United States,

TABLE 1 Currently approved and commercially available bacteriophage-based products to reduce the presence of foodborne pathogen and spoilage bacteria in foods and food animals[a]

Bacteriophage product	Target microorganism	Regulatory approval	Application
ListShield	*Listeria monocytogenes*	FDA, Israel Ministry of Health	RTE meats, fruits, vegetables, dairy, fish
Listex	*L. monocytogenes*	FDA, FSANZ (New Zealand), Health Canada	RTE foods, RTE meat, meat, dairy, fish, seafood
Finalyse	*Escherichia coli* O157:H7	USDA	Hides of livestock
BacWash	*Salmonella* spp.	USDA	Hides of livestock
Omnilytics	*Salmonella* spp.	USDA	Feathers of live poultry
EcoShield	*E. coli* O157:H7	FDA	Ground beef
SalmoFresh	*Salmonella* spp.	FDA, Health Canada, Israel Ministry of Health	Poultry, fish, fruits, vegetables
Salmonelex	*Salmonella*	FDA	Pork and poultry
Agriphage	*Xanthomonas campestris*, *Pseudomonas syringae*	EPA	Vegetables

[a]Adapted from Woolston and Sulakvelidze (46).
[b]RTE = ready to eat.

animal drugs are approved by the FDA Center for Veterinary Medicine (CVM) (50), and the process begins with the drug sponsor, who collects information regarding the safety and effectiveness of the new animal drug in question. Next, the sponsor submits a new animal drug application (NADA) to the CVM. The NADA includes all the information about the drug and the proposed label. A team of CVM personnel, including veterinarians, animal scientists, biostatisticians, chemists, microbiologists, pharmacologists, and toxicologists, reviews the NADA. If the CVM team agrees with the sponsor's conclusion that the drug is safe and effective (if used according to the proposed label), the NADA is approved and the drug sponsor can legally sell the drug (50).

The current FDA CVM drug approval process raises several questions with respect to phage-based antimicrobials. For example, given the FDA definition of an animal drug, and since foodborne pathogens do not tend to cause disease in animals, is a phage treatment developed to reduce the presence of foodborne pathogens in food animals considered to be a drug? And if a phage product is developed to modulate the immune system, will it be considered to be a drug or a biologic?

Another challenge of the current regulatory approval process is the fact that the regulations do not take into consideration the differences between phages and antibiotics or other antimicrobials. One such difference is that different phages with differing host range specificities can be quickly incorporated into new cocktails for use in the treatment of infections. However, the current drug approval process requires that every phage component of the cocktail must undergo individual safety trials and that the composition of the approved phage cocktail cannot be changed without reapproval. One potential opportunity for approval of phage therapy products is the adoption of a regulatory framework similar to that of influenza virus vaccines. These vaccines are formulated annually from a cocktail of three or four influenza strains that the FDA has approved to be reformulated according to the circulating flu strain (48). Strategies for continuous product development within a streamlined regulatory framework could be described in terms of a personalized drug approach in which individual phages or phage cocktails would be optimized to the current and also potentially regional characteristics of the bacterial agent being treated. If successful, this approach could serve as a mechanism by which the agricultural utility of phages would be aligned with the constant need for safe and abundant antibacterial drugs with specific mechanisms of action (51).

CHALLENGES

The many opportunities for the application of phages as antimicrobials to control infections and the presence of foodborne pathogens in food-producing animals is tempered by several challenges to this approach. Virulent phages are abundant and have proven to be effective *in vitro*. Clinical trials in animals show promising results but also show that the treatments are not completely effective. This is partly due to the studies being carried out with only a few phages, and with limited experimental groups, but is also because phage therapy has limitations *in vivo*. For example, in comparison with antibiotics, an individual phage can only infect one or a few bacterial strains. Therefore, a large number of different phages are needed to treat infections because these are caused by genetically different strains and species of bacteria. In addition, phages are only effective if enough of them can reach the site of bacterial colonization and increase in number *in situ*, which means that the phages will need to avoid clearance by the immune system upon introduction to the animal. Even if the phages can reach the sight of the infection, difficulties may arise in attaching to and infecting the target bacteria, which may be embedded in host tissue and therefore not be accessible, or in many cases bacteria may reside intracellularly. Alternatively, bacterial resistance to the phages may develop, resulting in treatment failure. Finally, a successful phage infection and lysis of its host still has consequences, because large amounts of endotoxin are released upon lysis of Gram-negative bacteria, which can lead to local and systemic complications. Taken together, overcoming these challenges requires careful design and development of phage cocktails, including comprehensive characterization of phage host range and assessment of immunological risks associated with phage treatment.

Phage Clearance by the Immune System

One potential obstacle to using phages for the treatment of systemic disease is that phages can be rapidly removed from the circulatory system when introduced orally or injected intravenously into healthy animals. For example, one characteristic feature of oral administration is the production of secretory IgA in the gut. In one long-term study (52) of antibody induction in mice, following 100 days of phage T4 oral introduction and 112 days without phage treatment, the authors introduced a second application of phage up to day 240. Serum and gut antibodies (IgM, IgG, secretory IgA) were analyzed, and the oral application of T4 phage induced anti-phage antibodies on day 36 (IgG) and day 79 (IgA). The antibody production was related to the high dosage of the phage. Cessation of phage treatment resulted in a decrease of IgA to basal levels, while a second administration of phage induced secretory IgA sooner than that induced by the first administrations. Other studies have shown that the initial antibody response to phage is much less efficient than secondary exposure in primed individuals (44). The researchers concluded that the immunological response is a major factor determining phage survival in the GIT (52). Intravenous inoculation of phages may also result in decreased efficacy of phage therapy. For example, repeated intravenous injections of phages may result in the production of neutralizing antibodies. The effect of such antibodies on the efficacy of phage therapy depends largely on the target antigen. For example, the attachment of antibodies to the base plate of *Myoviridae* phages such as phage T4 can prevent attachment to the bacterium and reduce therapeutic efficacy. The ability of antibodies to inhibit phage therapy also depends on the phages used. Antigenicity varies considerably between phage families. Phage T4 is a good immunogen, but phages T1 and T5 are not. Furthermore, antigenic cross-reactivity has not been observed among *Podoviridae*, *Myoviridae*, and *Siphoviridae* phages, and even the closely related immunogenic T4-like phages show little or no neutralization cross-reactivity (53).

Innate immunity also leads to clearance of phages. Merril et al. (54) demonstrated that a virulent λ phage injected into healthy mice was rapidly sequestered by the reticuloendothelial system (RES); this reinforces that phage survival following application can be a concern during phage therapy, especially given exposure to systemic circulation. Extension of circulation and phage survival, through phage design, can be accomplished via genetic as well as strictly phenotypic means. To that end, Merril et al. (54) developed a serial-passage technique in mice to select for phage mutants able to remain in the circulatory system for longer periods of time. The method entailed the initial intraperitoneal injection into each mouse of 10^{11} PFU of phage λW60 grown on a mutator *E. coli* strain (CRM2), which allowed for an increased rate of mutation in phage λW60, increasing the chances that one or more of the phage offspring would have properties that permitted evasion of the RES. The serial passage of phage, which was followed by injection into animals, isolation, and regrowth in bacteria, was repeated 10 times. Following the 10th cycle of the selection process, two phages, designated Argo1 and Argo2, were isolated and displayed similar enhanced capacity to avoid RES clearing. For example, the 18-hour survival following intraperitoneal injection of Argo1 was 16,000-fold higher and that of Argo2 was 13,000-fold higher than that of the wild-type λW60 strain. A similar selection process was used to isolate long circulating variants of the *Salmonella* phage R34. Analysis of the capsid proteins of the wild-type (λW60) and mutant phages (Argo1 and Argo2) revealed an alkaline shift in the 38-kDa major viral protein in Argo1 compared to λW60. The same electrophoretic protein shift was observed in Argo2. The protein was subsequently identified as the major λ capsid head protein E. Sequence analysis of the protein E gene in Argo1 and Argo2 revealed a G to A transition mutation at nucleotide 6606 in both phages. This transition mutation resulted in the substitution of the basic amino acid lysine for the acidic amino acid glutamic acid at position 158 of the λ capsid E protein in both Argo strains. Argo2 protein profiles displayed the presence of an additional altered protein, which also had an alkaline shift and corresponded to the major capsid head protein D of λ. Capparelli et al. (55) used similar techniques to obtain a bacteriophage (ϕD) which was capable of persisting in mice for 38 days. This phage was effective in clearing an experimental *E. coli* O157 infection within 48 hours.

Alternative (physical) approaches have been employed in an attempt to decrease the immunogenicity of phages and decrease clearance by the RES. Kim and colleagues (56) chemically modified phages by conjugation of monomethoxy-polyethylene glycol (mPEG) to phage proteins. This approach was based on the fact that, since mPEG polymer is nonimmunogenic, conjugation to phage structural components would mask immunogenic proteins, thereby increasing the blood circulation time of the phages. The *Salmonella*-specific phage Felix-O1 and the *Listeria*-specific phage A511 were modified with mPEG. Loss of phage infectivity following PEGylation was observed and related to the degree of modification, which could be controlled by adjusting the mPEG concentration. When injected into naive mice, PEGylated phages showed a strong increase in circulation half-life, but challenge of immunized mice did not reveal a significant difference. The results of the study suggest that the prolonged half-life is due to decreased susceptibility to innate immunity as well as avoidance of cellular defense mechanisms.

The studies aimed at increasing the circulation time of phages used as antibacterials confirm related work that shows that the time of clearance of phages in mammalian organisms depends on their surface protein properties (53). In addition, these studies also show that it is possible, using genomic and physical means, to attenuate immunogenicity of surface proteins via alterations to these surface antigens and, in so doing, decrease their recognition as foreign antigens and avoid rapid clearance by the RES.

Stability of Phages in the GIT

The oral route represents the most common method of phage introduction into animals during phage therapy. This makes sense from a practical standpoint in that phages can be delivered to animals via drinking water and feed. However (and depending on the animal species to be treated), for the phages to be delivered to the lower GIT, they must first traverse various organs which contain hydrochloric acid and digestive enzymes needed for the breakdown of feeds. Because phages tend to become rapidly inactivated at low pH, the viability of orally administered phage may be rapidly reduced under the acidic conditions of organs such as the bovine abomasum (57) and in the presence of enzymes and other digestive compounds such as bile. Without protection, phages may not survive gastric passage and may become inactive in the GIT.

To address this issue, several groups have investigated liposome-based encapsulation of phages as a way to protect phage viability. In one study (58), several *Salmonella*-specific phages were encapsulated in liposomes, and their efficacy in reducing *Salmonella* in poultry was assessed. In simulated gastric fluid (pH 2.8), the titer of unencapsulated phages decreased by 5.7 to 7.8 log units, whereas encapsulated phages were more stable, with a 99% increase in viability compared to the unencapsulated phages. The liposomes also improved the retention of phages in the chicken intestinal tract. For example, when cocktails of the encapsulated and unencapsulated phages were administered to broilers, the encapsulated phages were observed in 38.1% of the animals after 72 hours, while the unencapsulated phages were detected in only 9.5% of the broilers. In experimentally infected broilers, daily administration of the two cocktails over a period of 6 days postinfection with *Salmonella* led to similar levels of protection against *Salmonella* colonization. However, upon cessation of treatment, protection by the unencapsulated phages disappeared, while the encapsulated phages protected poultry for 1 week against *Salmonella*. In a related study, antiphage antibodies were evaluated for their ability to neutralize free and liposome-encapsulated phages. Liposomes completely protected encapsulated phages from neutralizing antibodies, while unencapsulated phages were neutralized within 3 hours by the antibodies. When compared with the inability of unencapsulated phages to enter macrophages, liposomes were able to deliver phages into macrophages and reduce the concentration of intracellular *Klebsiella pneumoniae* by approximately 1.5 logs. In addition, an attempt to compare the efficacy of unencapsulated and encapsulated phages to destroy biofilms alone or in conjunction with amikacin was assessed. The liposome-encapsulated phages exhibited synergistic activity with the amikacin in reducing *K. pneumoniae* biofilms. The authors concluded that liposome-encapsulated phages are effective at reducing bacterial concentrations by overcoming the majority of the hurdles related to clinical use of phages.

In another approach, Ma et al. (59) microencapsulated the phage Felix O1 using a chitosan-alginate-$CaCl_2$ system. *In vitro* studies were used to determine the effects of simulated gastric fluid and bile salts on the viability of free and encapsulated phage. Phage Felix O1 was found to be extremely sensitive to acidic environments and was not detectable after a 5-min exposure to pH values below 3.7, in comparison to microencapsulated phage, which only decreased by 0.67 log units, even at pH 2.4, for the same period of incubation. After 3 hours of incubation in 1% and 2% bile solutions, the unencapsulated phage count decreased by 1.29 and 1.67 log units, respectively, while the concentration of encapsulated phage did not decrease. Also, encapsulated phages were completely released from the microspheres upon exposure to simulated intestinal fluid (pH 6.8) within 6 hours. In the presence of trehalose, which was employed as a stabilizing agent, the microencapsulated phage had a 12.6% survival rate after storage for 6 weeks.

Collectively these studies highlight that the use of encapsulation techniques allows phages to remain viable for longer periods in the GIT and demonstrate that encapsulation allows for efficient delivery of phages to the GIT.

Development of Bacterial Insensitive Mutants (BIMs)

A major concern regarding the use of phage therapy in the treatment of infections is the development of phage-resistant bacteria. In early *in vitro*-based phage resistance work, Luria and Delbrück (60) made the general observation that while a decrease in turbidity occurred within several hours of phage infection of pure bacterial cultures, continued incubation of the culture resulted in the growth of a bacterial variant (a BIM) which was resistant to the attacking phage. The susceptibility of a bacterium to phage infection is primarily dependent on whether the phage can attach to receptors on the cell (61), and mutation of phage receptors represents a primary mechanism by which BIMs can be established. Many cell-surface-exposed components act as receptors for phage adsorption (reviewed by Goodridge [62]) including pili, flagella, capsule, and Gram-specific constituents such as teichoic acid of Gram-positive bacteria components of the O-antigen and outer core oligosaccharide of the LPS in Gram-negative bacteria. Phages also employ various outer membrane proteins (OmpA, OmpC, OmpF, LamB,Tsx, OmpP, FadL) as receptors.

Several groups have developed strategies for developing phage cocktails that are designed to address the issue of BIMs. Tanji et al. (63) described a method to develop a phage cocktail against *E. coli* O157:H7, which was based on the combination of phages that used different receptors to infect *E. coli* O157:H7. For example, the researchers showed that deletion of OmpC from the *E. coli* O157:H7 cells facilitated the emergence of resistant bacterial cells after 8 hours to one phage, designated SP21. Alternatively, alteration of the LPS profile of *E. coli* O157:H7 facilitated cell resistance to a second phage, SP22, which was observed following 6 hours of incubation. However, when a mixture of both phages was used to infect the *E. coli* O157:H7 cells, the emergence of phage-resistant cells was not observed for 30 hours. This work indicated that the combination of two phages employing different receptors was able to significantly delay the emergence of phage resistance cells when compared to either phage used alone (63).

Following up on this work, Chase et al. (64) used a receptor modeling procedure to produce a phage cocktail that contained phages that used multiple different receptors on the cell surface for reduction of *E. coli* O157:H7 in cattle. A total of 56 different phages, with varying degrees of specificity for *E. coli* O157:H7, were screened using isogenic *E. coli* strains (with respect to outer membrane protein [OMP] genes) to determine the receptors used by each phage. In addition to the OMP strains, 12 *E. coli* LPS outer core oligosaccharide mutant strains were used to assess the usefulness of *E. coli* LPS as a receptor for the phages. Each phage was individually tested against each bacterial isolate. The specificity data was used to generate a database of the receptors that each phage used. Some phages used more than one receptor. The database was used to construct a phage cocktail consisting of 37 different phages. When screened against 58 *E. coli* O157:H7 isolates, the cocktail produced complete clearing on all of the isolates, and no resistant colonies were observed. Next, the cocktail was examined for its ability to reduce the presence of *E. coli* O157:H7 in bovine fecal slurries and in calves. In an anaerobic *in vitro* model, the phage cocktail completely eliminated a strain mixture of 10^4 CFU/ml of *E. coli* O157:H7 from bovine fecal slurries within 4 hours. To evaluate the phage cocktail in live animals, 14 black Angus calves ranging from 4 to 6 months of age were orally inoculated with 10^8 CFU *E. coli* O157:H7. The phage cocktail initially decreased the numbers of *E. coli* O157:H7 in the calves, but an increase in the *E. coli* O157:H7 concentration was observed at 36 hours postinoculation. However, the increases in cell concentration were associated with a decrease in phage concentration (meaning there were not enough phages present to kill the remaining *E. coli* cells) and were not due to development of bacterial resistance.

While the above studies suggest that phage resistance can be overcome through the use of well-designed cocktails, it should be noted that BIMs can be produced even when phage cocktails are employed to reduce the presence of a given bacterium.

Release of Endotoxin

One of the side effects of phage therapy to control Gram-negative bacteria is the release of the lipid A portion of LPS (endotoxin) which mediates the general pathological aspects of Gram-negative septicemia. To address this issue, several groups have engineered phages so that they kill without effecting bacterial lysis, by eliminating lysis genes from phages or through use of naturally lysis-deficient phages such as filamentous phages which are released continuously through the cell membrane. Hagens and Bläsi (65) modified phage M13 to encode the restriction endonuclease *Bgl*II gene (phage M13R), resulting in phages that were efficient at killing their hosts while leaving them structurally intact. When a culture of *E. coli* MC4100F cells were infected with M13R, a rapid decline in the cell concentration was observed, resulting in a 2 log decrease of the bacteria after 6 hours. The effects of phage M13R on endotoxin release were compared with those of λcI^- (a lytic λ phage), and the data indicated that the endotoxin levels increased 18-fold after 1 hour of λcI^- infection, while only a 2-fold increase was observed after 1 hour of infection with M13R. After 4 hours, a 27-fold increase in endotoxin concentration was observed in the λcI^--infected culture, compared with a 6-fold increase with M13R.

In a related study, Hagens et al. (66) developed a genetically engineered, nonreplicating, nonlytic phage (Pf3R) based on the filamentous *Pseudomonas aeruginosa* phage Pf3. Phage Pf3R efficiently killed its wild-type host *in vitro* while reducing endotoxin release. In a murine model of *P. aeruginosa* infection, treatment with Pf3R or with a virulent phage resulted in comparable survival rates upon challenge, but in the presence of increased *P. aeruginosa* densities, the survival rate following phage therapy with Pf3R increased, and the inflammatory response decreased compared to that of the virulent phage. Therefore, this study suggests that the increased survival rate of Pf3R-treated mice likely resulted from reduced endotoxin release.

CONCLUSION

The current movement away from antibiotic use in food animal production provides an opportunity for alternative approaches to improve and maintain food animal health. Phage therapy represents an emerging class of natural antimicrobials that could be potentially employed as growth promoters and to control and treat zoonotic diseases in food animals. Depending on which approach is used, phage cocktails may be used to eliminate specific pathogens while leaving the rest of the microflora intact. Alternatively, where there is a need to eliminate multiple bacterial species (as is the case for use as a growth promoter), prophages could be induced from within these species through the use of natural chemicals. These statements highlight the fact that to apply phages to their full potential, it is likely that a combination of approaches will be required.

Nevertheless, for phages to become recognized as effective antimicrobials to control bacterial diseases in animals, many of the challenges and limitations to the use of phages will need to be addressed. The continuing emergence of genomics-based approaches to manipulate bacterial and bacteriophage chromosomes such as Crisper/Cas will enable elegant and specific methods to address concerns such as rapid clearance by the immune system and bacterial resistance to phages. Should the opportunities presented for the use of phage therapy in food animals be realized, the postantibiotic era, described by the WHO (67), in which common infections which have been treatable for decades will once again predominate, may be avoided.

CITATION

Colavecchio A, Goodridge LD. 2017. Phage therapy approaches to reducing pathogen persistence and transmission in animal production environments: opportunities and challenges. Microbiol Spectrum 5(3):PFS-0017-2017.

REFERENCES

1. **World Health Organization.** 2015. WHO estimates of the global burden of foodborne diseases. Foodborne diseases burden epidemiology reference group 2007–2015. http://www.who.int/foodsafety/publications/foodborne_disease/fergreport/en/.
2. **Centers for Disease Control and Prevention.** 2013. National Antimicrobial Resistance Monitoring System: Enteric Bacteria 2013. Human Isolates Final Report. https://www.cdc.gov/narms/pdf/2013-annual-report-narms-508c.pdf.
3. **Scallan E, Hoekstra RM, Angulo FJ, Tauxe RV, Widdowson M-A, Roy SL, Jones JL, Griffin PM.** 2011. Foodborne illness acquired in the United States—major pathogens. *Emerg Infect Dis* **17:**7–15.
4. **Lederberg J, Harrison PF.** 1998. *Antimicrobial Resistance: Issues and Options*. National Academies Press, Washington, DC.
5. **Food and Agriculture Organization of the UN.** 2016. FAO calls for international action on antimicrobial resistance. http://www.fao.org/news/story/en/item/382636/icode/.
6. **Levy S.** 2014. Reduced antibiotic use in livestock: how Denmark tackled resistance. *Environ Health Perspect* **122:**A160–A165.
7. **d'Herelle F.** 1917. Sur un microbe invisible antagoniste des bacilles dysentériques. *CR Acad Sci Paris* **165:**373–375.
8. **Aminov RI.** 2010. A brief history of the antibiotic era: lessons learned and challenges for the future. *Front Microbiol* **1:**134.
9. **Hughes P, Heritage J.** 2004. Antibiotic growth-promoters in food animals. *FAO Anim Prod Health Pap* 129–152. http://www.fao.org/docrep/007/y5159e/y5159e08.htm.
10. **Cadieux B, Colavecchio A, Goodridge L.** 2016. Control of bacterial foodborne pathogens on fresh produce: a Trojan horse tale, abst. T7-O7. Annu. Meet. Int. Assoc. Food Protection, St. Louis, MO.
11. **Zhang X, McDaniel AD, Wolf LE, Keusch GT, Waldor MK, Acheson DW.** 2000. Quinolone antibiotics induce Shiga toxin-encoding bacteriophages, toxin production, and death in mice. *J Infect Dis* **181:**664–670.
12. **Pricer WE Jr, Weissbach A.** 1964. The effect of lysogenic induction with Mitomycin C on the deoxyribonucleic acid polymerase of *Escherichia coli* K12λ. *J Biol Chem* **239:**2607–2612.
13. **Łoś JM, Łoś M, Węgrzyn A, Węgrzyn G.** 2010. Hydrogen peroxide-mediated induction of the Shiga toxin-converting lambdoid prophage ST2-8624 in *Escherichia coli* O157:H7. *FEMS Immunol Med Microbiol* **58:**322–329.
14. **Colomer-Lluch M, Jofre J, Muniesa M.** 2014. Quinolone resistance genes (qnrA and qnrS) in bacteriophage particles from wastewater samples and the effect of inducing agents on packaged antibiotic resistance genes. *J Antimicrob Chemother* **69:**1265–1274.
15. **Nilsson AS.** 2014. Phage therapy: constraints and possibilities. *Ups J Med Sci* **119:**192–198.
16. **Canchaya C, Proux C, Fournous G, Bruttin A, Brüssow H.** 2003. Prophage genomics. *Microbiol Mol Biol Rev* **67:**238–276.
17. **Kang HS.** 2016. Comprehensive analysis of curated prophage genomes from PhiSpy for assessment of phage genome mosaicism and tRNA dependencies. M.S. thesis. San Diego State University, San Diego, CA.
18. **Arthur TM, Brichta-Harhay DM, Bosilevac JM, Guerini MN, Kalchayanand N, Wells JE, Shackelford SD, Wheeler TL, Koohmaraie M.** 2008. Prevalence and characterization of *Salmonella* in bovine lymph nodes potentially destined for use in ground beef. *J Food Prot* **71:**1685–1688.
19. **Enault F, Briet A, Bouteille L, Roux S, Sullivan MB, Petit M-A.** 2017. Phages rarely encode antibiotic resistance genes: a cautionary tale for virome analyses. *ISME J* **11:**237–247.
20. **Allen HK, Looft T, Bayles DO, Humphrey S, Levine UY, Alt D, Stanton TB.** 2011. Antibiotics in feed induce prophages in swine fecal microbiomes. *MBio* **2:**e00260-11.

21. **Labrie SJ, Samson JE, Moineau S.** 2010. Bacteriophage resistance mechanisms. *Nat Rev Microbiol* **8:**317–327.
22. **Irbe RM, Morin LM, Oishi M.** 1981. Prophage (phi 80) induction in *Escherichia coli* K-12 by specific deoxyoligonucleotides. *Proc Natl Acad Sci USA* **78:**138–142.
23. **Norris JS, Westwater C, Schofield D.** 2000. Prokaryotic gene therapy to combat multidrug resistant bacterial infection. *Gene Ther* **7:**723–725.
24. **Westwater C, Kasman LM, Schofield DA, Werner PA, Dolan JW, Schmidt MG, Norris JS.** 2003. Use of genetically engineered phage to deliver antimicrobial agents to bacteria: an alternative therapy for treatment of bacterial infections. *Antimicrob Agents Chemother* **47:**1301–1307.
25. **Wu K, Wood TK.** 1994. Evaluation of the hok/sok killer locus for enhanced plasmid stability. *Biotechnol Bioeng* **44:**912–921.
26. **Jensen RB, Gerdes K.** 1995. Programmed cell death in bacteria: proteic plasmid stabilization systems. *Mol Microbiol* **17:**205–210.
27. **Gerdes K, Gultyaev AP, Franch T, Pedersen K, Mikkelsen ND.** 1997. Antisense RNA-regulated programmed cell death. *Annu Rev Genet* **31:**1–31.
28. **Couturier M, Bahassi el-M, Van Melderen L.** 1998. Bacterial death by DNA gyrase poisoning. *Trends Microbiol* **6:**269–275.
29. **Engelberg-Kulka H, Glaser G.** 1999. Addiction modules and programmed cell death and antideath in bacterial cultures. *Annu Rev Microbiol* **53:**43–70.
30. **Lu TK, Collins JJ.** 2007. Dispersing biofilms with engineered enzymatic bacteriophage. *Proc Natl Acad Sci USA* **104:**11197–11202.
31. **Bikard D, Euler CW, Jiang W, Nussenzweig PM, Goldberg GW, Duportet X, Fischetti VA, Marraffini LA.** 2014. Exploiting CRISPR-Cas nucleases to produce sequence-specific antimicrobials. *Nat Biotechnol* **32:**1146–1150.
32. **Citorik RJ, Mimee M, Lu TK.** 2014. Sequence-specific antimicrobials using efficiently delivered RNA-guided nucleases. *Nat Biotechnol* **32:**1141–1145.
33. **Yosef I, Manor M, Kiro R, Qimron U.** 2015. Temperate and lytic bacteriophages programmed to sensitize and kill antibiotic-resistant bacteria. *Proc Natl Acad Sci USA* **112:**7267–7272.
34. **Pajtasz-Piasecka E, Rossowska J, Duś D, Weber-Dąbrowska B, Zabłocka A, Górski A.** 2008. Bacteriophages support anti-tumor response initiated by DC-based vaccine against murine transplantable colon carcinoma. *Immunol Lett* **116:**24–32.
35. **Górski A, Kniotek M, Perkowska-Ptasińska A, Mróz A, Przerwa A, Gorczyca W, Dąbrowska K, Weber-Dąbrowska B, Nowaczyk M.** Bacteriophages and transplantation tolerance. *Transport Proc* **38:**331–333.
36. **Miernikiewicz P, Kłopot A, Soluch R, Szkuta P, Kęska W, Hodyra-Stefaniak K, Konopka A, Nowak M, Lecion D, Kaźmierczak Z, Majewska J, Harhala M, Górski A, Dąbrowska K.** 2016. T4 phage tail adhesin Gp12 counteracts LPS-induced inflammation *in vivo*. *Front Microbiol* **7:**1112.
37. **Nishikawa M, Hashida M, Takakura Y.** 2009. Catalase delivery for inhibiting ROS-mediated tissue injury and tumor metastasis. *Adv Drug Deliv Rev* **61:**319–326.
38. **Kim K, Ingale S, Kim J, Lee S, Lee J, Kwon I, Chae B.** 2014. Bacteriophage and probiotics both enhance the performance of growing pigs but bacteriophage are more effective. *Anim Feed Sci Technol* **196:**88–95.
39. **Weber-Dabrowska B, Mulczyk M, Górski A.** 2003. Bacteriophages as an efficient therapy for antibiotic-resistant septicemia in man. *Transplant Proc* **35:**1385–1386.
40. **Hoshino K, Takeuchi O, Kawai T, Sanjo H, Ogawa T, Takeda Y, Takeda K, Akira S.** 1999. Cutting edge: Toll-like receptor 4 (TLR4)-deficient mice are hyporesponsive to lipopolysaccharide: evidence for TLR4 as the Lps gene product. *J Immunol* **162:**3749–3752.
41. **Yu F, Mizushima S.** 1982. Roles of lipopolysaccharide and outer membrane protein OmpC of *Escherichia coli* K-12 in the receptor function for bacteriophage T4. *J Bacteriol* **151:**718–722.
42. **Górski A, Weber-Dabrowska B.** 2005. The potential role of endogenous bacteriophages in controlling invading pathogens. *Cell Mol Life Sci* **62:**511–519.
43. **Górski A, Ważna E, Dąbrowska B-W, Dąbrowska K, Switała-Jeleń K, Międzybrodzki R.** 2006. Bacteriophage translocation. *FEMS Immunol Med Microbiol* **46:**313–319.
44. **Kaur T, Nafissi N, Wasfi O, Sheldon K, Wettig S, Slavcev R.** 2012. Immunocompatibility of bacteriophages as nanomedicines. *J Nanotechnol* **2012:**247427.
45. **Hendrix RW.** 2002. Bacteriophages: evolution of the majority. *Theor Popul Biol* **61:**471–480.
46. **Woolston J, Sulakvelidze A.** 2015. Bacteriophages and food safety. *eLS*.
47. **EFSA.** 2009. The use and mode of action of bacteriophages in food production. *EFSA J* **1076:**1–26.
48. **Sulakvelidze A.** 2011. The challenges of bacteriophage therapy. *Eur Ind Pharm* **10:**14–18.
49. **U.S. FDA.** 2015. How U.S. FDA's GRAS notification program works. https://www.fda.gov/

Food/IngredientsPackagingLabeling/GRAS/ucm083022.htm.

50. **U.S. FDA.** 2015. From an idea to the marketplace: the journey of an animal drug through the approval process. https://www.fda.gov/AnimalVeterinary/ResourcesforYou/Animal HealthLiteracy/ucm219207.htm.
51. **Chan BK, Abedon ST, Loc-Carrillo C.** 2013. Phage cocktails and the future of phage therapy. *Future Microbiol* **8:**769–783.
52. **Majewska J, Beta W, Lecion D, Hodyra-Stefaniak K, Kłopot A, Kaźmierczak Z, Miernikiewicz P, Piotrowicz A, Ciekot J, Owczarek B, Kopciuch A, Wojtyna K, Harhala M, Mąkosa M, Dąbrowska K.** 2015. Oral application of T4 phage induces weak antibody production in the gut and in the blood. *Viruses* **7:**4783–4799.
53. **Sulakvelidze A, Kutter E.** 2004. Bacteriophage therapy in humans, p. 381. *In* Kutter E, Sulakvelidze S (ed), *Bacteriophages: Biology and Applications.* CRC Press, Boca Raton, FL.
54. **Merril CR, Biswas B, Carlton R, Jensen NC, Creed GJ, Zullo S, Adhya S.** 1996. Long-circulating bacteriophage as antibacterial agents. *Proc Natl Acad Sci USA* **93:**3188–3192.
55. **Capparelli R, Ventimiglia I, Roperto S, Fenizia D, Iannelli D.** 2006. Selection of an *Escherichia coli* O157:H7 bacteriophage for persistence in the circulatory system of mice infected experimentally. *Clin Microbiol Infect* **12:**248–253.
56. **Kim KP, Cha JD, Jang EH, Klumpp J, Hagens S, Hardt WD, Lee KY, Loessner MJ.** 2008. PEGylation of bacteriophages increases blood circulation time and reduces T-helper type 1 immune response. *Microb Biotechnol* **1:**247–257.
57. **Smith HW, Huggins MB, Shaw KM.** 1987. Factors influencing the survival and multiplication of bacteriophages in calves and in their environment. *J Gen Microbiol* **133:**1127–1135.
58. **Colom J, Cano-Sarabia M, Otero J, Cortés P, Maspoch D, Llagostera M.** 2015. Liposome-encapsulated bacteriophages for enhanced oral phage therapy against *Salmonella* spp. *Appl Environ Microbiol* **81:**4841–4849.
59. **Ma Y, Pacan JC, Wang Q, Xu Y, Huang X, Korenevsky A, Sabour PM.** 2008. Microencapsulation of bacteriophage felix O1 into chitosan-alginate microspheres for oral delivery. *Appl Environ Microbiol* **74:**4799–4805.
60. **Luria SE, Delbrück M.** 1943. Mutations of bacteria from virus sensitivity to virus resistance. *Genetics* **28:**491–511.
61. **Lindberg AA.** 1973. Bacteriophage receptors. *Annu Rev Microbiol* **27:**205–241.
62. **Goodridge LD.** 2010. Design of phage cocktails for therapy from a host range point of view. *In* Villa TG, Veiga-Crespo P (ed), *Enzybiotics: Antibiotic Enzymes as Drugs and Therapeutics.* John Wiley, Hoboken, NJ.
63. **Tanji Y, Shimada T, Yoichi M, Miyanaga K, Hori K, Unno H.** 2004. Toward rational control of *Escherichia coli* O157:H7 by a phage cocktail. *Appl Microbiol Biotechnol* **64:**270–274.
64. **Chase J, Kalchayanand N, Goodridge LD.** 2005. Use of bacteriophage therapy to reduce *Escherichia coli* O157:H7 concentrations in an anaerobic digestor that stimulates the bovine gastrointestinal tract. Institute of Food Technologists Annual Meeting and Food Expo, New Orleans, Louisiana.
65. **Hagens S, Bläsi U.** 2003. Genetically modified filamentous phage as bactericidal agents: a pilot study. *Lett Appl Microbiol* **37:**318–323.
66. **Hagens S, Habel A, von Ahsen U, von Gabain A, Bläsi U.** 2004. Therapy of experimental *Pseudomonas* infections with a nonreplicating genetically modified phage. *Antimicrob Agents Chemother* **48:**3817–3822.
67. **World Health Organization.** 2014. WHO's first global report on antibiotic resistance reveals serious, worldwide threat to public health. http://www.who.int/mediacentre/news/releases/2014/amr-report/en/.

Regulatory Issues Associated with Preharvest Food Safety: United States Perspective

17

SHIRLEY A. MICALLEF[1,2] and ROBERT L. BUCHANAN[2,3]

INTRODUCTION

The microbial safety of agricultural products starts at the preharvest/preslaughter level of primary production, and this stage may be considered as one of the most crucial steps in enhancing safety along the entire farm-to-table continuum. In the United States, the safety of the food supply encompasses a variety of potential chemical, biological, microbiological, radiological, and immunological hazards that are managed by three federal agencies (U.S. Food and Drug Administration [FDA], U.S. Department of Agriculture Food Safety and Inspection Service [USDA-FSIS], and U.S. Environmental Protection Agency [EPA]) and various state agricultural, public health, and environmental protection agencies. This article will focus on hazards associated with bacteriological agents associated with fresh produce as an example of the evolving regulations for managing food safety risks. As such, the primary focus will be changes in regulations enforced by the FDA.

A safety breach early in the food production chain has the potential for broad ramifications and widespread ensuing consequences. As contaminated food fans out from its point of production, it passages through several other

[1]Department of Plant Science and Landscape Architecture, University of Maryland, College Park, MD 20742; [2]Center for Food Safety and Security Systems, University of Maryland, College Park, MD 20742; [3]Department of Nutrition and Food Science, University of Maryland, College Park, MD 20742.

Preharvest Food Safety
Edited by Siddhartha Thakur and Kalmia E. Kniel

doi:10.1128/microbiolspec.PFS-0016-2016

handling and processing steps, with the potential of coming in contact with other products and a multitude of surfaces as it travels along the production chain. Hence, the risk of cross-contamination may be amplified when pathogen introduction occurs early in the chain. Globalization and lengthening of the food supply chain further complicate and potentially augment the transmission of contamination. The risks of pathogenic microbial growth and dispersal are time-, contact surface-, food-, and temperature-dependent. Adding to the importance of establishing on-farm food safety policies, microbiological contamination that occurs at the preharvest stage may be recalcitrant to neutralization, due to the sporadic and ephemeral nature of the source, the volume of product that could be impacted, the duration of occurrence, and the success and fitness with which the contaminating pathogen associates with the food. The physiological state, low prevalence, or low population levels of microbial agents can also impede detection, further hindering control.

While the FDA has had on-farm initiatives in the past (e.g., control of *Salmonella enterica* serovar Enteritidis in layer flocks during the 1970s), a strategic shift in food safety program implementation starting at the preharvest stage has occurred in the United States over the past 20 years. The impetus came from increasing frequencies and more unusual, larger, and more geographically widespread foodborne illness outbreaks in recent years (1). Several outbreaks have been traced to contamination occurring at early points in the food supply chain (2–5). In general, food systems have shifted to become more global, with more centralized production, widespread distribution, and longer supply chains on national and international scales for many products (6). In parallel, local and regional food systems in the United States that operate under models emphasizing sustainable practices, community engagement, and local farm-to-table networks have grown. These systems are persisting and increased in popularity to almost 8% of all systems in 2012 as direct-to-consumer marketing channels have opened up (7). Farming system, food supply chain length, and distances traveled from production to points of sale may impact the microbial safety risk of food products. Shorter supply chains and direct-to-consumer food systems are less likely to be implicated in geographically widespread foodborne illness outbreaks. However, local outbreaks or sporadic disease may still occur. Direct marketing sales routes may thwart the identification of an outbreak due to limited resources at the local level, potentially diminishing the likelihood of pinpointing the source. Despite differences in production and management practices, and food supply chain length, food safety regulatory systems need to encompass the whole gamut of food system models. Protecting food at the point of primary production can help provide assurance for a more uniformly safe product reaching the consumer as a raw commodity or entering secondary agriculture in the manufacturing of foods that provide value addition to producers and should be independent of the scale of production and the target market.

In spite of the importance of maintaining microbial safety at the farm level, guaranteeing food safety at this step is most challenging. Safety is dependent on a multitude of external factors that can be very difficult to identify, ranging from environmental to managerial, and including unpredictable and variable parameters such as climate, extreme weather events, resource availability and ecological interactions in the agro-environment. Historically, food safety problems were dealt with as they occurred, resulting in a concentration of resources being poured into a response. This resulted in the development of robust epidemiological and microbiological tools, data repositories, and a regulatory agency laboratory network called Pulsenet, which utilizes pulsed-field gel electrophoresis and is now transitioning to include whole-genome sequencing (8–11). Pulsenet revolutionized foodborne illness outbreak investigations, allowing for identification of etiologic agents implicated in an outbreak and any contaminated foods containing these agents, serving to implement swift action that results in more rapid outbreak control. Lives have been saved by averting exposure to contaminated food as a result of Pulsenet. Estimates of avoided illness have been determined at over 250,000 salmonellosis cases, over 9,000 pathogenic *Escherichia coli* infections, and over 50 listeriosis cases annually for the period between 1994 and 2009 (12). The associated reduction in productivity and public health costs was calculated at $507 million (12).

Improved surveillance and outbreak identification paved the way for another impactful change that has occurred in recent years in the United States' food safety system: focused attention to safety at the primary production and handling steps. With the advent of the Food Safety Modernization Act (FSMA), signed into law in January 2011 (U.S. Public Law 111-353), the food safety philosophy shifted from responsive to preventive. The law emphasizes the industry's responsibility to provide a safe food supply and the need for standards backed by scientific data and risk assessments, while still recognizing the importance of an effective and prompt response in the event of food contamination. The FSMA is compatible with CODEX ALIMENTARIUS and other standards and provides flexibility for regional and production-specific differences in practices. Moreover, the FSMA directs agencies involved in food safety to work in collaboration and coordination, making the best use of resources and allowing for recognition of inspections among agencies.

PREHARVEST FOOD SAFETY REGULATION IN THE UNITED STATES

As mentioned above, federal oversight of food safety is shared by more than one federal agency, including the FDA (U.S. Department of Health and Human Services), the FSIS (USDA), and the EPA. The FSIS oversees meat, poultry, and "cracked" egg products, working closely with state departments of agriculture. Other foods fall under the purview of the FDA. The FDA inspects and regulates preharvest food safety through the Office of Foods and Veterinary Medicine's Center for Food Safety and Applied Nutrition and Center for Veterinary Medicine, and the Office of Global Regulatory Operations and Policy's Office of Regulatory Affairs, addressing human food and animal feed, as well as both domestically produced food and imports. The EPA has specific responsibility over a number of food safety-related standards such as pesticide use and drinking water standards which impact on-farm food safety, subsequently enforced by the FDA or the FSIS. Both the USDA and the FDA work with other federal agencies; state, local, tribal, and international regulatory partners; and industry and other stakeholders to conduct inspections, ensure compliance, or issue regulations. A number of voluntary quality assurance programs also exist for livestock and poultry, typically driven by industry requirements, which cover preharvest food safety issues.

In raw agricultural crop production, harvesting, and handling, the implementation of practices recommended to minimize risks of crop contamination as outlined by the FDA (13) and industry groups was, until recently, only guidance. The FSMA, however, has directed a significant shift from this standpoint. The FSMA was a sweeping act that provided the FDA with increased authority in multiple areas. As is normal with such changes in legal authority, the FSMA is being implemented by modifying the Federal Food, Drug, and Cosmetic Act to include the requirements outlined in the FSMA, including establishing the factors that would lead to a food being declared adulterated. In relation to on-farm regulations, the key regulatory changes are the standards established under the FSMA's Final Rule on Produce Safety, effective January 2016, after consideration of scientific data and stakeholder input (80 FR 74354, 27 November 2015). This rule mandates compliance with the established standards for growing, harvesting, packing, and holding of fresh produce intended for human consumption. All growers grossing raw produce annual sales exceeding $25,000 are required to comply, although compliance is being phased in over the 4 years following the rule's effective date, based on operation size. Exemptions apply to agricultural commodities that have not been identified as intended for raw consumption and for farms with annual produce sales below $25,000 for the 3 preceding years. The rule offers additional flexibility in the form of qualified exemptions and variances. Qualified exemptions are based on total annual sales and selling to specific end users, including consumers and restaurants, while variances from the requirements may be sought if growing conditions are specific to a locality as long as the risk of produce contamination is not augmented.

BEST PRACTICES, CERTIFICATIONS, AND FSMA COMPLIANCE

Prior to the issuance of the FSMA's final produce safety standards, there was no umbrella system of oversight encompassing fresh produce production at the preharvest level. Instead, participation in voluntary industry programs was sought, often including third-party auditing to verify compliance to guidelines required by distributors and wholesalers. In 1998, the FDA issued the "Guide to Minimize Microbial Food Safety Hazards for Fresh Fruits and Vegetables" to outline nonbinding recommendations for good agricultural practices (GAPs) and good handling practices to reduce the likelihood of contaminating fresh produce crops with foodborne pathogens at the pre- and postharvest stages of production (13). The guide summarizes broad areas in the production of fruits and vegetables, offering recommendations designed to minimize food safety hazards associated with agricultural water use, soil amendments, manure handling and composting, equipment sanitation, wildlife exclusion, worker training, and hygiene and farm planning.

Commodity-specific guidelines were also developed by a number of commodity groups who felt the need to regulate the industry due to product-specific food safety risks, partly motivated by the collective damage incurred by the industry as a result of even a single foodborne illness outbreak. Many of the current commodity-specific guidelines were developed following the call for issuance of commodity-specific and practice-specific guidance in the FDA Food Safety Action Plan of 2004, in collaboration with government, research scientists, and industry, often requiring stricter regulatory mandates. Since 2008, the Florida Department of Agriculture and Consumer Services has enforced Tomato-Good Agricultural Practices, with the aim of reducing food safety contamination risk in tomato production, complemented with post-harvest Tomato Best Management Practices (Florida Administrative Code, 2007). The multistate *E. coli* O157:H7 outbreak associated with contaminated spinach grown in California in 2006 (14, 15) spurred the Leafy Greens Marketing Agreement (LGMA, 2 August 2013) in that state. The agricultural practices outlined in the LGMA are overseen by the California Department of Food and Agriculture through a government verification program that ensures the mandatory compliance of leafy greens safety practices. The California LGMA is regarded as a robust program and has served as a model for guidance specific to other regions, such as the Arizona Leafy Greens Products Shipper Marketing Agreement. Other examples include the California Cantaloupe Marketing Order and state GAPs programs such as the Maryland GAPs and Massachusetts GAPs. Meat and poultry guidelines also include practices that ensure animal health and welfare while encompassing food safety such as the United Egg Producers Animal Husbandry Guidelines for egg-laying flocks (2010).

Best practices verification programs offered by third-party certifiers emanated from these general and commodity-specific guidelines. Meat, poultry, and fresh produce crop industries all participate in third-party verification programs for compliance with recommended practices. Many of these verification programs remain voluntary, though certification requirements have been steadily increasing in recent years, driven mainly by wholesalers and food processing companies that demand guarantees of reduced food safety risk. As mentioned above, once compliance dates for the Final Rule on Produce Safety are reached, the standards therein will also become mandatory for all farms covered under the law. At this time, the FDA does not recognize any food safety program as an alternative to FSMA's Rule on Produce Safety. More partnership between industry and government is expected in the future and may pave the way for recognition of alternative programs by the FDA.

THIRD-PARTY AUDITS AND GOVERNMENT VERIFICATION

Before the enactment of the FSMA, participation in food safety verification programs was voluntary, although it was often mandated by wholesalers or other customers such as food processing plants and retailers. Any certification or verification program depends on regular third-party audits for proof of compliance and involves extensive record-keeping, microbial testing, worker training programs, and development of food safety plans. The industry-driven model has resulted in several commodity-specific guides that may be more stringent than federal requirements. Third-party audits have been able to supplement an otherwise spotty preharvest inspection program conducted by federal agencies. The Global Food Safety Initiative lists recognized schemes—such as Global G.A.P. and PrimusGFS for produce, Global Aquaculture Alliance for seafood, and the Global Red Meat Standard—all based on internationally identified minimum food safety requirements by Global Food Safety Initiative stakeholders (www.myGFSI.com). These are mainly driven by industry rather than by government. Other nonglobal food safety audit schemes also exist, such as the USDA Agricultural Marketing Service-supported Harmonized GAP program.

Prior to the FSMA, most USDA resources for meat and poultry inspections were diverted to slaughtering and processing plants, and FDA inspections of produce farms typically followed the trail of suspected violations. Although the model was generally effective in ensuring compliance, some exceptions, including the listeriosis outbreak linked to cantaloupe production at Jensen Farms in Colorado in 2011 (16) subsequent to successfully passing an audit, revealed possible deficiencies and ambiguities that could be improved. Regular comprehensive auditor training to ensure uniformity and thorough knowledge of the reporting requirements to federal and state agencies is crucial. Moreover, an accreditation program for third-party certifiers would guarantee equal standards among private certifiers. The FSMA is implementing a rigorous Foreign Supplier Verification Program for imported foods, alongside an Accredited Third-Party Certification program to ensure reliable, adequate, and objective inspections and certifications. Such a system of oversight still does not exist for audits performed in the United States. Although no system can be expected to be perfect even with oversight, having a mechanism that oversees auditing bodies could at least ensure that every accrediting body is keeping records of success and failure rates, investigating errors when they arise, and having a system in place that responds to deficiencies to improve the process.

There will be a staggered phasing-in period for compliance of the FSMA's Final Rule on Produce Safety, dependent on production scale, with large producers expected to comply within a year of issuance of the final rule and the smallest producers within 3 years. This shift in the approach of how preharvest food safety is guaranteed is seen as a necessary next step in the U.S. food safety system to ensure that all producers are held to the same standards and are liable under the law to ensure adherence to on-farm safety standards meant to minimize or eliminate microbiological contamination of fresh produce during the cultivation phase. The FDA, however, will face challenges to ensure comprehensive implementation: inspections are time-consuming and expensive and rely on specific expertise in various areas. As explained, although the FDA is currently not accepting third-party-verified food safety programs as alternatives to its rule, it appears that future recognition, following a process ensuring that guidelines within these programs will align with the rule, has not been ruled out. The California LGMA, for example, does not mandate membership, but members agree to compulsory government inspections as a means to verify the standard of this program.

ADVANTAGES AND CHALLENGES OF THE PREVENTATIVE APPROACH

The preventative approach is particularly suited for the protection of food from contamination at the preharvest stage. With the farm being the primary step in food production, implementing preventative measures that reduce pathogen occurrence and prevalence early in the farm-to-fork continuum can have the greatest public health impact and economic benefits. Paybacks are not always immediately apparent to producers, however. Despite the benefits, motivating producers to implement on-farm changes with the aim of reducing contamination of products with foodborne pathogens can be very challenging. This is especially true if the changes are perceived as financially burdensome and time-consuming, particularly in industries where foodborne illness is historically low or in regions that have not been entangled in large or publicized food safety problems. The following is a discussion of factors that can serve as obstacles to farmers, affecting decision-making and presenting difficulties in implementing the law.

Economies of Scale

A recommended first step in implementing on-farm produce safety measures is to develop a food safety plan. Although not an FDA requirement, major purchasers are almost guaranteed to demand a food safety plan as a means of limiting the liability of the retail buyer. Growers and primary producers may perceive the increasing pressures and demands to develop and implement food safety plans and acquire and maintain food safety certification as a burden, the magnitude of which can be greater for smaller businesses with limited resources. Large producers employ food safety staff dedicated to handling all required food safety plan development, record-keeping, testing, compliance, and maintenance of certification while keeping up to date with guidelines, shifting the onerous record-keeping burden from farm staff. Smaller operations may not have the resources for such a solution. Scale of production can impact the perceived burden, the motivation, as well as the ability to change practices or identify and remediate recognized problems (17).

In fact, scale of production does impact the true burden associated with implementation, with decreasing proportional costs for larger operations, leading to impediments in smaller operations (18). A survey of farms adopting the California LGMA found that even medium-sized farms incurred a higher average cost per acre for farm modifications needed for food safety compliance compared to larger farms (19). This was mostly attributed to scale effects. The FDA considered a blanket implementation approach controversial during the development of the Rule on Produce Safety and hence staggered compliance dates and allowed exemptions based on annual sales. However, nonexempt small-scale producers may still experience difficulties in absorbing the costs associated with the installation of food safety modifications on their farms and the employment of food safety staff, as a result of limited resources and the need to diversify crop production. Despite the identified scale effect for the required farm modifications, cost of implementation was ranked very low as a possible impediment to developing a food safety plan in a survey conducted on Mid-Atlantic farms in 2010 and again in 2013 (20). The need to understand grower behavior remains, because financial considerations may not be the only barriers to on-farm food safety policy implementation.

Market Channels

In spite of some small business protections, such as exemption from the FSMA Rule on Produce Safety and other industry programs based on sales, small growers are increasingly required to participate in food safety programs to sell to retail markets (21). Wholesale market channels are more likely than direct market channels to attract farms with GAP certification (20). However the belief exists that small growers are disadvantaged due to a lack of resources to employ food safety staff, implement food safety measures, or perform required testing. The likelihood of testing whether irrigation water used on the farm met microbial standards was higher for farms utilizing wholesale marketing channels (20). Difficulty in the ability to absorb the financial costs and added work needed to successfully develop and implement a food safety plan and keep detailed records can be an impediment to using this market channel. Mid-Atlantic farmers utilizing a wholesale market channel were significantly more likely to have a written food safety plan for their farm (20). By contrast, a Pennsylvania study found no relationship between farm size and writing a plan (22). Selling at farmer's markets, at roadside stands, and through community-supported agricultural programs may be an alternative market channel for growers in this category, but these direct channels should also have safeguards for consumers, making the FSMA exemptions based on sales a controversial issue. A study by Marine et al. found that financial constraints were not the primary reason for not having a written food safety plan; rather, farmers omitted to develop a plan because they were not required to do so under the FSMA (20). Recognition of state programs would help small farmers and consumers. State departments of agriculture and university extension programs can bridge this divide by providing education, support, and mock audit opportunities that help small growers understand their obligations.

The One-Size-Fits-All Approach in Educational Programs

As crucial as food safety programs are to effectively and consistently reduce risk across all agricultural production scales, including small-scale production, the imposition of burdensome regulations runs the risk of antagonizing small growers, hurting small business, and negatively impacting local communities engaged in efforts to promote local and regional production of fruits and vegetables. Incentivizing programs tailored to fit specific needs can be very effective in motivating small producers to seek information, initiate the development of a food safety plan, and learn about log maintenance and worker training tools available through university extension programs. In Maryland, a partnership between the Maryland Department of Agriculture and the University of Maryland has created a program aimed at providing support to growers and has been very successful in motivating growers to pursue food safety education through group training and one-on-one visits. This type of support can be crucial for small growers to navigate published guidance, understand standards, and maintain adequate records for multiple activities from manure composting logs, to water testing dates and results, to worker training records. Hands-on training that incorporates activity-based learning with inclusion of food safety plan development, organization, and writing can be extremely effective in ensuring that knowledge gained has an impact beyond the day of the training or farm visit. The Maryland Department of Agriculture-supported food safety training delivered by food safety educators at the University of Maryland has been very effective in increasing GAP certification through their state GAPs and good handling practices auditing and certification program, providing a model example of how compliance can be incentivized through ongoing and annual educational programs.

These programs are particularly important for farmers who may be exempt from the FSMA's Produce Rule. A Pennsylvania study found that several growers were self-motivated to provide a safe product for their customers (22). This may be the key to expanding GAP action on farms that are exempt from FSMA rules due to their production scale or are not required by their buyers to provide written food safety policies or GAP certifications. Understanding the drivers for GAP implementation that go beyond financial reasons and access to information and education customized to specific sectors of agriculture will ensure the safety of produce that is locally grown on a small scale, benefiting public health, local economies, and agriculture as a whole.

Assessing the Benefits of GAPs

With renewed focus to ensure compliance through the FSMA's direction, resistance to on-farm behavioral change should be approached through educational programs customized to specific grower groups. It is to be expected that some of these changes will take time, effort, and creativity, including for instance, making farmers aware of several added benefits of GAP implementation other than the direct effect of reducing foodborne illness. Several of the measures that can be implemented to reduce contamination risk at the preharvest level also contribute to environmental preservation, worker safety, long-term economic benefits to producers, and produce quality, which could be added incentives for behavioral changes on the farm. Grant funding for programs at the state level has served as a good model to successfully launch educational programs aimed at increasing compliance, but also to educate growers on the benefits of GAPs beyond food safety.

Assessing the impact of specific preharvest food safety mitigation interventions by measuring public health outcomes is crucial to ensure that implemented measures are having the intended effect in a cost-effective manner (23). A good example is the decline in *S.* Enteritidis infections from eggs coinciding with implementation of control measures to reduce infections in egg-laying flocks (24). These types of analyses are harder to conduct when illnesses are underreported, sporadic, or linked to small or singular outbreaks, making trend analysis unviable. This is the case with assessing the cost-effectiveness of GAPs implemented in the production of fresh fruits and vegetables. There is a dearth of data on the benefits of GAPs for crop production, and critics claim that the FSMA will increase costs without providing public health benefits. Studies have shown that, as a general rule, contamination of produce is low, sporadic, or present at undetectable levels (25–29), making it hard to assess the impact of GAPs by measuring the presence of an on-farm enteric pathogen. An alternative approach may be to assess managerial practices such as manure use (30) on regional and local effects such as soil type (31) and monitoring of water bodies (32). All these factors may contribute to persistent or sporadic occurrence of foodborne pathogens on farms and therefore provide a means by which GAP effectiveness may be evaluated indirectly. In turn, correlation analyses between implemented control measures on farms and ensuing health outcomes or pathogen prevalence outcomes can help identify ineffective measures and provide the rationale for devising intervention improvements, ensuring the optimal result for investments.

REGULATED AREAS, CURRENT ISSUES, AND FUTURE DIRECTIONS

The Final Rule for Produce Safety issued in 2015 represents a remarkable example of a rule developed by a regulatory body with substantial input from stakeholders through an extensive effort of public engagement. The final rule encompasses the integration of regulation proposed by the FDA that is risk-based and supported by scientific data, with consideration of stakeholder-raised issues, including some important concerns such as alignment with the National Organic Program and environmental regulations. The rule covers raw agricultural commodities intended for human consumption and covers both domestically grown and imported produce. Requirements for the main areas included in the rule and related limitations and future questions that need to be addressed are discussed below.

Agricultural Water

Until the FSMA-proposed Food Safety Standards were first published in January 2014, no federal standards for agricultural water had been defined by regulatory agencies. Prior to this date, EPA standards for recreational water use were adopted as *de facto* standards by industry and commodity groups such as the California Leafy Greens marketing agreement and the Florida Tomato-GAPs. The metrics in the Final Rule for Produce Safety have not deviated significantly from this initial standard, requiring a geometric mean of generic *E. coli* levels at or below 126 CFU/100 ml of four or more sequential water samples and a statistical threshold value representing acceptable variability of water quality of 410 CFU/100 ml of water for preharvest agricultural applications that will result in water coming in contact with produce, such as irrigation, evaporative cooling, frost protection, and pesticide mixing. Buchanan and Schaffner (32) pointed out that the effectiveness of such a performance-based verification testing program is highly dependent on the frequency of testing both in terms of identifying a loss of control and demonstrating that control has been reestablished. Measures that need to be taken in response to inadequate water quality, however, allow for much more flexibility. The Rule now allows for continued use of water not meeting the standards as long as an adequate time interval is applied between water use and harvest, calculated at a rate of decline of bacterial populations at 0.5 log CFU/day until the standard is met, taking remediating measures to improve water quality or treating the water. Water that will not contact produce does not fit the definition of agricultural water under the law and does not need to meet the requirements.

Conceptually, the definition of a microbiological standard to assess water quality and signal potential fecal contamination is sound and advantageous. By utilizing a single nonpathogenic bacterial species as a sentinel for several pathogenic agents, testing and interpretation of test results are facilitated and costs are kept within reason. Moreover, the *E. coli* standard provides a relatively easy guideline for growers while being independent of region, climate, and farming system. Unfortunately, the actual implementation of the standard is not free of caveats. The Rule does acknowledge the complex and seasonally dependent nature of the relationship between the *E. coli* indicator and several bacterial and viral pathogens and protozoan parasites. This is well documented, with reports of positive and significant correlations being weak or even nonexistent (32–34). This has led to widespread confusion in understanding the rationale for the metric's use. The FDA intends the standard to signal fecal contamination, not presence of enteric pathogens. Fecal indicators do serve an important purpose, pointing to a possible new source of foodborne pathogens when fecal indicator populations spike above background levels. They do little, however, to address another important food safety issue: the presence of pathogens that may be endemic or ephemerally present as part of the natural ecology of a water body being used as an agricultural water source.

An example of this is the widespread occurrence of *S. enterica* in surface waters along the east coast of the United States (25, 31, 32, 35, 36). The bidirectional nature of this weak association can mean that water bodies harboring high *E. coli* levels which exceed the standard may be harboring no pathogens. By contrast, high pathogen prevalence may coexist with low levels of generic *E. coli*, whereby water would pass the current metric in spite of dangerous pathogen loads (35). The former situation exerts an undue burden on growers and the environment. Growers must either discontinue water use for remedial action, which may lead to crop losses, or be forced to use other water sources such as groundwater. This practice in itself is not without consequence: increased demand on groundwater reservoirs can be detrimental to this important resource. On the other hand, farmers may continue to use water that passes the *E. coli* standard even when pathogen status is unknown. However, a lack of microbiological records does not exonerate the farmer from liability in the case of illness being traced back to a farm that is found to harbor pathogens in irrigation water, despite there being no requirement to test for pathogens. Growers may be held liable under a requirement by the Federal Food, Drug, and Cosmetic Act (U.S. Code Title 21) that food cannot be produced under unsanitary conditions. Foodborne illness outbreaks linked to produce contaminated by irrigation water, such as a recurrent *S. enterica* serovar Newport outbreak associated with tomatoes (3), have been reported, but such examples of direct connections between contaminated irrigation water as the source of pathogens found on food and causing outbreaks are rare. On the other hand, evidence that plant epiphytic communities are not influenced by microbiota introduced through foliar applications of water is mounting. Frequently, correlations between *E. coli* in water and the irrigated produce are weak to undetectable (26, 37, 38), while metagenomic analysis of phyllosphere communities exhibited little commonality with communities in water used for pesticide mixing for foliar spraying (39).

The need for risk-based standards remains. The currently implemented standards are an important step in the right direction whereby farmers are required to monitor the microbiological quality of water bodies being used as agricultural water sources. However, more needs to be done to protect both the consumer and the grower. A one-size-fits-all standard may not be adequate. In areas known to harbor specific endemic pathogens, growers should be required to test for that pathogen. Moreover, enough evidence exists to warrant a search for a superior standard, which may require a deviation from the current dogma. Standards need to be rooted in rigorous scientific data and validated in various regions, under varying climatic conditions, based on assessments that evaluate risk of transmission of pathogens, integrating crop type, application method, and time of application. One indicator species may seem convenient and cheap but appears to be unfeasible for endemic environmental pathogens. Although allowing the use of water high in *E. coli* levels would be irresponsible and is rightly not allowed, current metrics may permit the continued use of agricultural waters which pass microbial standards while still carrying significant pathogen risks. In this era of metagenomic tools, we may need to begin implementing ecological approaches to solve this complex issue. Microbial community shifts in response to environmental cues may be better indicators of the conditions that may favor pathogen persistence.

Organic Fertilization of Soil

The establishment of standards for the application of organic fertilizers to soil to improve and maintain soil health and sustain crop health and yield has been problematic. The FDA has adopted the term "biological soil amendments of animal origin" to refer to fertilizers deemed to be high-risk manure products. Data on survival of pathogens in soil amended with untreated manure reveal that pathogen persistence is dependent on a multitude of factors such as bacterial species, soil texture, soil temperature, moisture levels, season, protozoal predation pressures, and pathogen motility and other traits, with survival ranging from a few weeks to over 300 days (40–45). Moreover, there is the added complexity of establishing a standard that aligns with environmental regulation related to soil amendments, as well as other established standards such as the National Organic Program requirements for certified organic production of produce. Recognizing the need for more in-depth risk assessment on the use of animal manure, in lieu of establishing specific metrics for soil amendment application timelines, the FDA has opted to accept the National Organic Program requirements for soil amendments and instead is strictly regulating the process of treating the amendments themselves. Requirements and protocols for composting or manure treatments are clear and sound, as is the definition for what is considered fully composted manure. Simultaneously, the Rule allows for the use of other types of soil amendments, as long as those amendments are proven to be safe and meet the standards that are established for the pathogens *S. enterica*, *E. coli* O157:H7, and *Listeria monocytogenes*. The approach of regulating composting processes while allowing the flexibility of utilizing a variety of fertilizer products may be highly amenable, and future work should continue to dissect the factors that contribute to pathogen survival, as well as continue to map out pathogen dispersal dynamics in a variety of farm settings.

Other Requirements and Considerations: Workers, Wildlife, Equipment, and Environmental Impacts

Little controversy revolves around rules relating to worker health, training, and hygiene, and great strides have been made in this area, now largely accepted as absolutely critical for on-farm food safety. Reducing the possibility of transmission of contamination to produce at harvest has been further enhanced by the Final Rule for Produce Safety requirement to use potable-quality water for worker handwashing. The Rule also establishes requirements for washing equipment, tools, and facilities, including greenhouse structures, seed germination rooms, and any other structure utilized before harvest and for holding produce after harvest. The challenge may still lie with imported foods, for which sewage contamination or other human waste and hygiene-related issues may arise, as was seen in a 2013 outbreak of *Cyclospora cayetanensis*, a human coccidial parasite, traced back to imported cilantro from Mexico (46). Similar outbreaks continued in the spring and summer months of 2014 and 2015, prompting an import alert being imposed on a specific commodity from a specific region outside the United States (Import Alert 24-23, 27 August 2015).

The more contentious aspects of the Rule revolve around livestock, domesticated animals, and wildlife. Source tracking of foodborne illness outbreak strains to fields, livestock, feral animals, and bird populations inhabiting surrounding forested and riparian areas have directly incriminated animals in field contamination of fresh produce. Examples include an *E. coli* O157:H7 outbreak associated with spinach grown in California in 2006, where the outbreak strain was also identified in feral swine and cattle samples (47), and a *Campylobacter jejuni* outbreak linked to peas grown in Alaska in 2008, where the source was tracked to sandhill crane feces (48). Although separation of livestock can be managed on mixed animal-crop farms, directing run-off, capturing aerosols, and excluding wildlife are challenging and expensive, if not in some cases unachievable, goals.

Adding to the complexity of the issue is the need to balance ecosystem, wildlife conservation and water quality conservation, and preservation needs with the obligation to comply with FSMA rules and the desire to minimize preharvest contamination of produce. The Final Rule for Produce Safety does attempt to be flexible in relation to farm practices, and cognizant of cultural diversity and environmental concerns, allows provisions, for instance, for the presence of livestock and working animals. Although the Rule does not require habitat removal, a study assessing the environmental outcomes following the *E. coli* O157:H7 outbreak measured the destruction or degradation of over 13% of the surrounding riparian areas over the subsequent 5-year period (49). The risk here is that fear of becoming implicated in food contamination concerns pushes growers to take extreme action on their farm that imposes dire and detrimental environmental impacts, possibly driven by misinformation about FSMA. Educating growers about the law and its requirements is an important part of implementing this law, to prevent unnecessary and nonmandated animal and habitat destruction. Another potential risk relates to the decisions that growers will have to make in determining how to allocate financial and human resources. The message needs to be clear that agriculture bears a responsibility to balance both food safety and environmental protection. Employment of trade-off management schemes that balance these various needs are being proposed; growers could, for example, be required to assess the cost-benefit and the environmental impact of their food safety measures to reach a more holistic decision that balances food safety and environmental goals (50). This may become of increasing importance as consumers demand not only safe food, but also food that is produced in an environmentally sustainable manner.

CONCLUSION

The regulatory issues, difficulties, and concerns associated with preharvest food safety in the United States are complex. The FSMA has established rules and standards in a way that integrates food safety prevention with economic and environmental considerations. The new law is going into effect over the next 4 years, bringing with it the development of new FDA resources in the form of guides to assist in implementation of the law, strong GAPs educational programs through new collaborations such as the Produce Safety Alliance, and research that fills current gaps in scientific knowledge and investigates socioeconomic and behavioral aspects of the law's implementation. The FSMA and associated rules are a step in the right direction but should continue to be viewed as baseline rules subject to adaptation and change. Future research should continue to validate the proposed metrics, seek improved standards, and evaluate cost-effectiveness and public health outcomes of food safety policy implementation, farmer motivation and behavior, buyer requirements, and consumer opinions. The next phase of food safety policy implementation should pay particular attention to environmental goals while safeguarding agricultural economies and ensuring the best public health outcomes. This can be achieved through awareness and education, by engaging entire communities from grower to consumer. Our goal should be to have food safety plans for all farms, regardless of requirements, through the dissemination of information, public engagement, and the desire to provide and maintain a safe food supply for all.

ACKNOWLEDGMENTS

Some of the research studies described here were supported by Specialty Crop Research Initiative grant 2011-51181-30767 from the USDA's National Institute of Food and Agriculture (NIFA). The sponsor played no role in the study design and implementation. Any opinions, findings, and conclusions expressed in this material are those of the authors and do not necessarily reflect the views of the USDA-NIFA.

The authors declare no conflict of interest.

CITATION

Micallef SA, Buchanan RL. 2017. Regulatory issues associated with preharvest food safety:

United States perspective. Microbiol Spectrum 5(4):PFS-0016-2016.

REFERENCES

1. **Dewey-Mattia D, Roberts VA, Vieira A, Fullerton KE.** 2016. Foodborne (1973–2013) and waterborne (1971–2013) disease outbreaks: United States. *MMWR Morb Mortal Wkly Rep* **63:**79–84.
2. **Buss BF, Joshi MV, Dement JL, Cantu V, Safranek TJ.** 2015. Multistate product traceforward investigation to link imported romaine lettuce to a US cyclosporiasis outbreak: Nebraska, Texas, and Florida, June–August 2013. *Epidemiol Infect* [Epub ahead of print.]
3. **Greene SK, Daly ER, Talbot EA, Demma LJ, Holzbauer S, Patel NJ, Hill TA, Walderhaug MO, Hoekstra RM, Lynch MF, Painter JA.** 2008. Recurrent multistate outbreak of *Salmonella* Newport associated with tomatoes from contaminated fields, 2005. *Epidemiol Infect* **136:**157–165.
4. **Buchholz U, Bernard H, Werber D, Böhmer MM, Remschmidt C, Wilking H, Delere Y, an der Heiden M, Adlhoch C, Dreesman J, Ehlers J, Ethelberg S, Faber M, Frank C, Fricke G, Greiner M, Höhle M, Ivarsson S, Jark U, Kirchner M, Koch J, Krause G, Luber P, Rosner B, Stark K, Kühne M.** 2011. German outbreak of *Escherichia coli* O104:H4 associated with sprouts. *N Engl J Med* **365:**1763–1770.
5. **Cooley M, Carychao D, Crawford-Miksza L, Jay MT, Myers C, Rose C, Keys C, Farrar J, Mandrell RE.** 2007. Incidence and tracking of *Escherichia coli* O157:H7 in a major produce production region in California. *PLoS One* **2:** e1159.
6. **Fabiosa JF.** 2012. Globalization and trends in world food consumption, p 591–611. *In* Lusk JL, Roosen J, Shogren JF (ed), *The Oxford Handbook of the Economics of Food Consumption and Policy*. Oxford University Press, Oxford, United Kingdom.
7. **Low SA, Adalja A, Beaulieu E, Key N, Martinez S, Melton A, Perez A, Ralston K, Stewart H, Suttles S, Vogel S.** 2015. *Trends in U.S. Local and Regional Food Systems: Report to Congress*. USDA, Washington, DC.
8. **Swaminathan B, Barrett TJ, Hunter SB, Tauxe RV, CDC PulseNet Task Force.** 2001. PulseNet: the molecular subtyping network for foodborne bacterial disease surveillance, United States. *Emerg Infect Dis* **7:**382–389.
9. **Kwong JC, Mercoulia K, Tomita T, Easton M, Li HY, Bulach DM, Stinear TP, Seemann T, Howden BP.** 2016. Prospective whole-genome sequencing enhances national surveillance of *Listeria monocytogenes*. *J Clin Microbiol* **54:**333–342.
10. **Swaminathan B, Gerner-Smidt P, Ng L-K, Lukinmaa S, Kam K-M, Rolando S, Gutiérrez EP, Binsztein N.** 2006. Building PulseNet International: an interconnected system of laboratory networks to facilitate timely public health recognition and response to foodborne disease outbreaks and emerging foodborne diseases. *Foodborne Pathog Dis* **3:**36–50.
11. **Taylor AJ, Lappi V, Wolfgang WJ, Lapierre P, Palumbo MJ, Medus C, Boxrud D.** 2015. Characterization of foodborne outbreaks of *Salmonella enterica* serovar Enteritidis with whole-genome sequencing single nucleotide polymorphism-based analysis for surveillance and outbreak detection. *J Clin Microbiol* **53:** 3334–3340.
12. **Scharff RL, Besser J, Sharp DJ, Jones TF, Peter G-S, Hedberg CW.** 2016. An economic evaluation of PulseNet: a network for foodborne disease surveillance. *Am J Prev Med* **50**(Suppl 1): S66–S73.
13. **U.S. Food and Drug Administration.** 1997. *Guide to Minimize Microbial Food Safety Hazards for Fresh Fruits and Vegetables.* Center for Food Safety and Applied Nutrition, U.S. Food and Drug Administration, Washington, DC.
14. **Centers for Disease Control and Prevention (CDC).** 2006. Ongoing multistate outbreak of *Escherichia coli* serotype O157:H7 infections associated with consumption of fresh spinach: United States, September 2006. *MMWR Morb Mortal Wkly Rep* **55:**1045–1046.
15. **Grant J, Wendelboe AM, Wendel A, Jepson B, Torres P, Smelser C, Rolfs RT.** 2008. Spinach-associated *Escherichia coli* O157:H7 outbreak, Utah and New Mexico, 2006. *Emerg Infect Dis* **14:**1633–1636.
16. **Centers for Disease Control and Prevention.** 2011. Multistate outbreak of listeriosis associated with Jensen Farms cantaloupe: United States, August–September 2011. *MMWR Morb Mortal Wkly Rep* **60:**1357–1358.
17. **Parker J, Wilson R, LeJeune J, Doohan D.** 2012. Including growers in the "food safety" conversation: enhancing the design and implementation of food safety programming based on farm and marketing needs of fresh fruit and vegetable producers. *Agric Human Values* **29:** 303–319.
18. **Lichtenberg E, Tselepidakis Page E.** 2016. Prevalence and cost of on-farm produce safety measures in the Mid-Atlantic. *Food Control* **69:**315–323.

19. **Hardesty SD, Kusunose Y.** 2009. *Growers' Compliance Costs for the Leafy Greens Marketing Agreement and Other Food Safety Programs.* University of California Davis UC Small Farm Program Research Brief. http://sfp.ucdavis.edu/files/143911.pdf.
20. **Marine SC, Martin DA, Adalja A, Mathew S, Everts KL.** 2016. Effect of market channel, farm scale, and years in production on mid-Atlantic vegetable producers' knowledge and implementation of good agricultural practices. *Food Control* **59:**128–138.
21. **Tobin D, Thomson J, LaBorde L, Bagdonis J.** 2011. Developing GAP training for growers: perspectives from Pennsylvania supermarkets. *J Ext* **49:**5RIB7. http://www.joe.org/joe/2011october/rb7.php.
22. **Tobin D, Thomson J, LaBorde L, Radhakrishna R.** 2013. Factors affecting growers' on-farm food safety practices: evaluation findings from Penn State Extension programming. *Food Control* **33:** 73–80.
23. **International Commission on Microbiological Specifications for Foods.** 2006. Use of epidemiologic data to measure the impact of food safety control programs. *Food Control* **17:**825–837.
24. **Olsen SJ, Bishop R, Brenner FW, Roels TH, Bean N, Tauxe RV, Slutsker L.** 2001. The changing epidemiology of *Salmonella*: trends in serotypes isolated from humans in the United States, 1987–1997. *J Infect Dis* **183:**753–761.
25. **Micallef SA, Rosenberg Goldstein RE, George A, Kleinfelter L, Boyer MS, McLaughlin CR, Estrin A, Ewing L, Jean-Gilles Beaubrun J, Hanes DE, Kothary MH, Tall BD, Razeq JH, Joseph SW, Sapkota AR.** 2012. Occurrence and antibiotic resistance of multiple *Salmonella* serotypes recovered from water, sediment and soil on mid-Atlantic tomato farms. *Environ Res* **114:**31–39.
26. **Pagadala S, Marine SC, Micallef SA, Wang F, Pahl DM, Melendez MV, Kline WL, Oni RA, Walsh CS, Everts KL, Buchanan RL.** 2015. Assessment of region, farming system, irrigation source and sampling time as food safety risk factors for tomatoes. *Int J Food Microbiol* **196:**98–108.
27. **Gorski L, Parker CT, Liang A, Cooley MB, Jay-Russell MT, Gordus AG, Atwill ER, Mandrell RE.** 2011. Prevalence, distribution, and diversity of *Salmonella enterica* in a major produce region of California. *Appl Environ Microbiol* **77:**2734–2748.
28. **Mukherjee A, Speh D, Dyck E, Diez-Gonzalez F.** 2004. Preharvest evaluation of coliforms, *Escherichia coli, Salmonella,* and *Escherichia coli* O157:H7 in organic and conventional produce grown by Minnesota farmers. *J Food Prot* **67:**894–900.
29. **Mukherjee A, Speh D, Jones AT, Buesing KM, Diez-Gonzalez F.** 2006. Longitudinal microbiological survey of fresh produce grown by farmers in the upper midwest. *J Food Prot* **69:** 1928–1936.
30. **Strawn LK, Gröhn YT, Warchocki S, Worobo RW, Bihn EA, Wiedmann M.** 2013. Risk factors associated with *Salmonella* and *Listeria monocytogenes* contamination of produce fields. *Appl Environ Microbiol* **79:**7618–7627.
31. **Haley BJ, Cole DJ, Lipp EK.** 2009. Distribution, diversity, and seasonality of waterborne salmonellae in a rural watershed. *Appl Environ Microbiol* **75:**1248–1255.
32. **McEgan R, Mootian G, Goodridge LD, Schaffner DW, Danyluk MD.** 2013. Predicting *Salmonella* populations from biological, chemical, and physical indicators in Florida surface waters. *Appl Environ Microbiol* **79:**4094–4105.
33. **Wilkes G, Edge T, Gannon V, Jokinen C, Lyautey E, Medeiros D, Neumann N, Ruecker N, Topp E, Lapen DR.** 2009. Seasonal relationships among indicator bacteria, pathogenic bacteria, *Cryptosporidium* oocysts, *Giardia* cysts, and hydrological indices for surface waters within an agricultural landscape. *Water Res* **43:**2209–2223.
34. **Economou V, Gousia P, Kansouzidou A, Sakkas H, Karanis P, Papadopoulou C.** 2013. Prevalence, antimicrobial resistance and relation to indicator and pathogenic microorganisms of *Salmonella enterica* isolated from surface waters within an agricultural landscape. *Int J Hyg Environ Health* **216:**435–444.
35. **Luo Z, Gu G, Ginn A, Giurcanu MC, Adams P, Vellidis G, van Bruggen AHC, Danyluk MD, Wright AC.** 2015. Distribution and characterization of *Salmonella enterica* isolates from irrigation ponds in the southeastern United States. *Appl Environ Microbiol* **81:**4376–4387.
36. **Bell RL, Zheng J, Burrows E, Allard S, Wang CY, Keys CE, Melka DC, Strain E, Luo Y, Allard MW, Rideout S, Brown EW.** 2015. Ecological prevalence, genetic diversity, and epidemiological aspects of *Salmonella* isolated from tomato agricultural regions of the Virginia Eastern Shore. *Front Microbiol* **6:**415.
37. **Won G, Schlegel PJ, Schrock JM, LeJeune JT.** 2013. Absence of direct association between coliforms and *Escherichia coli* in irrigation water and on produce. *J Food Prot* **76:**959–966.
38. **Pahl DM, Telias A, Newell M, Ottesen AR, Walsh CS.** 2013. Comparing source of agricultural contact water and the presence of fecal

indicator organisms on the surface of 'juliet' grape tomatoes. *J Food Prot* **76:**967–974.

39. **Telias A, White JR, Pahl DM, Ottesen AR, Walsh CS.** 2011. Bacterial community diversity and variation in spray water sources and the tomato fruit surface. *BMC Microbiol* **11:**81.
40. **Holley RA, Arrus KM, Ominski KH, Tenuta M, Blank G.** 2006. *Salmonella* survival in manure-treated soils during simulated seasonal temperature exposure. *J Environ Qual* **35:**1170–1180.
41. **García R, Baelum J, Fredslund L, Santorum P, Jacobsen CS.** 2010. Influence of temperature and predation on survival of *Salmonella enterica* serovar Typhimurium and expression of invA in soil and manure-amended soil. *Appl Environ Microbiol* **76:**5025–5031.
42. **You Y, Rankin SC, Aceto HW, Benson CE, Toth JD, Dou Z.** 2006. Survival of *Salmonella enterica* serovar Newport in manure and manure-amended soils. *Appl Environ Microbiol* **72:**5777–5783.
43. **McLaughlin HP, Casey PG, Cotter J, Gahan CG, Hill C.** 2011. Factors affecting survival of *Listeria monocytogenes* and *Listeria innocua* in soil samples. *Arch Microbiol* **193:**775–785.
44. **Nicholson FA, Groves SJ, Chambers BJ.** 2005. Pathogen survival during livestock manure storage and following land application. *Bioresour Technol* **96:**135–143.
45. **Côté C, Quessy S.** 2005. Persistence of *Escherichia coli* and *Salmonella* in surface soil following application of liquid hog manure for production of pickling cucumbers. *J Food Prot* **68:**900–905.
46. **desVignes-Kendrick M, Reynolds K, Lee T, Gaul L, Klein K, Irvin K, Wellman A, Hardin A, Williams I, Wiegand R, Harris J, Parise M, Abanyie F, Reid Harvey R, Centers for Disease Control and Prevention (CDC).** 2013. Outbreaks of cyclosporiasis: United States, June–August 2013. *MMWR Morb Mortal Wkly Rep* **62:**862.
47. **Jay MT, Cooley M, Carychao D, Wiscomb GW, Sweitzer RA, Crawford-Miksza L, Farrar JA, Lau DK, O'Connell J, Millington A, Asmundson RV, Atwill ER, Mandrell RE.** 2007. *Escherichia coli* O157:H7 in feral swine near spinach fields and cattle, central California coast. *Emerg Infect Dis* **13:**1908–1911.
48. **Gardner TJ, Fitzgerald C, Xavier C, Klein R, Pruckler J, Stroika S, McLaughlin JB.** 2011. Outbreak of campylobacteriosis associated with consumption of raw peas. *Clin Infect Dis* **53:**26–32.
49. **Gennet S, Howard J, Langholz J, Andrews K, Reynolds MD, Morrison SA.** 2013. Farm practices for food safety: an emerging threat to floodplain and riparian ecosystems. *Front Ecol Environ* **11:**236–242.
50. **Pollans MJ.** 2015. Regulating farming: balancing food safety and environmental protection in a cooperative governance regime (2015). *Wake Forest Law Rev* **50:**399–460.

Regulatory Issues Associated with Preharvest Food Safety: European Union Perspective

18

LIS ALBAN[1]

FOOD SAFETY AND THE NEED FOR SURVEILLANCE AND CONTROL IN THE EUROPEAN UNION

Free movement of safe and wholesome food is an essential aspect of the internal market within the European Union. This contributes significantly to the health and well-being of citizens and to their social and economic interests. A high level of protection of human life and health should be assured in the pursuit of community policies while allowing for flexibility when appropriate (1). These lines describe the overall policy in the European Union. The policy is practiced through different parts of the European Union regulatory framework, which is described below.

The responsibility for the progress of food safety is split between the European Union (EU) Commission, which makes legislation as risk managers, and The European Food Safety Authority (EFSA), which provides independent scientific support and advice to the EU Commission as risk assessors. EFSA provides support and advice to the EU Commission on the risks to human and animal health related to zoonotic hazards in the environment, in the food chain, and in animal populations. EFSA takes an integrated approach to its work, involving a number of its scientific panels and units if there is a concern about the food chain. The scientific panels assess the risk related to a

[1]Danish Agriculture and Food Council, DK-1609 Copenhagen V, Denmark.

Preharvest Food Safety
Edited by Siddhartha Thakur and Kalmia E. Kniel

doi:10.1128/microbiolspec.PFS-0003-2014

given activity or hazard and produce opinions about questions received from the EU Commission. Moreover, independent scientists contribute through a number of working groups, for instance, a working group on antimicrobial resistance. Input is also received and discussed with a number of stakeholders such as the European Livestock and Meat Trading Union (UECBV), the Liaison Center for the Meat Processing Industry (CLITRAVI), the European Poultry Industry (AVEC), and Copa-Cogeca (the union of European farmers and their cooperatives).

Based on data collected and reported by the individual member states, European Union summary reports on zoonotic infections, foodborne outbreaks, and antimicrobial resistance are produced in collaboration between EFSA and the European Center for Disease Prevention and Control (ECDC). EFSA's scientific panels review the annual reports and make recommendations on prevention and measures. The output of all EFSA's work is presented on their website (http://www.efsa.europa.eu).

According to EFSA (2, 3), several zoonotic hazards are capable of causing disease in humans. These are *Campylobacter*, *Salmonella*, *Yersinia*, verotoxigenic *Escherichia coli* (VTEC) (also known as shigatoxigenic *E. coli* [STEC]), *Toxoplasma gondii*, *Listeria monocytogenes*, *Coxiella burnetii* (Q-fever), *Brucella*, *Trichinella*, West Nile fever, bovine tuberculosis, and lyssavirus (rabies). *Campylobacter*, *Salmonella*, and *Yersinia* account for the majority of human cases (Fig. 1). All hazards on this list are foodborne except for *C. burnetii*, lyssavirus, and West Nile virus.

The importance of a zoonosis as a human infection does not depend solely on the incidence in the human population. The severity of the disease and the case fatality are also important factors affecting the relevance. Therefore, despite the relatively low number of cases caused by VTEC/STEC, *Listeria*, *Echinococcus*, *Trichinella*, and lyssavirus (rabies), these hazards are considered important in the European Union due to the severity of the illnesses they cause and thereby higher case fatality risks (2, 3).

Generally speaking, poultry is considered the source of the majority of human campylobacteriosis cases—consumption of poultry meat attributing to 20 to 30% of cases,

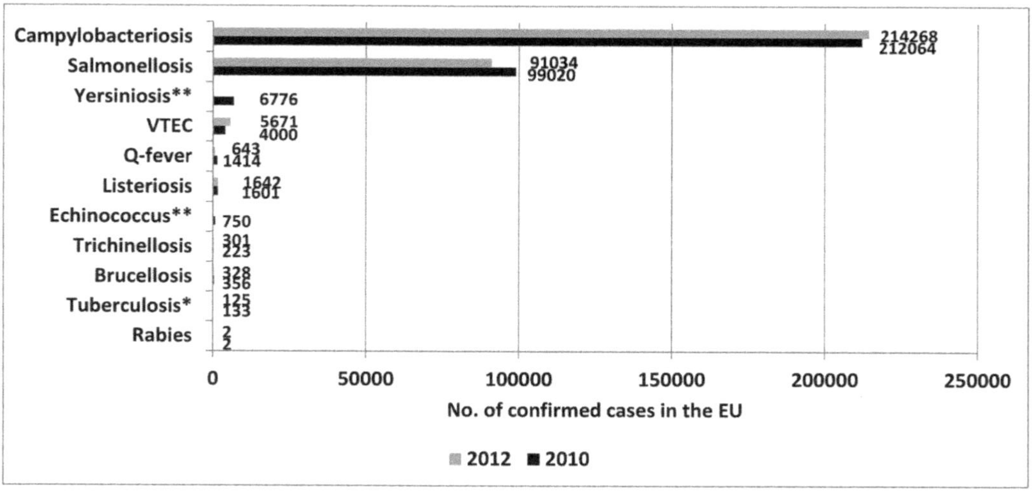

FIGURE 1 Reported number of confirmed human cases of zoonoses in the European Union in 2010 and 2012. Modified from references 2 and 3. *Tuberculosis caused by *Mycobacterium bovis*. **No data available for 2012.

while 50 to 80% may be attributed to the poultry reservoir seen as a whole in the European Union (4). Poultry meat and eggs are also a significant source of human *Salmonella* infections, although improvements are being seen as a result of setting targets for *Salmonella* in poultry production in the European Union (2). Furthermore, a non-negligible proportion of the human cases of infection with *Salmonella*, *Yersinia*, and *Trichinella* can be ascribed to pork. For *Toxoplasma*, pork is one among several sources, and the proportion of human cases caused by pork is unknown. Finally, VTEC/STEC is primarily ascribed to cattle/beef and other ruminants as well as vegetables.

Apart from the zoonotic infections listed in Fig. 1, there is concern in the European Union about the development of antimicrobial resistance due to the use of antimicrobials in livestock. It is feared that this might result in treatment failure and/or prolonged treatment in humans infected with resistant bacteria. Drug-resistant bacteria are estimated to be responsible for 25,000 human deaths annually and cost European Union member states more than €1.5 billion annually (5). In most of these cases the specific impact of the resistance is unknown, because the fatalities are related to a combination of advanced age, comorbidity, lack of immune competence, and the disease course itself (e.g., dehydration after diarrhea). An example of this is, e.g., seen in humans infected with macrolide-resistant *Campylobacter*, as described by Alban et al. (6).

Despite of a plethora of actions taken to ensure food safety in the European Union, a large number of human cases of foodborne illness occur annually. In times of economic crisis, resources are even more limited than usual, hampering implementation of additional preharvest control programs. Hence, the feasibility of improvements in areas such as food safety is related to the economy in the livestock industry—unless governments can or will pay. A global economic crisis will not motivate farmers or abattoirs to implement costly changes that will further minimize the risk related to the animals or the products thereof. Hence, it is a challenge to ensure food safety in a cost-effective way. Still, this is aimed for continuously within the European Union.

The following sections present the most important parts of the regulatory framework in the European Union and discuss actions taken in member states that are forerunners within food safety. In addition, actions in the pipeline will be presented and discussed. Ultimately, food safety policy is a result of a combination of consumers' perception and what is feasible, and in the European Union the latter means what the member states can agree to.

The focus will be on pig/pork, cattle/beef, poultry meat, and eggs. Bovine spongiform encephalopathy (BSE) will also be covered.

REGULATORY FRAMEWORK IN THE EUROPEAN UNION

In the European Union, most legislation is framed as either a regulation or a directive. Regulations have general application; they are binding and directly applicable in all member states. In contrast, a European Union directive sets out the objective or the policy which should be attained. Each member state must then pass the relevant legislation to give effect to the terms of the directive within a certain time period.

The General Food Law and the Associated Parts of Legislation

Regulation 178/2002—called the General Food Law—deals with the general principles, requirements, and procedures regarding food. The core of the entire European Union food safety legislation can be found in article 14, where it is stated that food shall not be placed on the market if it is unsafe (1). Because it is a regulation, it applies directly in all member states.

The General Food Law was introduced to ensure free movement of safe and wholesome food while avoiding distortion of competition. Previous to the adoption of this regulation, the food laws differed between the individual member states with respect to concepts, principles, and procedures. It was the intention to approximate these concepts, principles, and procedures to form a common basis for the measures governing food and feed. Moreover, it was found necessary to ensure that consumers, other stakeholders, and trading partners had confidence in the decision-making process underpinning food law, including the scientific basis and the structures and independence of the institutions protecting health and other interests. An example of this is that the establishment and duties of EFSA are stipulated in the General Food Law. The BSE crisis taking place during the years up to 2002 probably also contributed to the need for a common, clear, fair, and effective food law.

The parts of the regulatory framework that deals specifically with preharvest food safety are the Zoonosis Directive 2003/99/EC, the Zoonosis Regulation 2160/2003, and the Hygiene Package (Fig. 2). The following will describe the Zoonosis Directive, the Zoonosis Regulation, and selected parts of the Hygiene Package—as well as the food law.

Control of *Salmonella* and Other Foodborne Zoonotic Agents

The Zoonosis Directive 2003/99/EC sets the general requirement for monitoring zoonoses and zoonotic agents (7), whereas the Zoonosis Regulation 2160/2003 states that common targets and control programs for the reduction of *Salmonella* in poultry and pig production will be set on the European Union

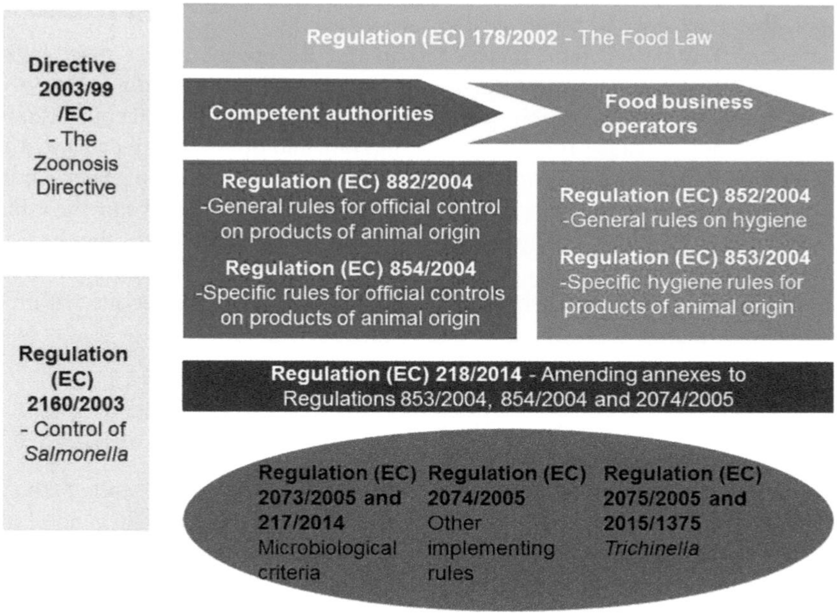

FIGURE 2 Graphical description of the European Union regulatory framework for food safety: the Food Law, the Zoonosis Directive, the *Salmonella* Regulation, and the Hygiene Package (Regulations 852, 853, 854, and 882); other relevant legislation (Regulations 2073, 2074, and 2075 from 2005); and the new regulations 216, 217 and 218 from 2014, which have updated the older parts of the legislation.

level (Annex I of the regulation) (8). Other foodborne zoonotic agents or other animal populations should be addressed, if necessary. The regulation foresees that target setting should take place at the level of primary production and/or, where necessary, at other appropriate stages of the food chain. In other words, testing should be performed at the herd level or of carcasses at the slaughterhouse (8). It is stated in the regulation that the member states are to implement the control program. Annex II lists general requirements for national control programs.

For pigs, no preharvest target has been set so far. It is specified that testing should cover all *Salmonella* serotypes with public health significance for both breeding and finishing pigs. Even though this is a regulation, not all member states have established a preharvest monitoring program for finishing pigs. However, at least Belgium, Finland, Germany, The Netherlands, Sweden, and the United Kingdom have monitoring programs in place; see Baptista (9) for a thorough presentation of the various programs in the European Union. Only few member states apart from the Scandinavian countries have a surveillance program implying that actions are taken on results obtained from preharvest surveillance. In Denmark, finishing pigs from herds with a high prevalence of *Salmonella* are subjected to sanitary slaughter, which mainly consists of hot water decontamination (10). Moreover, an industry-driven penalty scheme is in place for finishing pig herds, in which a certain percentage of the carcass value is deducted from the finishers originating from herds with a moderate or a high level of *Salmonella* (10).

For cattle, no specific regulation on or requirement for preharvest testing for *Salmonella* is in place at the European Union level. Moreover, the prevalence of *Salmonella* in beef is sporadic to low and of limited concern in the European Union in general (3). However, this may change due to the severity of the disease in humans caused by the host-adapted *Salmonella enterica* serotype Dublin, which accounts for the majority of isolates from cattle and beef (11). Reductions in prevalence among cattle are possible, but it is difficult to keep the infection at a low within-herd prevalence without risking new outbreaks and consequential food safety issues and production losses related to outbreaks. Thus, eradication of *Salmonella* in cattle is aimed for in the Scandinavian countries, which are the only member states which have nationally organized control efforts in place. In Sweden, eradication is aimed for through very strict handling of infected premises by veterinary authorities, followed by repeated bacteriological culturing of samples from all animals on the premises before the herd can be considered free from infection (12, 13).

S. Dublin mainly causes disease in calves but also causes abortions, and it has been shown that veal producers who purchased calves from *Salmonella* test–positive dairy herds were more likely to deliver infected animals to slaughter (14). The Danish approach therefore aims at eradicating *S.* Dublin from the cattle population through stepwise regional eradication with strict movement restrictions and mandatory, documented effective action plans for elimination of *S. Dublin* in test-positive herds (15). All cattle herds are classified by use of serology. Owners of test-positive herds have to implement a risk-based action plan to eliminate *S. Dublin* and have to document the effect of this plan by serological testing of indicator groups of animals. The Danish Veterinary and Food Administration imposes official restrictions on infected herds. This requires, among other things, that the animals are accompanied for slaughter by an official passport and slaughtered late in the day under tightened hygienic precautions, at lowered line-speed, and carcasses are subjected to microbiological testing; *S.* Dublin–positive carcasses are either heat-treated or condemned (16).

For poultry and eggs, testing should cover all *Salmonella* serotypes with public health significance (Regulation 2160/2003). Regulation 646/2007 has set a target for broilers of 1% prevalence of *S. enterica* serotype Enteritidis and *S. enterica* serotype Typhimurium, which constitute the majority of human cases ascribed to poultry. The target for table egg layer flocks is 2% or below as laid down in Regulation 1168/2006 (17). The target had to be reached by 31 December 2011 (18). Moreover, Regulation 200/2010 specifies that the maximum percentage of adult breeding flocks of *Gallus gallus* remaining positive for *S.* Enteritidis, *S. enterica* serotype Infantis, *S. enterica* serotype Hadar, *S.* Typhimurium, and *S. enterica* serotype Virchow (the relevant *Salmonella* spp.) is to be 1% or less (19). European Union data show a continued decrease in the numbers of human salmonellosis cases, which is likely to be mainly related to the successful *Salmonella* control programs in fowl (3).

Denmark has had intensive *Salmonella* control programs since the 1990s, and a zero strategy (eradication) is in place for production of eggs and broilers, implying actions against all serotypes. The target of 1% *Salmonella*-positive broiler flocks was reached in 2000, and the prevalence in table egg laying flocks has been lower than 2% since 2004 (20). In 2007, Denmark applied to the European Union for special guarantees for *Salmonella* in table eggs, a position attained by Sweden and Finland at the time of their accession into the European Union in 1995 (21). In 2012, special guarantees for *Salmonella* in table eggs were granted to Denmark, implying that eggs placed on the market should be of the same high standard (22). Hence, in the European Union a member state will be able to benefit from actions taken to deal effectively with a food safety challenge.

The Precautionary Principle

The General Food Law (Regulation [EC] 178/2002) specifies that risk assessments should be undertaken to ensure confidence in the scientific basis for the food law in general, except where this is not appropriate to the circumstances (1). Moreover, according to Article 6 in the regulation, risk assessment will be based on the available scientific evidence and undertaken in an independent, objective, and transparent manner (1).

Article 7 in the General Food Law sets rules for how and when the precautionary principle should be used:

> 1. In specific circumstances where, following an assessment of available information, the possibility of harmful effects on health is identified but scientific uncertainty persists, provisional risk management measures necessary to ensure the high level of health protection chosen in the Community may be adopted, pending further scientific information for a more comprehensive risk assessment.
>
> 2. Measures adopted on the basis of paragraph 1 shall be proportionate and no more restrictive of trade than is required to achieve the high level of health protection chosen in the Community, regard being had to technical and economic feasibility and other factors regarded as legitimate in the matter under consideration. The measures shall be reviewed within a reasonable period of time, depending on the nature of the risk to life or health identified and the type of scientific information needed to clarify the scientific uncertainty and to conduct a more comprehensive risk assessment.

Hence, the precautionary principle makes it possible to introduce preliminary actions that should be in place until sufficient knowledge is collected to assess the risk. There might not always be agreement about what sufficient knowledge is.

Compared to earlier regulations, the General Food Law induced a shift in primary responsibility for food safety: from veterinary authorities to the so-called food business operators (FBO), who can be farmers, meat processors including abattoir owners, and retailers. Moreover, focus has shifted from control of food safety at the company level to the supply chain level, which implies the production chain from stable to table (23). However, the shift in responsibility to the FBO is somewhat contradicted in Regulation (EC) 854/2004, which specifies in detail what the official control should consist of in relation to meat inspection. According to this regulation, almost no responsibility is delegated to the FBO (24).

General Rules on Hygiene: for the FBOs

Regulation (EC) 852/2004 lays down the general rules on hygiene to be followed by the FBOs (25). According to the regulation, FBOs who carry out any stage of production, processing, and distribution of food after primary production must implement and maintain procedures based on hazard analysis and critical control point (HACCP) principles. Such an HACCP program is based on the following elements: (i) conduct a hazard analysis, (ii) identify critical control points, (iii) set critical limits, (iv) establish monitoring procedures, (v) establish corrective actions, (vi) establish record-keeping procedures, and (vii) verify that the HACCP works (Table 1).

As a part of an HACCP, a measure of compliance is conducted. This might consist of a test determining the number of coliform bacteria (indicating growth or not of a specific bacteria), or more commonly it could be an indirect measure such as the temperature. To identify where in the food production process interventions can be done, a risk analysis needs to be conducted. A risk analysis consists of the following four elements: (i) hazard identification, (ii) risk assessment, (iii) risk management, and (iv) risk communication. Figure 3 shows how the different elements of an HACCP relate to risk analysis, as suggested by Mellor (26).

Specific Rules on Hygiene for the FBOs

Regulation (EC) 853/2004 states the specific hygiene rules for products of animal origin (21). This includes the concept of food chain information. In brief, this relates to information about the animal such as:

1. The status of the holding or the regional animal health status
2. The health status of the animals
3. Veterinary medicinal products and compliance with withdrawal periods
4. Occurrence of disease that may affect the safety of the meat
5. Results of any analysis taken that may be of relevance for protection of public health
6. Relevant reports about previous ante- and postmortem inspections of animals from the same holding, including reports from the official veterinarian
7. Production data, when this might indicate the presence of disease
8. Name and address of the private veterinarian normally attending the farm

The slaughterhouse operator may be provided with the main part of the information listed above through a standing arrangement or a quality assurance scheme. Record-keeping of relevant food safety information is a part of an HAACP system, as described in Table 1.

TABLE 1 Relations between various elements of hazard analysis of critical control points (HACCP) and risk analysis – according to Mellor (26)

	Element of risk analysis			
Element of HACCP	**Hazard identification**	**Risk assessment**	**Risk management**	**Risk communication**
Conduct a hazard analysis	X	X		
Identify critical control points	X		X	
Set critical limits		X	X	
Establish monitoring procedures			X	
Establish corrective actions			X	
Establish record-keeping procedures			X	X
Verification that HACCP works			X	X

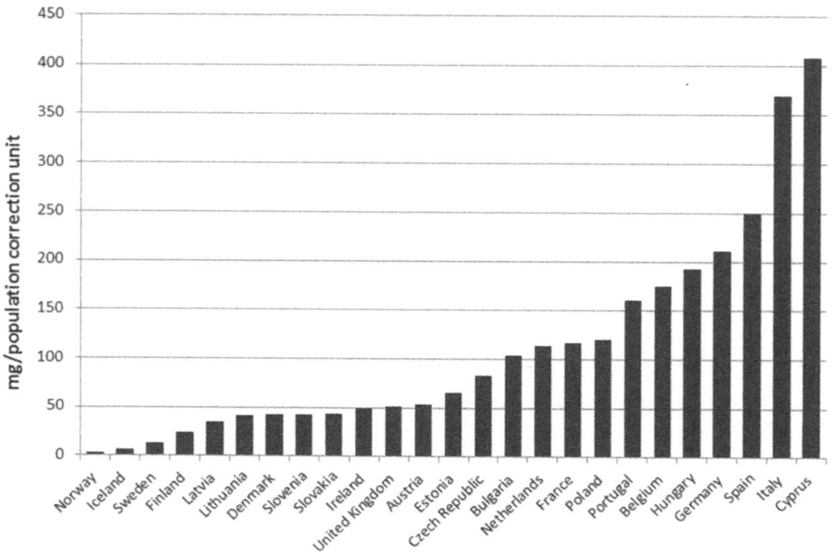

FIGURE 3 Population-corrected sales of antimicrobials for food-producing animals, including horses, for 25 European Union/European Economic Area countries in 2012 (41).

According to Regulation (EC) 853/2004, slaughterhouse operators must not accept animals onto the slaughterhouse premises unless they have been provided with relevant food chain information. Slaughterhouse operators must be provided with the information no less than 24 hours before the arrival of animals at the slaughterhouse, except in specific circumstances such as emergency slaughter, slaughter of horses, or where antemortem inspection has taken place at the farm (21). However, today the competent authority may permit the information to arrive less than 24 hours before the animal arrives. This makes it possible for smaller abattoirs to receive the information by paper upon the arrival of the animal (27).

The regulation also describes requirements for dairies, abattoirs, and cutting plants and for maintenance of hygiene, e.g., by setting maximum limits for the temperature during production.

Microbiological Criteria and *Salmonella*

Regulation (EC) 2073/2005, concerning microbiological criteria, sets the requirement for the level of food safety that the FBO should provide by use of his or her own control program. The regulation specifies how surveillance and monitoring should be conducted for carcasses and the meat products thereof as well as a variety of other food items of animal origin (28). The regulation operates with two kinds of criteria: process hygiene criteria and food safety criteria. For each criterion, it is specified how many samples should be taken and how often. This includes a description of the sampling plan (n = number of samples, and c = number of samples above m), acceptable limits (a maximum of c/n values are observed between m and M, and no value exceeds M), analytical reference methods, stage where a specific criterion applies, as well as action in case of unsatisfactory results. Moreover, the reactions to findings above the specified limits are described; for process hygiene criteria, corrective measures must be taken in the production line, whereas for food safety criteria, recall of products is required (28).

For pig carcasses, five swab samples are to be taken and analyzed for *Salmonella* per week, and the results are evaluated over

10 weeks. Until June 2014, the limit for this process criterion was five positive out of the 50 samples. Today, the limit is three out of 50 (29). If more positives are found, corrective measures must be taken along the production line. This implies that prevalence up to 6% (3 positive out 50 samples) is accepted on the carcass.

For poultry and turkeys, the initial process criterion was 7 out of 50 samples, reflecting a generally higher prevalence in poultry compared to pigs. In 2011, the criterion was changed to a food safety criterion requiring the absence of *S.* Enteritidis and *S.* Typhimurium in 25 g fresh poultry meat.

For cattle carcasses, the process criterion is 2 out of 50 samples, reflecting a generally low prevalence of *Salmonella* in beef.

For minced meat, the criterion is a food safety criterion. Here, absence in the five weekly samples is required for both pig, beef, and poultry meat. If *Salmonella*-positive samples are found, the food product should be recalled. This also applies to minced meat that is to be heat-treated prior to consumption. This reflects that eating habits vary in the European Union; in some member states minced pork would never be eaten raw, whereas in other member states raw minced pork is considered a delicacy.

Salmonella is known to be present occasionally on the carcass (which is why prevalence up to 6% is accepted in the European Union). However, *Salmonella* does not disappear when the meat is being processed, e.g., into minced meat. Hence, the criterion set for carcasses may not be considered that ambitious, whereas the criterion set for processed meat such as minced meat intended to be heat-treated prior to consumption is so ambitious that it is difficult for the processors to comply.

Trichinella

Until 2014, Regulation (EC) 2075/2005 laid down the specific rules on official controls for *Trichinella* in meat (30). Susceptible livestock includes pigs and horses as well as wild boar and other game species that are destined for human consumption. Until June 2014, it was mandatory to test all these animals for *Trichinella* as a part of meat inspection. Exemptions could be made for a holding or a category of holdings that had been officially recognized by the competent authority as free from *Trichinella* or for a region where the risk of *Trichinella* in domestic swine was officially recognized as negligible. Belgium and Denmark were the only member states which were granted the status of areas with negligible risk (31, 32).

In most parts of the world, pig production has changed. Today many pigs are raised indoors under high levels of biosecurity, hampering the transmission of *Trichinella.* This is supported by European Union and U.S. data (33, 34). Eventually, this led to a change in European Union regulation toward lifting the requirement for testing pigs raised under so-called controlled housing conditions in integrated production systems. This calls for a very high degree of biosecurity and requires a type of animal husbandry where after weaning, swine are kept at all times under conditions controlled by the FBO with regard to feeding and housing. The animals must be reared indoors after weaning, there must be no contact with wildlife, and effective rodent control must be in place. A set of procedures laid out in Chapter I of Annex IV of the regulation must be followed (32) (Table 2).

All carcasses from holdings not officially recognized as applying controlled housing conditions will be systematically examined for *Trichinella.* Moreover, at least 10% of slaughtered carcasses from each holding officially recognized as applying controlled housing conditions will be examined for *Trichinella.* This will last until the member state has demonstrated that historical data on continuous testing carried out on the slaughtered swine population provide at least 95% confidence that the prevalence of *Trichinella* does not exceed 1 per million in that population.

TABLE 2 List of requirements that food business operators must meet to obtain official recognition of holdings as applying controlled housing conditions – modified from Annex IV, Chapter 1 in Regulation (EC) 2015/1375

Issue	Specific requirement
Housing	The operator must have taken **all practical precautions** with regard to **building construction and maintenance** in order **to prevent rodents**, any other kind of mammals, and large carnivorous birds **from having access to buildings where animals are kept.**
Rodent control	The operator must apply a **pest-control program, in particular for rodents**, effectively to prevent infestation of pigs. The operator must keep records of the program to the satisfaction of the competent authority.
Buying feed	The operator must ensure that all **feed** has been obtained from a facility that produces feed in accordance with the principles described in Regulation (EC) No 183/2005 of the European Parliament of 12 January 2005 and of the Council laying down requirements for feed hygiene.
Storing feed	The operator must **store feed** intended for *Trichinella*-susceptible species **in closed silos or other containers that are impenetrable to rodents**. All other feed supplies must be heat-treated or produced and stored to the satisfaction of the competent authority.
Disposal of dead animals	The operator must ensure that **dead animals are collected, identified, and transported without undue delay.**
Rubbish dump	If a **rubbish dump** is located in the neighborhood of the holding, the operator must inform the competent authority. Subsequently, the competent authority must assess the risks involved and decide whether the holding is to be recognized as applying controlled housing conditions.
Purchase of piglets	The operator must ensure that **piglets** coming onto the holding from outside and pigs purchased are born and bred under **controlled housing conditions.**
Identification of pigs	The operator must ensure that **pigs are identified** so each animal can be traced back to the holding.
Introduction of new animals	The operator may **introduce new animals** onto the holding only if they come from holdings also officially recognized as applying controlled housing conditions.
Access to outdoor facilities	**None of the animals has access to outdoor facilities** unless the food business operator can show by a risk analysis to the satisfaction of the competent authority that the time period, facilities, and circumstances of outdoor access do not pose a danger for introduction of *Trichinella* in the holding.

Denmark and Belgium are exempted from the 10% testing because of their negligible risk status in accordance with Regulation 2075/2005 (32).

The testing of pigs from noncontrolled housing can be interpreted as a sentinel because these pigs are believed to be of relatively high risk. Testing the high-risk pigs will form part of both a continuous documentation of disease freedom and an early warning system allowing fast and effective risk mitigation in case of possible findings (31). The possibility of stopping testing pigs from controlled housing will most likely lead to a decrease in the number of *Trichinella* tests. However, the full implementation will await acceptance from important trade partners from countries outside the European Union. Therefore, a country like Denmark, which exports >90% of its pork production will continue testing almost all pigs. Denmark's challenge shows the importance of international recognition regarding surveillance for animal health. Currently, work is being conducted in the international organization Codex Alimentarius to identify internationally accepted methods of surveillance for *Trichinella* (see later section).

BSE

BSE is under control in the European Union. By June 2015, 22 member states had obtained the status of negligible BSE risk by the World Organization for Animal Health, and the remaining 6 member states had the status of controlled risk (35). This is a result of having effective bans in place for many years regarding the use of mammalian protein as feed for livestock. This has opened up for discussion of the need to continue removing specified risk material (SRM). The position that it is unnecessary to remove SRM from cattle in areas or member states with negligible BSE risk was prevalent at the time the original version of the European Union TSE Regulation 999/2001 was adopted (Annex V, paragraph 1) (36). However, this was changed later, which can be seen in the consolidated version of Regulation 999/2001 (37). So today, SRM must still be removed in member states with negligible risk of BSE—but not in countries outside the European Union exporting to the European Union.

The EU Commission has developed a so-called roadmap for BSE (38), according to which a review of the necessity to remove SRM is on the agenda. Until recently, the intestines from duodenum to rectum (both inclusive) and of all ages of cattle were defined as SRM. In May 2015, new legislation (European Union Regulation 2015/728) was adopted that reduces the SRM from the intestine to the last 4 m of the small intestines. This brings the European Union list of SRM closer to international standards. Furthermore, a new European Union regulation (Regulation 2015/1162) came into force (August 2015) revising the definition of SRM for member states with a negligible risk of BSE. Hence, from now on, these member states only have to remove the skull including the eyes, the brain, and the spinal cord as SRM from cattle over 12 months of age. This means the intestines will no longer have to be dealt with as SRM for these member states.

The economic gain related to lifting the feed ban cannot be quantified easily, because it depends on the commercial possibilities. Most likely, the gain will be substantial; currently, there is a large number of products that need to be rendered by use of a resource-demanding and costly process. If these products can be manufactured and sold instead, then they will represent both an economic gain and a cost saving. For example, if the spinal column is no longer considered SRM, then T-bone steaks can be sold again. This used to be a highly valued cut. Cattle intestines have previously been used for human consumption, and this would again be considered a possibility if the SRM requirements are lifted. There used to be a demand for human consumption for cattle brains, in particular, in Southern Europe, but it is doubtful whether this interest can be re-established (39). In fact, cattle brains will continue to be considered SRM.

Antimicrobial Use and Resistant Bacteria

In 2001, the EU Commission launched its first European Union strategy to combat the threat of antimicrobial resistance. This led to the banning of the use of antimicrobials as growth promoters in 2006 (40). Moreover, it set requirements for data collection and monitoring and supported actions in the areas of research and raising awareness. For example, a specification for the harmonized monitoring of antimicrobial resistance in *Salmonella* and *Campylobacter* was published by EFSA in 2007. Since then these data have been collected and published on EFSA's website.

In November 2011, the EU Commission outlined a 5-year action plan against the rising threats from antimicrobial resistance (5). According to the action plan, policymakers need to protect consumers from risks related to the food chain and to establish the best control options to reduce such risks. Moreover, the EU Commission considers that the ongoing efforts are insufficient and that a holistic approach is needed to mitigate the risk. A total of 12 key actions are aimed for in the action plan. Among these are promoting the appropriate use of antimicrobials in humans and animals, focusing on medicated feed, promoting microbiological diagnosis and follow-up report, putting in place effective prevention, and developing new effective antimicrobials or alternatives for treatment of infections in humans and animals. Moreover, the monitoring systems for antimicrobial use and resistance should be strengthened in all member states—in both human and animal medicine—and prevention and control of infections in animals should be enhanced. The European Medicine Agency is running the European Surveillance of Veterinary Antimicrobial Consumption, which calculates the consumption of antimicrobials in milligram per kilogram of animal produced for each member state (41). These data show that there is a substantial variation in the consumption of antimicrobials across the member states (Fig. 3).

ACTIONS IN THE PIPELINE IN THE EUROPEAN UNION

As noted in Fig. 1, *Salmonella* and *Campylobacter* constitute the main foodborne hazards in the European Union. It is being increasingly acknowledged in the European Union member states that *Salmonella* and *Campylobacter* require risk management.

Salmonella

The European Union strategy for *Salmonella* began with poultry, which was ascribed to the highest number of human cases. No further actions are expected for poultry, because of the recently set production targets. The coming years will probably show a positive impact on human health related to achieving these targets.

Currently, the focus is on pigs, which are the second-largest source of *Salmonella* in humans in the European Union. The discussion is currently dealing with the question of where to go for control: preharvest, postharvest, or both. Initially, the focus was on preharvest measures and therefore also on monitoring or surveillance of pig herds. The idea was to increase awareness among farmers about the status of the animals.

However, as highlighted by Alban et al. (10), if there is no focus on the abattoir, the actions taken against *Salmonella* preharvest might easily be wasted. This was clearly shown in the EFSA baseline study for finishing pigs, where for two member states the prevalence of *Salmonella* was higher in the pork leaving the abattoir (measured as the proportion of *Salmonella*-positive carcasses) than the prevalence in the pigs entering the abattoir (measured as the proportion of lymph-node-positive samples) (42). This probably reflects a lack of focus on what can be done at the abattoir to prevent *Salmonella* contamination and cross-contamination.

The EU Commission is currently considering how to move on. Scientific cost–benefit analyses of actions taken preharvest have indicated that the feasibility of managing and reducing *Salmonella* in breeding and slaughter pigs was neither easy nor clear, and did not show larger benefits than costs (43, 44). A subsequent cost–benefit analysis of postharvest measures also showed that none of the actions considered were cost-effective (45). So far, the only step taken politically has been a tightening of the microbiological process hygiene criterion for carcasses as described in "Regulatory Framework in the European Union," above.

As stated earlier, only a few member states have plans in place for cattle, probably because beef and milk are already dealt with sufficiently to prevent human infection. Moreover, no targets are being discussed regarding the prevalence of *Salmonella* in cattle in the European Union.

Campylobacter

No regulation is in place for *Campylobacter* in the European Union. A recently conducted cost–benefit analysis pointed to the importance of improved on-farm biosecurity, such as fly screens as suggested by Hald et al. (46), and best practice hygiene at the slaughterhouse as feasible ways of reducing the exposure of humans to *Campylobacter* in broilers (47). However, such requirements depend on continued commitment from the producers and the slaughterhouses. Political discussions are underway concerning how to move on. Regulatory actions will most likely involve the use of process criteria, and such criteria will be set by the EU Commission. However, flexibility might be sought because the prevalence of *Campylobacter* varies enormously between European Union member states; in some member states between-flock prevalence above 70% is seen. When setting a criteria for *Campylobacter*, the concentration might be used instead of the prevalence, e.g., by use of a limit of 1,000 CFU/g. Before setting such a target, the impact on production will be estimated, i.e., the number of batches that will be subjected to heat treatment or freezing. Meanwhile, individual member states have regulations in place aiming at reducing consumer exposure.

In Denmark, a national mandatory surveillance program is in place. The Danish Action Plan describes measures implemented preharvest, such as biosecurity which is regulated in details, not only in the national legislation but also by a quality assurance system, owned and driven by the industry, called KIK. The KIK scheme is based on national and European Union legislation as well as international published literature and practical inputs from the farmers (http://www.danskslagtefjerkrae.dk/Aktiviteter/Kvalitet_i_Kyllingeproduktionen.aspx). *Campylobacter* flock prevalence is monitored by cloacal swab samples collected at the slaughterhouse. Each flock is sampled. Moreover, national surveillance of fresh, chilled broiler meat at slaughterhouses is in place involving weekly sampling of whole legs that are analyzed quantitatively. The Action Plan also covers surveillance of fresh broiler meat intended for retail, both produced in Denmark and imported. This is covered by the so-called case-by-case program, which was initiated in 2006 as an intensified nondiscriminatory control of *Salmonella* and *Campylobacter* in fresh meat (48).

Other sources and routes may contribute to the number of people acquiring a *Campylobacter* infection. Source account models are being developed to increase the knowledge about the relative importance of the various reservoirs. According to Boysen and Hald (49), the cattle reservoir has been found to be the second most important reservoir in Denmark. High *Campylobacter jejuni* prevalence has been reported in cattle, but very low prevalence has been found in Danish beef. Thus, if cattle should bear the second highest responsibility in relation to human campylobacteriosis, routes other than meat should be considered.

Trichinella and Other Zoonotic Parasites

The need for surveillance for *Trichinella* in pigs to ensure food safety and international trade is being debated internationally. Various approaches to surveillance are suggested or in place in the European Union, the United States, the World Organization for Animal Health, and Switzerland. The difference relates to the choice of test (serology/digestion), the target prevalence by which freedom from infection should be demonstrated, and the requirements for housing conditions for the pigs that do not need to be tested. According to Schuppers et al. (50), the programs do not recognize each other as providing an equivalent level of protection for consumers. However, a common set of guidelines would be helpful to ensure free trade. The question is whether it is possible to identify a globally agreed level of consumer protection.

The Codex Committee on food hygiene in collaboration with the European Union and the World Organization for Animal Health is currently developing a set of guidelines for *Trichinella* surveillance. An electronic working group for *Trichinella* was organized in March 2012 and is led by the European Union and New Zealand, and this working group is close to finding a common standard for *Trichinella* testing about which there will be consensus. Please see the website of Codex Alimentarius for the latest version of this work (http://www.codexalimentarius.org).

It is being debated to what extent *Taenia saginata/Cysticercus bovis* should be considered a food safety hazard. Viable cysticerci present in beef may result in infection of humans with tapeworm (51). Although this condition is not associated with pain or discomfort, it is perceived as disgusting. The infection is not notifiable in the European Union, and no data are available that can be used to assess the number of human cases with sufficient precision. Still, in some member states the prevalence of *C. bovis* found at meat inspection is considered nonnegligible. Due to the European habit of consuming raw or less than well-done beef, there is a risk of becoming infected with tapeworm. How to deal with this risk is currently being discussed at the Codex level—in the same working group that is discussing *Trichinella*. One solution may be risk-based surveillance aimed at targeting animals, herds, or cuts with the highest risk of harboring *C. bovis* (51).

Toxoplasma is known to be associated with rodents and cats, and therefore outdoor-reared pigs have a higher risk of *Toxoplasma* than pigs reared indoors on farms with rodent control and indoor-living cats. A risk assessment has shown that meat destined for production of raw or lightly cured products intended to be consumed without further heat treatment (such as salami) represents a risk to humans (52). It may be suggested to use the concept of controlled housing to divide pig herds into controlled (low-risk) and noncontrolled (high-risk) subpopulations. That would make it possible to freeze meat from noncontrolled housing destined for such ready-to-eat products as a way to mitigate the potential risk related to *Toxoplasma*.

Other Zoonotic Hazards

In the European Union, no definite actions have been decided upon regarding *Yersinia*, *Toxoplasma*, VTEC/STEC, *L. monocytogenes*, and viruses. The considerations regarding feasible control are presented in the following.

Although *Yersinia* is number 3 on the list of zoonotic hazards causing human illness (Fig. 1), no specific risk-mitigating regulations are in place in the European Union. Recently, scientific investigations have been made into the use of multidiagnostics whereby a single sample of blood taken from a pig at slaughter can be used to divide herds into among others *Yersinia*-positive and -negative. However, *Yersinia* is commonly found in pig farms, and eradication from the pig population does not seem to be realistic. The presence of *Yersinia* on a carcass is related to fecal contamination. The focus could therefore be on the hygienic measures taken in the abattoir. For example, there is discussion concerning to what extent we are dealing with *Yersinia* when we are preventing the spreading of *Salmonella* during slaughter. EFSA has suggested introducing a 4-year survey for *Yersinia* in the European Union. Such European Union surveys would ensure collection of comparable data enabling trend analyses as well as risk factor studies—similar to what has been undertaken for *Salmonella* (53). However, so far there is no acceptance of such a survey in the European Union.

VTEC/STEC is associated with ruminants. Within the European Union, outbreaks related to consumption of raw milk, among other sources, have occurred in England and Scotland, although the largest recent outbreak occurred in 2011 and was related to fenugreek sprouts (54). It is not known how to control the spread of VTEC/STEC among cattle or other livestock. Therefore, the focus is on how to control the hazard postharvest. Here, pasteurization of milk (55) and cooling of meat resulting in a low surface temperature combined with a dry surface have a documented risk-mitigating effect (56). According to Regulation 853/2004, an unbroken cooling chain is required (25). Since 2013, operators of bovine slaughterhouses can use lactic acid on whole carcasses, half-carcasses, or parts of bovines with the objective being to reduce the microbiological surface contamination. This is in line with U.S. postharvest procedures for decontamination (57).

In 2014, the EU Commission initiated a work aimed at developing guidelines for how to deal with food contaminated with VTEC/STEC. The discussion between the Commission and the meat-producing stakeholders (represented by the European Livestock and Meat Trading Union) concerned the criteria for when to submit meat for heat treatment and the practical difficulties arising due to lack of rapid analytical methods. If the finding of *stx* genes in meat destined for ready-to-eat products or dishes that are not fully heat-treated would become the foundation for the risk assessment, then a large quantity of meat would have to be destined for heat treatment for no reason because *stx* genes alone will not cause outbreaks; not just *stx* genes, but also *eae* genes—limited to serogroups O157, O111, O26, O103, O145, and O104—have to be present in the same live bacteria. Therefore, the first step of the analysis would be bacteriological enrichment and detection of *stx* and *eae* genes associated with the aforementioned serogroups. The next step will be confirmation of the simultaneous presence of the genes in the same isolated live bacteria (58). Such an analysis is at present so time-consuming that it will create substantial losses to the abattoirs and make it nearly impossible to put freshly minced meat on the market. Other classification approaches were therefore also being discussed. However, the EU Commission have recently given up the guidelines due to these technical constraints.

In the European Union, the preharvest regulation of *L. monocytogenes* is indirect and uses indicators. For milk, this includes requirements set to (i) somatic cell counts aimed at controlling mastitis, (ii) general hygiene practices, and (iii) measures directed against microbial development in the milk, including requirements for temperature, time, and total viable counts, both upon collection and prior to heat treatment (21). The remaining actions are industry-driven such as good silage production methods, good milking techniques, correct equipment, and cleaning of equipment. Moreover, *Listeria* is regulated postharvest; for meat products, milk, cheese, fish, etc. microbiological criteria are defined (28).

Foodborne viruses are the second most important cause of foodborne outbreaks in the European Union after *Salmonella*. In 2009, they were responsible for 19% of all outbreaks in the European Union, causing over 1,000 outbreaks and affecting more than 8,700 individuals. There is no legislation in place to mitigate this risk. EFSA has provided advice on possible measures to control and prevent the spread of these viruses. Focus is on the prevention of contamination rather than removal of the virus from contaminated food. More specifically, norovirus and hepatitis A virus play a role in fresh produce, ready-to-eat foods, and bivalve mollusks such as oysters, mussels, and scallops. Thorough cooking is currently the only efficient way to remove or inactivate norovirus or hepatitis A virus from these products. Moreover, hepatitis E is highly prevalent in pigs across Europe, and there is some evidence of transmission through food, although human clinical cases are rare in the European Union. Meat or liver should also be completely cooked to ensure effective removal of the potential presence of hepatitis E virus. To prevent hepatitis E infections, EFSA also recommends that people with liver diseases or immune deficiencies and pregnant women should be advised against eating undercooked meat and liver from wild boar and pork (59).

Antimicrobials and the Associated Risk of Development of Resistant Bacteria

The European Union action plan against the rising threats from antimicrobial resistance that is described in the section "Regulatory Framework in the European Union" will lead to initiatives in the various member states at a varying speed because of the different situations and approaches in the member states. The EU Council conclusions from 2012 regarding antimicrobial resistance calls on the EU Commission and the member states to take the necessary steps to implement the action plan (60).

To get an idea of what measures, more specifically, will be put in place, attention might be directed to Denmark, which—together with Sweden—has been ahead of the rest of the European Union regarding mitigation of the risk related to the use of antimicrobials in livestock. These experiences, good as well as bad, will probably act as an inspiration for the European Union with respect to how to mitigate the risk related to antimicrobials in livestock. In the following, the legal framework in place in Denmark is described in brief.

- Restricted use of extemporaneously prepared medicines (the cascade rule; imposing mandatory first priority to medicinal products approved for the relevant species, subsidiary approved for other species).
- The veterinarians' profit when distributing medicines is limited to a maximum of 5 to 10% at sales.
- Livestock herds are recommended to have a veterinary advisory service contract with a veterinary practitioner.
- Prescriptions made by veterinarians are limited to a maximum of 5 days of treatment in production animals. Exceptions are granted only if a veterinary advisory service contract is made between the veterinarian and the farmer. In such cases, up to 35 days of treatment is allowed for a diagnosed disease or a disease that was expected in the pigs, calves, and poultry, on the basis of the veterinarian's knowledge of the herd.
- Treatment is allowed only in diseased animals or animals in a well-defined incubation period (metaphylaxis); prophylactic use is illegal.
- Mandatory recording of medicines used, and drugs delivered and prescribed by the veterinary practitioners to farmed animals. This information must be available for inspection by veterinary officials for 3 years.
- Mandatory reporting to the Vetstat database.
- Pharmacies and the pharmaceutical industry are prohibited from offering economic incentives to veterinarians or others for the purpose of increasing product sales.
- Restricted use of fluoroquinolones and cephalosporins in livestock either officially or through industry-driven bans. This includes among others that the injection is restricted to use by the veterinary practitioner only. Furthermore, it is mandatory to conduct susceptibility testing in relation to use for production animals, documenting the need. It is mandatory to notify authorities of the use of fluoroquinolones.
- Treatment guidelines have been developed regarding use of antimicrobial agents in cattle and pigs.
- The Yellow Card scheme endorsed by the Danish Veterinary and Food Administration imposes preventive measures in pig herds using more than twice the average national consumption. Permit limits are gradually being adjusted. Please see Alban et al. (61) for an extended description of the Yellow Card scheme. (Source: modified from DANMAP [62].)

Meat Safety Assurance Programs

The primary purpose of meat inspection is to provide consumers with healthy meat. The meat inspection procedures were developed about 100 years ago, when tuberculosis and brucellosis played a major role in human health. Since then, the picture has changed as *Campylobacter*, *Salmonella*, and *Yersinia* today represent the most significant causes of zoonotic infections in humans, as noted in Fig. 1. In the European Union it is being asked whether the existing meat inspection process provides the optimal protection, what the costs are, and how we can do better. This implies a political and scientific focus on the modernization of meat inspection in the member states, the European Commission, and EFSA. The EU Council asked the EU Commission to look at how a risk-based approach could be incorporated increasingly into the legislation. This applies to which hazards are to be covered by meat inspection and how such hazards should be monitored/surveyed. The EU Commission then asked EFSA to clarify these two issues scientifically. EFSA sent out two reports regarding pigs in September 2011. The first report concerned the identification of infectious agents (hazards) that are relevant for food safety in relation to pigs and pork (63). The next report identified so-called harmonized epidemiological indicators (metrics) for the hazards that are identified as being relevant to food safety (64). Similar reports have been made for poultry (in 2012), ruminants (in 2013), sheep and goats (in 2013), and solipeds (in 2013). Please go to http://www.efsa.europa.eu/ for all reports.

According to these reports, the main food safety hazards associated with pork are *Salmonella*, *Yersinia*, *Toxoplasma*, and *Trichinella*. None of these can be dealt with macroscopically at meat inspection (63), and only *Salmonella* and *Trichinella* are covered by existing regulations as described above. As also stated above, EFSA developed a list of epidemiological indicators, meaning ways an indication of the presence (or prevalence) of each of the individual hazards can be obtained (64). The list was primarily based on scientific considerations, and not on economic and practical concerns. The approach taken by EFSA was to look broadly at surveillance and thus not only on what happens in a slaughterhouse in connection with the meat inspection. The indicators were divided according to location in the production line: in the herd, during transport to the slaughterhouse, at the slaughterhouse, or by biosecurity audits of the herds. The latter can be seen as similar to the concept of controlled housing used in the *Trichinella* regulation and referring to top biosecurity. EFSA recommends the establishment of pork safety assurance frameworks that cover the entire production chain from stable to table (63). A similar line of thinking has been used for cattle/beef and poultry/poultry meat.

The concept of food chain information can be used for such safety assurance programs. For example, currently, all cattle must be inspected for *C. bovis* by making incisions into the masseter muscles even in regions where the prevalence is very low. A recent Danish study has shown that female cattle are at higher risk of *C. bovis* compared to males, probably because females are grazed more often and live longer than males, resulting in more exposure to the parasite. If only females are inspected, almost the same number of positive cattle are found as when all cattle are inspected. Hence, gender can be used as food chain information determining the kind of meat inspection to which the carcass is subjected (51). The future of meat inspection for bovines is currently being discussed between the EU Commission and the member states. As of mid-2015 it is unknown whether changes will be introduced. In particular, *C. bovis* and bovine tuberculosis are being discussed. It is feared that a visual-only inspection will lead to poorer detection of these zoonotic agents. This can be compensated for by the use of risk-based sampling of high-risk subpopulations as well as using several years of surveillance, as described by Calvo-Artavia et al. (51, 65) and Foddai et al. (66).

By June 2014, the European Union meat inspection regulation for pigs was changed, making visual-only the customary way of conducting inspection if food chain information is exchanged routinely between the farmer and the abattoir (67). Palpation will now take place only upon suspicion. Suspicion could arise if any abnormalities are seen during antemortem or postmortem inspection or if epidemiological data or food chain information indicates that palpation or incision is necessary. It is expected that visual-only inspection will be gradually implemented during the coming years in most European Union member states. Again, concern about reactions from importing trade partners outside the European Union may play a role in the speed of the implementation.

Private Industry Standards

Private standards might act as food safety assurance frameworks as suggested by EFSA. Such standards are gaining increasing importance by combining different aspects of food production that are of interest to customers and consumers. Private standards are built on top of international and national legislation (Fig. 4). They constitute an effective way of dealing with food safety, because they ensure implementation of legislation and new requirements while also providing documentation for trade partners. The specific requirements in the standard are customer-driven. One possible drawback is that use of standards is more costly when herds and abattoirs are numerous and small compared to few and large.

In Denmark, a private standard has been developed for pig production. This standard is called the DANISH Product Standard, and it defines the requirements for the production of Danish pigs. It is accredited to an international standard called EN45011. Compliance with the standard is checked by an independent third party, who visits every pig farm at least once every 3 years. During the visit, compliance with all requirements is checked in detail. The main focus of the DANISH Product Standard is the key areas affecting animal welfare, meat safety, and traceability in the primary production of pigs.

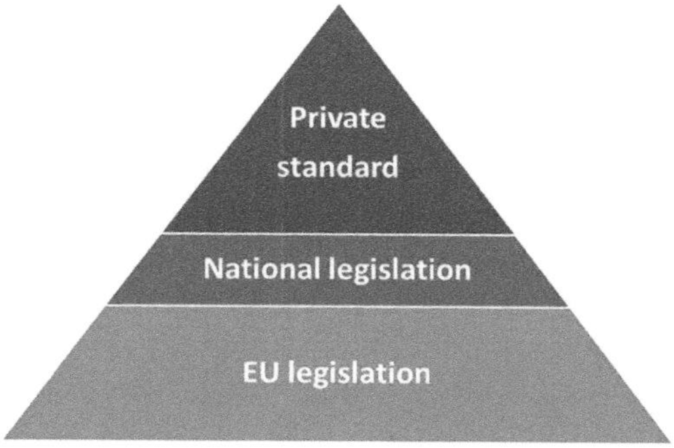

FIGURE 4 Graphical display of how private standards relate to national and international legislation.

Detailed requirements are described for each of the following areas, and an example of what it covers is given for each area:

1. **Pig identification and traceability:** Pigs for slaughter must be tattooed and ear-marked if exported.
2. **Feed:** Antimicrobial growth promoters are not allowed.
3. **Herd health and use of medicine:** Treated animals must be clearly identified, and withdrawal periods after antimicrobial treatment should be complied with.
4. **Treatment of sick or injured pigs:** Diseased pigs are placed in hospital pens.
5. **Housing and equipment:** Bedding must not be harmful to the pigs. Therefore, it is forbidden to use peat unless it has been heat-treated or specifically approved by the Danish SPS-Health status department.
6. **Outdoor production:** Pigs kept outdoors must be fenced with an inner and outer fence.
7. **Feed and water provision:** There should be free access to clean and fresh drinking water for pigs >14 days old.
8. **Management:** All pigs >28 days old have to be inspected at least once a day.
9. **Delivery of pigs:** Pigs may be held for a maximum of 2 hours in mobile collection pens.

If the inspector finds conditions not in accordance with the above-listed requirements, he might take one of the following three actions, depending upon how serious the lack of compliance is:

1. Record a critical comment in the auditing report with a requirement for immediate action.
2. Ask producer to submit follow-up documentation in the form of a photo or a note showing that the condition is corrected.
3. Revisit within 3 months. If conditions have not been put right by this time, the approval is lost.

More information and appendices for the DANISH Product Standard can be found at http://vsp.lf.dk/~/media/Files/DANISH/DANISH%20produktstandard/Produkt_Standard_UK.pdf.

CONCLUSION

Within the European Union, it is being recognized that *Salmonella* and *Campylobacter* need to be addressed specifically. For *Salmonella* in poultry, the experience is that setting targets in the primary production arena results in a lower number of human cases, whereas for *Salmonella* in pigs it is crucial to focus on the abattoir. For *Campylobacter*, no easy risk-mitigating strategies exist. The focus here should first be on identification of feasible instruments. Moreover, the role of the environment in the transmission of *Campylobacter* is being increasingly acknowledged, although not much is known about the exact pathways leading to exposure of humans.

For *Trichinella*, testing has to take place in only the noncontrolled housing compartment, meaning outdoor-reared pigs or pigs reared on farms with poor biosecurity. For BSE, changes related to how to deal with SRM are in the pipeline, because the vast majority of member states in the European Union are considered to have a negligible risk of BSE.

There is no direct preharvest regulation of the zoonotic hazards *C. bovis/T. saginata*, *Toxoplasma*, *Yersinia*, STEC/VTEC, *Listeria*, and foodborne viruses. Discussion of how to deal with these hazards in the future is ongoing.

To ensure food safety, further focus should be on the entire production chain. For each commodity, it should be carefully considered which hazards need to be covered by surveillance and how. Subsequently, control points should be identified. This should involve measurements of either the hazards themselves or the production systems, where this makes sense and is most cost-effective.

To combat the risk related to antimicrobial resistance, the focus in the European Union will be on prudent use, treatment guidelines, and restriction of use of critically important antimicrobials such as fluoroquinolones and cephalosporins. Detailed monitoring of use in individual animal species and age groups is expected because without such knowledge no targeted effort can be made.

Traditional meat inspection provides limited public health compared to the resources invested. Meat inspection is therefore being modernized within the European Union, emphasizing a simple and targeted approach to ensure cost-effectiveness.

All these needs can be addressed by the use of quality assurance schemes based on private standards incorporating public requirements and make use of appropriate food chain information. Such schemes should be accepted internationally.

The precautionary principle can and should be used where needed, but actions taken should be of limited duration and of a proportionate dimension in order not to distort trade. Moreover, risk managers including politicians should be willing to lift restrictions when the risk is declining (as noted for BSE) to ensure cost-effective protection of food safety implying no unnecessary waste of resources while safeguarding consumer confidence.

The European Union consists of 28 member states, and the rules that are set are minimum rules, which can be complied with by most member states. In general, flexibility is allowed as an acceptance of the wide variety of productions systems, climatic conditions, and traditions that exists in the different member states. This flexibility also makes it possible for individual member states to set higher standards, where this is judged reasonable.

Finally, it should be emphasized that it is the FBO's responsibility to place safe products on the market. Agreements on targets and method freedom will act as incentives to identify cost-effective and feasible means.

ACKNOWLEDGMENTS

Several people have contributed to this chapter: Annette Cleveland Nielsen, Annette Lychau Petersen, Gudrun Sandø (the Danish Veterinary and Food Administration), Maj-Britt Albrechtsen, Mie Nielsen Blom, Jan Dahl, Claus Heggum, Vibeke Møgelmose, Annette Dresling, Marianne Sandberg, Lene Lund Sørensen, Flemming Thune-Stephensen (the Danish Agriculture and Food Council), Erik Rattenborg (SEGES), Lene Trier Olesen (Arla Foods Amba), and Liza Rosenbaum Nielsen (University of Copenhagen).

The author declares a conflict of interest: The author is working for an organization that gives advice to the agricultural sector of Denmark, including farmers and abattoirs.

CITATION

Alban L. 2016. Regulatory issues associated with preharvest food safety: European Union perspective. Microbiol Spectrum 4(5):PFS-0003-2014.

REFERENCES

1. **Anonymous.** 2002. Regulation (EC) No 178/2002 laying down the general principles and requirements of food law, establishing the European Food Safety Authority and laying down procedures in matters of food safety. http://eur-lex.europa.eu/LexUriServ/LexUriServ.do?uri=OJ:L:2002:031:0001:0024:EN:PDF.
2. **EFSA.** 2012. The European Union Summary Report on Trends and Sources of Zoonoses, Zoonotic Agents and Foodborne Outbreaks in 2010. *EFSA J* **10:**2597.
3. **EFSA.** 2014. The European Union Summary Report on Trends and Sources of Zoonoses, Zoonotic Agents and Foodborne Outbreaks in 2012. http://www.efsa.europa.eu/en/efsajournal/pub/3547.htm.
4. **EFSA.** 2011. Scientific opinion on *Campylobacter* in broiler meat production: control options and performance objectives and/or targets at different stages of the food chain. *EFSA J* **9:**2105. http://www.efsa.europa.eu/en/efsajournal/doc/2105.pdf.
5. **Anonymous.** 2011. Communication from the Commission to the European Parliament and the Council. Action plan against the rising threats from antimicrobial resistance. http://ec.europa.eu/dgs/health_consumer/docs/communication_amr_2011_748_en.pdf.
6. **Alban L, Nielsen EO, Dahl J.** 2008. A human health risk assessment for macrolide-resistant *Campylobacter* associated with the use of macrolides in Danish pig production. *Prev Vet Med* **83:**115–129.
7. **Anonymous.** 2003. Directive 2003/99/EC on the monitoring of zoonoses and zoonotic agents. http://eur-lex.europa.eu/LexUriServ/LexUriServ.do?uri=OJ:L:2003:325:0031:0040:EN:PDF.
8. **Anonymous.** 2003. Regulation (EC) No 2160/2003 on the control of *Salmonella* and other specified foodborne zoonotic agents. http://eur-lex.europa.eu/LexUriServ/LexUriServ.do?uri=OJ:L:2003:325:0001:0015:EN:PDF.
9. **Baptista FM.** 2011. *Salmonella* in pigs and pork in Denmark and Portugal: use of risk-based principles to achieve cost-effective surveillance strategies. Ph.D. thesis. Faculty of Life Sciences, University of Copenhagen. pp. 177
10. **Alban L, Baptista FM, Møgelmose V, Sørensen LL, Christensen H, Aabo S, Dahl J.** 2011. *Salmonella* surveillance and control for finisher pigs and pork in Denmark: a case study. *Food Res Int* **45:**656–665.
11. **Helms M, Vastrup P, Gerner-Smidt P, Mølbak K.** 2003. Short and long term mortality associated with foodborne bacterial gastrointestinal infections: registry based study. *BMJ* **326:**357–361.
12. **Boqvist S, Vågsholm I.** 2005. Risk factors for hazard of release from *Salmonella*-control restriction on Swedish cattle farms from 1993 to 2002. *Prev Vet Med* **71:**35–44.
13. **Lewerin SS, Skog L, Frössling J, Wahlström H.** 2011. Geographical distribution of *Salmonella* infected pig, cattle and sheep herds in Sweden 1993–2010. *Acta Vet Scand* **53:**51–58.
14. **Nielsen LR, Baggesen DL, Aabo S, Moos MK, Rattenborg E.** 2011. Prevalence and risk factors for *Salmonella* in veal calves at Danish cattle abattoirs. *Epidemiol Infect* **139:**1075–1080.
15. **Nielsen LR, Dohoo I.** 2012. Survival analysis of factors affecting incidence risk of *Salmonella* Dublin in Danish dairy herds during a 7-year surveillance period. *Prev Vet Med* **107:**160–169.
16. **Anonymous.** 2012. Danish order no. 143 about *Salmonella* in cattle and others. (In Danish.) https://www.retsinformation.dk/Forms/R0710.aspx?id=140575.
17. **Anonymous.** 2006. Regulation 1168/2006 implementing Regulation 2160/2003 as regards a community target for the reduction of the

prevalence of certain *Salmonella* serotypes in laying hens of *Gallus gallus* and amending Regulation 1003/2005. http://eur-lex.europa.eu/legal-content/EN/TXT/PDF/?uri=CELEX:32006R1168&rid=2.

18. **Anonymous.** 2007. Regulation 646/2007 implementing Regulation 2160/2003 as regards a community target for the reduction of the prevalence of *Salmonella* Enteritidis and *Salmonella* Typhimurium in broilers and repealing Regulation 1091/2005. http://eur-lex.europa.eu/LexUriServ/LexUriServ.do?uri=CONSLEG:2007R0646:20080701:EN:PDF.
19. **Anonymous.** 2010. Regulation 200/2010 implementing regulation 2160/2003 as regards a union target for the reduction of the prevalence of *Salmonella* serotypes in adult breeding flocks of *Gallus gallus*. http://eur-lex.europa.eu/legal-content/EN/TXT/PDF/?uri=CELEX:32010R0200&rid=2.
20. **Sandøe G.** 2014. International topics, p 32. *In* Sørensen AIV, Helwigh B, Müller L (ed), *Annual Report on Zoonoses in Denmark 2013*. National Food Institute, Technical University of Denmark. http://www.food.dtu.dk.
21. **Anonymous.** 2004. Regulation (EC) No 853/2004 laying down specific hygiene rules for food of animal origin. http://eur-lex.europa.eu/LexUriServ/LexUriServ.do?uri=OJ:L:2004:226:0022:0082:EN:PDF.
22. **Anonymous.** 2012. Commission implementing regulation (EU) no. 427/2012 on the special guarantees concerning *Salmonella* laid down in Regulation (EC) No 853/2004 to eggs intended for Denmark. http://eur-lex.europa.eu/legal-content/EN/TXT/PDF/?uri=CELEX:32012R0427&from=EN.
23. **Van Wagenberg CPA.** 2010. Incentive mechanisms for food safety control in pork supply chains: a study on the relationship between finishing pig producers and slaughterhouses in the Netherlands. Ph.D. thesis. Wageningen University, The Netherlands. pp. 148
24. **Anonymous.** 2004. Regulation (EC) No 854/2004 laying down specific rules for the organisation of official controls on products of animal origin intended for human consumption. http://eurlex.europa.eu/LexUriServ/LexUriServ.do?uri=OJ:L:2004:226:0083:0127:EN:PDF.
25. **Anonymous.** 2004. Regulation (EC) No 852/2004 on the hygiene of foodstuffs. http://eur-lex.europa.eu/LexUriServ/LexUriServ.do?uri=OJ:L:2004:226:0003:0021:EN:PDF.
26. **Mellor D.** 2011. The trouble with epidemiology: the tyranny of numbers. Honorary presentation at the annual meeting of the Society of Veterinary Epidemiology and Preventive Medicine (SVEPM). Leipzig, Germany. 22–25 March, 2011.
27. **Anonymous.** 2009. Commission Regulation (EC) No 1161/2009 amending Annex II to Regulation (EC) no 853/2004 as regards food chain information to be provided to food business operators operating slaughterhouses. http://eur-lex.europa.eu/LexUriServ/LexUriServ.do?uri=OJ:L:2009:314:0008:0009:EN:PDF.
28. **Anonymous.** 2005. Commission Regulation (EC) No 2073/2005 on microbiological criteria for foodstuffs. http://eur-lex.europa.eu/LexUriServ/LexUriServ.do?uri=OJ:L:2005:338:0001:0026:EN:PDF.
29. **Anonymous.** 2014. Commission Regulation (EC) No 217/2014 amending Regulation (EC) No 2073/2005 as regards *Salmonella* on pig carcases. http://eur-lex.europa.eu/legal-content/EN/TXT/PDF/?uri=CELEX:32014R0217&rid=1.
30. **Anonymous.** 2005. Commission Regulation (EC) No 2075/2005 laying down specific rules on official controls for *Trichinella* in meat. http://eur-lex.europa.eu/legal-content/EN/TXT/PDF/?uri=CELEX:32005R2075&from=EN.
31. **Alban L, Boes J, Kreiner H, Petersen JV, Willeberg P.** 2008. Towards a risk-based surveillance for *Trichinella* spp. in Danish pig production. *Prev Vet Med* **87:**340–357.
32. **Anonymous.** 2015. Commission Implementing Regulation (EC) No 2015/1375 laying down specific rules on official controls for *Trichinella* in meat. http://eur-lex.europa.eu/legal-content/EN/TXT/PDF/?uri=CELEX:32015R1375&rid=1.
33. **Alban L, Pozio E, Boes J, Boireau P, Boué F, Claes M, Cook AJC, Dorny P, Enemark H, van der Giessen J, Hunt KR, Howell M, Kirjusina M, Nöckler K, Rossi P, Smith GC, Snow L, Taylor MA, Theodoropoulos G, Vallée I, Viera-Pinto MM, Zimmer IA.** 2011. Towards a standardised surveillance for *Trichinella* in the European Union. *Prev Vet Med* **99:**148–160.
34. **CDC.** 2014. Trichinellosis: epidemiology and risk factors. Centers for Disease Control and Prevention, Atlanta, GA. http://www.cdc.gov/parasites/trichinellosis/epi.html.
35. **OIE.** 2015. List of bovine spongiform encephalopathy risk status of member countries. According to Resolution No. 21 (83rd general session May 2015). http://www.oie.int/animal-health-in-the-world/official-disease-status/bse/list-of-bse-risk-status/.
36. **Anonymous.** 2001. Regulation (EC) 999/2001 laying down rules for the prevention, control and eradication of certain transmissible spongiform encephalopathies. http://eur-lex.europa.eu/legal-content/EN/TXT/PDF/?uri=CELEX:02001R0999-20010701&rid=38.

37. **Anonymous.** 2001. Regulation (EC) 999/2001 laying down rules for the prevention, control and eradication of certain transmissible spongiform encephalopathies. Consolidated version. http://eur-lex.europa.eu/LexUriServ/LexUriServ.do?uri=CONSLEG:2001R0999:20110318:EN:PDF.
38. **Anonymous.** 2010. The TSE Roadmap 2. http://ec.europa.eu/food/food/biosafety/tse_bse/docs/roadmap_2_en.pdf.
39. **Alban L, Thune-Stephensen F.** 2013. Successfully controlling BSE. *Fleischwirtschaft Int* **3:** 16–22.
40. **Anonymous.** 2003. Regulation 1831/2003 on additives for use in animal nutrition prohibiting the use of antibiotic growth promoters. http://eur-lex.europa.eu/LexUriServ/LexUriServ.do?uri=CONSLEG:2003R1831:20100901:EN:PDF.
41. **ECDC/EFSA/EMA.** 2015. ECDC/EFSA/EMA first joint report on the integrated analysis of the consumption of antimicrobial agents and occurrence of antimicrobial resistance in bacteria from humans and food-producing animals. *EFSA J* **13:**4006.
42. **EFSA.** 2008. Report of the Task Force on Zoonoses Data Collection on the analysis of the baseline survey on the prevalence of *Salmonella* in slaughter pigs. Part A: *Salmonella* prevalence estimates (1–111). http://www.efsa.europa.eu/en/efsajournal/pub/135r.htm.
43. **FCC.** 2010. Analysis of the costs and benefits of setting a target for the reduction of *Salmonella* in slaughter pigs. Report by FFC Consortium for the European Commission Directorate-General Health and Consumers SANCO/2008/E2/036. http://ec.europa.eu/food/food/biosafety/salmonella/docs/fattening_pigs_analysis_costs.pdf.
44. **FCC.** 2011. Analysis of the costs and benefits of setting a target for the reduction of *Salmonella* in breeding pigs: report by FFC Consortium for the European Commission Directorate-General Health and Consumers.
45. **FCC.** 2014. Analysis of the costs and benefits of setting a target for the reduction of *Salmonella* in pigs at slaughterhouse level. Report by FFC Consortium for the European Commission Directorate-General for Health and Consumers SANCO/E2/2009/SI2.534057. http://ec.europa.eu/food/food/biosafety/salmonella/docs/br_pigs_repfor_submission_16032011_en.pdf.
46. **Hald B, Sommer HM, Skovgård H.** 2007. Use of fly screens to reduce *Campylobacter* spp. introduction in broiler houses. *Emerg Infect Dis* **13:**1951–1953.
47. **Elliott J, Lee D, Erbilgic A, Jarvis A.** 2012. Analysis of the costs and benefits of setting certain control measures for reduction of *Campylobacter* in broiler meat at different stages of the food chain. Project report by ICF GHK and ADAS for DG SANCO of the European Commission. http://ec.europa.eu/food/food/biosafety/salmonella/docs/campylobacter_cost_benefit_analysis_en.pdf.
48. **Helwigh B, Müller L (ed).** 2012. *Annual Report on Zoonoses in Denmark 2011.* National Food Institute, Technical University of Denmark. http://www.food.dtu.
49. **Boysen L, Hald L.** 2012. Trends and sources in human campylobacteriosis, p 10. *In* Helwigh B, Müller L (ed), *Annual Report on Zoonoses in Denmark 2011.* National Food Institute, Technical University of Denmark. http://www.food.dtu.
50. **Schuppers ME, Pyburn DG, Alban L.** 2011. Alternative monitoring and surveillance scenarios for *Trichinella* spp. and their impact on public health. *Epidémiol Santé Animal* **59–60:**120–122.
51. **Calvo-Artavia FF, Nielsen LR, Alban L.** 2013. Epidemiologic and economic evaluation of risk-based meat inspection for bovine cysticercosis in Danish cattle. *Prev Vet Med* **108:**253–261.
52. **Boes J, Alban L, Sørensen LL, Nersting L.** 2007. Risk assessment for *Toxoplasma gondii* in the Danish pig industry. 7th International Symposium on the Epidemiology and Control of Foodborne Pathogens in Pork. 7-9 May 2007. Verona, Italy. P 195–198. http://lib.dr.iastate.edu/cgi/viewcontent.cgi?article=1086&context=safepork.
53. **EFSA.** 2009. Technical specifications for harmonised national surveys on *Yersinia enterocolitica* in slaughter pigs. *EFSA J* **7:**1374.
54. **EFSA.** 2014. Shiga toxin-producing *E. coli* outbreak(s). http://www.efsa.europa.eu/en/topics/topic/ecolioutbreak2011.htm.
55. **Pearce LE, Smythe BW, Crawford RA, Oakley E, Hathaway SC, Shepherd JM.** 2012. Pasteurization of milk: the heat inactivation kinetics of milk-borne dairy pathogens under commercial-type conditions of turbulent flow. *J Dairy Sci* **95:**20–35.
56. **Anonymous.** 2004. Decline of VTEC during cooling with and without blast chilling. –(In Danish.) The Danish Veterinary and Food Administration. http://www.foedevarestyrelsen.dk/SiteCollectionDocuments/25_PDF_word_filer%20til%20download/04kontor/henfald%20vtec%20slutrapport.pdf.
57. **EFSA.** 2013. Scientific opinion on VTEC-seropathotype and scientific criteria regarding pathogenicity assessment. *EFSA J* **11:**3138.
58. **Anonymous.** 2013. Commission Regulation (EU) No 101/2013 concerning the use of lactic

acid to reduce microbiological surface contamination on bovine carcasses. http://eur-lex.europa.eu/LexUriServ/LexUriServ.do?uri=OJ:L:2013:034:0001:0003:EN:PDF.

59. **EFSA.** 2011. Scientific opinion on an update on the present knowledge on the occurrence and control of foodborne viruses. *EFSA J* **9:**2190.

60. **Anonymous.** 2012. Council conclusions on the impact of antimicrobial resistance in the human health sector and in the veterinary sector: a "One Health" perspective. http://www.consilium.europa.eu/uedocs/cms_data/docs/pressdata/en/lsa/131126.pdf.

61. **Alban L, Dahl J, Andreasen M, Petersen JV, Sandberg M.** 2013. Possible impact of the "yellow card" antimicrobial scheme on meat inspection lesions in Danish finisher pigs. *Prev Vet Med* **108:**334–341.

62. **DANMAP.** 2010. Use of antimicrobial agents and occurrence of antimicrobial resistance in bacteria from food animals, food and humans in Denmark. pp. 160 http://www.danmap.org/Downloads/~/media/Projekt%20sites/Danmap/DANMAP%20reports/Danmap_2010.pdf.

63. **EFSA.** 2011. Scientific opinion on the public health hazards to be covered by inspection of meat (swine). *EFSA J* **9:**2351.

64. **EFSA.** 2011. Technical specifications on harmonised epidemiological indicators for public health hazards to be covered by meat inspection of swine. *EFSA J* **9:**2371.

65. **Calvo-Artavia FF, Alban L, Nielsen LR.** 2013. Evaluation of surveillance for documentation of freedom from bovine tuberculosis. *Agriculture* **3:**310–326.

66. **Foddai A, Nielsen LR, Willeberg P, Alban L.** 2015. Comparison of output-based approaches used to substantiate bovine tuberculosis free status in Danish cattle herds. *Prev Vet Med* **121:**21–29.

67. **Anonymous.** 2014. Commission Regulation (EC) No 218/2014 amending annexes to Regulation (EC) No 853/2004, (EC), No 854/2004 and (EC) Regulation 2074/2005. http://www.fsai.ie/uploaded/Files/Reg218_2014.pdf.

Current Status of the Preharvest Application of Pro- and Prebiotics to Farm Animals to Enhance the Microbial Safety of Animal Products

19

ROLF D. JOERGER[1] and ARPEETA GANGULY[1]

HISTORY AND DEFINITIONS

Elie Metchnikoff, who is "regarded as the grandfather of modern probiotics" (1) mentioned in his book *The Prolongation of Life*, published in 1907, that a researcher at the Pasteur Institute, Dr. Belonowsky, had shown that administration of the "Bulgarian bacillus cures a special intestinal disease known as mouse typhus" (2). Although likely impossible to prove, this passage in a book might have been one of the first to describe experimental probiotic action against an intestinal pathogen. Whatever one might think today about Metchnikoff's ideas and his preoccupation with "putrefaction" in the digestive tract, he provided what could still be considered the basis of the modern definition of a probiotic when he wrote with reference to lactic bacilli, "The latter become acclimatized in the human digestive tube as they find there the sugary material required for their subsistence, and by producing disinfecting bodies benefit the organism which supports them" (2). With the term "disinfecting bodies," Metchnikoff was referring primarily to lactic acid, but he was also aware, based on Belonowsky's research, that more than lactic acid was involved in the probiotic action of the "Bulgarian bacillus." Nowadays,

[1]Department of Animal and Food Sciences, University of Delaware, Newark, DE 19716.

Preharvest Food Safety
Edited by Siddhartha Thakur and Kalmia E. Kniel

doi:10.1128/microbiolspec.PFS-0012-2016

most authors have settled on a broad definition of probiotics as "live microorganisms that, when administered in adequate amounts, confer a health benefit to the host" (3); however, with respect to food safety, this definition might not be broad enough. Certainly, a food animal host that is healthier as a consequence of probiotic administration would be less likely to be a food safety concern, but would live microorganisms that reduce a human pathogen such as *Campylobacter jejuni* in the chicken's intestinal tract without any noticeable health benefits to the host not also be a probiotic? Similarly, a product that would reduce the *Escherichia coli* 0157:H7 carrier state in cattle would also fall into that category.

Likely unintentionally, Metchnikoff also embedded a reference to prebiotics in his statement on lactic bacilli by mentioning "sugary materials." He apparently recognized that a suitable substrate for beneficial bacteria was the basis of desired effects of these microorganisms. It was not until 1995 that dietary compounds that would be able to modulate the microbiota were given the name "prebiotics" (4). Today, prebiotics are defined as "a nonviable food component that confers a health benefit on the host associated with modulation of the microbiota" (5). Usually there are some limitations to which food components actually count as prebiotics. The compounds need to be resistant to hydrolysis and absorption by the upper gastrointestinal tract so that they can reach the target organisms in the lower gastrointestinal tract. It is desirable that these compounds be substrates more or less only for those microorganisms that one intends to support. It has been argued that only fructooligosaccharides and inulin meet the criteria (6); however, numerous other compounds have been included in lists of prebiotics such as galactooligosaccharides, soy-oligosaccharides, xylooligosaccharides, pyrodextrins, isomaltooligosaccharides, lactulose, pectinoligosaccharides, lactosucrose, sugar alcohols, glucooligosaccharides, levans, resistant starch, and xylosaccharides (5).

MICROBIOTA MANIPULATIONS AND FOOD SAFETY

Current products designed to manipulate the microbiota of food animals with live microorganisms (or with products directly derived from the culture of these organisms) fall into two rather lopsided categories. The predominant type attempts to improve or maintain the animals' health status under the conditions encountered in modern animal husbandry practices without making specific claims to target pathogens that are of concern to human health. The smaller group claims to establish or modify intestinal microbiota that have a direct measurable effect on pathogens of concern to humans, such as *Salmonella enterica* and *C. jejuni*. The two categories are not mutually exclusive because healthier animals are expected to be less susceptible to colonization with certain human pathogens or to carry fewer of these pathogens. Similarly, microbiota changes designed to inhibit foodborne pathogen colonization can also improve overall animal health and, for example, lead to increased weight gains. As will be shown later in this article, the number of probiotic products sold in the United States that claim to be directed against pathogens or are competitive exclusion products is exceedingly small compared to products that claim to improve feed conversion efficiency, growth, immune system function, or resistance to stressful events.

The strategy of increasing food safety by microbiota interventions was the first to be addressed experimentally, primarily in chickens. In 1952 Milner and Schaffer (7) observed that a mature microbiota can confer resistance to infection of chicks by *Salmonella*, but it was not until the 1970s that effective and commercially viable microbial preparations were developed that, when administered to newly hatched chicks, were capable of reducing *Salmonella* infection (8). This approach to food safety enhancement in the poultry industry is sometimes termed the "Nurmi concept," named after the author who first developed the concept. It is assumed that this approach works by the bacteria preparation competitively excluding *Salmonella* from sites that this pathogen would occupy if no other bacteria were present. Generally it is assumed that a combination of factors work together to exclude unwanted organisms. These factors include effects on the immune system, interference with adhesion to intestinal surfaces, competition for nutrients and perhaps oxygen, as well as production of inhibitory molecules such as volatile fatty acids, lactic acid, bacteriocins, and perhaps extracellular enzymes. Years of research showed that the more complex the mixture of bacteria that was administered to chicks, the more successful was the exclusion of *Salmonella*. Thus, administering bacterial mixtures from fecal or cecal sources was more protective than administering single bacterial isolates or a combination of just a few isolates (9).

The competitive exclusion concept was later also tested on pigs with undefined cultures of porcine origin plus *Bacillus thetaiotaomicron* (10). Interestingly, it appears that complex or undefined cultures were not prepared for use in cattle. Here, the route chosen was primarily to find single isolates that would be inhibitory to *E. coli* 0157:H7, for which cattle have long been considered the primary reservoir (11). For example, Brashears et al. (12) isolated numerous lactic acid bacteria and tested them for inhibition of *E. coli* 0157:H7 inoculated into manure and rumen fluid. The authors also determined the effect of the lactic acid bacteria on *E. coli* 0157:H7 carriage in live cattle (12, 13). The reported effects were generally positive, indicating that single strains or combinations of a couple of strains were capable of reducing *E. coli* 0157:H7 shedding and carcass contamination.

REGULATION OF PRO- AND PREBIOTICS

The trend for using single strains or a combination of several, characterized strains to improve animal well-being and growth, and directly or indirectly to improve food safety, has continued. Rather than calling these products "probiotics," the term "direct-fed microbial products" is used in the United States for products that are given to animals. According to the Bovine Alliance on Management and Nutrition, which is comprised of representatives from the American Association of Bovine Practitioners, the American Dairy Science Association, the American Feed Industry Association, and the U.S. Department of Agriculture, the terms "probiotic" and "direct-fed microbial" (DFM) can be used interchangeably (14).

The primary reason for suppliers of such products to stay away from complex, partially, or completely undefined probiotics is based on regulations for the use of such products in animals that require spelling out the microbial composition of the products. For this reason, the commercial product originally developed by Nurmi and coworkers, is only available in countries that do not require product labeling that includes a list of the microorganisms included in the product. Furthermore, some countries have established lists of microorganisms that are allowed to be used in such products. For example, the European Union and the United States have published lists of acceptable organisms (Table 1). In the United States, Compliance Policy Guidelines (Sec. 689.100 Direct-Fed Microbial Products) (15) specify how probiotic products have to be labeled and what claims can be made. A particular influence on what kind of probiotics can be marketed is Policy 3 of the guidelines, which states that a product that "contains one or more microorganisms not listed in the AAFCO Official Publication is a food additive and is adulterated under Section 402(a)(2)(C) unless it is the subject of a food additive regulation." Therefore, a substantial investment will have to be made by anyone who wants to bring to market a product that contains microorganisms not currently on the list. The previously approved poultry competitive exclusion product, Preempt (NADA 141-101 PREEMPT TM) (16), has been removed from the list of FDA-approved products as of 2013 (17).

TABLE 1 List of microorganisms approved for use in DFBs in the United States and the European Union[a]

Microorganisms approved by FDA and AAFCO (61)		Microorganisms with QPS[b] status by EFSA[c]		
Bacillus coagulans	*Lactobacillus curvatus*	*Bacillus amyloliquefaciens*	*Lactobacillus coryniformis*	*Pediococcus acidilactici*
Bacillus lentus	*Lactobacillus delbrueckii*	*Bacillus atrophaeus*	*Lactobacillus crispatus*	*Pediococcus dextrinicus*
Bacillus licheniformis	*Lactobacillus fermentum*	*Bacillus clausii*	*Lactobacillus curvatus*	*Pediococcus pentosaceus*
Bacillus pumilus	*Lactobacillus helveticus*	*Bacillus coagulans*	*Lactobacillus delbrueckii*	*Propionibacterium acidipropionici*
Bacillus subtilis	*Lactobacillus lactis*	*Bacillus fusiformis*	*Lactobacillus farciminis*	*Propionibacterium freudenreichii*
Bacteroides amylophilus	*Lactobacillus plantarum*	*Bacillus lentus*	*Lactobacillus fermentum*	*Streptococcus thermophiles*
Bacteroides capillosus	*Lactobacillus reuteri*	*Bacillus licheniformis*	*Lactobacillus gallinarum*	*Xanthomonas campestris*[b]
Bacteroides ruminicola	*Leuconostoc mesenteroides*	*Bacillus megaterium*	*Lactobacillus gasseri*	Yeasts:
Bacteroides suis	*Megasphaera elsdenii* (cattle only)	*Bacillus mojavensis*	*Lactobacillus helveticus*	*Candida cylindracea*[b]
Bifidobacterium adolescentis	*Pediococcus acidilactici*	*Bacillus pumilus*	*Lactobacillus hilgardii*	*Debaryomyces hansenii*
Bifidobacterium animalis	*Pediococcus cerevisiae (damnosus)*	*Bacillus subtilis*	*Lactobacillus johnsonii*	*Hanseniaspora uvarum*
Bifidobacterium bifidum	*Pediococcus pentosaceus*	*Bacillus. vallismortis*	*Lactobacillus kefiranofaciens*	*Kluyveromyces lactis*
Bifidobacterium infantis	*Propionibacterium acidipropiconici* (cattle only)	*Bifidobacterium adolescentis*	*Lactobacillus kefiri*	*Kluyveromyces marxianus*
Bifidobacterium longum	*Propionibacterium freudenreichii*	*Bifidobacterium animalis*	*Lactobacillus mucosae*	*Komagataella pastoris*[b]
Bifidobacterium thermophilum	*Propionibacterium shermanii*	*Bifidobacterium bifidum*	*Lactobacillus panis*	*Lindnera jadinii*[b]
Enterococcus cremoris	*Rhodopseudomonas palustris (broiler chickens only)*	*Bifidobacterium breve*	*Lactobacillus paracasei*	*Ogataea angusta*[b]
Enterococcus diacetylactis	Yeasts and molds:	*Bifidobacterium longum*	*Lactobacillus paraplantarum*	*Saccharomyces bayanus*
Enterococcus faecium	*Aspergillus niger*	*Carnobacterium divergens*	*Lactobacillus pentosus*	*Saccharomyces cerevisiae*
Enterococcus intermedius	*Aspergillus oryzae*	*Corynebacterium glutamicum*[b]	*Lactobacillus plantarum*	*Saccharomyces pastorianus*
Enterococcus lactis	*Saccharomyces cerevisiae*	*Geobacillus stearothermophilus*	*Lactobacillus pontis*	*Schizosaccharomyces pombe*
Enterococcus thermophilus		*Gluconobacter oxydans*[b]	*Lactobacillus reuteri*	*Wickerhamomyces anomalus*[b]
Lactobacillus acidophilus		*Lactobacillus acidophilus*	*Lactobacillus rhamnosus*	*Xanthophyllomyces dendrorhous*
Lactobacillus brevis		*Lactobacillus amylolyticus*	*Lactobacillus sakei*	
Lactobacillus buchneri (cattle only)		*Lactobacillus amylovorus*	*Lactobacillus salivarius*	
Lactobacillus bulgaricus		*Lactobacillus alimentarius*	*Lactobacillus sanfranciscensis*	
Lactobacillus casei		*Lactobacillus aviaries*	*Lactococcus lactis*	
Lactobacillus cellobiosus		*Lactobacillus brevis*	*Leuconostoc citreum*	
Lactobacillus farciminis (swine only)		*Lactobacillus buchneri*	*Leuconostoc lactis*	
		Lactobacillus casei	*Leuconostoc mesenteroides*	
		Lactobacillus cellobiosus	*Microbacterium imperiale*[b]	
		Lactobacillus collinoides	*Oenococcus oeni*	

[a]Abbreviations: AAFCO, Association of American Feed Control Officials; QPS, qualified presumption of safety; EFSA, European Food Safety Authority.
[b]QPS status subject to qualifications.
[c]Adapted from Table 1 in ref. 62.

DFMs AND THEIR COMPOSITION

Despite the regulatory constraints imposed on probiotic products in the United States and the European Union, numerous products are currently on the market. Based on listings provided in the Microbial Compendium (18), nearly 130 products were being offered in the United States that contained either live bacteria or bacterial culture-derived components that may or may not include live bacteria. Of all these products, only two make the claim that the product has competitive exclusion functions. Two additional products mention activity against pathogens. One product consisting of *Lactobacillus acidophilus* live cells mentions "food safety." The remainder of the products either do not make specific claims about the effects of the product or claim benefits to the target animal only. Some producers or distributors describe the bacteria in the product as "beneficial," as promoting or enhancing a "balanced" microbiota, as improving feed digestion and feed conversion, and as being helpful during or following stressful events. The obvious absence of specific claims, especially food safety claims, stems from the requirements that would have to be met to substantiate such claims. Challenge trials with pathogens under controlled conditions are frequently carried out to test the efficacy of microbial interventions, because such trials are easier and cheaper to conduct than actual field trials; however, challenge trials have drawbacks, and it has been argued that "the public health benefits of preharvest interventions to reduce zoonotic pathogens in livestock might be best served by field trial results alone" (19).

The scope of bacteria and products from these bacteria is limited as a consequence of regulations and agreements between the U.S. government and the organization representing the DFM industry (Fig. 1). *L. acidophilus* leads the list of bacteria included in DFMs, closely followed by *Enterococcus faecium*. *L. acidophilus*, in particular, has a long history of use in food and feed and has been evaluated for its probiotic properties such as survival under gastrointestinal conditions, adhesion to tissue, immunomodulatory properties, and production of bacteriocins and other antimicrobial activities in numerous trials (reviewed in 20, 21). The bacterium features prominently in studies aimed at reducing the prevalence of fecal shedding of *E. coli* 0157 in beef cattle that showed significant reductions in shedding (22).

A similar wealth of studies has attempted to elucidate the probiotic properties of *E. faecium*, and such studies have shown properties of this bacterium similar to those of *L. acidophilus*. Recently, the impact of the administration of *E. faecium* NCIMB 11181 on the fecal microbiota of swine has been studied using high-throughput 16S rRNA gene pyrosequencing. The bacterium, when added to feed over a 2-week period, produced various changes in the proportions of the phyla and genera of the fecal bacteria (23). The authors pointed out that the level of *E. coli* decreased, perhaps indicating a beneficial effect on food safety. The application of a commercial product containing *E. faecium*, *Lactobacillus salivarius*, and *Bifidobacter animalis* to broilers housed on floor pens and challenged with three *Eimeria* spp. resulted in lower *Eimeria* counts in the probiotic-treated birds than in the untreated controls (24). Since associations between *Eimeria* infection and infection with other pathogens such as *Salmonella* have been established (25, 26), decreased carriage of *Eimeria* could be interpreted as a positive influence of the probiotic on food safety. A similar commercial preparation consisting of *E. faecium*, *Lactobacillus* spp., and *Bifidobacterium thermophilum* showed evidence for causing reduced *Salmonella* levels in poultry litter from flocks given the preparation (27). Such a reduction potentially correlates with improved microbial safety of birds after processing, although such a connection was not established.

The third most common bacterial component of DFMs, *Bacillus subtilis*, has a shorter history of use than lactic acid bacteria, but

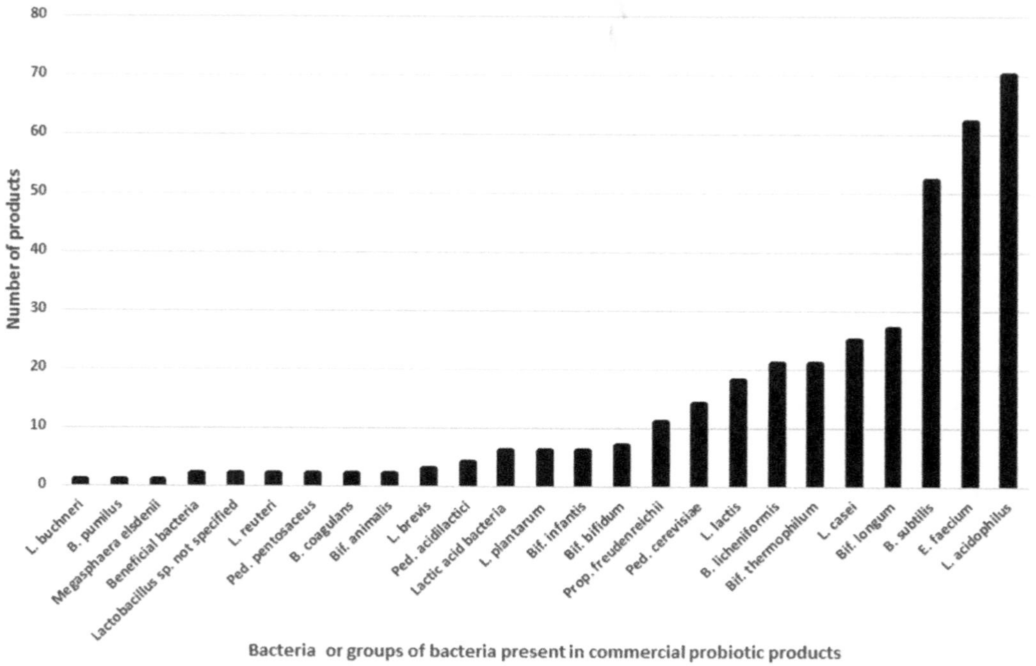

FIGURE 1 Number of products entered in the Microbial Compendium (18) that contain particular bacteria or groups of bacteria. Abbreviations: L., *Lactobacillus*; B., *Bacillus*; Ped., *Pediococcus*; Bif., *Bifidobacterium*; Prop., *Propionibacterium*; E., *Enterococcus*.

because of its obvious advantages over other bacteria with respect to storage and survival, it is now frequently included in probiotic products. Several studies have shown that the spores of various *Bacillus* spp. are able to germinate in the intestinal tract (for example, reviewed in 28–30); thus, they can be expected to produce metabolites such as acids, extracellular enzymes, bacteriocins, or antibiotics able to influence intestinal microbiota. The vast majority of studies examining the antipathogenic effect of *Bacillus* spp. so far have been done *in vitro*, but a number of challenge trials have been carried out, primarily with poultry (reviewed in reference 29), and immunomodulation by *Bacillus* spp.-containing preparations has also been observed in pigs and cattle (for example, references 31, 32). A commercial preparation containing a *B. subtilis* strain reduced *Clostridium perfringens* counts when administered to *C. perfringens*-challenged chicks (33). Since some level of association between C. *perfringens* and *Salmonella* infection may exist (25), the probiotic might be able to contribute to food safety of chickens.

Various species of *Bifidobacterium* and *Pediococcus* and *Propionibacterium freudenreichii* are also represented among the bacteria in DFMs. The probiotic effects of *Bifidobacterium* spp. have been studied extensively (reviewed, for example, in reference 34), including their protective effects against *Salmonella* and *E. coli* 0157:H7. Studies that showed protective effects against these pathogens were primarily carried out with animals under laboratory conditions such as those with streptomycin-treated (35) or germ-free mice (36), and it is obviously not possible to make direct connections to animals in farm settings. The same is true for studies using bioreactor models simulating the proximal colon of pigs (37). Animal trials involving *Bifidobacterium longum* in conjunction with fructo- and galactooligosaccharides revealed that this combination reduced *Campylobacter* spp. (38).

A commercial probiotic preparation containing *Pediococcus parvulus* and *L. salivarius* was administered to broiler chicks to study its effect on *Salmonella* colonization under conditions simulating commercial practices. Significant differences were found in the *Salmonella* counts in the ceca of birds receiving the probiotic supplement (39). The same product had earlier shown similar results with respect to *S. enterica* serovar Heidelberg colonization of broiler chickens and poults (40). Propionibacteria have a history of use as probiotics (41), and their probiotic properties such as adhesion to cells, survival at low pH, and survival in the presence of bile, as well as their antimicrobial activities are the subject of ongoing research (42). Antimicrobial activity has primarily been studied *in vitro*, but *P. freudenreichii* in conjunction with *L. acidophilus* was successfully utilized in trials with cattle carrying *E. coli* 0157:H7 (22).

A sizable number of DFM products include not just bacteria and/or their culture products, but also yeasts and molds. Of the approximately 130 products currently listed in the Microbial Compendium (18), more than 70 contain these eukaryotes as well as bacteria. Some DFMs are composed entirely of yeasts, although none of those products make any claim about food safety or antipathogen activity. The number of eukaryotic species allowed is limited, as can be seen in Table 1. The reason why yeasts and mold species and their culture products are included in DFMs is likely 2-fold. Based on studies with biomass from yeast fermentations fed to animals, it can be concluded that such preparations have performance-enhancing properties that might be related to the provision of additional nutrients such as vitamins, but also substrates that can be utilized by intestinal bacteria or enzymes that aid in the breakdown of plant polymers. The nutritional benefits of DFMs can be enhanced by including yeasts grown on inorganic selenium sources. The resulting organic selenium fraction is considered more nutritionally available to consuming animals than inorganic sources fed directly to the animals (43, 44).

In addition to the nutritional support provided by yeasts and molds, these organisms can potentially also act as antagonists of certain bacteria or as supporters of bacteria considered beneficial. Antagonism can be exerted as envisioned for bacteria simply through competition for nutrients and production of inhibitory metabolites. Obviously, the production of ethanol by *Saccharomyces* and other organisms has to be high on the list. Blocking of attachment sites on intestinal surfaces has also been proposed to be a factor that could prevent the establishment of pathogens (45). Due to their relatively large size and ability to aggregate, yeasts and molds can also be attachment sites for bacteria and provide environments that are conducive for the bacteria or that prevent them from interacting with the animal host. The presence of mannan oligosaccharides in yeast cell walls and therefore also in many preparations derived from yeasts has been proposed as providing structures to which enterobacteria can adhere and be prevented from adherence to intestinal surfaces; however, mechanisms and extent of attachment have not been fully elucidated (46).

PREBIOTICS

For virtually all DFM products, the more than 50 prebiotic products listed in the Microbial Compendium (18) also do not make any claims about food safety enhancement. Usually, the products emphasize contributions to animal health. Many of the products are not solely prebiotic preparations, but also contain live bacteria and/or yeasts. These products would fall into the category of synbiotics. Yeasts and molds are added not just in live form, but also in the form of autolysates, hydrolysates, or extracts. In this form, the components of these organisms are not only nutritional supplements, but are also a potential source of prebiotics in the form of mannan oligosaccharides or other oligosaccharides originating from the cell walls. Almost half of the prebiotic products listed in the compendium contain these types of oligosaccharides. An only slightly lower number lists fructo-oligosaccharides (FOS) and inulin as part of their content, and about 10 mention beta glucans. Rarely is the source of these oligosaccharides provided. Beta glucans might originate from yeasts, but also from plant material. FOS including inulin are likely exclusively of plant origin, and some products list yucca or chicory as sources. A review of the literature regarding the use of FOS-type prebiotics made it clear that studies of the effect of these probiotics have not yielded definitive proof of effectiveness (46–48). In part, this outcome can be attributed to the potential for differences in the FOS products utilized due to differences in origin and processing.

By definition, prebiotics are intended to be unavailable to the host but be a substrate for beneficial bacteria resulting in the exertion of inhibitory effects on pathogens. Many, but not all, lactic acid bacteria (LAB) and bifidobacteria have the ability to utilize prebiotic oligosaccharides (48), but it is difficult to document in animal trials that reductions in pathogens such as *Salmonella* and *Campylobacter* are indeed correlated with effects on these bacteria. Since not all LABs or bifidobacteria are able to utilize prebiotic oligosaccharides (49, 50), and not all of them can be expected to be equally effective as antipathogen agents, the presence of these bacteria and changes in their abundance do not automatically result in increased pathogen reduction. These circumstances might explain the observed heterogeneity in results with prebiotics, and they might also give impetus to combine prebiotic with live bacteria known to be inhibitory to pathogens. A recent review of prebiotic-related studies involving poultry (51) again highlighted the paucity of experiences with prebiotics as efficient contributors to food safety; however, the authors expressed their conviction that high-throughput approaches to intestinal microbiology will advance knowledge of the effects of prebiotics on intestinal microbiota and reveal suitable applications of prebiotics for the promotion of food safety. Perhaps microarray-based instrumentation such as the FDA's GutProb (52), which allows rapid detection and analyses of probiotic strains, will also accelerate development of pro- and prebiotic products. Without approaches that allow a comprehensive analysis of the microbial composition of intestinal sites, it is difficult to detect changes in the microbiota in response to the application of prebiotics. For example, a recent study on the effects of a mannanoligosaccharide-based commercial prebiotic on the cecal microbiota of broiler chickens that utilized denaturing gradient gel electrophoresis was not able to reveal obvious and consistent differences in the cecal microbiota of birds fed the prebiotic and control birds (53). The study also highlighted the difficulty in demonstrating potential food safety benefits of prebiotic application in large-scale settings because the frequency of *Salmonella* isolations from the ceca of the treated and control birds was low regardless of the treatment.

FUTURE PROSPECTS

Prebiotic research in the past was clearly focused on how these compounds can support well-known intestinal populations such as LABs and bifidobacteria, but recently, evidence was provided, albeit in a zebra fish model, that even the minority component of the gut microbiota can have a strong effect on intestinal immune responses (54). Provided these findings are true not just in a model system, but apply to complex intestinal microbiota, it will be necessary to evaluate means by which even minor components of the microbiota might have to be supported in the future.

It is difficult to predict at what pace applications of pro- and prebiotics for the purpose of improved preharvest food safety will move forward. As this chapter has attempted to outline, a relatively conservative regulatory environment has been established in the United States and Europe that makes it necessary to invest in considerable research and other efforts prior to the introduction of novel types of bacteria as DFMs. Considerable gaps in the understanding of the interactions of intestinal microorganisms with each other and with the host organism still exist, but these gaps are expected to disappear due to high-throughput sequencing and other molecular and genetic techniques. In the food safety field, a strong impetus toward improved applications of pro- and prebiotics is not only the need for improved animal health and performance, but also the movement away from the application of antibiotics in animal feed (55, 56). Pro- and prebiotics are viewed as components of strategies to reduce or even eliminate routine antibiotic use in animal agriculture. Obviously, the concerns about the spread of antibiotic resistance also pose one more prerequisite for the selection of suitable probiotic strains because it is absolutely essential that such strains do not carry antibiotic resistance genes. Other genes that should, of course, be absent in probiotic strains are those that code for the production of toxins or compounds that in any way will interfere with an animal's well-being and productivity.

In contrast, genes that should be present are those that enable the organism to grow in inexpensive media for cost-effective production, to survive processing and storage, to pass through the stomach or other inhospitable parts of an animal's digestive tract without being overly decimated, to persist and perhaps grow under the conditions of the lower intestinal tract, and to produce compounds that inhibit unwanted bacteria and perhaps also those that stimulate the animal's natural defenses and performance. Finding organisms that possess all these traits is not easy, and not surprisingly, the genetic modification of probiotic strains has been considered and even tested. Lactic acid bacteria, in particular, have been the target of genetic modifications, with an eye on ultimately utilizing them for human applications such as combatting or preventing colitis (57). A few genetically modified strains have been tested with the aim of improving animal performance. For example, *Lactococcus lactis* was engineered to express the epidermal growth factor EGF-LL in an effort to boost the performance of early-weaned piglets (58), and the yeast *Pichia pastoris* was modified by the introduction of the *C. perfringens* alpha toxin gene in an attempt to induce immunity against *C. perfringens* in broiler chickens (59).

If and when such genetically modified organisms will ever be utilized commercially depends on scientific and economic factors but most importantly on whether or not the farming community and the public will be willing to accept the widespread release of genetically modified microorganisms in the environment and the food supply. As a way to avoid the direct introduction of genetically engineered microorganisms but still be able to utilize the benefits of genetic engineering, the use of culture supernatants is an option. This method is now being pursued, for example, in the case of the *L. lactis* expressing EGF-LL (60).

CITATION

Joerger RD, Ganguly A. 2017. Current status of the preharvest application of pro- and prebiotics to farm animals to enhance the microbial safety of animal products. Microbiol Spectrum 5(1):PFS-0012-2016.

REFERENCES

1. **Anukam KC, Reid G.** 2007. Probiotics: 100 years (1907–2007) after Elie Metchnikoff's observation. *Commun Curr Res Educ Top Trends Appl Microbiol* **1:**466–474.
2. **Metchnikoff E.** 1907. *The Prolongation of Life: Optimistic Studies.* p 171. Translated and edited by Mitchell PC. Heinemann, London, United Kingdom.

3. **Hill C, Guarner F, Reid G, Gibson GR, Merenstein DJ, Pot B, Morelli L, Canani RB, Flint HJ, Salminen S, Calder PC, Sanders ME.** 2014. Expert consensus document: The International Scientific Association for Probiotics and Prebiotics consensus statement on the scope and appropriate use of the term probiotic. *Nat Rev Gastroenterol Hepatol* **11:**506–514.
4. **Gibson GR, Roberfroid MB.** 1995. Dietary modulation of the human colonic microbiota: introducing the concept of prebiotics. *J Nutr* **125:**1401–1412.
5. **Pineiro M, Asp N-G, Reid G, Macfarlane S, Morelli L, Brunser O, Tuohy K.** 2008. FAO technical meeting on prebiotics. *J Clin Gastroenterol* **42**(Suppl 3 Pt 2)**:**S156–S159.
6. **Roberfroid M.** 2007. Prebiotics: the concept revisited. *J Nutr* **137:**830S–837S.
7. **Milner KC, Shaffer MF.** 1952. Bacteriologic studies of experimental *Salmonella* infections in chicks. *J Infect Dis* **90:**81–96.
8. **Nurmi E, Rantala M.** 1973. New aspects of *Salmonella* infection in broiler production. *Nature* **241:**210–211.
9. **Kerr AK, Farrar AM, Waddell LA, Wilkins W, Wilhelm BJ, Bucher O, Wills RW, Bailey RH, Varga C, McEwen SA, Rajio A.** 2013. A systematic review-meta-analysis and meta-regression on the effect of selected competitive exclusion products on *Salmonella* spp. prevalence and concentration in broiler chickens. *Prev Vet Med* **111:**112–25.
10. **Anderson RC, Nisbet DJ, Buckley SA, Genovese KJ, Harvey RB, Deloach JR, Keith NK, Stanker LH.** 1998. Experimental and natural infection of early weaned pigs with *Salmonella choleraesuis*. *Res Vet Sci* **64:**261–262.
11. **Rasmussen MA, Casey TA.** 2001. Environmental and food safety aspects of *Escherichia coli* 0157:H7 infections in cattle. *Crit Rev Microbiol* **27:**57–73.
12. **Brashears MM, Jaroni D, Trimble J.** 2003. Isolation, selection, and characterization of lactic acid bacteria for a competitive exclusion product to reduce shedding of *Escherichia coli* 0157:H7 in cattle. *J Food Prot* **66:**355–363.
13. **Brashears MM, Galyean ML, Loneragan GH, Mann JE, Killinger-Mann K.** 2003. Prevalence of *Escherichia coli* 0157:H7 and performance by beef feedlot cattle given *Lactobacillus* direct-fed microbials. *J Food Prot* **66:**748–754.
14. **APHIS.** 2011. Direct-fed microbials (probiotics) in calf diets. https://www.aphis.usda.gov/animal_health/nahms/dairy/downloads/bamn/BAMN11_Probiotics.pdf.
15. **FDA.** 2015. CPG Sec. 689.100 direct-fed microbial products. http://www.fda.gov/ICECI/ComplianceManuals/CompliancePolicyGuidanceManual/ucm074707.htm.
16. **FDA.** 2015. NADA 141-101 PREEMPT™: original approval. http://www.fda.gov/AnimalVeterinary/Products/ApprovedAnimalDrugProducts/FOIADrugSummaries/ucm117130.htm.
17. **FDA.** 2013. Actions taken by FDA center for veterinary medicine. http://www.fda.gov/downloads/animalveterinary/products/approvedanimaldrugproducts/ucm346319.pdf.
18. **Penton Agriculture.** 2015. Microbial compendium. http://microbialcompendium.com/.
19. **Wisener LV, Sargeant JM, O'Connor AM, Faires MC, Glass-Kaastra SK.** 2014. The evidentiary value of challenge trials for three pre-harvest food safety topics: a systematic assessment. *Zoonoses Public Health* **61:**449–476.
20. **Bull M, Plummer S, Marchesi J, Mahenthiralingam E.** 2013. The life history of *Lactobacillus acidophilus* as a probiotic: a tale of revisionary taxonomy, misidentification and commercial success. *FEMS Microbiol Lett* **349:**77–87.
21. **Sanders ME, Klaenhammer TR.** 2001. Invited review: the scientific basis of *Lactobacillus acidophilus* NCFM functionality as a probiotic. *J Dairy Sci* **84:**319–331.
22. **Elam NA, Gleghorn JF, Rivera JD, Galyean ML, Defoor PJ, Brashears MM, Younts-Dahl SM.** 2003. Effects of live cultures of *Lactobacillus acidophilus* (strains NP45 and NP51) and *Propionibacterium freudenreichii* on performance, carcass, and intestinal characteristics, and *Escherichia coli* strain 0157 shedding of finishing beef steers. *J Anim Sci* **81:**2686–2698.
23. **Pajarillo EAB, Chae JP, Balolong MP, Kim HB, Park C-S, Kang D-K.** 2015. Effects of probiotic *Enterococcus faecium* NCIMB 11181 administration on swine fecal microbiota diversity and composition using barcoded pyrosequencing. *Anim Feed Sci Technol* **201:**80–88.
24. **Ritzi MM, Abdelrahman W, Hohnl M, Dalloul R.** 2014. Effects of probiotics and application methods on performance and response of broiler chickens to an *Eimeria* challenge. *Poult Sci* **93:**2772–2778.
25. **Shivaramaiah S, Wolfenden RE, Barra JR, Morgan MJ, Wolfenden AD, Hargis BM, Tellez G.** 2011. The role of early *Salmonella* Typhimurium infection as a predisposing factor for necrotic enteritidis in a laboratory challenge model. *Avian Dis* **55:**319–323.
26. **Volkova VV, Wills RW, Hubbard SA, Magee D, Byrd JA, Bailey RH.** 2011. Associations between vaccinations against protozoal and viral infections and *Salmonella* in broiler flocks. *Epidemiol Infect* **139:**206–215.

27. **Pedroso AA, Hurley-Bacon AL, Zedek AS, Kwan TW, Jordan APO, Avellaneda G, Hofacre CL, Oakley BB, Collett SR, Maurer JJ, Lee MD.** 2013. Can probiotics improve the environmental microbiome and resistome of commercial poultry production? *Int J Environ Res Public Health* **10:**4534–4559.
28. **Cutting SM.** 2011. *Bacillus* probiotics. *Food Microbiol* **28:**214–220.
29. **Ricke SC, Saengkerdsub S.** 2015. *Bacillus* probiotics and biologicals for improving animal and human health: current applications and future prospects, p 341–360. *In* Rai VR, Bai JA (ed), *Beneficial Microbes in Fermented and Functional Foods*. CRC Press, Boca Raton, FL.
30. **Ripamonti B, Stella S.** 2009. Bacterial spore formers as probiotics for animal nutrition. *Large Anim Rev* **15:**7–12.
31. **Novak KN, Davis E, Wehnes CA, Shields DR, Coalson JA, Smith AH, Rehberger TG.** 2012. Effect of supplementation with an electrolyte containing a *Bacillus*-based direct-fed microbial on immune development in dairy calves. *Res Vet Sci* **92:**427–434.
32. **Zhou D, Zhu Y-H, Zhang W, Wang M-L, Fan W-Y, Song D, Yang G-Y, Jensen BB, Wang J-F.** 2015. Oral administration of a select mixture of *Bacillus* probiotics generates Tr1 cells in weaned F4ab/acR(−) pigs challenged with an F4 (+) ETEC/VTEC/EPEC strain. *Vet Res* **46:**95–110.
33. **Abudabos AM, Alyemni AH, Al Marshad MBA.** 2013. *Bacillus subtilis* PB6-based probiotic (CloSTATTM) improves intestinal morphological and microbiological status of broiler chickens under *Clostridium perfringens* challenge. *Int J Agric Biol* **15:**978–982.
34. **Picard C, Fioramonti J, Francois A, Robinson T, Neant F, Matuchansky C.** 2005. Review article: bifidobacteria as probiotic agents: physiological effects and clinical benefits. *Aliment Pharmacol Ther* **22:**495–512.
35. **Asahara T, Shimizu K, Nomoto K, Hamabata T, Ozawa A, Takeda Y.** 2004. Probiotic bifidobacteria protect mice from lethal infection with shiga toxin-producing *Escherichia coli* 0157:H7. *Infect Immun* **72:**2240–2247.
36. **Yoshimura K, Matsui T, Itoh K.** 2010. Prevention of *Escherichia coli* 0157:H7 infection in gnotobiotic mice associated with *Bifidobacterium* strains. *Anton Leeuw Int J Gen Mol Microbiol* **97:**107–117.
37. **Tanner SA, Chassard C, Zihler Berner A, Lacroix C.** 2014. Synergistic effect of *Bifidobacterium thermophilum* RBL67 and selected prebiotics on inhibition of *Salmonella* colonization in the swine proximal colon PolyFermS model. *Gut Pathogens* **6:**44–55.
38. **Baffoni L, Gaggia F, Di Gioia D, Santini C, Mogna L, Biavati B.** 2012. A Bifidobacterium-based synbiotic product to reduce the transmission of C. *jejuni* along the poultry food chain. *Int J Food Microbiol* **157:**156–161.
39. **Biloni A, Quintana CF, Menconi A, Kallapura G, Latorre J, Pixley C, Layton S, Dalmagro M, Hernadnez-Velasco X, Wolfenden A, Hargis BM, Tellez G.** 2013. Evaluation of effects of EarlyBird associated with FloraMax-B11 on *Salmonella* Enteritidis, intestinal morphology, and performance of broiler chickens. *Poult Sci* **92:**2337–2346.
40. **Menconi A, Wolfenden AD, Shivaramaiah S, Rerraes JC, Urbano T, Kuttel J, Kremer C, Hargis BM, Tellex G.** 2011. Effect of lactic acid bacteria probiotic culture for the treatment of *S. enterica* serovar Heidelberg in neonatal broiler chickens and turkey poults. *Poult Sci* **90:**561–565.
41. **Mantere-Alhonen S.** 1995. Propionibacteria used as probiotics: a review. *Le Lait* **75:**447–452.
42. **Campaniello D, Bevilacqua A, Sinigaglia M, Altieri C.** 2015. Screening of *Propionibacterium* spp. for potential probiotic properties. *Anaerobe* **34:**169–173.
43. **Cozzi G, Prevedello P, Stefani AL, Piron A, Contiero B, Lante A, Gottardo F, Chevaux E.** 2011. Effect of dietary supplementation with different sources of selenium on growth response, selenium blood levels and meat quality of intensively finished Charolais young bulls. *Animal* **5:**1531–1531.
44. **Delezie E, Rovers M, Van der Aa A, Ruttens A, Wittocx S, Segers L.** 2014. Comparing responses to different selenium sources and dosages in laying hens. *Poult Sci* **93:**3083–3090.
45. **Ganner A, Schatzmayr G.** 2012. Capability of yeast derivatives to adhere enteropathogenic bacteria and to modulate cells of the innate immune system. *Appl Microbiol Biotechnol* **95:** 289–297.
46. **Ricke SC.** 2015. Potential of fructooligosaccharide prebiotics in alternative and nonconventional poultry production systems. *Poult Sci* **94:**1411–1418.
47. **Jordan K, Dalmasso M, Zentek J, Mader A, Bruggeman G, Wallace J, De Medici D, Fiore A, Prukner-Radovcic E, Lukac M, Axelsson L, Holck A, Ingmer H, Malakauskas M.** 2014. Microbes versus microbes: control of pathogens in the food chain. *J Sci Food Agric* **94:**3079–3089.
48. **Totton SC, Farrar AM, Wilkins W, Bucher O, Waddell LA, Wilhelm BJ, McEwen SA, Rajic A.** 2012. The effectiveness of selected feed and

water additives for reducing *Salmonella* spp. of public health importance in broiler chickens: a systematic review, meta-analysis, and meta-regression approach. *Prev Vet Med* **106:**197–213.

49. **Hopkins MJ, Cummings JH, Macfarlane GT.** 1998. Inter-species differences in maximum specific growth rates and cell yields of bifidobacteria cultured on oligosaccharides and other simple carbohydrate sources. *J Appl Microbiol* **85:**381–386.
50. **Grimoud J, Durand H, Courtin C, Monsan P, Ouarne F, Theodorouo V, Rogues C.** 2010. *In vitro* screening of probiotic lactic acid bacteria and prebiotic glucooligosaccharides to select effective symbiotics. *Anaerobe* **16:**493–500.
51. **Pourabedin M, Zhao X.** 2015. Prebiotics and gut microbiota in chickens. *FEMS Microbiol Lett* **362:**fnv122.
52. **Patro JN, Ramachandran P, Lewis JL, Mammel MK, Barnab T, Pfeiler EA, Elkins CA.** 2015. Development and utility of the FDA 'GutProbe' DNA microarray for identification, genotyping and metagenomic analysis of commercially available probiotics. *J Appl Microbiol* **118:**1478–1488.
53. **Lee SI, Park SH, Ricke SC.** 2015. Assessment of cecal microbiota, integrin occurrence, fermentation responses, and *Salmonella* frequency in conventionally raised broilers fed a commercial yeast-based prebiotic compound. *Poult Sci* **95:**144–153.
54. **Rolig AS, Parthasarathy R, Burns AR, Bohannan BJM, Guillemin K.** 2015. Individual members of the microbiota disproportionately modulate host innate immune responses. *Cell Host Microbe* **18:** 613–620.
55. **European Union.** 2003. Regulation (EC) No 1831/2003 of the European Parliament and of the Council of 22 September 2003 on additives for use in animal nutrition. *Off J Europ Union* **46:**29–43.
56. **FDA.** 2015. Guidance for industry #213. http://www.fda.gov/downloads/AnimalVeterinary/GuidanceComplianceEnforcement/Guidanceforindustry/UCM299624.pdf.
57. **de LeBlanc AD, del Carmen S, Chatel JM, Miyoshi A, Azevedo V, Langella P, Bermudez-Humaran LG, LeBlanc JG.** 2015. Current review of genetically modified lactic acid bacteria for the prevention and treatment of colitis using murine models. *Gastorenterol Res Pract.* **2015:**146972.
58. **Bedford A, Li Z, Li M, Ji S, Liu W, Huai Y, de Lange CEM, Li J.** 2012. Epidermal growth factor-expressing *Lactococcus lactis* enhances growth performance of early-weaned pigs fed diets devoid of blood plasma. *J Anim Sci* **90:**4–6.
59. **Gil de los Santos JR, Storch OB, Fernandes CG, Gil-Turnes C.** 2012. Evaluation in broilers of the probiotic properties of *Pichia pastoris* and a recombinant *P. pastoris* containing the *Clostridium perfringens* alpha toxin gene. *Vet Microbiol* **156:**448–451.
60. **Bedford A, Chen T, Huynh E, Zhu C, Medeiros S, Wey D, de Lange C, Li J.** 2015. Epidermal growth factor containing culture supernatant enhances intestine development of early-weaned pigs *in vivo*: potential mechanisms involved. *J Biotechnol* **196:**9–19.
61. **AAFCO.** 2015. *2015 Official Publication*. Association of American Feed Control Officials, Champaign, IL.
62. **EFSA.** 2013. Scientific opinion on the maintenance of the list of QPS biological agents intentionally added to food and feed (2013 update). *EFSA Journal* **11**(11)**:**3449.

Molecular Tools To Study Preharvest Food Safety Challenges

20

DEEPAK KUMAR[1] and SIDDHARTHA THAKUR[2]

INTRODUCTION

There has been a marked shift in consumer preferences about food choices. Today, consumers not only demand safe, inexpensive, tasty, and healthy food but also expect animal welfare and environmental safety. Consumers have become more curious about the source of meat—whether it comes from a conventional or an antibiotic-free farm (ABF) and if biosecurity requirements were met. To ensure the supply of safe meat to consumers, preventive measures need to be taken at the farm when animals are alive. This is where the concept of preharvest food safety comes from. Preharvest food safety is the combination of measures and interventions adopted at the farm to detect harmful pathogens and to reduce pathogen load in the food chain. Preharvest food safety is not only about the food animals and the microbes they carry but also includes the surrounding environment and human activities. The farm is a dynamic environment where pathogens, food animals, human interventions, environmental factors, and other animal species interact closely.

[1]Department of Veterinary Public Health & Epidemiology, College of Veterinary and Animal Sciences, Govind Ballabh Pant University of Agriculture & Technology, Pantnagar, Uttarakhand-263145, India; [2]Department of Population Health and Pathobiology, College of Veterinary Medicine, North Carolina State University, Raleigh, NC 27607.

Preharvest Food Safety
Edited by Siddhartha Thakur and Kalmia E. Kniel

doi:10.1128/microbiolspec.PFS-0019-2017

Safe meat production has to be a holistic approach instead of a single-point inspection at slaughter or during processing. Earlier, all food safety interventions were largely focused on postmortem inspection of the carcass and other quality checks administered before the finished product could reach the market. These quality checks would categorize meat into categories such as acceptable, unacceptable, and unsafe for human consumption. It was believed that because the postharvest stages (slaughter and processing) in meat production are closest to the final product, by controlling the incidence of pathogens in the finished product, the risk of foodborne diseases in humans could be reduced. Nevertheless, postharvest food safety measures such as meat inspection and detection of meat that is unfit for human consumption do contribute to consumer protection but do not completely prevent major food safety issues in the final product; i.e., these are only quality checkpoints at the end of an on-farm production phase.

Another reason is that the controlled environment of a slaughterhouse or processing unit is unlike an open and complex farm environment, which comprises different pathogens moving freely and looking for an opportunity to colonize food animals. Classical zoonotic diseases such as tuberculosis and brucellosis could be recognized at the farm stage, and their lesions could be detected during postmortem inspection. However, foodborne pathogens such as multidrug-resistant *Salmonella* and *Campylobacter*, enterohemorrhagic *Escherichia coli* (*E. coli* O157:H7), *Toxoplasma*, and *Yersinia* are only detectable through targeted monitoring systems, because they cause neither clinical symptoms nor lesions to be detected during postmortem inspection of the carcass (1). The traditional mandatory meat inspection procedures are essential, but they are unable to prevent new/emerging foodborne pathogens posing risks to human health (2, 3).

Moreover, issues such as the emergence of antimicrobial-resistant (AMR) bacterial pathogens (*Salmonella* and *Campylobacter*) and the presence of antimicrobial and chemical residues in meat have forced researchers and the meat industry to focus on the preharvest stages of meat production. In particular, the use of antimicrobials and the emergence of multidrug resistance bacteria are the most important preharvest food safety issues affecting public health. The modern food animal production system depends a lot on the use of antimicrobials for prophylaxis and growth promotion in addition to the treatment of animals. Domestic sales and distribution of antimicrobials approved for use in food animals in the United States in 2015 was approximately 15.6 million kg (4). Subtherapeutic doses of antimicrobials are used in food animal production for rapid growth. Such doses, however, are not able to kill microbes and lead to the emergence of antimicrobial resistance in microbes. AMR bacteria can pass through the food chain to the consumers. Extensive usage of antimicrobials in production animals creates selective pressure on bacteria that eventually develop resistance to survive (5, 6). These resistant bacteria originating from production systems could travel to the slaughter and processing units and transfer antimicrobial-resistant attributes to the native microbial population.

Preharvest food safety research and activities have advanced over time with the recognition of the importance and complicated nature of the preharvest phase of food production. In developed nations, implementation of preharvest food safety procedures along with strict monitoring and containment at various postharvest stages such as slaughter, processing, storage, and distribution have remarkably reduced the burden of foodborne pathogens in humans (1). Early detection and adequate surveillance of pathogens at the preharvest stage is of the utmost importance to ensure a safe meat supply. There is an urgent need to develop rapid, cost-effective, and point-of-care diagnostics which could be used at the preharvest stage and which complement postmortem and other quality checks performed at the postharvest stage.

With newer methods and technologies, more effort needs to be directed toward developing rapid, sensitive, and specific methods for detection or screening of foodborne pathogens at the preharvest stage. In this review, we will discuss the molecular methods available for detection and molecular typing of bacterial foodborne pathogens at the farm. Such methods include conventional techniques such as endpoint PCR, real-time PCR, DNA microarray, restriction fragment length polymorphism (RFLP), random amplification of polymorphic DNA (RAPD), and more advanced techniques such as matrix-assisted layer desorption ionization–time of flight mass spectrometry (MALDI-TOF MS), whole-genome sequencing (WGS), and multilocus sequence typing (MLST).

MOLECULAR DETECTION TECHNIQUES

The development and application of new tools for the rapid identification of microorganisms is of paramount importance. Culture and isolation of bacterial pathogens is time-consuming and typically takes 36 to 48 h from sample processing to complete results (7). This time is considerably increased if the growth of the pathogen is slow or delayed (8). The use of molecular methods is a key approach for the rapid and accurate identification of foodborne pathogens. One of the major reasons for the wide acceptance of molecular techniques in pathogen detection is their ability to generate results within a short time compared to culture-based methods. In this section, we discuss molecular methods available for pathogen detection.

Endpoint PCRs

The rapid and accurate detection of foodborne pathogens became possible with the arrival of PCR. PCR involves the amplification of a specific segment of DNA of any pathogen under restricted conditions of reagents and time-temperature combinations. PCR is divided into two categories, endpoint PCR and real-time PCR, on the basis of the time to result interpretation and the detection of the amount of amplified DNA. Endpoint PCR is further divided into simplex and multiplex PCR, the details of which are given below.

Simplex PCR

The easiest molecular approach to detecting pathogens is the amplification of genomic DNA using PCR with oligonucleotide primers that are specific for the genus or species of the bacteria. Simplex PCR can identify only one gene/target in a single PCR reaction and requires designing highly specific primers against the gene (9). PCR amplification is followed by endpoint visualization of the amplification product(s) in an agarose gel (10, 11). Simplex PCRs can identify the known microorganisms at the genus or species level in a very short time. Early detection of foodborne pathogens in farm settings is particularly important because it provides sufficient time to set up measures to prevent further spread and to strengthen biosecurity. However, it has limited utility in outbreak investigations involving new pathogens with no sequence information available in GenBank/NCBI. In such cases, PCR needs to be coupled with gene sequencing or restriction digestion of the amplified product for identification and differentiation of the new pathogens involved (9). Simplex PCR has been extensively utilized in prevalence studies for detection of an array of pathogens at the production (preharvest), slaughter, and processing stages (12–14).

Multiplex PCR

The inability of simplex PCR to amplify multiple targets/genes led to the development of multiplex PCR. Multiplex PCR can be used for simultaneous detection of multiple pathogens in a single PCR reaction. However, designing primers for a multiplex reaction is a challenge. Primers should be designed such that each primer sequence is able to produce a unique amplicon for the target DNA. Moreover, identification of a common annealing temperature is a prerequisite for successful design of a multiplex PCR (15).

Multiplex PCR has been highly commercialized, and several multiplex PCR detection kits are available. To date, several multiplex PCRs able to detect different genera or species of foodborne pathogens have been developed (16–21). For example, multiplex PCR was developed for concurrent detection of five major foodborne pathogens, namely *Staphylococcus aureus*, *Listeria monocytogenes*, *E. coli* O157:H7, *Salmonella enterica* serovar Enteritidis, and *Shigella flexneri*, targeting 16S rDNA, listeriolysin O (*hlyA*), the intimin (*eaeA*) gene, *invA*, and the invasion plasmid antigen H (*ipaH*) genes, respectively (22). This multiplex PCR assay was able to simultaneously detect all five of these organisms in artificially contaminated pork samples. Multiplex PCR has been successfully applied to the differentiation of species, serotypes, subtypes, toxin genes, and antimicrobial resistance genes of bacteria. Recently, Rawool and coworkers reported a multiplex PCR for simultaneous detection of the *Listeria* genus, *L. monocytogenes*, and three important lineages (LI, LII, and LIII) of *L. monocytogenes* (23). A multiplex PCR was reported for rapid identification of multiple serovars of *Salmonella*, namely Typhimurium, Choleraesuis, Infantis, Hadar, Enteritidis, Dublin, and Gallinarum (24). Application of such multiplex PCRs could be very useful in disease investigation because sometimes multiple serovars of *Salmonella* are found in food or food animals. This will not only help to significantly reduce the cost of diagnosis but will also reduce the time required for serotyping the isolates. Such PCRs could be very useful in detecting multiple pathogens from samples and in outbreak investigations involving several etiological agents. These can be successfully used in low-resource countries for rapid identification of pathogens.

Real-Time PCR

The most remarkable development in PCR technology has been the introduction of real-time monitoring of DNA amplification during a PCR reaction (25, 26). Also known as quantitative PCR, real-time PCR is a well-recognized method for the detection and quantification of foodborne pathogens. Real-time PCR involves target amplification and detection in a single step (27). Post-PCR procedures such as gel electrophoresis and gel imaging are not required in real-time PCR. Real-time PCR merges PCR with the use of fluorescent reporter molecules to monitor the production of amplification products during each cycle of the PCR reaction. Foodborne pathogens such as *Salmonella*, *Campylobacter*, *E. coli*, and *Listeria* are the leading causes of foodborne infections and outbreaks worldwide. Rapid assays are needed to identify these pathogens quickly from fecal and environmental samples at food production sites to minimize the infection across the food chain.

Real-time PCR amplification can be mainly categorized into a non-amplicon sequence-specific method and sequence-specific methods which are based on fluorescent labeled probes (28). The nonspecific method is based on the use of double-stranded intercalating DNA binding dyes, such as SYBR green I and EvaGreen. Amplicon sequence-specific detection methods such as TaqMan, molecular beacons, and Scorpions are based on the use of oligonucleotide probes labeled with a donor fluorophore and an acceptor dye known as quencher (29). Both of these approaches use some kind of fluorescent marker which binds to the DNA, and as the number of DNA copies increases during the PCR reaction, the fluorescence signal also increases.

SYBR green is a double-stranded DNA binding fluorescent dye (30). This nonspecific intercalating dye emits slight fluorescence, and the fluorescence signal is enhanced when bound to the minor groove of the DNA double helix (31, 32). Comparison with reference samples of known concentration allows the quantification of the initial concentration of the target DNA. In initial cycles the fluorescence is too low to be distinguishable from the background. However, the point at which the fluorescence intensity increases above the detectable level corresponds proportionally to the initial number of template DNA molecules in the sample. The SYBR green real-time PCR cannot differentiate between specific target and nonspecific amplifications or primer dimers (33). The advantage of using SYBR green is that a DNA melting curve can be generated after PCR along with the calculation of the T_m value of the amplified products. SYBR green lacks specificity and binds to all double-stranded DNA; thus, it can be used to detect any PCR product (34).

TaqMan probes are short oligonucleotides (normally 10 bp long—10-mer) specific to the target sequence between the two primers used in the PCR. TaqMan probes carry a fluorescent reporter molecule at one end and a quencher molecule capable of quenching the fluorescence of the reporter molecule at the other end. However, during the PCR the TaqMan probe binds to the target sequence and is cleaved by the polymerase. Therefore, the reporter and quencher molecules are physically separated, and the fluorescence increases. The fluorescence, measured after each cycle in a real-time PCR, is proportional to the amount of the specific target amplification product and does not include PCR artifacts (35). TaqMan probes and molecular beacons are sequence-specific probes and only bind to the target sequence (36).

The major advantages of real-time PCR include (i) fast and high-throughput detection and quantification of target DNA sequences, (ii) simultaneous amplification and visualization of newly synthesized DNA amplicons, (iii) no post-PCR processing or gel imaging, (iv) very low chances of cross-contamination after PCR amplification, and (v) the ability to multiplex several targets in a single reaction (37). Real-time PCR has become an indispensable tool in the detection of viral pathogens because isolating viruses in the laboratory is difficult and time-consuming. Moreover, methods like enzyme-linked immunosorbent assay have low sensitivity, specificity, and reproducibility. For RNA viruses, real-time PCR is used in conjunction with reverse transcription to synthesize cDNA, which works as a template for PCR. Several real-time reverse transcriptase PCRs have been reported for the detection of RNA viruses (38, 39). However, a major limitation of real-time PCR is its inability to differentiate between live and dead cells. This may lead to false-positive results when estimating the presence of viable pathogens in a sample. Detection of dead cells may overestimate the risk associated with the foodborne pathogens. Also, real-time PCR has limited utility in the detection of new or unreported pathogens (40).

In a farm setting, food animals can be infected by several pathogens at a time. Therefore, it is desirable to develop molecular assays that can detect multiple pathogens concurrently. Multiplex PCR has been widely used for amplifying multiple targets by using multiple primer pairs in a single reaction. However, developing multiplex PCR is difficult and requires expertise during standardization. Moreover, including multiple targets in a PCR reaction significantly decreases sensitivity, and hence, the number of targets is kept to a minimum (28). Compared to conventional PCR, real-time PCR offers better options for multiplexing several targets in a single reaction. Recently, there has been an increase in the development of real-time multiplex PCRs for the detection of multiple pathogens including foodborne pathogens (41–46). In one such study, Hu et al. developed a molecular beacon-based multiplex real-time PCR for the detection of various foodborne pathogens, namely *S. enterica* subsp. *enterica*, *L. monocytogenes*, *E. coli* O157, *Vibrio parahaemolyticus*, *Vibrio vulnificus*, *Campylobacter jejuni*, *Enterobacter sakazakii*, and *Shigella* spp. (46). Specific genes were targeted for simultaneous detection of eight foodborne pathogens. Fukushima et al. used a SYBR green-based multiplex real-time PCR for simultaneous identification of 24 genes of several foodborne pathogens (42). This assay was validated for the detection of foodborne pathogens in 33 out of 35 cases of foodborne outbreaks (42).

Another study reported a rapid real-time multiplex PCR assay using four primer sets and four TaqMan probes for the detection of multiple *Salmonella* serotypes. The assay was found to be sensitive and correctly identified different *Salmonella* serotypes in artificially contaminated chicken skin and chicken meat samples (41). Fratamico et al. detected Shiga toxin-producing *E. coli* serogroups O26, O45, O103, O111, O121, and O145 with a TaqMan-based real-time multiplex PCR with primers targeting the *stx1, stx2, eae,* and *wzx* genes (43). Apart from pathogen detection, real-time PCR has been commonly utilized for the detection of antimicrobial resistance genes in foodborne pathogens (47–49). For instance, Roschanski et al. reported a multiplex real-time PCR for the detection of class A beta-lactamase genes *bla*CTX-M, *bla*SHV, and *bla*TEM and CIT-type AmpCs in *Enterobacteriaceae* (50). The authors correctly identified resistance genes in previously identified resistance gene subtypes and a few animal and environmental isolates (50).

DNA Microarray

DNA microarray is considered a rapid and sensitive pathogen detection technique that overcomes the limitations of PCR-based methods (9, 51). Microarrays allow a large number of specific DNA sequences to be detected simultaneously. A microarray consists of a microscopic glass slide (also known as a chip) containing thousands of chemically synthesized short sequences/probes of 25 to 80 bases adhered to the surface of the slide (35, 52). These probes have nucleotide sequences that are complementary to one or more target genes/organisms. Nucleic acids are extracted from the sample and labeled with a fluorescent or radioactive dye. DNA is denatured into single-stranded DNA fragments and incubated over the surface of the glass slide for several hours for hybridization to take place. DNA fragments hybridize to their complementary or near complementary DNA probes on the glass slide. The array is then washed to get rid of the unbound additional DNA and scanned to capture and quantify the fluorescence signal probe-sample complex. The fluorescence intensity is directly proportional to the concentration of fluorescent dye-loaded nucleic acid (53).

Several microarrays have been developed for the detection of foodborne pathogens (51, 54–62). For instance, Guo et al. developed a microarray system for simultaneous identification of 46 *Salmonella* O serogroups targeting O antigen-specific genes. The assay accurately identified 40 serogroups. The remaining six serogroups were identified in the combination of three pairs and needed further differentiation using PCR or conventional serotyping. The assay correctly identified 98% of the *Salmonella* strains tested, with a low detection sensitivity of 50 ng genomic DNA. This was the first time a comprehensive microarray system was developed for simultaneous identification of as many as 46 *Salmonella* O serogroups (51). In a similar study, Braun et al. developed a microarray assay to identify multiple *Salmonella* serovars prevalent in Europe and North America (59). This assay differentiated O antigens and 86 H antigens of *Salmonella.* The authors successfully evaluated their microarray system in 117 of 132 reference strains of *Salmonella.* The other 15 serovars gave a similar band pattern shared by multiple serovars, which could be due to high homogeneity between serogroups A and D1. Further, this assay was also used to identify a panel of *Salmonella* field isolates; all were correctly identified at the genus level, and the results of 88.6% of the isolates correctly matched the conventional serotyping data (59).

In addition to *Salmonella,* the microarray method has also been used for the identification and differentiation of other foodborne pathogens. For example, Wang et al. reported an oligonucleotide microarray for the detection of 22 foodborne bacterial pathogens. Their assay identified most of the major foodborne pathogens, namely *Salmonella* spp., *C. jejuni, Clostridium perfringens, S. aureus,*

L. monocytogenes, *Bacillus cereus*, *Proteus vulgaris*, *V. parahaemolyticus*, and *Clostridium botulinum*, to name a few (54). This assay possessed good specificity and a low sensitivity of 10^2 CFU/ml of bacterium. Moreover, the assay was also validated using field samples (54). Chiang et al. developed a nylon chip-based microarray system targeting 16S rRNA for simultaneous identification of *Bacillus* spp., *E. coli*, *Salmonella* spp., *Staphylococcus* spp., and *Vibrio* spp. (63). Microarrays have also been used for the detection of resistance genes in bacterial pathogens. In one such study, Perreten et al. reported a microarray system for simultaneous detection of 90 antibiotic resistance genes in Gram-positive bacteria (64). The chip contained 137 oligonucleotide probes and was hybridized with bacterial strains carrying specific antibiotic resistance genes. Microarray results were in accordance with the phenotypic resistance (64).

Microarrays allow simultaneous detection of multiple targets in a single assay. Several identification assays can be run in parallel, with each probe representing a specific small section of a genome or a sequence common to multiple genomes (65). A major limitation of this method is the complexity and the amount of time invested to design genetic regions which are diverse enough to allow discrimination of several species (66).

Loop-Mediated Isothermal Amplification (LAMP) Assay

LAMP is an innovative nucleic acid amplification technique used for rapid, sensitive, and specific detection of foodborne pathogens (53, 67). LAMP is considered an ideal technique for use in field conditions because it does not require costly laboratory equipment (thermal cycler) for DNA amplification. LAMP is an isothermal nucleic acid amplification technique in which amplification is carried out at a single constant temperature of 60° to 65°C. For amplification, LAMP uses two inner and two outer primers to target six distinct regions within the target DNA. A pair of "loop primers" is also used to accelerate the amplification process. A polymerase enzyme with high displacement activity (usually *Bst* DNA polymerase) facilitates DNA amplification at a constant temperature, unlike conventional PCR, where amplification is carried out in an alternating time-temperature combination. Amplified DNA can be visually detected as increased turbidity resulting from the accumulation of magnesium pyrophosphate (an amplification byproduct) (68), by measuring the turbidity over time, or by measuring fluorescence using intercalating dyes such as SYBR green I (69). Several variants of the LAMP assay are in vogue, namely multiplex LAMP (70), reverse-transcription LAMP (71), real-time LAMP (72), RFLP LAMP (73), multiple inner primers LAMP (74), and *in situ* LAMP (75).

The LAMP assay has several advantages over conventional molecular detection methods. Due to the high specificity of the primers, the amount of DNA produced in the LAMP assay is significantly higher and the detection limit is lower than conventional PCRs (53, 76). Moreover, a LAMP assay takes only 30 to 60 min for amplification (77, 78), which is considerably less than the time taken by conventional PCRs (4 to 6 hours). One of the major limitations of LAMP is its poor ability to multiplex (70). Multiplexing in LAMP increases the probability of false-positive amplification because of the presence of several pairs of primers targeting different regions.

LAMP has been applied for the detection of various foodborne pathogens due to its rapidity, high sensitivity, and ease of producing results. Dong et al. developed a LAMP assay targeting the *hip*O gene to detect *C. jejuni* in samples from cattle on farms (79). The assay was found to be specific (100% inclusivity and exclusivity), sensitive (100 fg/μl detection limit), and quantifiable. The assay took less than 30 min to produce amplification for all *C. jejuni* isolates (79). Yamazaki et al. compared the LAMP assay with conventional isolation methods for the detection of

C. jejuni and *C. coli* in naturally contaminated chicken meat samples (80). The LAMP assay showed high sensitivity and specificity of 98.5% and 97.4%, respectively. Moreover, LAMP required only 23.5 to 25.5 hours for the complete detection of the bacteria starting from the enrichment culture compared to 3 to 4 days for conventional culture methods (80). A six-primer LAMP system targeting eight regions of the *hlyA* gene has been reported for the detection of *L. monocytogenes* (81). This LAMP assay was rapid and produced results within 40 min. Sensitivity of the LAMP method for the detection of *L. monocytogenes* in pure cultures was 2.0 CFU per reaction, and the sensitivity was 100-fold higher compared to conventional PCR (81).

The LAMP assay has also been used for the simultaneous detection of antimicrobial resistance genes and the genus or species of the bacteria. In one such instance, the LAMP assay simultaneously detected staphylococcal *mecA* (methicillin resistance) and *spa* (*S. aureus*) genes (82). The LAMP assay amplified both the genes within 60 min with high specificity and sensitivity. In a recent study, Chen et al. developed a LAMP system for identification of methicillin-resistant *S. aureus* by simultaneously amplifying the *mecA* (methicillin resistance specific), *nuc* (*S. aureus* specific), and *femB* (virulence marker) genes (83). The LAMP assay identified all three genes in less than 60 min. There was uniformity in the results of LAMP and PCR assays (83).

MALDI-TOF MS

MALDI-TOF MS is a rapid, sensitive, and cost-effective technique currently used for pathogen detection. It generates unique mass-spectral fingerprint signatures for individual microorganisms, which helps in genus and species identification (84). In this technique, a sample is mixed with an organic compound known as a matrix. Matrix components vary according to the type of sample analyzed and the type of laser used (84, 85). The most commonly used matrix components are α-cyano-4-hydroxycinnamic acid, 2,5-dihydroxy benzoic acid, 3,5-dimethoxy-4-hydroxycinnamic acid (sinapinic acid), ferulic acid, and 2,4-hydroxy-phenyl benzoic acid. These matrix compounds have strong optical absorption capabilities within the range of laser wavelength used. The samples are mixed with the matrix and deposited on a metal plate. This mixture is then left to dry, which results in crystallization of the matrix along with the sample. The sample embedded in the matrix is then bombarded with a pulsed laser beam, which results in desorption of the analytes in the matrix. The analytes are vaporized and ionized into the gaseous phase. Desorption and ionization generates singly charged ions from the analytes by either gain or loss of more than one proton. The soft ionization approach used in MALDI is beneficial because it does not lead to a loss of sample integrity. The ionized analyte molecules are accelerated through an electrostatic field at a fixed potential and made to travel through a metal tube toward a detector. The detector determines the TOF taken by gaseous analyte molecules to travel through the flight tube and reach the detector (86). Detection of each bioanalyte molecule depends on the molecular mass (m), the charge (z), the ratio mass/charge (m/z), and the relative intensity of the signal (87). Small bioanalyte molecules travel faster than larger molecules (higher molecular weight) and are detected early. Different biomolecules in a sample with different mass (m) to charge (z) ratios form a mass spectrum on the basis of the TOF. Currently, various applications of MALDI-TOF MS are being utilized for AMR detection in bacterial pathogens. These include detection of enzymes/antibiotics and/or their degraded products, direct MALDI-TOF MS analysis of bacterial extracts, and mini-sequencing-primer extension followed by MALDI-TOF assay.

During the past few years, a surge has been observed in publications using the MALDI approach for pathogen detection, including foodborne pathogens. MALDI-TOF MS has been developed and validated for several bacterial foodborne pathogens, namely *Salmonella* (88, 89), *Campylobacter* (90), *S. aureus* (91, 92), and *E. coli* (93). Dieckmann and Malorny successfully used MALDI-TOF MS for identification of the five most frequently isolated *S. enterica* serovars, namely Enteritidis, Typhimurium, Virchow, Infantis, and Hadar (89). They also reported 100% inclusivity and exclusivity of serovar-identifying biomarker ions for these *Salmonella* serovars (89). A study validated MALDI-TOF MS as a powerful tool for the identification of clinical isolates of coagulase-negative staphylococci (94). Mandrell et al. reported the identification and differentiation of multiple strains of *Campylobacter coli, C. jejuni, Campylobacter helveticus, Campylobacter lari, Campylobacter sputorum*, and *Campylobacter upsaliensis* isolated from animal, clinical, or food samples using a MALDI-TOF MS system (90).

MALDI-TOF MS is a rapid, cost-effective, sensitive, and specific technique (7, 95). The expected result turnaround time for MALDI-TOF MS is less than 5 min, which is significantly faster than other molecular techniques used for pathogen detection (77). Another advantage is its ability to detect unknown organisms, which is not limited to the prespecified targets (7). Mass spectra obtained from unknown organisms are compared with the already available MS database to identify the organism. Once identified, the mass spectra of unidentified species can be added to the MS database after identification by sequencing for future reference (96). Disadvantages of MALDI-TOF MS include its initial high setup cost, which limits its use as a routine diagnostic method in resource-limited countries. However, in comparison to conventional methods, MALDI is much less labor intensive, with a low overall cost. Very few reagents and laboratory supplies are needed for a MALDI experiment; these include a metal slide, matrix solution, pipette tips, and loops for sample application. A research group reported that use of MALDI-TOF MS resulted in net savings of 87.8% in reagent costs annually compared to conventional methods (97). The authors further stated that the routine use of MALDI in pathogen detection could offset the initial high cost of MALDI equipment in approximately 3 years (97). Because of its ease of operation, rapid result turnaround time, and very low overall cost per sample, the MALDI technique is revolutionizing the field of pathogen detection.

MOLECULAR TYPING TECHNIQUES

Several PCR-based molecular typing methods have been developed and applied to study the population structure and molecular epidemiology of foodborne pathogens at the preharvest and postharvest stages. These methods are based on PCR amplification and subsequent analysis of the banding pattern in gel electrophoresis. One of the main advantages of typing techniques is their utility in the investigation of foodborne outbreaks and the complete understanding of the epidemiology of foodborne infections (98). Rapid typing methods can significantly reduce costs associated with detection, containment, and decontamination. Working principles, advantages, disadvantages, and the application of such methods in food safety research are discussed in the following subsections.

PCR-RFLP

RFLP is a molecular typing technique which involves amplification of genomic DNA and digestion with suitable restriction enzymes. Restriction fragments are then separated using gel electrophoresis. Restriction enzymes, also known as restriction endonucleases, cut the bacterial DNA at specific sites and produce restriction products of specific sizes. The presence or absence of these restriction fragments of different sizes constitutes a specific banding pattern for each target that can be visualized over gel electrophoresis. RFLP is a simple method to detect known mutations that alter recognition sites of restriction endonucleases (99). However, with the advent of pulsed-field gel electrophoresis (PFGE) and other inexpensive techniques for molecular typing, PCR-RFLP has become obsolete (100, 101). The advantages of using PCR-RFLP include its low cost and lack of requirement of expensive and advanced instruments. It can be performed in a laboratory with basic PCR facilities, and no specialized personnel training is required. PCR-RFLP is a very sensitive technique for strain identification and differentiation of microbes. The disadvantages include the requirement for specific restriction enzymes and difficulties in identifying the perfect combination of enzymes targeting different sites in the target sequence. Moreover, the process is time-consuming and involves several steps, such as PCR amplification, gel extraction, restriction digestion using endonucleases, and final gel electrophoresis for result interpretation. PCR-RFLP is not a suitable technique for preharvest food safety research because simultaneous detection of multiple pathogens is difficult due to the requirement of specific primers and restriction enzymes for each target. This technique can be used for strain differentiation when time is not a constraint and there is no urgency to get rapid results.

RAPD

RAPD is a molecular subtyping technique that is widely used for typing for various bacterial pathogens (102). It involves PCR amplification of random segments of genomic DNA with a single primer of arbitrary nucleotide sequence (9 to 10 bp). Unlike conventional PCR, the size of the target PCR products is unknown in RAPD. The arbitrary primer may or may not amplify the segment of the target genome, depending on the positions that are complementary to the primer sequence. Moreover, the arbitrary primer can simultaneously anneal to multiple sites in the whole genome and produce amplification products of varied sizes; this difference in the banding pattern is the basis of strain differentiation. PCR products can be visualized over gel electrophoresis, and banding patterns of different strains/isolates can be compared.

RAPD has been extensively used for molecular subtyping of various foodborne pathogens. For instance, Lin et al. developed and optimized a RAPD protocol for molecular subtyping of *S.* Enteritidis strains. They used a panel of primers to characterize isolates of *S.* Enteritidis, which had been previously characterized by phage typing, ribotyping, and PFGE. RAPD distinguished isolates into different RAPD subtypes. Moreover, RAPD successfully differentiated eight *S.* Enteritidis isolates which were previously untyped by the above three methods. It was also reported that RAPD had more discriminatory power than any of the other three subtyping methods applied individually (103). Two years later, a RAPD method was developed to differentiate strains of *Salmonella* Typhi and other *Salmonella* spp. (104). Five-primer RAPD successfully differentiated several strains of *Salmonella* Typhi into 21 distinct RAPD profiles. Moreover, RAPD was also able to differentiate 65 other *Salmonella* isolates representing 42 serotypes. Yoshida et al. used four-primer RAPD to characterize 20 unrelated *L. monocytogenes* strains isolated from different animals and locations at different points in time (105). All strains were classified into 18 unique subtypes. The authors further characterized seven epidemiologically related *L. monocytogenes* strains isolated from raw milk and a bulk tank collected from a dairy farm and found that all *L. monocytogenes* strains had the same RAPD profile (105). Hernandez et al. studied a set of *C. jejuni* and *C. coli* isolates recovered from human feces, seawater, and poultry products and reported a total of 118 different RAPD profiles, with each profile having 4 to 11 bands (106). These authors found the RAPD method to be highly discriminatory for *Campylobacter* subtyping. The RAPD protocol reported by Hernandez et al. (106) was later validated by Açik and coworkers (107) in *C. jejuni* and *C. coli* isolated from healthy cattle and sheep. The authors found a high degree of heterogeneity among *Campylobacter* isolates using RAPD (107). Another study successfully used RAPD to distinguish 20 reference strains of *Clostridium difficile* (108).

One of the major drawbacks of RAPD is low reproducibility. Highly standardized experimental conditions with consistent reagents and cycling conditions are required to get consistent results. The quality and concentration of DNA and PCR cycling and reaction conditions can affect the reproducibility of the RAPD-PCR (102). Another limitation is the requirement of very-high-quality DNA for the reaction. Precautions need to be taken to avoid contamination of DNA because the short random primers (9 to 10 bp) used for target amplification can cross-amplify similar sequences of other pathogens.

Amplified Fragment Length Polymorphism (AFLP-PCR)

AFLP-PCR combines the advantages of RFLP and PCR assays. This technique involves the digestion of entire bacterial DNA by one or more restriction enzymes, commonly MseI and EcoRI (109). The genomic DNA is incubated with restriction enzymes for complete digestion of the bacterial DNA into restriction fragments. Digestion is followed by the ligation of the resulting fragments to a double-stranded oligonucleotide adapter complementary to the base sequence of the restriction site. The adapters are designed such that the original restriction site is not restored after ligation, thus preventing further restriction digestion. Selective amplification of these fragments is done by using primers targeting specific sequences ligated to either end of restriction endonuclease-digested genomic DNA rather than genomic DNA itself; this generates a high number of fragments for analysis. The resulting patterns of PCR-amplified DNA fragments are then analyzed by gel electrophoresis to assign subtypes (110, 111). AFLP-PCR is considered reliable and robust because it uses stringent reaction conditions for primer annealing (112). The high discriminatory ability of this assay is due to the generation of a higher number of fragments for analysis (113). Prior knowledge of the bacterial DNA sequence is not required for analysis. Moreover, highly restrictive amplification conditions due to high specificity of primers toward adaptors minimizes nontarget amplification. The discriminatory power of AFLP is considered to be equal to PFGE, the current gold standard in molecular typing.

This whole-genome fingerprinting technique has been used for high-resolution typing of several bacteria in epidemiological studies (113). Guerra et al. developed a single-enzyme (EcoR1) AFLP assay for molecular typing of *L. monocytogenes* (111). They evaluated this method with 84 *L. monocytogenes* cultures and compared AFLP-PCR results with serotyping, phage-typing, and cadmium and arsenic resistance typing. The technique was found to be reproducible, and a range of different banding patterns was obtained. AFLP produced indistinguishable banding patterns for *L. monocytogenes* isolates related by origin and serovar. AFLP, phage typing, and cadmium/arsenic typing produced indistinguishable patterns for the majority of epidemiologically related groups of *L. monocytogenes* cultures (111).

Since its inception, a few variants of AFLP technique have been reported. An example of these is fluorescent AFLP (FAFLP). FAFLP uses fluorescent oligonucleotide primers, which enables accurate and high-resolution detection of the amplified digested fragments. Several studies have used FAFLP for high-resolution genotyping of the foodborne pathogens (109, 114, 115). Recently, Roussel and coworkers compared FAFLP and PFGE for molecular subtyping of *L. monocytogenes* isolated from humans, food, food processing environments, and animals (109). Both the techniques were able to produce unique types for epidemiologically associated strains. The discriminatory power of FAFLP was similar to that of PFGE. The authors concluded that as a less labor-intensive assay, FAFLP could be used for outbreak investigations and source tracking of the pathogens (109). However, they also highlighted the necessity of complete standardization of FAFLP protocols and reproducibility assessment through trials before using FAFLP for routine typing (109). In a similar study, FAFLP and PFGE were compared for molecular typing of *Salmonella* Typhimurium DT126 to differentiate between outbreak-associated and epidemiologically unrelated DT126 isolates in Australia. In this study, the discriminatory power of AFLP was found to be greater than that of PFGE. Both techniques differentiated between isolates from separate outbreaks. However, neither assay could separate epidemiologically unrelated isolates from the outbreak isolates (116).

MLST

MLST was first proposed in 1998 as a typing approach that enables the unambiguous characterization of bacterial isolates in a standardized, reproducible, and portable way (117). Since its discovery in 1998, MLST has been widely applied to a large number of pathogens of public health and food safety importance. MLST involves the extraction of genomic DNA and PCR amplification of highly conserved housekeeping genes (preferably seven). After amplification, amplicons are sequenced, aligned, and investigated using bioinformatics software. Housekeeping genes are distributed in different loci around the chromosome of microorganisms (118, 119). For each housekeeping gene, different sequences are assigned as alleles, and the alleles at the multiple loci constitute an allelic profile, which unambiguously defines the sequence type of each isolate. Reasons for targeting housekeeping genes in MLST include their uniform presence in all organisms, their conserved nature, and their ability to provide adequate discrimination for molecular typing. Moreover, sequence data submitted by researchers located anywhere in the world are freely available online via the internet. Research laboratories can easily compare their sequences with other laboratories. Online collection of sequence data helps with the rapid comparison of sequences of bacteria obtained from any part of the world (120).

MLST data available online have been applied in several food animal farm-based epidemiological and evolutionary studies (121–127). For example, a study used MLST to compare *C. coli* isolates selected from fecal, environmental, and carcass samples of antibiotic-free and conventional swine production systems (128). The authors reported a genotypically diverse *C. coli* population with the presence of *C. coli* isolates sharing a common ancestry in both production systems. They concluded that a common population structure between *C. coli* from conventional and antibiotic-free farm systems may be the reason for persistence of antimicrobial-resistant *C. coli* in antibiotic-free farms (128).

MLST has several advantages over other genotyping techniques. First, it is a DNA-based technique and does not require processing of live cultures. Second, MLST detects changes at the genotypic level (DNA), which may not be apparent by phenotypic typing methods such as serotyping. MLST targets coding regions and uses variation that accumulates very slowly in the population, in contrast to PFGE. Moreover, the online availability of the sequence data from various parts of the world makes it easy to compare sequences from different places. These qualities make MLST an excellent molecular tool for long-term epidemiological investigations (117). Although MLST is easy, rapid, and less expensive than WGS, it is costlier than other genotyping methods such as PFGE and PCR-RFLP.

PFGE

PFGE is the most widely used molecular typing system for bacterial pathogens. It is considered one of the most discriminatory subtyping methods and thus is regarded as a gold standard in molecular epidemiological studies (129, 130). In 1996, the Centers for Disease Control and Prevention (CDC) established a nationwide molecular subtyping network called PulseNet for molecular surveillance of foodborne diseases to facilitate early outbreak detection and capacity building in disease investigation. PulseNet uses DNA fingerprints of the foodborne bacteria to detect an outbreak. Once a PFGE pattern is generated, the laboratory analyzes the pattern using bioinformatics software (BioNumerics). Then the DNA fingerprint is uploaded to the PulseNet database, where it is investigated and analyzed to determine if it is causing an outbreak. The PulseNet PFGE database helps with the quick identification of outbreaks that are geographically separated and caused by a similar pathogen or strain across states. PulseNet facilitates rapid web-based sharing of PFGE fingerprints of the bacterial pathogens. Online sharing of DNA fingerprints helps with detection of temporally and spatially related foodborne disease outbreaks. PFGE is based on the digestion of bacterial genomic DNA with restriction enzymes, which allows analysis of fragments up to 10,000 kb (131). Large fragments of DNA molecules are separated by applying an electric field that changes direction periodically in a gel matrix. The comparison of banding patterns of genomic DNA fragments after digestion with a restriction enzyme is the basis of PFGE. Briefly, bacterial cells are incubated overnight, immobilized in agarose, and digested with a rare restriction enzyme. The enzyme-treated plugs are then loaded onto an agarose gel, and the DNA fragments are separated based on size using an alternating electric field. This results in unique DNA fingerprints/PFGE patterns of the bacterial pathogens. The PFGE electrophoretic banding patterns are exceptionally reproducible and specific for various bacterial organisms (132). The major limitations of PFGE are that it is labor intensive and requires 3 to 4 days to generate complete results. Moreover, some strains of bacteria cannot be typed using PFGE.

However, PFGE has been reported as an efficient, time-saving, and cost-effective method for serogrouping of *Salmonella* isolates (133). In a study from Taiwan, 45 *S.* Typhimurium isolates which caused food-poisoning diarrhea during 1991 to 1994 were subjected to restriction digestion with XbaI, AvrII, and SpeI, followed by PFGE, resulting in 26 PFGE patterns. Since isolates of the same patterns were recovered from unrelated food-poisoning cases, it was suggested that these isolates could be prevalent and circulating in Taiwan (134). McCullagh et al. used PFGE for molecular typing of 109 *S. aureus* isolates in Northern Ireland, which included 47 isolates from broilers with clinical conditions and 62 strains from hatcheries (135). PFGE analysis showed a similarity between 85% of the strains from clinical sources and 71% of the hatchery isolates. On the basis of the PFGE data, the authors concluded that the hatchery was a potential source of infection for clinical broiler disease (135). Another study used PFGE for identification and differentiation of various *Salmonella* serotypes, namely Heidelberg, Javiana, Typhimurium, Newport, Enteritidis, Dublin, Pullorum, and Choleraesuis (136). Molla et al. used PFGE to investigate the occurrence and genotypic relatedness of *S. enterica* isolated from feed and fecal samples in commercial swine farms (137). They reported that more than 50% of the *Salmonella* isolates recovered from feed shared genotypic clonality with those detected in swine fecal samples, suggesting the dissemination of *Salmonella* via feed (137). The presence of highly clonal *Salmonella* isolates indicated an epidemiological link between *Salmonella* in feed and in feces (137). PFGE can be used for short-term as well as for long-term epidemiological surveillance programs. Refsum et al. performed PFGE to analyze 142 isolates of *S.* Typhimurium obtained from passerines (n = 46), gulls (n = 26), domestic animals (n = 50), and the environment (n = 13) collected over a period of 30 years and revealed that passerines are the major source of human infection in Norway (138).

WGS

WGS of bacterial pathogens using next-generation techniques (NGS) can provide comprehensive genetic information, including genus and species identification and sequence information on drug resistance and virulence determinants (139). The term "next-generation sequencing" specifically refers to the strategies that have succeeded the old sequencing method, i.e., Sanger sequencing. Popular platforms available for NGS include pyrosequencing (454/Roche), Illumina (Hiseq, Miseq), ion-torrent sequencing, SOLiD, PacBio RS, and Oxford Nanopore (140, 141). Each NGS platform, however, follows the same workflow sequence, which involves nucleic acid extraction, library preparation, sequencing, and data analysis using bioinformatics software. Illumina sequencing involves DNA synthesis by a DNA polymerase incorporating deoxynucleoside triphosphates labeled with fluorescent terminators which are detected in real time (142). Illumina offers a range of sequencing instruments which includes low-throughput benchtop machines as well as ultra-high-throughput machines. The most extensively used Illumina platforms are MiSeq and HiSeq (143). MiSeq has been shown to be the best regarding both throughput per run and error rates compared to pyrosequencing and ion-torrent sequencing platforms (77). Ion-torrent sequencing utilizes the "sequencing-by-synthesis" approach. It is performed on a semiconductor chip and is based on the detection of hydrogen ions that are released during DNA synthesis (9, 142).

The huge amount and comprehensiveness of the sequence data and the low cost per base makes NGS an interesting alternative to current PCR and culture-based detection methods. Moreover, apart from molecular typing, NGS also has utility in virtual screening/detection of antimicrobial resistance and virulence determinants in the whole genome of the microorganism. Once a microorganism is sequenced, the whole genome of the bacteria is screened for the presence of antimicrobial resistance and virulence determinants/markers using gene prediction programs (ResFinder). This approach helps in predicting future resistance. However, the reliability of WGS as a tool for predicting antimicrobial resistance is critically dependent on the availability of a regularly updated resistance database and prior knowledge of the target sequence. WGS data in the absence of any phenotypic support from traditional tests might lead to false-positive or -negative results. For example, the absence of a known gene sequence in the whole genome does not necessarily mean that the isolate is susceptible. There might be other novel or unreported molecular mechanisms that confer the same resistance, which is not characterized or sequenced yet. Similarly, NGS may not be able to identify and interpret the presence of an unreported/unknown sequence. NGS analysis depends a lot on the annotation interpretation of the genome sequence using prior knowledge of such sequences (144). In our opinion, until a technological advancement is made which can accurately predict antimicrobial resistance, at least some support from the phenotypic tests will be required. Another major limitation of NGS is the requirement of a highly trained professional for comprehensive analysis and interpretation of the WGS data. Relevant information needs to be skillfully extracted from the vast NGS data sets using bioinformatics software packages. Also, the high initial setup cost and the scarcity of user-friendly bioinformatics tools curtails its use in resource-limited countries.

However, these limitations could be overcome by the fact that NGS methods provide high-resolution sequence data in a short time frame of only 2 to 5 days. Moreover, the cost of performing WGS is expected to go down in the near future due to improvement in the methodology and increasing competition among the commercial houses to provide affordable and cost-effective genome sequencing options. Keeping all these facts in mind, it can be safely assumed that WGS will become an indispensable tool in molecular diagnostics, pathogen typing, and detection of antimicrobial resistance determinants in the near future.

CONCLUSION

Current methods for pathogen detection and identification are mainly culture based. Although they are accurate and complete, culture methods are time-consuming and laborious, which seriously hampers the in-time implementation of biosecurity measures on farms. To overcome the limitations of bacterial isolation, rapid molecular detection methods have been developed. In particular, DNA base detection methods such as PCR, real-time PCR, multiplex PCR, and microarray have been extensively used for pathogen detection at the pre- and postharvest stages. However, these methods still require a prior culture step and a trained individual to perform the tests. LAMP is a unique nucleic acid amplification technique used for rapid, sensitive, and specific detection of foodborne pathogens in a single step. LAMP is called "equipment-free technology" because it does not require an expensive thermal cycler for DNA amplification, and visualization with the naked eye is sufficient for result interpretation. No post-PCR gel electrophoresis is required. Moreover, result turnaround time for a LAMP protocol is only 30 min to 1 h. These features make LAMP a method of choice for use in field conditions. However, the only limitation of LAMP is its inability to multiplex various targets. The introduction of techniques like MALDI-TOF MS has completely revolutionized the field of pathogen detection. MALDI is based on the production of mass-spectral fingerprints for each microorganism, which helps in precise genus and species identification. This technique is very fast and sensitive and provides results within 5 min from a single isolated bacterial colony. Another advantage of MALDI is its ability to detect unknown organisms. The initial equipment cost of MALDI-TOF MS is very high, but research suggests that routine use of this technique could save 87.8% in reagent costs annually compared to conventional methods. Once the equipment is installed, the cost of processing samples becomes very low.

For molecular typing of microorganisms, several methods are in vogue. However, PFGE is considered the gold standard for molecular typing and is used extensively worldwide for epidemiological typing of foodborne pathogens. NGS, with its high multiplexing capability, is rapidly becoming the method of choice for pathogen and AMR detection. In addition to pathogen detection, WGS can accurately predict the presence of antimicrobial resistance and virulence determinants within the whole-genome sequence. MLST, although it is a good typing technique, is less likely to be the method of choice because of the high cost of sequencing multiple housekeeping genes for a single isolate. Moreover, MLST is not considered an ideal technique for serovar differentiation. MLST is more suited for long-term epidemiological investigations. Although a lot of progress has been made in developing rapid and sensitive methods for pathogen detection; culture methods still remain essential. Culture methods provide useful information about phenotypic characteristics such as antimicrobial resistance, which is crucial in deciding on preventive measures. We strongly believe that recent molecular-based diagnostic methods will not be able to completely replace conventional microbiological and biochemical methods for the detection of genus/species and antimicrobial resistance. We foresee in-tandem use of traditional phenotypic methods and the latest molecular methods, at least for some time to come.

CITATION

Kumar D, Thakur S. 2018. Molecular tools to study preharvest food safety challenges. Microbiol Spectrum 6(1):PFS-0019-2017.

REFERENCES

1. **Blaha T.** 1997. Public health and pork: pre-harvest food safety and slaughter perspectives. *Rev Sci Tech* **16:**489–495.
2. **Blaha TH.** 1996. What's coming in food safety and pork quality, p 136–138. Proceedings of the 23rd Allen D. Leman Conference, 21–24 September, St Paul, MN.
3. **Buntain B.** 1997. The role of the food animal veterinarian in the HACCP era. *J Am Vet Med Assoc* **210:**492–495.
4. **Food and Drug Administration (FDA).** 2015. Summary report on antimicrobials sold or distributed for use in food-producing animals. https://www.fda.gov/downloads/forindustry/userfees/animaldruguserfeeactadufa/ucm534243.pdf. Accessed July 22, 2017.
5. **Turnidge J.** 2004. Antibiotic use in animals: prejudices, perceptions and realities. *J Antimicrob Chemother* **53:**26–27.
6. **Singer RS, Finch R, Wegener HC, Bywater R, Walters J, Lipsitch M.** 2003. Antibiotic resistance: the interplay between antibiotic use in animals and human beings. *Lancet Infect Dis* **3:**47–51.
7. **Angeletti S.** 2017. Matrix assisted laser desorption time of flight mass spectrometry (MALDI-TOF MS) in clinical microbiology. *J Microbiol Methods* **138:**20–29.
8. **Dixon P, Davies P, Hollingworth W, Stoddart M, MacGowan A.** 2015. A systematic review of matrix-assisted laser desorption/ionisation time-of-flight mass spectrometry compared to routine microbiological methods for the time taken to identify microbial organisms from positive blood cultures. *Eur J Clin Microbiol Infect Dis* **34:**863–876.
9. **Lupo A, Papp-Wallace KM, Sendi P, Bonomo RA, Endimiani A.** 2013. Non-phenotypic tests to detect and characterize antibiotic resistance mechanisms in *Enterobacteriaceae*. *Diagn Microbiol Infect Dis* **77:**179–194.
10. **Mullis KB, Faloona FA.** 1987. Specific synthesis of DNA *in vitro* via a polymerase-catalyzed chain reaction. *Methods Enzymol* **155:**335–350.
11. **Predari SC, Ligozzi M, Fontana R.** 1991. Genotypic identification of methicillin-resistant coagulase-negative staphylococci by polymerase chain reaction. *Antimicrob Agents Chemother* **35:**2568–2573.
12. **Keelara S, Scott HM, Morrow WM, Gebreyes WA, Correa M, Nayak R, Stefanova R, Thakur S.** 2013. Longitudinal study of distributions of similar antimicrobial-resistant *Salmonella* serovars in pigs and their environment in two distinct swine production systems. *Appl Environ Microbiol* **79:**5167–5178.
13. **Paião FG, Arisitides LGA, Murate LS, Vilas-Bôas GT, Vilas-Boas LA, Shimokomaki M.** 2013. Detection of *Salmonella* spp, *Salmonella* Enteritidis and Typhimurium in naturally infected broiler chickens by a multiplex PCR-based assay. *Braz J Microbiol* **44:**37–41.
14. **Frana TS, Beahm AR, Hanson BM, Kinyon JM, Layman LL, Karriker LA, Ramirez A, Smith TC.** 2013. Isolation and characterization of methicillin-resistant *Staphylococcus aureus* from pork farms and visiting veterinary students. *PLoS One* **8:**e53738.
15. **Markoulatos P, Siafakas N, Moncany M.** 2002. Multiplex polymerase chain reaction: a practical approach. *J Clin Lab Anal* **16:**47–51.
16. **Radhika M, Saugata M, Murali HS, Batra HV.** 2014. A novel multiplex PCR for the simultaneous detection of *Salmonella enterica* and *Shigella* species. *Braz J Microbiol* **45:**667–676.
17. **Alvarez J, Sota M, Vivanco AB, Perales I, Cisterna R, Rementeria A, Garaizar J.** 2004. Development of a multiplex PCR technique for detection and epidemiological typing of *Salmonella* in human clinical samples. *J Clin Microbiol* **42:**1734–1738.
18. **Kawasaki S, Horikoshi N, Okada Y, Takeshita K, Sameshima T, Kawamoto S.** 2005. Multiplex PCR for simultaneous detection of *Salmonella* spp., *Listeria monocytogenes*, and *Escherichia coli* O157:H7 in meat samples. *J Food Prot* **68:**551–556.
19. **Park SH, Hanning I, Jarquin R, Moore P, Donoghue DJ, Donoghue AM, Ricke SC.** 2011. Multiplex PCR assay for the detection and quantification of *Campylobacter* spp., *Escherichia coli* O157:H7, and *Salmonella* serotypes in water samples. *FEMS Microbiol Lett* **316:**7–15.
20. **Klena JD, Parker CT, Knibb K, Ibbitt JC, Devane PM, Horn ST, Miller WG, Konkel ME.** 2004. Differentiation of *Campylobacter coli, Campylobacter jejuni, Campylobacter lari,* and *Campylobacter upsaliensis* by a multiplex PCR developed from the nucleotide sequence of the lipid A gene lpxA. *J Clin Microbiol* **42:**5549–5557.
21. **Persoons D, Van Hoorebeke S, Hermans K, Butaye P, de Kruif A, Haesebrouck F, Dewulf J.** 2009. Methicillin-resistant *Staphylococcus aureus* in poultry. *Emerg Infect Dis* **15:**452–453.
22. **Chen J, Tang J, Liu J, Cai Z, Bai X.** 2012. Development and evaluation of a multiplex PCR for simultaneous detection of five foodborne pathogens. *J Appl Microbiol* **112:**823–830.
23. **Rawool DB, Doijad SP, Poharkar KV, Negi M, Kale SB, Malik SV, Kurkure NV, Chakraborty

T, Barbuddhe SB. 2016. A multiplex PCR for detection of *Listeria monocytogenes* and its lineages. *J Microbiol Methods* **130:**144–147.

24. **Akiba M, Kusumoto M, Iwata T.** 2011. Rapid identification of *Salmonella enterica* serovars, Typhimurium, Choleraesuis, Infantis, Hadar, Enteritidis, Dublin and Gallinarum, by multiplex PCR. *J Microbiol Methods* **85:**9–15.
25. **Kralik P, Ricchi M.** 2017. A basic guide to real time PCR in microbial diagnostics: definitions, parameters, and everything. *Front Microbiol* **8:**108.
26. **Higuchi R, Dollinger G, Walsh PS, Griffith R.** 1992. Simultaneous amplification and detection of specific DNA sequences. *Biotechnology (N Y)* **10:**413–417.
27. **Malorny B, Paccassoni E, Fach P, Bunge C, Martin A, Helmuth R.** 2004. Diagnostic real-time PCR for detection of *Salmonella* in food. *Appl Environ Microbiol* **70:**7046–7052.
28. **Alemu K.** 2014. Real-time PCR and its application in plant disease diagnostics. *Adv Life Sci Technol* **27:**39–49.
29. **Navarro E, Serrano-Heras G, Castaño MJ, Solera J.** 2015. Real-time PCR detection chemistry. *Clin Chim Acta* **439:**231–250.
30. **Hein I, Lehner A, Rieck P, Klein K, Brandl E, Wagner M.** 2001. Comparison of different approaches to quantify *Staphylococcus* aureus cells by real-time quantitative PCR and application of this technique for examination of cheese. *Appl Environ Microbiol* **67:**3122–3126.
31. **Fukushima H, Tsunomori Y, Seki R.** 2003. Duplex real-time SYBR green PCR assays for detection of 17 species of food- or waterborne pathogens in stools. *J Clin Microbiol* **41:**5134–5146.
32. **Singh J, Batish VK, Grover S.** 2009. A molecular beacon-based duplex real-time polymerase chain reaction assay for simultaneous detection of *Escherichia coli* O157:H7 and *Listeria monocytogenes* in milk and milk products. *Foodborne Pathog Dis* **6:**1195–1201.
33. **Jamnikar Ciglenecki U, Grom J, Toplak I, Jemersić L, Barlič-Maganja D.** 2008. Real-time RT-PCR assay for rapid and specific detection of classical swine fever virus: comparison of SYBR Green and TaqMan MGB detection methods using novel MGB probes. *J Virol Methods* **147:**257–264.
34. **Madani M, Subbotin SA, Moens M.** 2005. Quantitative detection of the potato cyst nematode, *Globodera pallida*, and the beet cyst nematode, *Heterodera schachtii*, using real-time PCR with SYBR green I dye. *Mol Cell Probes* **19:**81–86.
35. **Lauri A, Mariani PO.** 2009. Potentials and limitations of molecular diagnostic methods in food safety. *Genes Nutr* **4:**1–12.
36. **Levin RE.** 2005. The application of real-time PCR to food and agricultural systems. A review. *Food Biotechnol* **18:**97–133.
37. **Klein D.** 2002. Quantification using real-time PCR technology: applications and limitations. *Trends Mol Med* **8:**257–260.
38. **Marthaler D, Raymond L, Jiang Y, Collins J, Rossow K, Rovira A.** 2014. Rapid detection, complete genome sequencing, and phylogenetic analysis of porcine deltacoronavirus. *Emerg Infect Dis* **20:**1347–1350.
39. **Marthaler D, Homwong N, Rossow K, Culhane M, Goyal S, Collins J, Matthijnssens J, Ciarlet M.** 2014. Rapid detection and high occurrence of porcine rotavirus A, B, and C by RT-qPCR in diagnostic samples. *J Virol Methods* **209:**30–34.
40. **Rodriguez-Lazaro D, Cook N, Hernandez M.** 2013. Real-time PCR in food science: PCR diagnostics. *Curr Issues Mol Biol* **15:**39–44.
41. **O'Regan E, McCabe E, Burgess C, McGuinness S, Barry T, Duffy G, Whyte P, Fanning S.** 2008. Development of a real-time multiplex PCR assay for the detection of multiple *Salmonella* serotypes in chicken samples. *BMC Microbiol* **8:**156.
42. **Fukushima H, Kawase J, Etoh Y, Sugama K, Yashiro S, Iida N, Yamaguchi K.** 2010. Simultaneous screening of 24 target genes of foodborne pathogens in 35 foodborne outbreaks using multiplex real-time SYBR green PCR analysis. *Int J Microbiol* **2010:**1–18.
43. **Fratamico PM, Bagi LK, Cray WC Jr, Narang N, Yan X, Medina M, Liu Y.** 2011. Detection by multiplex real-time polymerase chain reaction assays and isolation of Shiga toxin-producing *Escherichia coli* serogroups O26, O45, O103, O111, O121, and O145 in ground beef. *Foodborne Pathog Dis* **8:**601–607.
44. **Denis E, Bielińska K, Wieczorek K, Osek J.** 2016. Multiplex real-time PCRs for detection of *Salmonella, Listeria monocytogenes*, and verotoxigenic *Escherichia coli* in carcasses of slaughtered animals. *J Vet Res (Pulawy)* **60:**287–292.
45. **Kim HJ, Lee HJ, Lee KH, Cho JC.** 2012. Simultaneous detection of pathogenic *Vibrio* species using multiplex real-time PCR. *Food Control* **23:**491–498.
46. **Hu Q, Lyu D, Shi X, Jiang Y, Lin Y, Li Y, Qiu Y, He L, Zhang R, Li Q.** 2014. A modified molecular beacons-based multiplex real-time PCR assay for simultaneous detection of eight foodborne pathogens in a single reaction and its application. *Foodborne Pathog Dis* **11:**207–214.
47. **Hanemaaijer NM, Nijhuis RHT, Slotboom BJ, Mascini EM, van Zwet AA.** 2014. New

screening method to detect carriage of carbapenemase-producing *Enterobacteriaceae* in patients within 24 hours. *J Hosp Infect* **87:** 47–49.
48. **Lowman W, Marais M, Ahmed K, Marcus L.** 2014. Routine active surveillance for carbapenemase-producing *Enterobacteriaceae* from rectal swabs: diagnostic implications of multiplex polymerase chain reaction. *J Hosp Infect* **88:**66–71.
49. **Chen L, Chavda KD, Mediavilla JR, Zhao Y, Fraimow HS, Jenkins SG, Levi MH, Hong T, Rojtman AD, Ginocchio CC, Bonomo RA, Kreiswirth BN.** 2012. Multiplex real-time PCR for detection of an epidemic KPC-producing *Klebsiella pneumoniae* ST258 clone. *Antimicrob Agents Chemother* **56:**3444–3447.
50. **Roschanski N, Fischer J, Guerra B, Roesler U.** 2014. Development of a multiplex real-time PCR for the rapid detection of the predominant beta-lactamase genes CTX-M, SHV, TEM and CIT-type AmpCs in *Enterobacteriaceae*. *PLoS One* **9:**e100956.
51. **Guo D, Liu B, Liu F, Cao B, Chen M, Hao X, Feng L, Wang L.** 2013. Development of a DNA microarray for molecular identification of all 46 *Salmonella* O serogroups. *Appl Environ Microbiol* **79:**3392–3399.
52. **Sibley CD, Peirano G, Church DL.** 2012. Molecular methods for pathogen and microbial community detection and characterization: current and potential application in diagnostic microbiology. *Infect Genet Evol* **12:**505–521.
53. **Law JW, Ab Mutalib NS, Chan KG, Lee LH.** 2015. Rapid methods for the detection of foodborne bacterial pathogens: principles, applications, advantages and limitations. *Front Microbiol* **5:**770.
54. **Wang XW, Zhang L, Jin LQ, Jin M, Shen ZQ, An S, Chao FH, Li JW.** 2007. Development and application of an oligonucleotide microarray for the detection of food-borne bacterial pathogens. *Appl Microbiol Biotechnol* **76:**225–233.
55. **Li Y, Liu D, Cao B, Han W, Liu Y, Liu F, Guo X, Bastin DA, Feng L, Wang L.** 2006. Development of a serotype-specific DNA microarray for identification of some *Shigella* and pathogenic *Escherichia coli* strains. *J Clin Microbiol* **44:** 4376–4383.
56. **Bang J, Beuchat LR, Song H, Gu MB, Chang HI, Kim HS, Ryu JH.** 2013. Development of a random genomic DNA microarray for the detection and identification of *Listeria monocytogenes* in milk. *Int J Food Microbiol* **161:**134–141.
57. **Scaria J, Palaniappan RUM, Chiu D, Phan JA, Ponnala L, McDonough P, Grohn YT, Porwollik S, McClelland M, Chiou CS, Chu C, Chang YF.** 2008. Microarray for molecular typing of *Salmonella* enterica serovars. *Mol Cell Probes* **22:**238–243.
58. **Porwollik S, Boyd EF, Choy C, Cheng P, Florea L, Proctor E, McClelland M.** 2004. Characterization of *Salmonella enterica* subspecies I genovars by use of microarrays. *J Bacteriol* **186:**5883–5898.
59. **Braun SD, Ziegler A, Methner U, Slickers P, Keiling S, Monecke S, Ehricht R.** 2012. Fast DNA serotyping and antimicrobial resistance gene determination of *Salmonella enterica* with an oligonucleotide microarray-based assay. *PLoS One* **7:**e46489.
60. **Keramas G, Bang DD, Lund M, Madsen M, Bunkenborg H, Telleman P, Christensen CBV.** 2004. Use of culture, PCR analysis, and DNA microarrays for detection of *Campylobacter jejuni* and *Campylobacter coli* from chicken feces. *J Clin Microbiol* **42:**3985–3991.
61. **Keramas G, Bang DD, Lund M, Madsen M, Rasmussen SE, Bunkenborg H, Telleman P, Christensen CBV.** 2003. Development of a sensitive DNA microarray suitable for rapid detection of *Campylobacter* spp. *Mol Cell Probes* **17:**187–196.
62. **Laksanalamai P, Jackson SA, Mammel MK, Datta AR.** 2012. High density microarray analysis reveals new insights into genetic footprints of *Listeria monocytogenes* strains involved in listeriosis outbreaks. *PLoS One* **7:** e32896.
63. **Chiang YC, Yang CY, Li C, Ho YC, Lin CK, Tsen HY.** 2006. Identification of *Bacillus* spp., *Escherichia coli*, *Salmonella* spp., *Staphylococcus* spp. and *Vibrio* spp. with 16S ribosomal DNA-based oligonucleotide array hybridization. *Int J Food Microbiol* **107:**131–137.
64. **Perreten V, Vorlet-Fawer L, Slickers P, Ehricht R, Kuhnert P, Frey J.** 2005. Microarray-based detection of 90 antibiotic resistance genes of Gram-positive bacteria. *J Clin Microbiol* **43:**2291–2302.
65. **De Boer SH, López MM.** 2012. New grower-friendly methods for plant pathogen monitoring. *Annu Rev Phytopathol* **50:**197–218.
66. **Everett KR, Rees-George J, Pushparajah IPS, Janssen BJ, Luo Z.** 2010. Advantages and disadvantages of microarrays to study microbial population dynamics: a mini review. *N Z Plant Prot* **63:**1–6.
67. **Notomi T, Okayama H, Masubuchi H, Yonekawa T, Watanabe K, Amino N, Hase T.** 2000. Loop-mediated isothermal amplification of DNA. *Nucleic Acids Res* **28:**E63.
68. **Mori Y, Nagamine K, Tomita N, Notomi T.** 2001. Detection of loop-mediated isothermal

amplification reaction by turbidity derived from magnesium pyrophosphate formation. *Biochem Biophys Res Commun* **289:**150–154.

69. **Tomita N, Mori Y, Kanda H, Notomi T.** 2008. Loop-mediated isothermal amplification (LAMP) of gene sequences and simple visual detection of products. *Nat Protoc* **3:**877–882.
70. **Liu N, Zou D, Dong D, Yang Z, Ao D, Liu W, Huang L.** 2017. Development of a multiplex loop-mediated isothermal amplification method for the simultaneous detection of *Salmonella* spp. and *Vibrio parahaemolyticus*. *Sci Rep* **7:** 45601.
71. **Chen HT, Zhang J, Sun DH, Ma LN, Liu XT, Cai XP, Liu YS.** 2008. Development of reverse transcription loop-mediated isothermal amplification for rapid detection of H9 avian influenza virus. *J Virol Methods* **151:**200–203.
72. **Wang D, Wang Y, Xiao F, Guo W, Zhang Y, Wang A, Liu Y.** 2015. A comparison of in-house real-time LAMP assays with a commercial assay for the detection of pathogenic bacteria. *Molecules* **20:**9487–9495.
73. **Shao Y, Zhu S, Jin C, Chen F.** 2011. Development of multiplex loop-mediated isothermal amplification-RFLP (mLAMP-RFLP) to detect *Salmonella* spp. and *Shigella* spp. in milk. *Int J Food Microbiol* **148:**75–79.
74. **Wang Y, Wang Y, Ma A, Li D, Luo L, Liu D, Hu S, Jin D, Liu K, Ye C.** 2015. The novel multiple inner primers-loop-mediated isothermal amplification (MIP-LAMP) for rapid detection and differentiation of *Listeria monocytogenes*. *Molecules* **20:**21515–21531.
75. **Ye Y, Wang B, Huang F, Song Y, Yan H, Alam MJ, Yamasaki S, Shi L.** 2011. Application of *in situ* loop-mediated isothermal amplification method for detection of *Salmonella* in foods. *Food Control* **22:**438–444.
76. **Xu Z, Li L, Chu J, Peters BM, Harris ML, Li B, Shi L, Shirtliff ME.** 2012. Development and application of loop-mediated isothermal amplification assays on rapid detection of various types of staphylococci strains. *Food Res Int* **47:**166–173.
77. **Frickmann H, Masanta WO, Zautner AE.** 2014. Emerging rapid resistance testing methods for clinical microbiology laboratories and their potential impact on patient management. *Biomed Res Int* **2014:**375681.
78. **Hordijk J, Wagenaar JA, Kant A, van Essen-Zandbergen A, Dierikx C, Veldman K, Wit B, Mevius D.** 2013. Cross-sectional study on prevalence and molecular characteristics of plasmid mediated ESBL/AmpC-producing *Escherichia coli* isolated from veal calves at slaughter. *PLoS One* **8:**e65681.
79. **Dong HJ, Cho AR, Hahn TW, Cho S.** 2014. Development of a loop-mediated isothermal amplification assay for rapid, sensitive detection of *Campylobacter jejuni* in cattle farm samples. *J Food Prot* **77:**1593–1598.
80. **Yamazaki W, Taguchi M, Kawai T, Kawatsu K, Sakata J, Inoue K, Misawa N.** 2009. Comparison of loop-mediated isothermal amplification assay and conventional culture methods for detection of *Campylobacter jejuni* and *Campylobacter coli* in naturally contaminated chicken meat samples. *Appl Environ Microbiol* **75:**1597–1603.
81. **Tang MJ, Zhou S, Zhang XY, Pu JH, Ge QL, Tang XJ, Gao YS.** 2011. Rapid and sensitive detection of *Listeria monocytogenes* by loop-mediated isothermal amplification. *Curr Microbio* **63:**511–516.
82. **Koide Y, Maeda H, Yamabe K, Naruishi K, Yamamoto T, Kokeguchi S, Takashiba S.** 2010. Rapid detection of *mecA* and *spa* by the loop-mediated isothermal amplification (LAMP) method. *Lett Appl Microbiol* **50:**386–392.
83. **Chen C, Zhao Q, Guo J, Li Y, Chen Q.** 2017. Identification of methicillin-resistant *Staphylococcus aureus* (MRSA) using simultaneous detection of mecA, nuc, and femB by loop-mediated isothermal amplification (LAMP). *Curr Microbiol* **74:**965–971.
84. **Croxatto A, Prod'hom G, Greub G.** 2012. Applications of MALDI-TOF mass spectrometry in clinical diagnostic microbiology. *FEMS Microbiol Rev* **36:**380–407.
85. **Fenselau C, Demirev PA.** 2001. Characterization of intact microorganisms by MALDI mass spectrometry. *Mass Spectrom Rev* **20:** 157–171.
86. **Singhal N, Kumar M, Kanaujia PK, Virdi JS.** 2015. MALDI-TOF mass spectrometry: an emerging technology for microbial identification and diagnosis. *Front Microbiol* **6:**791.
87. **Freiwald A, Sauer S.** 2009. Phylogenetic classification and identification of bacteria by mass spectrometry. *Nat Protoc* **4:**732–742.
88. **Kang L, Li N, Li P, Zhou Y, Gao S, Gao H, Xin W, Wang J.** 2017. MALDI-TOF mass spectrometry provides high accuracy in identification of *Salmonella* at species level but is limited to type or subtype *Salmonella* serovars. *Eur J Mass Spectrom* **23:**70–82.
89. **Dieckmann R, Malorny B.** 2011. Rapid screening of epidemiologically important *Salmonella enterica* subsp. *enterica* serovars by whole-cell matrix-assisted laser desorption ionization-time of flight mass spectrometry. *Appl Environ Microbiol* **77:**4136–4146.

90. **Mandrell RE, Harden LA, Bates A, Miller WG, Haddon WF, Fagerquist CK.** 2005. Speciation of *Campylobacter coli, C. jejuni, C. helveticus, C. lari, C. sputorum,* and *C. upsaliensis* by matrix-assisted laser desorption ionization-time of flight mass spectrometry. *Appl Environ Microbiol* **71:**6292–6307.
91. **Cameron M, Barkema HW, De Buck J, De Vliegher S, Chaffer M, Lewis J, Keefe GP.** 2017. Identification of bovine-associated coagulase-negative staphylococci by matrix-assisted laser desorption/ionization time-of-flight mass spectrometry using a direct transfer protocol. *J Dairy Sci* **100:**2137–2147.
92. **Wolters M, Rohde H, Maier T, Belmar-Campos C, Franke G, Scherpe S, Aepfelbacher M, Christner M.** 2011. MALDI-TOF MS fingerprinting allows for discrimination of major methicillin-resistant *Staphylococcus* aureus lineages. *Int J Med Microbiol* **301:**64–68.
93. **Egli A, Tschudin-Sutter S, Oberle M, Goldenberger D, Frei R, Widmer AF.** 2015. Matrix-assisted laser desorption/ionization time of flight mass-spectrometry (MALDI-TOF MS) based typing of extended-spectrum β-lactamase producing *E. coli*: a novel tool for real-time outbreak investigation. *PLoS One* **10:**e0120624.
94. **Carbonnelle E, Beretti JL, Cottyn S, Quesne G, Berche P, Nassif X, Ferroni A.** 2007. Rapid identification of staphylococci isolated in clinical microbiology laboratories by matrix-assisted laser desorption ionization-time of flight mass spectrometry. *J Clin Microbiol* **45:** 2156–2161.
95. **Emonet S, Shah HN, Cherkaoui A, Schrenzel J.** 2010. Application and use of various mass spectrometry methods in clinical microbiology. *Clin Microbiol Infect* **16:**1604–1613.
96. **Urwyler SK, Glaubitz J.** 2016. Advantage of MALDI-TOF-MS over biochemical-based phenotyping for microbial identification illustrated on industrial applications. *Lett Appl Microbiol* **62:**130–137.
97. **Tran A, Alby K, Kerr A, Jones M, Gilligan PH.** 2015. Cost savings realized by implementation of routine microbiological identification by matrix-assisted laser desorption ionization–time of flight mass spectrometry. *J Clin Microbiol* **53:**2473–2479.
98. **Trindade PA, McCulloch JA, Oliveira GA, Mamizuka EM.** 2003. Molecular techniques for MRSA typing: current issues and perspectives. *Braz J Infect Dis* **7:**32–43.
99. **Chroma M, Kolar M.** 2010. Genetic methods for detection of antibiotic resistance: focus on extended-spectrum β-lactamases. *Biomed Pap Med Fac Univ Palacky Olomouc Czech Repub* **154:**289–296.
100. **Mohran ZS, Guerry P, Lior H, Murphy JR, el-Gendy AM, Mikhail MM, Oyofo BA.** 1996. Restriction fragment length polymorphism of flagellin genes of *Campylobacter jejuni* and/or *C. coli* isolates from Egypt. *J Clin Microbiol* **34:**1216–1219.
101. **Babalola OO.** 2003. Molecular techniques: an overview of methods for the detection of bacteria. *Afr J Biotechnol* **2:**710–713.
102. **Chen Y, Son I.** 2014. Polymerase chain reaction-based subtyping methods, p 3–26. *In* Oyarzabal OA, Kathariou S (ed), *DNA Methods in Food Safety. Molecular Typing of Foodborne and Waterborne Bacterial Pathogens.* Wiley-Blackwell, Oxford, United Kingdom.
103. **Lin AW, Usera MA, Barrett TJ, Goldsby RA.** 1996. Application of random amplified polymorphic DNA analysis to differentiate strains of *Salmonella enteritidis*. *J Clin Microbiol* **34:** 870–876.
104. **Shangkuan YH, Lin HC.** 1998. Application of random amplified polymorphic DNA analysis to differentiate strains of *Salmonella typhi* and other *Salmonella* species. *J Appl Microbiol* **85:** 693–702.
105. **Yoshida T, Takeuchi M, Sato M, Hirai K.** 1999. Typing *Listeria monocytogenes* by random amplified polymorphic DNA (RAPD) fingerprinting. *J Vet Med Sci* **61:**857–860.
106. **Hernandez J, Fayos A, Ferrus MA, Owen RJ.** 1995. Random amplified polymorphic DNA fingerprinting of *Campylobacter jejuni* and *C. coli* isolated from human faeces, seawater and poultry products. *Res Microbiol* **146:** 685–696.
107. **Açik MN, Cetinkaya B.** 2006. Random amplified polymorphic DNA analysis of *Campylobacter jejuni* and *Campylobacter coli* isolated from healthy cattle and sheep. *J Med Microbiol* **55:**331–334.
108. **Barbut F, Mario N, Delmée M, Gozian J, Petit JC.** 1993. Genomic fingerprinting of *Clostridium difficile* isolates by using a random amplified polymorphic DNA (RAPD) assay. *FEMS Microbiol Lett* **114:**161–166.
109. **Roussel S, Félix B, Grant K, Dao TT, Brisabois A, Amar C.** 2013. Fluorescence amplified fragment length polymorphism compared to pulsed field gel electrophoresis for *Listeria monocytogenes* subtyping. *BMC Microbiol* **13:**14.
110. **Singh DV, Mohapatra H.** 2008. Application of DNA-based methods in typing *Vibrio cholerae* strains. *Future Microbiol* **3:**87–96.
111. **Guerra MM, Bernardo F, McLauchlin J.** 2002. Amplified fragment length polymorphism

(AFLP) analysis of *Listeria monocytogenes*. *Syst Appl Microbiol* **25:**456–461.

112. **Vos P, Hogers R, Bleeker M, Reijans M, Lee T, Hornes M, Friters A, Pot J, Paleman J, Kuiper M, Zabeau M.** 1995. AFLP: a new technique for DNA fingerprinting. *Nucleic Acids Res* **23:**4407–4414.
113. **Savelkoul PHM, Aarts HJM, de Haas J, Dijkshoorn L, Duim B, Otsen M, Rademaker JL, Schouls L, Lenstra JA.** 1999. Amplified-fragment length polymorphism analysis: the state of an art. *J Clin Microbiol* **37:**3083–3091.
114. **Duim B, Vandamme PA, Rigter A, Laevens S, Dijkstra JR, Wagenaar JA.** 2001. Differentiation of *Campylobacter* species by AFLP fingerprinting. *Microbiology* **147:**2729–2737.
115. **Aarts HJ, Hakemulder LE, Van Hoef AM.** 1999. Genomic typing of *Listeria monocytogenes* strains by automated laser fluorescence analysis of amplified fragment length polymorphism fingerprint patterns. *Int J Food Microbiol* **49:**95–102.
116. **Ross IL, Heuzenroeder MW.** 2005. Use of AFLP and PFGE to discriminate between *Salmonella enterica* serovar Typhimurium DT126 isolates from separate food-related outbreaks in Australia. *Epidemiol Infect* **133:**635–644.
117. **Maiden MC, Bygraves JA, Feil E, Morelli G, Russell JE, Urwin R, Zhang Q, Zhou J, Zurth K, Caugant DA, Feavers IM, Achtman M, Spratt BG.** 1998. Multilocus sequence typing: a portable approach to the identification of clones within populations of pathogenic microorganisms. *Proc Natl Acad Sci USA* **95:**3140–3145.
118. **Cooper JE, Feil EJ.** 2004. Multilocus sequence typing: what is resolved? *Trends Microbiol* **12:**373–377.
119. **Enright MC, Spratt BG.** 1999. Multilocus sequence typing. *Trends Microbiol* **7:**482–487.
120. **Knabel J.** 2014. Multilocus sequence typing: an adaptable tool for understanding the global epidemiology of bacterial pathogens, p 47–64. *In* Oyarzabal OA, Kathariou S (ed), *DNA Methods in Food Safety. Molecular Typing of Foodborne and Waterborne Bacterial Pathogens*. Wiley-Blackwell, Oxford, United Kingdom.
121. **Thakur S, Morrow WE, Funk JA, Bahnson PB, Gebreyes WA.** 2006. Molecular epidemiologic investigation of *Campylobacter coli* in swine production systems, using multilocus sequence typing. *Appl Environ Microbiol* **72:**5666–5669.
122. **Thakur S, White DG, McDermott PF, Zhao S, Kroft B, Gebreyes W, Abbott J, Cullen P, English L, Carter P, Harbottle H.** 2009. Genotyping of *Campylobacter coli* isolated from humans and retail meats using multilocus sequence typing and pulsed-field gel electrophoresis. *J Appl Microbiol* **106:**1722–1733.
123. **Zhao X, Gao Y, Ye C, Yang L, Wang T, Chang W.** 2016. Prevalence and Characteristics of *Salmonella* isolated from free-range chickens in Shandong Province, China. *BioMed Res Int* **2016:**8183931.
124. **Thakur S, Gebreyes WA.** 2010. Phenotypic and genotypic heterogeneity of *Campylobacter coli* within individual pigs at farm and slaughter in the US. *Zoonoses Public Health* **57** (Suppl 1)**:**100–106.
125. **Stone D, Davis M, Baker K, Besser T, Roopnarine R, Sharma R.** 2013. MLST genotypes and antibiotic resistance of *Campylobacter* spp. isolated from poultry in Grenada. *BioMed Res Int* **2013:**794643.
126. **Vidal AB, Colles FM, Rodgers JD, McCarthy ND, Davies RH, Maiden MC, Clifton-Hadley FA.** 2016. Genetic diversity of *Campylobacter jejuni* and *Campylobacter coli* isolates from conventional broiler flocks and the impacts of sampling strategy and laboratory method. *Appl Environ Microbiol* **82:**2347–2355.
127. **Molla B, Byrne M, Abley M, Mathews J, Jackson CR, Fedorka-Cray P, Sreevatsan S, Wang P, Gebreyes WA.** 2012. Epidemiology and genotypic characteristics of methicillin-resistant *Staphylococcus aureus* strains of porcine origin. *J Clin Microbiol* **50:**3687–3693.
128. **Quintana-Hayashi MP, Thakur S.** 2012. Phylogenetic analysis reveals common antimicrobial resistant *Campylobacter coli* population in antimicrobial-free (ABF) and commercial swine systems. *PLoS One* **7:**e44662.
129. **Ray M, Schwartz DC.** 2014. Pulsed-field gel electrophoresis and the molecular epidemiology of foodborne pathogens, p 27–46. *In* Oyarzabal OA, Kathariou S (ed), *DNA Methods in Food Safety. Molecular Typing of Foodborne and Waterborne Bacterial Pathogens*. Wiley-Blackwell, Oxford, United Kingdom.
130. **Foley SL, White DG, McDermott PF, Walker RD, Rhodes B, Fedorka-Cray PJ, Simjee S, Zhao S.** 2006. Comparison of subtyping methods for differentiating *Salmonella enterica* serovar Typhimurium isolates obtained from food animal sources. *J Clin Microbiol* **44:**3569–3577.
131. **Anand R.** 1986. Pulsed field gel electrophoresis: a technique for fractionating large DNA molecules. *Trends Genet* **2:**278–283.
132. **Gautom RK.** 1997. Rapid pulsed-field gel electrophoresis protocol for typing of *Escherichia coli* O157:H7 and other Gram-negative organisms in 1 day. *J Clin Microbiol* **35:**2977–2980.
133. **Bopp DJ, Baker DJ, Thompson L, Saylors A, Root TP, Armstrong L, Mitchell K, Dumas**

NB, Musser KA. 2016. Implementation of *Salmonella* serotype determination using pulsed-field gel electrophoresis in a state public health laboratory. *Diagn Microbiol Infect Dis* **85:**416–418.

134. **Tsen HY, Hu HH, Lin JS, Huang CH, Wang TK.** 2000. Analysis of the *Salmonella* Typhimurium isolates from food-poisoning cases by molecular subtyping methods. *Food Microbiol* **17:**143–152.

135. **McCullagh JJ, McNamee PT, Smyth JA, Ball HJ.** 1998. The use of pulsed field gel electrophoresis to investigate the epidemiology of *Staphylococcus aureus* infection in commercial broiler flocks. *Vet Microbiol* **63:**275–281.

136. **Zou W, Lin W-J, Foley SL, Chen C-H, Nayak R, Chen JJ.** 2010. Evaluation of pulsed-field gel electrophoresis profiles for identification of *Salmonella* serotypes. *J Clin Microbiol* **48:** 3122–3126.

137. **Molla B, Sterman A, Mathews J, Artuso-Ponte V, Abley M, Farmer W, Rajala-Schultz P, Morrow WEM, Gebreyes WA.** 2010. *Salmonella enterica* in commercial swine feed and subsequent isolation of phenotypically and genotypically related strains from fecal samples. *Appl Environ Microbiol* **76:**7188–7193.

138. **Refsum T, Heir E, Kapperud G, Vardund T, Holstad G.** 2002. Molecular epidemiology of *Salmonella enterica* serovar Typhimurium isolates determined by pulsed-field gel electrophoresis: comparison of isolates from avian wildlife, domestic animals, and the environment in Norway. *Appl Environ Microbiol* **68:** 5600–5606.

139. **Roetzer A, Diel R, Kohl TA, Rückert C, Nübel U, Blom J, Wirth T, Jaenicke S, Schuback S, Rüsch-Gerdes S, Supply P, Kalinowski J, Niemann S.** 2013. Whole genome sequencing versus traditional genotyping for investigation of a *Mycobacterium tuberculosis* outbreak: a longitudinal molecular epidemiological study. *PLoS Med* **10:**e1001387.

140. **Dark MJ.** 2013. Whole-genome sequencing in bacteriology: state of the art. *Infect Drug Resist* **6:**115–123.

141. **Jünemann S, Sedlazeck FJ, Prior K, Albersmeier A, John U, Kalinowski J, Mellmann A, Goesmann A, von Haeseler A, Stoye J, Harmsen D.** 2013. Updating benchtop sequencing performance comparison. *Nat Biotechnol* **31:**294–296.

142. **Liu L, Li Y, Li S, Hu N, He Y, Pong R, Lin D, Lu L, Law M.** 2012. Comparison of next-generation sequencing systems. *J Biomed Biotechnol* **2012:**251364.

143. **Schürch AC, van Schaik W.** 2017. Challenges and opportunities for whole-genome sequencing-based surveillance of antibiotic resistance. *Ann N Y Acad Sci* **1388:**108–120.

144. **Dunne WM Jr, Westblade LF, Ford B.** 2012. Next-generation and whole-genome sequencing in the diagnostic clinical microbiology laboratory. *Eur J Clin Microbiol Infect Dis* **31:**1719–1726.

Mathematical Modeling Tools to Study Preharvest Food Safety

21

CRISTINA LANZAS[1] and SHI CHEN[1]

INTRODUCTION

Preharvest food safety is a complex system because pathogen transmission and dissemination within a farm environment are determined by multiple interrelated factors, including ecological, evolutionary, environmental, and management drivers that act on different scales of time, space, and organizational complexity. The nonlinear dynamics of pathogen transmission and the complexity of the systems involved pose challenges in understanding key determinants of preharvest food safety and in identifying critical points and designing effective mitigation strategies. Mathematical modeling provides tools to explicitly represent the variability, interconnectedness, and complexity of such systems (1). In biology, mathematical models have contributed to numerous insights and theoretical advances, as well as to changes in public policy, health practice, and management (2). Mathematical biology has, indeed, become one of the most prominent interdisciplinary areas of research, but the use of mathematical models in preharvest food safety is recent (3). Results from modeling research in preharvest food safety have been published mostly since the 2000s. Most of these publications describe the development of epidemiological models to represent foodborne pathogen transmission in animal

[1]Department of Population Health and Pathobiology, College of Veterinary Medicine, North Carolina State University, Raleigh, NC 27607.

Preharvest Food Safety
Edited by Siddhartha Thakur and Kalmia E. Kniel

doi:10.1128/microbiolspec.PFS-0001-2013

farming, following the longstanding tradition of applying mathematical modeling in epidemiology (4, 5).

In this article, we provide an overview of the emerging field of mathematical modeling in preharvest food safety. The first section provides an overview of the discipline of mathematical modeling. We describe the steps involved in developing mathematical models, different types of models, and their multiple applications. The article focuses on mechanistic models, which describe the underlying processes driving the spatial and temporal patterns that we aim to comprehend. We do not comprehensively discuss statistical (phenomenological) models. The sections that follow introduce the most common modeling approaches used in preharvest systems. The article concludes with an outline of the potential future directions for the field.

PRINCIPLES OF MATHEMATICAL MODELING

Modeling Steps

Mathematical modeling is the process of constructing, testing, analyzing, verifying, and validating mathematical models. Mathematical models are representations and abstractions of real systems or hypotheses in a mathematical language. The steps in mathematical modeling parallel those of the scientific method: we formulate the research question or model objectives, propose falsifiable hypotheses, develop the mathematical model, analyze and validate the model, and finally relate the model to the objectives (2). This process is often iterative. The analysis of the model can help us to further crystallize the modeling objectives, or it may lead to revisions in the model structure. Usually, several versions of a model are developed prior to obtaining the final model.

Figure 1 shows the modeling process in more detail. The process starts by defining objectives for the project. A clearly defined set of objectives is the most important element for a successful modeling project. The usefulness of the model relies precisely on the ability to represent the key components of the system while ignoring the less important ones. The modeling objectives act as criteria to determine which components of the system need to be included, which ones can be ignored, and what modeling approach to use. Once the modeling objectives are clear, the modeler can generate a proper hypothesis and review the available data and information to develop a conceptual model that incorporates the key components and interactions of the system to model. A conceptual model may not necessarily contain a mathematical description of the system; it could be only a graphic representation of the system. Regardless, the interaction between modelers and subject matter experts is important in this step to ensure that the theory and assumptions underlying the conceptual model are appropriate for addressing the intended modeling objectives and hypotheses.

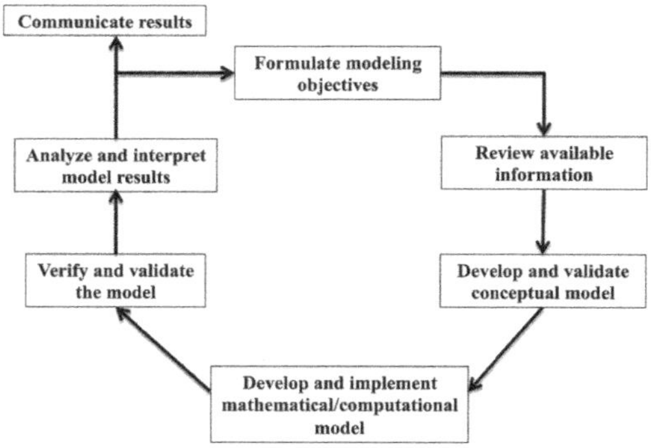

FIGURE 1 The modeling cycle.

To translate the conceptual model into a mathematical or computational model, we first need to decide what modeling approach is the most suitable to the question in hand (discussed in the next subsection). For example, the conceptual model can be translated into a population-based, ordinary differential equations model or into a spatially explicit, individual-based computational model, where each individual is tracked across the space. Advances in computational power have allowed us to consider modeling approaches such as individual-based models that a few years ago were computationally infeasible. The availability of data to parameterize the model is another factor to consider when deciding on the model approach and detail level. The availability of appropriate data to parameterize models has always been one of the most limiting factors in advancing the use of mathematical models. In recent years, advancements in the area of computational statistics (e.g., bootstrap method, Markov process, Monte Carlo method, various filtering techniques) have substantially improved the integration of mathematical models and empirical data (6).

Model verification and validation are integrated components of the overall model development process and need to be addressed before the model is applied (Fig. 1). The models are often complex enough that deriving the analytical solution is not feasible; therefore, we rely on computational implementation to explore the model through simulations. Model verification ensures that the model is implemented properly and that no errors are present in the computer code, so the model should work as intended. However, model verification does not guarantee that the model will predict the system accurately. Once the model is verified, it needs to be validated. Researchers often equate model validity with the ability of the model to replicate some observed data. Yet model validation is a much broader process that involves ensuring the appropriateness of the model at different levels and steps along the modeling process (7). First, the theories and assumptions underlying the conceptual model should be deemed reasonable and sound for the intended model purpose (7). Several techniques can be used to evaluate the conceptual validity of the model (7, 8). One such technique is the structure assessment test, which evaluates the level of aggregation, whether the model meets basic conservation laws (e.g., total energy or mass does not change over time in a closed population), dimensional consistency (dimension on the left side of an equation equals that of the right side), and the realism of the incorporated decision rules (8). Finally, to evaluate model accuracy (i.e., how close predicted values are to observed values), we compare model outcomes with real data using statistical techniques (9). The availability of appropriate data to both parameterize and validate models is often limited, but when possible, different data sets should be used to build and validate the model.

The type of model analysis and simulations depends on the objectives of the model. One of the most common objectives is to evaluate scenarios with potential control interventions. We often simulate a control strategy under scenarios that differ in management and risk profiles. When communicating the model results to stakeholders, it is important to explain the model assumptions, potential outcomes, and limitations in a transparent manner.

Model Classifications

Models can be classified based on several criteria, such as (i) the treatment of time or space, (ii) whether they account for variability, or (iii) whether we track changes in the overall population or in each individual.

1. Dynamic models describe how a system changes over time and are the most common models in mathematical biology. We can further subdivide dynamic models into discrete-time models and continuous-time models. Discrete-time models track changes to variables in discrete time steps (e.g., a day, a year, or a generation). Continuous-time models track the variables continuously over any period of time.
2. Models are defined as deterministic or stochastic depending on their treatment of variability. Deterministic models do not involve stochasticity or randomness

and therefore do not address variability. The outcome in a deterministic model is predicted entirely by the model equations and initial values. Given the same model system and same initial values, we should get exactly the same result every time we simulate the model. Stochastic models incorporate the effect of chance. They predict the probability of various outcomes in the future. In each model realization, the model prediction is different even when we use the same initial conditions and parameters (Fig. 2). Statistical analysis of a large number of replicates is often necessary to summarize the results of stochastic models. For the same system, if the population is large enough, stochastic realizations of the model will converge to the deterministic counterpart. For small populations, stochastic models are better suited than deterministic models because demographic stochasticity has an important role in the system. In small populations (or when the number of infectious individuals is very low), each individual carries a significant weight in the population, and therefore the probabilistic nature of individual processes such as birth, death, transmission, and recovery cannot be ignored.

3. Models can also be classified based on their treatment of the population and individuals. In a compartmental model, the individuals in the population are subdivided into broad subgroups (compartments), and the model tracks changes over time for individuals in the same compartment collectively. In individual-based models (also referred to as agent-based models), each individual unit is simulated. Individual-based models have gained popularity recently because of their capability of handling individual variability. They can incorporate complicated patterns of interaction among individuals in the population, and individuals (also called agents) can be characterized by a given number of flexible attributes (e.g., level of colonization, antimicrobial use, susceptibility, etc.). The main disadvantages of individual-

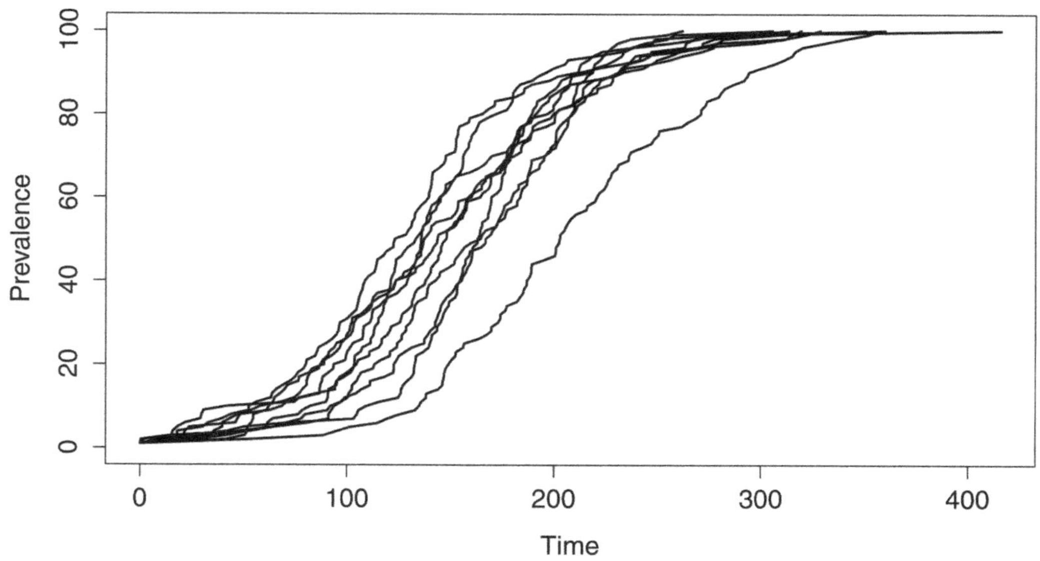

FIGURE 2 Time series (days) of infection prevalence simulated by a stochastic susceptible-infectious model. Ten realizations of the model are presented.

based models are that they require a more comprehensive understanding of the system at hand and usually require more effort to formulate, run, and analyze.

In summary, we should choose the modeling approach based on the purpose of the study, the resources available, and what accuracy and level of detail the model requires. For a modeling project, we may develop several models using different modeling approaches.

APPLICATIONS OF MATHEMATICAL MODELS IN PREHARVEST FOOD SAFETY

Model applications have research-oriented purposes or more practical uses, such as developing decision support systems. Mathematical models can be used throughout all the steps involved in the research process (10). We can use models to refine our study questions, select the most sensitive variables, and design sample strategies. One of the most important applications of models in research is to clarify our understanding of the system under study. Mathematical models describe our hypotheses in the rigorous language of mathematics. In mathematical language, we make explicit the underlying assumptions, mechanisms, and connections among the components of a system. For example, mathematical models have provided numerous insights into the relationships between antimicrobial use and resistance (11). Field studies of associations between risk factors and measured outcomes can provide information regarding the existence or strength of an association, but they do not provide information regarding the mechanisms underlying the relationship. In models, changes in measures of association such as odds ratios can be linked to specific mechanisms to investigate how the strength and direction of the association change under different scenarios (12). This is especially useful to evaluate field studies with conflicting results.

Mathematical models can support the experimental design of field studies and data collection in several ways. Models help to identify areas where more data collection is necessary through the identification of the key components of the system and existent knowledge gaps. This is most effectively accomplished by conducting a global sensitivity analysis of the model (13). In a global sensitivity analysis, inputs and model parameters are both considered as distributions. Parameters are sampled according to the distributions and fed to the model in a Monte Carlo–type simulation, which is a commonly used numerical method to compute results by random sampling. Statistical methods are then used to relate parameter variations, determined by specific distributions, to the model outputs. Improving our knowledge of highly influential parameters can increase model accuracy and yield better results to support decisions. For example, several models have highlighted the limited knowledge of the epidemiological effects of immunity against *Salmonella* (14–16). A global sensitivity analysis of an epidemiological model describing the transmission of *Salmonella* in dairy farms identified that endemic prevalence was sensitive to the duration of immunity, a parameter that is uncertain (14). Similarly, Lurette et al. performed a comprehensive sensitivity analysis for a model representing *Salmonella* spread within a pig batch (15). The protective effect of maternal immunity was identified as an influential parameter and a gap in the current knowledge of *Salmonella* infection in pigs (15). Models can also assist in the selection of a sampling strategy, as well as in determining the statistical power associated with different sample sizes (10).

Additional uses of mathematical models in preharvest food safety include characterizing the production modules in risk assessments and identifying, designing, and optimizing mitigation strategies and surveillance efforts. The production module of a farm-to-table risk assessment can be as simple as a description of the foodborne pathogen prevalence at the farm level as a probability distribution, with the main objective being to feed an input to the next phase of the food chain. Recent risk assessments for foodborne pathogens use a mechanistic modeling approach at the farm level (17, 18). These models are stochastic to account for variability and uncertainty. The use of a mechanistic approach has numerous advantages: critical control points at the farm level can be identified, and scenarios where the effect of mitigations is simulated can be evaluated to assess the effect of preharvest interventions on human disease.

The most common application of mathematical models preharvest is the prediction of the effect and efficacy of preharvest control strategies. Testing multiple levels and combinations of control measures in experimental and field settings is often logistically and economically unfeasible. In addition, on-farm studies of preharvest interventions can be difficult to interpret due to potentially confounding factors, and they may be disruptive to a farm's day-to-day practices (19). Mathematical models are capable of simulating the effect of control measures under a wide range of management conditions, and they can provide both qualitative and quantitative results under various controlling efforts, which are otherwise hard to achieve due to usual constraints on resources and timing. Model predictions can also guide the design of subsequent field studies to further test interventions. Models can be integrated with economic approaches to perform cost-benefit analysis and determine economically optimal control strategies. Finally, mathematical modeling is a suitable approach to identify the best sampling strategies to implement in surveillance programs (20).

COMPARTMENTAL MODELS FOR FOODBORNE PATHOGENS

Farm animals are considered the main reservoir and host for most foodborne pathogens (21), and therefore considerable efforts have been devoted to understanding the epidemiology and ecology of these pathogens at the animal reservoir level. The most common mathematical modeling framework used to describe the ecology of foodborne pathogens is compartment modeling for infectious diseases. These models capture the process of infection and transmission of the pathogens. The animal (host) population is described based on its infection state (e.g., susceptible, represented as S; exposed, E; infectious, I; and recovered, R). Susceptible animals are those not infected but at risk of becoming infected if exposed to the pathogen. Exposed animals are infected but cannot immediately infect other animals because they do not yet shed the pathogen. Infectious animals are infected animals that shed the pathogen and therefore can infect other animals. Recovered animals have immunity against the pathogen. Furthermore, multiple types of infectious individuals can be included in the model depending on whether infection results in clinical or subclinical disease or based on differences in the duration and intensity of pathogen shedding (14). Most foodborne pathogens colonize animals asymptomatically. Within asymptomatically colonized animals, the level of shedding can be variable (22). Luckily, models can include different infectious states (e.g., high- and low-level shedders) based on distinct levels of pathogen shedding. A carrier state may also be included to describe animals carrying the pathogen but not shedding it after being infectious (23). Carrier animals can become infectious again when they start shedding the pathogen again. The onset of the pathogen re-excretion is linked to stressful events such as transport from farm to slaughter house (17).

Acronyms for the epidemiological models are based on the flow patterns between the states included in the compartment model (e.g., SIR, SEIR, SIRS, SI, etc.). The simplest case is an SI model, where no immunity is assumed, so animals are either susceptible or infectious. The SI model is often used for pathogens that behave as commensals (e.g.,

Campylobacter jejuni in poultry) (24). A more complicated SIRS model indicates that immunity lasts for a limited period of time, after which animals become susceptible again. Models with waning immunity are often used to characterize *Salmonella* (14, 25). *Salmonella* induces both humoral and cellular responses, but long and full protective immunity is not always induced after natural infection (26). The choice of which epidemiological states to include in a model depends on the host-pathogen being modeled and the purpose of the model. A foodborne pathogen may require different states depending on the host. For example, *Salmonella* may cause clinical disease or systemic or gastrointestinal infections depending on the different interactions with the host. Figure 3 shows examples of different *Salmonella*-host systems. The equations behind the epidemiological models describe the transitions between the epidemiological states represented in the model flowchart (see Fig. 4 for an example). The transmission term of an epidemiological model describes how animals acquire the pathogen and become infected. Several transmission formulations can be used depending on the transmission pathways and the assumptions underlying the transmission term (27, 28). Transmission of a pathogen directly from animal to animal depends on the prevalence of infectious animals (I/N, where I = the number of infectious individuals and N = the total population size), the population contact structure, and the probability of transmission given contact between a susceptible and an infectious animal (4). If the rate of contact between animals increases or decreases directly with the population size change, transmission is said to be density-dependent.

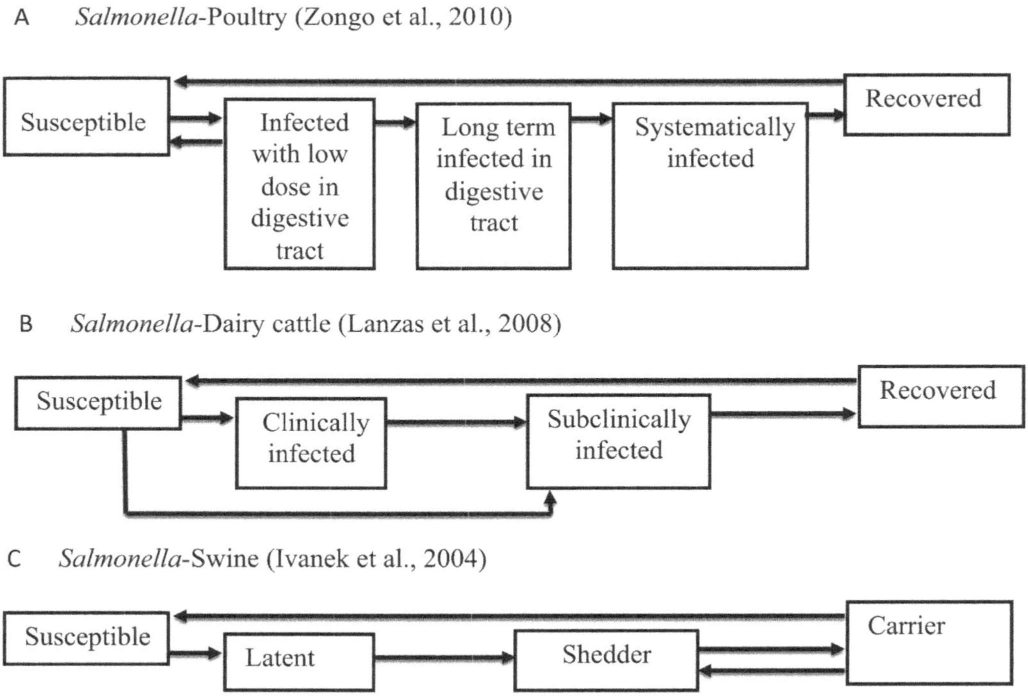

FIGURE 3 Examples of flow charts for epidemiological models commonly used for modeling foodborne pathogens in the animal host. (A) *Salmonella*-poultry (25). (B) *Salmonella*-dairy cattle (14). (C) *Salmonella*-swine (72).

A

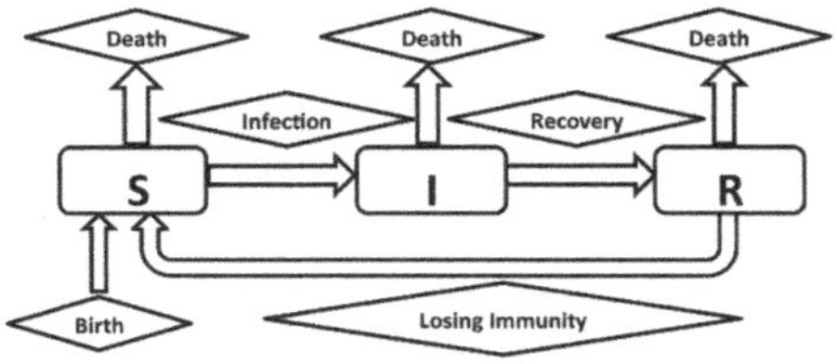

B

$$\frac{dS}{dt} = \mu N + rR - \frac{\beta IS}{N} - \mu S$$

$$\frac{dI}{dt} = \frac{\beta IS}{N} - \gamma I - \mu I$$

$$\frac{dR}{dt} = \gamma I - rR - \mu R$$

C

$$R_0 = \frac{\beta}{\gamma + \mu}$$

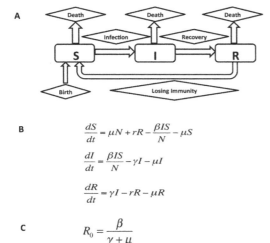

FIGURE 4 An example of a deterministic compartmental infectious disease model. (A) The flowchart represents an SIRS model. Compartments S, I, and R track the number of individuals who are susceptible, infectious, and recovered, respectively. The triangles represent change in the number of individuals in the compartments. Models that contain both demographic (e.g., births and deaths) and infection flows are called endemic models. (B) Compartmental models are often described mathematically as ordinary differential equations. The ordinary differential equations here describe the inflows and outflows by which the number of individuals in each epidemiological state (S, I, and R) changes over time. The total population, N is equal to S + I + R; υ = birth and death rate; γ = recovery rate of infectious individuals; β = transmission coefficient (transmission is frequency dependent); and *r* = immunity loss rate. The inflow and outflow of a compartment is reflected in the equation. For instance, the S compartment has four arrows associated with it: two inflows and two outflows. The two inflows are birth and loss of immunity from recovered compartment (hence the + sign in the first equation); the two outflows are natural death and infection to infectious compartment (hence the – sign in the first equation). (C) The basic reproduction number of this model.

On the other hand, if the contact rate does not change with population size, transmission is frequency-dependent (4).

Foodborne pathogens can often survive and even grow in nonhost environments at the farm (e.g., water troughs and feed bunks) (29). Indirect transmission through the environment to the animal and the persistence of the pathogens in the environment are then important determinants of foodborne pathogen ecology. Indirect transmission is related to free-living pathogens in the environment (3). These pathogens can be represented in a single, homogenous farm environment or in different nonhost habitats where the pathogen can survive and grow (30). For example, Ayscue et al. (30) investigated the roles of three specific nonbovine habitats (i.e., water tanks, feed bunks, surrounding pen environment) on persistence and loads of *Escherichia coli* O157 in feedlots. Water troughs and contaminated pen floors appeared to be particularly influential sources driving *E. coli* O157 population dynamics and thus were prime environmental targets for interventions to effectively reduce the *E. coli* O157 load at the pen level (30). The survival of pathogens in the environment is influenced by ambient factors such as temperature and humidity. However, the effect of environmental factors can be incorporated by modifying bacteria growth and survival parameters as functions of ambient factors (e.g., 31). These approaches of linking bacteria dynamics in the nonhost environment are closely related to the field of predictive microbiology, which mechanistically models microbial growth based on intrinsic and extrinsic determinants such as temperature, pH, and water activity (32). To date, the application of predictive microbiology preharvest has been small; Sauli et al. used predictive microbiology to predict the growth, survival, and inactivation of *Salmonella* during the processing of swine feed (33). One of the main limitations of this approach is the reliability of the predictions outside the data used to fit the parameters of the growth model.

In addition to direct and environmental transmission pathways, more complex transmission pathways can be included in the model. Transmission of foodborne pathogens can take place through contaminated fomites and mechanical vectors, such as workers and equipment. These indirect transmission pathways can be explicitly modeled by using the infectious disease modeling framework used for vector-borne diseases such as malaria (34). To complete a transmission cycle, animals (host population) contaminate the workers/equipment (vector population), and contaminated vectors then infect animals. Some foodborne pathogens such as *Toxoplasma gondii* have complex cycles with both a definitive host and several intermediate hosts. The models need to address the transmission paths between the different hosts (35).

Another important parameter in compartmental models of infectious disease is the rate at which individuals transit from one state/compartment to another. The recovery rate is the rate at which individuals leave the infectious compartment. The recovery rate is assumed to be the inverse of the infectious period (the time period during which an infectious individual sheds the pathogen). Data on the infectious period may be obtained from longitudinal studies that describe the shedding period of an agent, or data may be obtained from a few artificially infected hosts in a small-scale challenge experiment. For pathogens that transmit through the oral-fecal route, the duration of fecal shedding is often used as an estimate of the infectious period (14).

Basic Reproduction Number

The basic reproduction number (R_O) is often determined to evaluate whether a pathogen would propagate upon introduction in a farm (36). R_O is defined as the expected number of secondary cases produced by a single typical infectious individual during its infectious period in a completely susceptible population (5). R_O is considered a threshold quantity because if R_O is greater than 1, the number of cases will usually grow in the population because one case leads to more than one secondary case; if R_O is less than 1, on average, one case leads to less than one secondary case, and therefore the infection will die out in the population. R_O has been estimated for some foodborne pathogens; for *E. coli* O157, R_O has been estimated to be 4.3 in beef cattle populations (37). For *Salmonella*, R_O has been estimated to be 1.3 to 5.8 in dairy cow populations (38, 39), 2.4 in dairy calves (34), and 2.8 in laying hen populations (40). There are several methods to calculate R_O from epidemiological data (36). In general, all the methods involve formulating R_O in terms of compartment model parameters and using epidemiological data (e.g., incidence data, average age at infection, etc.) to estimate the parameters. Caution should be applied when comparing R_O from different studies because its value varies from population to population and with the assumed model and available data.

The basic reproduction number is also interpreted to represent the epidemiological fitness of a strain (41). Estimating R_O for strains that share an ecological niche or have some level of cross-immunity can aid in forecasting the effect of serotype-specific interventions (e.g., strain-specific vaccines). Epidemiological models reveal that when two strains with cross-immunity compete, the strain with the largest R_O is able to competitively exclude the other strain from the host population (4). *Salmonella Gallinarum* was eradicated from commercial flocks in the 1960s (42). In the mid-1980s, *Salmonella enterica* serovar Enteritidis (*S.* Enteritidis) emerged in poultry populations and become an important cause of human salmonellosis (42). Both are members of the same *Salmonella* serogroup D and share some antigens (43). It has been hypothesized that while *S. Gallinarum* was present in poultry populations ($R_O \sim 2.8$), *S.* Enteritidis ($R_O \sim 1.05$) was not present because of competitive exclusion (43). As immunity against *S. Gallinarum* declined in the flocks after its eradication, *S.* Enteritidis filled the ecological niche left empty by *S. Gallinarum*. The acquisition of antimicrobial resistance by the bacteria is often assumed to cause a decrease in strain fitness that may cause a reduction in transmissibility and/or duration of infection and therefore a reduction in the bacteria's R_O in the absence of antimicrobial selective pressure (44). However, some resistant clones may be highly transmissible due to reasons unrelated to the carriage of resistance and the presence of antimicrobial selective pressure (45); this anomaly may explain the presence of some foodborne pathogens in systems without an apparent antimicrobial selective pressure such as organic farms. Overall, scarce data exists regarding the R_O of different strains for foodborne pathogens.

Evaluation of Control Strategies

The effect of control strategies on foodborne pathogen transmission can be evaluated in mathematical models by predicting the effect of interventions in several model outcomes, such as R_O, the duration and size of outbreaks, or the change in prevalence or pathogen shedding loads over time in the presence and absence of interventions (34). As a general trend, models that evaluate intervention strategies are often more complex, with additional details compared to models built with the purpose of gaining a general understanding (4).

Typically, the greater the R_O, the greater are the efforts necessary to control a pathogen in a population. We can link R_O with intervention and secure vaccination coverage thresholds. In general, the intervention efficacy or vaccination coverage threshold should exceed $1 - 1/R_O$ to allow the infection to die out in the population. Intuitively, the larger the R_O value, the faster the disease will spread, and hence a higher proportion of the population needs to be immunized. The R_O is known as the effective reproductive number, R_e, when the population is partially immune or control strategies are applied. A change in R_e can be evaluated both before and after a control measure is applied to determine which measures, at what level, and in which combinations are able to reduce R_e to a value less than 1 (its threshold value) (36). One of the main limitations in using changes in R_e to evaluate interventions is that the measure does not capture the speed at which interventions are taking place (36). Changes in prevalence are the most common outcome we use to evaluate the efficacy of the interventions. Foodborne pathogens are widespread, and therefore the aim is often to reduce prevalence but not necessarily to achieve a zero prevalence.

The models can include more specific details and parameters that may be necessary to better capture and evaluate the effect of control measures in transmission. As examples, we discuss below how mathematical models are modified and used to address vaccination. The epidemiologic measure of protection induced by vaccination is known as vaccine efficacy (VE) and is expressed as a measure of relative risk (RR) in the vaccinated group compared with the unvaccinated group (46): $VE = 1 - RR$. The relative risk is defined as the ratio of infection in the vaccinated group versus the unvaccinated group. For example, if in the vaccinated group 2 of the total 100 animals get the disease while 10 are infected in the unvaccinated group with a population size of 75, then RR is computed as $(2/100)/(10/75) = 0.15$, meaning the risk of getting infected in the vaccinated group is only 15% of those in unvaccinated group. Consequently, the vaccine efficacy (VE) is $1 - RR = 0.85$, indicating that vaccination could reduce infection by 85%.

Vaccines have multiple biological effects. At the individual level, vaccines may reduce infectiousness or colonization or protect against infection (47). At the population level, the reduction in transmission due to widespread vaccination has indirect protective effects (i.e., herd immunity) for unvaccinated animals as well as for vaccinated animals (47). We can estimate different measures of VE to capture the multidimensional effect of vaccines (e.g., VE_{col} is vaccine efficacy for colonization; VE_I is vaccine efficacy for infectiousness). For foodborne pathogens, the desired outcomes often include the ability to decrease pathogen colonization and infectiousness simultaneously (i.e., fecal shedding). For example, some of the existent vaccines against *E. coli* O157 target the type III secretory proteins, which mediate

O157 colonization, and decrease colonization and pathogen shedding (48). Mathematical models can be used to assess multiple vaccine effects and different vaccine strategies (e.g., pulse and continuous vaccination). To model vaccination in a population, animals are classified based on both their epidemiological and vaccination states (e.g., susceptible and vaccinated susceptible), and vaccine effects are explicitly included in the model. Lu et al. (16) evaluated the potential impact of imperfect vaccines in *Salmonella* transmission on a dairy herd. The model included six states: susceptible, infectious, recovered, susceptible vaccinees, infected vaccinees, and recovered vaccinees. Vaccine effects on susceptibility, infectiousness, and duration of the infectious period were evaluated. The study showed that the vaccine had a larger impact on the reduction of endemic prevalence if the infectious period is reduced compared to the effect of the vaccine on susceptibility or infectiousness, and that vaccine coverage should be adjusted at different stages of infection (16).

SPATIAL AND INDIVIDUAL-BASED MODELS

We have discussed and demonstrated how mathematical modeling significantly enhances our understanding of the temporal dynamics of foodborne pathogens in preharvest food systems. The models described assume that pathogens and hosts are homogeneously distributed across the space. However, to understand the spread of the foodborne pathogens across space, both time and space need to be addressed in the models. Spatial structure is one type of heterogeneity. Other sources of heterogeneity that help to explain differences in the prevalence of foodborne pathogens are individual differences in infectiousness, susceptibility, age, and behavior, among others.

Metapopulation Models

The simplest and most fundamental spatial model is the metapopulation model (49). Metapopulation means "a population of populations." A metapopulation is a group of interacting populations, each with its own dynamics. Metapopulation models are readily used in studying pathogen transmission between farms. These models can represent the transmission between farms using two main approaches (4). Transmission between populations can be represented by describing the transmission in each population as a function of the weighted sum of prevalence in all populations and is related to the relative strength of interaction between the populations (4). The interaction between farms is also mediated by the movement of animals. In this case, metapopulation models represent explicitly the movement of animals, and transmission depends only on the animals present in the farm, but infected animals can introduce the infection in a farm. The movement of individuals creates a contact network for the spread of pathogens. This last approach has been commonly used to model spatial transmission of foodborne pathogens from farm to farm because the introduction of animals is a known risk factor, and data on the movements of animals have recently become available (50). For *E. coli* O157, metapopulation models identified cattle movement between farms as an important contributor to the observed prevalence of *E. coli* O157–positive farms in Scotland, but movement alone did not explain the persistence of *E. coli* O157 (50).

Network Models

Another popular spatial modeling method is network modeling. The models discussed so far often assumed random mixing of the individuals. Random mixing means that each individual has an equal chance of contacting other individuals in the whole population (e.g., farm) during a given time period (28). Nevertheless, individuals often only interact with a limited portion of the population, and hence random mixing is not an appropriate assumption of the models. Network models capture explicitly the contact structure of a population (51). Networks have evolved from the fields of graph theory and social sciences and are often represented as a graph in which hosts are represented as nodes, and the contacts between nodes are referred to as edges (see Fig. 5 for an example). Descriptions of the contact network among animals within farms have been possible through the tracking and recording of animal data with devices such as global positioning systems or proximity collars (52). Turner et al. (53) modeled the transmission of *E. coli* O157 in a dynamic contact network of a typical dairy farm to evaluate the effects of heterogeneity and clustering on the prevalence of infection within the herd. Heterogeneity in direct contacts decreased the persistence of the pathogen at the farm (53). The contact structure among farms can also be represented when animal movement data is available or the structure of the industry is known (54). In addition, human movements and livestock-wildlife interactions create contact networks among farms that contribute to the spread of foodborne pathogens in those farms and the introduction of pathogens in a naïve facility (55, 56). These two types of networks are more difficult to characterize because data to generate the networks are scarcer.

Network models focus more on the topology aspect of the infection transmission system. In addition, network analysis can provide important descriptive information regarding the underlying contact structure such as measurements of connectivity (how individual nodes are connected) and centrality (relative importance of nodes and edges in the network), and it can be integrated with individual-based models (51). In the example in Fig. 5, calf 1 is more active in the group: it contacts all other members (node 1 has four edges pointing out), while calf 2 contacts only one other calf. We can measure this difference as centrality, which is defined as the relative importance of a node in the graph. There are different variations of centrality. The easiest and most intuitive one is degree centrality, which is the number of the edges of a node. So the degree centrality of the five calves is 4, 1, 3, 3, 2, for calves 1 to 5, respectively. Other definitions include closeness centrality, betweenness centrality, and eigenvector centrality. All these measures of centrality have their

A

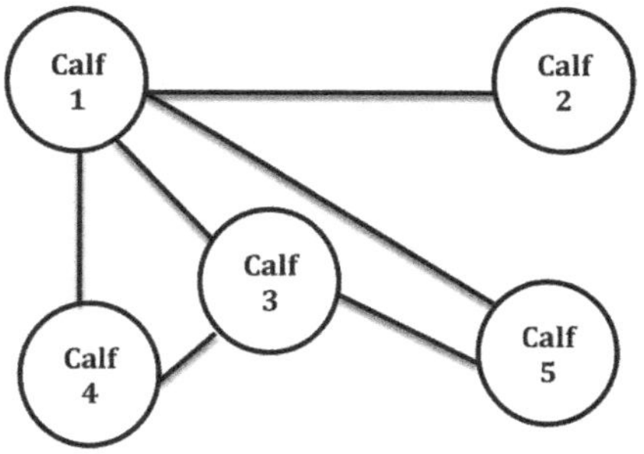

B

Calf No.	1	2	3	4	5
1	--	1	1	1	1
2	1	--	0	0	0
3	1	0	--	1	1
4	1	1	1	--	0
5	1	0	1	0	--

FIGURE 5 Network models are usually represented as graphs. A graph contains a set of nodes (individuals) and edges (contacts) that are associated with these nodes. Shown is an example with five interacting calves. The calves with contact with each other have a line (edge) between them to indicate their mutual relationship (A). The graph can be translated to an adjacency matrix, which contains as many rows and columns as there are nodes (B). The elements of the matrix record information about the edges between each pair of nodes, with 1 indicating a contact and 0 meaning no contact.

particular purpose. For instance, eigenvector centrality measures the influence of a node in the network. In the example in Fig. 4, the eigenvector centrality values are 1, 0.37, 0.90, 0.71, and 0.71 for the five calves, respectively, indicating that calf 1 is the most influential in the population. Studying the centrality properties of the nodes could provide important information on whether there are supercontacters in the population. Supercontacters can be defined as individuals that transmit more infection by making more contacts in the population (57).

Individual-Based Models

Individual-based models (also known as agent-based models) have become very popular in epidemiology and ecology (58). In individual-based models, each individual in the population is simulated explicitly, and a wide range of individual characteristics and interactions can be included. Individual-based models and network models (and stochastic models) are not exclusive alternatives; they are based on different modeling classifications. Most individual-based models are built around some network structure and therefore are also considered network models. The most important purpose of individual-based models is to provide more comprehensive insight into the heterogeneity of the system, thanks to the natural capability of individual-based models to incorporate and handle variability. Among all the heterogeneities, spatial heterogeneity is probably the most commonly encountered variability and is the cornerstone for many other behaviors or observed epidemiological variability.

Individual-based models provide a powerful approach to study mechanisms underlying observed patterns (58) because they allow for a more realistic description of the infection and transmission processes. Animals exhibit a variable degree of infectiousness as they shed variable amounts of the pathogen. In compartmental models, animals are assumed to have the same level of infectiousness (i.e., bacteria load shed) during the duration of their infectious period, which is often assumed to be exponentially distributed. This strict description of the infection process can be relaxed in individual-based models. Individual-based models provide a flexible modeling framework to study relationships between infection/transmission patterns and sources of variation and mechanisms underlying these patterns. For example, Chen et al. (59) developed an individual-based model that integrates individual animal data on temporal fecal shedding dynamics with pen-level *E. coli* O157 transmission to study how the temporal (and aggregation) patterns of *E. coli* O157 shedding loads and prevalence arise at the pen level. Model simulation indicated that scenarios without between-host variability (all the animals had the same shedding profile) were able to generate the proportion of animals with concentrations of *E. coli* O157 in feces that exceed 10^4 CFU/g, similar to that observed in cross-sectional field data (59). Thus, heterogeneity in the number of new infections caused by each infectious animal was produced even in scenarios with random contacts and no individual host variability. Therefore, caution should be applied in interpreting observed patterns such as cross-sectional shedding or prevalence data. In *E. coli* O157 epidemiology, super-shedders are loosely identified as cattle that shed high concentrations of *E. coli* O157 ($\geq 10^3$ or $\geq 10^4$ CFU/g of feces) at a single (or multiple) cross-section in time (60–62). The individual-based model simulations indicated that observed cross-sectional shedding patterns (including super-shedders) may arise without the presence of a subset of highly infectious individuals within the population. Therefore, testing efforts to truly identify highly infectious individuals might need to be longitudinal or require more complex sampling strategies.

WITHIN-HOST MODELS

Some of the strategies devised to reduce food-borne pathogen loads in animals aim to target the pathogen at the gut level by modulating the intestinal microbiome or interacting with the pathogen (63). Mathematical models can help to understand and design mitigation strategies for probiotics, prebiotics, and bacteriophages, among others. Compartmental models that represent the population dynamics of the pathogen are the most common modeling approaches used in within-host models (64–66). For example, Wood et al. (66) represented the population dynamics of *E. coli* O157 with four compartments: bacteria in the rumen, abomasum, feces, and environment. In each compartment, bacterial growth and death are represented. One of the main challenges in modeling the pathogens within host dynamics and the effect of mitigation strategies is the level of uncertainty on growth parameters and parameters related to the efficacy of the mitigation strategies because of the difficulty of sampling and quantifying the bacterial population within the host (67). Nonetheless, models can provide ranges for the potential effects of mitigation strategies and a qualitative understanding of the effects of mitigation strategies on foodborne pathogens.

Within-host models are often based on the existent theoretical body of population dynamics and interacting populations in the field of ecology. Lotka-Volterra models are a family of population biology models that describe interacting populations, and they represent an extension of the logistic growth model for a single population (2). The logistic growth model implicitly includes intraspecific competition in the form of a carrying capacity term. The intrapopulation competition is represented by density-dependent growth, $rN(1-NK)$, where r is the intrinsic growth rate, N is the population size, and K is the carrying capacity (maximum population that resources are able to sustain) (2). Competition among two populations influences the growth of both interacting populations as follows: for population N_1, growth is $r1N1(1-N1K1-\alpha12N2K1)$, and for population N_2, growth is $r2N2(1-N2K2-\alpha21N1K2)$. The sign and strength of the coefficients for α indicate the degree of interaction between the two populations. The two populations can be two competing strains, a probiotic and the target pathogen, or a bacteria and its associated bacteriophage. Predator-prey models (a type of Lotka-Volterra model) have been developed to characterize the dynamics of bacteria and their associated bacteriophages (64, 68). Hurley et al. (64) developed a mathematical model to assess the pharmacodynamics of the use of bacteriophages to modulate *Salmonella* load in chickens. Results from this model and other related models highlighted the importance of bacteria densities and the ratio of bacteriophage to bacteria in the efficacy of bacteriophages as mitigation strategies. When the target bacteria are sufficiently dense, few phage particles are able to replicate and effectively reduce bacterial populations (68). But at low bacteria densities, the self-replication of the phages is compromised.

Within-host models are also critical to understanding the emergence and dissemination of antimicrobial resistance in foodborne pathogens. Dissemination of antimicrobial resistance genes takes place on several scales (within bacterial populations through horizontal gene transmission and within and between hosts through clonal bacteria dissemination) (69). Mathematical models can specifically address the emergence and dissemination of antimicrobial resistance at the animal level during the use of antimicrobial treatments. The main factor in the emergence of antimicrobial resistance is the use of antimicrobials, and the administration of antimicrobial agents results in residual levels in the gastrointestinal tract in treated animals that may select for resistant bacteria at the gut level (70). Models that combine pharmacokinetics and pharmacodynamics with bacteria population dynamics can predict the effect of antimicrobial selection pressure on antimicrobial resistance dissemination within the host and subsequently can predict the load of resistant bacteria shed by an animal. The dynamics of plasmid-mediated resistance to cephalosporin ceftiofur in enteric commensals of cattle was assessed using a population dynamics model of ceftiofur-sensitive and resistant commensal enteric *E. coli* in the absence of and during parenteral therapy with ceftiofur (71). The model indicated that a low but stable fraction of resistant enteric *E. coli* could persist in the absence of immediate ceftiofur pressure. The presence of residual levels of ceftiofur in the gastrointestinal tract favored the selection and dissemination of ceftiofur-resistant bacteria. During parenteral therapy with ceftiofur, resistant enteric *E. coli* expanded in absolute number and relative frequency. The pharmacokinetic and pharmacodynamic parameters influenced the selection of antimicrobial resistance during treatment, but fitness cost determined the persistence of the resistant bacteria after the end of the treatment. The fitness cost of resistance is a key factor to successful displacement of resistant bacteria.

CONCLUSIONS

Mathematical models provide a framework to integrate and use information about the transmission and control of foodborne pathogens. The robustness of the modeling recommendations to control foodborne pathogens increases when modeling studies that differ in their underlying data, approaches, and assumptions reach similar conclusions. To date, there are a limited number of available models; therefore, we foresee a need to develop more models that fully use existent modeling methodologies and available data. Although most foodborne pathogens colonize multiple hosts, and a single host type carries multiple foodborne pathogens, the current models focus mostly on modeling one host–one pathogen systems. Steps toward developing models that account for multiple hosts, strains, and/or pathogens are necessary to understand and forecast the emergence of new foodborne pathogen threats. Additional applications for the models that could greatly benefit future research are used in the emergence and dissemination of antimicrobial resistance in food animals, as well as the dissemination of foodborne pathogens in produce. Both areas have received limited attention from a modeling point of view.

CITATION

Lanzas C, Chen S. 2016. Mathematical modeling tools to study preharvest food safety. Microbiol Spectrum 4(4):PFS-0001-2013.

REFERENCES

1. **Robinson S.** 2004. *Simulation. The Practice of Model Development and Use*. John Wiley & Sons, West Sussex, England.
2. **Otto SP, Day T.** 2007. *A Biologist's Guide to Mathematical Modeling in Ecology and Evolution*. Princeton University Press, Princeton, NJ.
3. **Lanzas C, Lu Z, Grohn YT.** 2011. Mathematical modeling of the transmission dynamics and control of foodborne pathogens and antimicrobial resistance at preharvest. *Foodborne Pathog Dis* **8:**1–10.
4. **Keeling MJ, Rohani P.** 2008. *Modeling Infectious Diseases in Humans and Animals*. Princeton University Press, Princeton, NJ.
5. **Anderson RM, May RM.** 1992. *Infectious Diseases of Humans: Dynamics and Control*. Oxford University Press, Oxford, England.
6. **O'Neill PD.** 2010. Introduction and snapshot review: relating infectious disease transmission models to data. *Stat Med* **29:**2069–2077.
7. **Sargent RG.** 2005. Verification and validation of simulation models. *In* Kuhl ME, Steiger NM, Armstrong FB, Joines JA (ed), *Proc. 2005 Winter Simulation Conference*. Orlando, FL.
8. **Sterman JD.** 2000. *Business Dynamics: System Thinking and Modeling for a Complex World*. Irwin McGraw-Hill, Boston, MA.
9. **Tedeschi LO.** 2006. Assessment of the adequacy of mathematical models. *Agric Syst* **89:** 225–247.
10. **Chubb MC, Jacobsen KH.** 2010. Mathematical modeling and the epidemiological research process. *Eur J Epidemiol* **25:**13–19.
11. **Lipsitch M, Samore MH.** 2002. Antimicrobial use and antimicrobial resistance: a population perspective. *Emerg Infect Dis* **8:**347–354.
12. **Samore MH, Lipsitch M, Alder SC, Haddadin B, Stoddard G, Williamson J, Sebastian K, Carroll K, Ergonul O, Carmeli Y, Sande MA.**

2006. Mechanisms by which antibiotics promote dissemination of resistant pneumococci in human populations. *Am J Epidemiol* **163:** 160–170.

13. **Saltelli A.** 2000. What is sensitivity analysis?, p 3–13. *In* Saltelli A, Chan K, Scott EM (ed), *Sensitivity Analysis*. John Wiley & Sons, Ltd, Chichester, England.
14. **Lanzas C, Brien S, Ivanek R, Lo Y, Chapagain PP, Ray KA, Ayscue P, Warnick LD, Grohn YT.** 2008. The effect of heterogeneous infectious period and contagiousness on the dynamics of *Salmonella* transmission in dairy cows. *Epidemiol Infect* **136:**1496–1510.
15. **Lurette A, Touzeau S, Lamboni M, Monod H.** 2009. Sensitivity analysis to identify key parameters influencing *Salmonella* infection dynamics in a pig batch. *J Theor Biol* **258:**43–52.
16. **Lu Z, Grohn YT, Smith RL, Wolfgang DR, Van Kessel JA, Schukken YH.** 2009. Assessing the potential impact of *Salmonella* vaccines in an endemically infected dairy herd. *J Theor Biol* **259:**770–784.
17. **Hill AA, Snary EL, Arnold ME, Alban L, Cook AJC.** 2008. Dynamics of *Salmonella* transmission on a British pig grower-finisher farm: a stochastic model. *Epidemiol Infect* **136:** 320–333.
18. **Katsma WEA, De Koeijer AA, Jacobs-Reitsma WF, Mangen M-JJ, Wagenaar JA.** 2007. Assessing interventions to reduce the risk of *Campylobacter* prevalence in broilers. *Risk Anal* **27:**863–876.
19. **Doyle MP, Erickson MC.** 2012. Opportunities for mitigating pathogen contamination during on-farm food production. *Int J Food Microbiol* **152:**54–74.
20. **Jordan D, Nielsen LR, Warnick LD.** 2008. Modelling a national programme for the control of foodborne pathogens in livestock: the case of *Salmonella* Dublin in the Danish cattle industry. *Epidemiol Inf* **136:**1521–1536.
21. **Doyle MP, Erickson MC.** 2006. Reducing the carriage of foodborne pathogens in livestock and poultry. *Poultry Sci* **85:**960–973.
22. **Robinson SE, Brown PE, Wright EJ, Hart CA, French NP.** 2009. Quantifying within- and between-animal variation and uncertainty associated with counts of *Escherichia coli* O157 occurring in naturally infected cattle faeces. *J R Soc Interface* **6:**169–177.
23. **Lurette A, Belloc C, Touzeau S, Hoch T, Ezanno P, Seegers H, Fourichon C.** 2008. Modelling *Salmonella* spread within a farrow-to-finish pig herd. *Vet Res* **39:**49.
24. **van Gerwe T, Miflin JK, Templeton JM, Bouma A, Wagenaar JA, Jacobs-Reitsma WF, Stegeman A, Klinkenberg D.** 2009. Quantifying transmission of *Campylobacter jejuni* in commercial broiler flocks. *Appl Environ Microbiol* **75:**625–628.
25. **Zongo P, Viet A-F, Magal P, Beaumont C.** 2010. A spatio-temporal model to describe the spread of *Salmonella* within a laying flock. *J Theor Biol* **267:**595–604.
26. **Dougan G, John V, Palmer S, Mastroeni P.** 2011. Immunity to salmonellosis. *Immunol Rev* **240:**196–210.
27. **McCallum H, Barlow N, Hone J.** 2001. How should pathogen transmission be modelled? *Trends Ecol Evol* **16:**295–300.
28. **Begon M, Bennett M, Bowers RG, French NP, Hazel SM, Turner J.** 2002. A clarification of transmission terms in host-microparasite models: numbers, densities and areas. *Epidemiol Infect* **129:**147–153.
29. **Winfield MD, Groisman EA.** 2003. Role of nonhost environments in the lifestyles of *Salmonella* and *Escherichia coli*. *Appl Environ Microbiol* **69:**3687–3694.
30. **Ayscue P, Lanzas C, Ivanek R, Grohn YT.** 2009. Modeling on-farm *Escherichia coli* O157: H7 population dynamics. *Foodborne Path Dis* **6:** 461–470.
31. **Gautam R, Bani-Yaghoub M, Neill WH, Döpfer D, Kaspar C, Ivanek R.** 2011. Modeling the effect of seasonal variation in ambient temperature on the transmission dynamics of a pathogen with a free-living stage: example of *Escherichia coli* O157:H7 in a dairy herd. *Prev Vet Med* **102:**10–21.
32. **Black DG, Davidson PM.** 2008. Use of modeling to enhance the microbiological safety of the food system. *Compr Rev Food Sci Food Saf* **7:**159–167.
33. **Sauli I, Danuser J, Geeraerd AH, Van Impe JF, Rüfenacht J, Bissig-Choisat B, Wenk C, Stärk KDC.** 2005. Estimating the probability and level of contamination with *Salmonella* of feed for finishing pigs produced in Switzerland: the impact of the production pathway. *Int J Food Microbiol* **100:**289–310.
34. **Lanzas C, Warnick LD, Ivanek R, Ayscue P, Nydam DV, Grohn YT.** 2008. The risk and control of *Salmonella* outbreaks in calf-raising operations: a mathematical modeling approach. *Vet Res* **39:**61.
35. **Lelu M, Langlais M, Poulle ML, Gilot-Fromont E.** 2010. Transmission dynamics of *Toxoplasma gondii* along an urban-rural gradient. *Theor Popul Biol* **78:**139–147.
36. **Heffernan JM, Smith RJ, Wahl LM.** 2005. Perspectives on the basic reproductive ratio. *J R Soc Interface* **2:**281–293.

37. **Laegreid WW, Keen JE.** 2004. Estimation of the basic reproduction ratio (R-0) for Shiga toxin-producing *Escherichia coli* O157 : H7 (STEC O157) in beef calves. *Epidemiol Infect* **132:**291–295.
38. **Chapagain PP, Van Kessel JS, Karns JK, Wolfgang DR, Hovingh E, Nelen KA, Schukken YH, Grohn YT.** 2008. A mathematical model of the dynamics of *Salmonella cerro* infection in a US dairy herd. *Epidemiol Infect* **136:**263–272.
39. **Van Schaik G, Klinkenberg D, Veling J, Stegeman A.** 2007. Transmission of *Salmonella* in dairy herds quantified in the endemic situation. *Vet Res* **38:**861–869.
40. **Thomas ME, Klinkenberg D, Ejeta G, Van Knapen F, Bergwerff AA, Stegeman JA, Bouma A.** 2009. Quantification of horizontal transmission of *Salmonella enterica* serovar Enteritidis bacteria in pair-housed groups of laying hens. *Appl Environ Microbiol* **75:**6361–6366.
41. **Luciani F, Sisson SA, Jiang H, Francis AR, Tanaka MM.** 2009. The epidemiological fitness cost of drug resistance in *Mycobacterium tuberculosis*. *Proc Natl Acad Sci USA* **106:**14711–14715.
42. **Bäumler AJ, Hargis BM, Tsolis RM.** 2000. Tracing the origins of *Salmonella* outbreaks. *Science* **287:**50–52.
43. **Rabsch W, Hargis BM, Tsolis RM, Kingsley RA, Hinz KH, Tschape H, Baumler AJ.** 2000. Competitive exclusion of *Salmonella* enteritidis by *Salmonella* gallinarum in poultry. *Emerg Infect Dis* **6:**443–448.
44. **Levin BR.** 2001. Minimizing potential resistance: a population dynamics view. *Clin Infect Dis* **33:**S161–S169.
45. **Davis MA, Hancock DD, Besser TE.** 2002. Multiresistant clones of *Salmonella enterica*: The importance of dissemination. *J Lab Clin Med* **140:**135–141.
46. **Halloran ME, Longini IM Jr, Struchiner CJ.** 1999. Design and interpretation of vaccine field studies. *Epidemiol Rev* **21:**73–88.
47. **Halloran ME, Longini IM Jr, Struchiner CJ.** 2009. *Design and Analysis of Vaccine Studies*. Springer, New York, NY.
48. **Smith DR, Moxley RA, Peterson RE, Klopfenstein TJ, Erickson GE, Bretschneider G, Berberov EM, Clowser S.** 2009. A two-dose regimen of a vaccine against type III secreted proteins reduced *Escherichia coli* O157:H7 colonization of the terminal rectum in beef cattle in commercial feedlots. *Foodborne Pathog Dis* **6:**155–161.
49. **Hanski I.** 1998. Metapopulation dynamics. *Nature* **396:**41–49.
50. **Liu W-C, Matthews L, Chase-Topping M, Savill NJ, Shaw DJ, Woolhouse MEJ.** 2007. Metapopulation dynamics of *Escherichia coli* O157 in cattle: an exploratory model. *J R Soc Interface* **4:**917–924.
51. **Keeling M, Eames K.** 2005. Review. Networks and epidemic models. *J R Soc Interface* **2:**295–307.
52. **Handcock R, Swain D, Bishop-Hurley G, Patison K, Wark T, Valencia P, Corke P, O'Neill C.** 2009. Monitoring animal behaviour and environmental interactions using wireless sensor networks, GPS collars and satellite remote sensing. *Sensors* **9:**3586–3603.
53. **Turner J, Bowers RG, Clancy D, Behnke MC, Christley RM.** 2008. A network model of *E. coli* O157 transmission within a typical UK dairy herd: the effect of heterogeneity and clustering on the prevalence of infection. *J Theor Biol* **254:**45–54.
54. **Lurette A, Belloc C, Keeling M.** 2011. Contact structure and *Salmonella* control in the network of pig movements in France. *Prev Vet Med* **102:**30–40.
55. **Böhm M, Hutchings MR, White PCL.** 2009. Contact networks in a wildlife-livestock host community: identifying high-risk individuals in the transmission of bovine TB among badgers and cattle. *PLoS One* **4:**e5016.
56. **Burns TE, Guerin MT, Kelton D, Ribble C, Stephen C.** 2011. On-farm study of human contact networks to document potential pathways for avian influenza transmission between commercial poultry farms in Ontario, Canada. *Transboundary Emerg Dis* **58:**510–518.
57. **McCaig C, Begon M, Norman R, Shankland C.** 2011. A symbolic investigation of super-spreaders. *Bull Math Biol* **73:**777–794.
58. **Grimm V, Revilla E, Berger U, Jeltsch F, Mooij WM, Railsback SF, Thulke H-H, Weiner J, Wiegand T, DeAngelis DL.** 2005. Pattern-oriented modeling of agent-based complex systems: lessons from ecology. *Science* **310:**987–991.
59. **Chen S, Sanderson M, Lanzas C.** 2013. Investigating effects of between- and within-host variability on *Escherichia coli* O157 shedding pattern and transmission. *Prev Vet Med* **109:**47–57.
60. **Chase-Topping M, Gally D, Low C, Matthews L, Woolhouse M.** 2008. Super-shedding and the link between human infection and livestock carriage of *Escherichia coli* O157. *Nat Rev Microbiol* **6:**904–912.
61. **Low JC, McKendrick IJ, McKechnie C, Fenlon D, Naylor SW, Currie C, Smith DGE, Allison L, Gally DL.** 2005. Rectal carriage of enterohemorrhagic *Escherichia coli* O157 in slaughtered cattle. *Appl Environ Microbiol* **71:** 93–97.

62. **Omisakin F, MacRae M, Ogden ID, Strachan NJC.** 2003. Concentration and prevalence of *Escherichia coli* O157 in cattle feces at slaughter. *Appl Environ Microbiol* **69:**2444–2447.
63. **Callaway TR, Edrington TS, Anderson RC, Byrd JA, Nisbet DJ.** 2008. Gastrointestinal microbial ecology and the safety of our food supply as related to *Salmonella*. *J Anim Sci* **86:** E163–E172.
64. **Hurley A, Maurer JJ, Lee MD.** 2008. Using bacteriophages to modulate *Salmonella* colonization of the chicken's gastrointestinal tract: lessons learned from *in silico* and *in vivo* modeling. *Avian Dis* **52:**599–607.
65. **Wood JC, McKendrick IJ, Gettinby G.** 2006. A simulation model for the study of the within-animal infection dynamics of *E. coli* O157. *Prev Vet Med* **74:**180–193.
66. **Wood JC, McKendrick IJ, Gettinby G.** 2006. Assessing the efficacy of within-animal control strategies against *E. coli* O157: a simulation study. *Prev Vet Med* **74:**194–211.
67. **Wood JC, Speirs DC, Naylor SW, Gettinby G, McKendrick IJ.** 2006. A continuum model of the within-animal population dynamics of *E. coli* O157. *J Biol Syst* **14:**425–443.
68. **Levin BR, Bull JJ.** 2004. Population and evolutionary dynamics of phage therapy. *Nat Rev Micro* **2:**166–173.
69. **Andersson DI, Hughes D.** 2010. Antibiotic resistance and its cost: is it possible to reverse resistance? *Nat Rev Microbiol* **8:**260–271.
70. **Yan SS, Gilbert JM.** 2004. Antimicrobial drug delivery in food animals and microbial food safety concerns: an overview of *in vitro* and *in vivo* factors potentially affecting the animal gut microflora. *Adv Drug Delivery Rev* **56:** 1497–1521.
71. **Volkova VV, Lanzas C, Lu Z, Gröhn YT.** 2012. Mathematical model of plasmid-mediated resistance to ceftiofur in commensal enteric *Escherichia coli* of cattle. *PLoS One* **7:**e36738.
72. **Ivanek R, Snary EL, Cook AJ, Grohn YT.** 2004. A mathematical model for the transmission of *Salmonella* Typhimurium within a grower-finisher pig herd in Great Britain. *J Food Protect* **67:**2403–2409.

Understanding the Complexities of Food Safety Using a "One Health" Approach

22

KALMIA E. KNIEL,[1] DEEPAK KUMAR,[2] and SIDDHARTHA THAKUR[3]

INTRODUCTION

The term "One Health" describes a discipline, a theory, and a way of thinking that bring together human, animal, and environmental health. A majority of infectious diseases critical to food safety in humans are zoonoses. In fact, the Centers for Disease Control and Prevention (CDC) states that scientists estimate that more than 6 out of every 10 known infectious diseases in people are spread from animals and that 3 out of every 4 new or emerging infectious diseases in people are spread from animals (https://www.cdc.gov/onehealth/index.html). Through the broad One Health concept, scientists can probe solutions and develop a better understanding of how to address the growing problems concerning human medicine, animal medicine, and environmental sciences. Educational advances in One Health are occurring quickly through many undergraduate, graduate, and professional programs across agricultural sciences and veterinary medicine, as well as in public and human health. Traditionally, human health has been managed separately from animal health, and the health of the environment has been considered less than the latter.

[1]Department of Animal and Food Science, University of Delaware, Newark, DE 19716; [2]Department of Veterinary Public Health & Epidemiology, College of Veterinary and Animal Sciences, Govind Ballabh Pant University of Agriculture & Technology, Pantnagar, Uttarakhand-263145, India; [3]Department of Population Health and Pathobiology, College of Veterinary Medicine, North Carolina State University, Raleigh, NC 27607.

Preharvest Food Safety
Edited by Siddhartha Thakur and Kalmia E. Kniel

doi:10.1128/microbiolspec.PFS-0021-2017

This is even more true in the recent past given the increase in specializations in human medicine, such as personalized medicine. Large government agencies and numerous private programs and companies work to protect human and animal health; therefore, clinicians, veterinarians, and environmentalists must all join together to fully address One Health.

One Health is certainly not a new concept: the Greek physician Hippocrates (460 to 370 BCE) recognized the impact of environmental conditions on human health in his *On Airs, Waters, and Places,* in which he wrote about how stagnant water and swampy environments were more likely to make people sick (1, 2); thousands of years later scientists would connect mosquitoes and malaria to those same environments. In the 1960s, Calvin Schwabe coined the term "One Medicine," which initiated the alignment of human and animal medicine (3). Over the past decade, in light of growing scientific interest in climate change and in the protection of global health, One Health has gained traction and recognition. While a myriad of chapters and books have covered various aspects of One Health, including political and social implications, here we discuss two issues relevant to preharvest food safety in connection with each other. The first is the daunting issue of antimicrobial resistance (AMR), and the second some consider equally as formidable, that is, disease spread and occurrence associated with *Salmonella* spp. The shift in thinking through the One Health paradigm is critical to our approach to targeting these two subjects. Not only does One Health bring together the necessary experts and scientists to solve the problems, but this multidisciplinary approach is critical to educational programs and for continued learning about these broad issues. Lonnie King described this interconnectivity of people, animals, and their products as being embedded in a threatened environment which results in an unprecedented 21st-century mixing bowl (4). Our impacts on each other now cannot be separated. The interaction of these forces, as described by King, are similar to the dynamics of Newton's third law of motion, which states that for every action there is an equal and opposite reaction. If this is true, then each action in any of the three domains has multiple consequences on the other.

ANTIMICROBIAL RESISTANCE SURVEILLANCE

To promote effective use of antibiotics, scientists and physicians are engaged in a One Health approach of integrated surveillance of antibiotic resistance in foodborne bacteria. This includes efficient reductions in use and only diligent use of antibiotics to treat humans and nonhuman animals raised for agriculture. Integrated surveillance of AMR in foodborne pathogens involves collection, validation, analysis, and reporting of relevant microbiological and epidemiological data on AMR in foodborne bacteria from humans, animals, and food and data on relevant antimicrobial use in humans and animals (http://www.who.int/foodsafety/publications/agisar_guidance2017/en/). The focus of an integrated approach is to monitor the patterns of AMR, including the development of novel resistance mechanisms in pathogens isolated from foods (including foods of animal and plant origin) destined for human consumption and food animals, and to monitor the use of antimicrobials in humans and food animals. The WHO Advisory Group on Integrated Surveillance of Antimicrobial Resistance (AGISAR), in its recent guidelines, has outlined the basic requirements for countries to establish integrated surveillance of antibiotic resistance in foodborne bacteria. AGISAR guidelines give a detailed view of various steps such as the need for standardized and validated antimicrobial susceptibility testing methods, harmonized zone or MIC interpretive criteria, and monitoring and collection of antimicrobial consumption and use data in humans and animals.

The first component of the WHO guidelines for integrated surveillance of AMR in foodborne bacteria involves monitoring and interpretation of AMR in pathogens isolated from food-producing animals, foods, and humans. The focus of this section is to develop comparable epidemiological and microbiological methods for unbiased comparison of antimicrobial susceptibility results between different areas, countries, and regions. Specific guidelines on sampling sources (humans, retail meats, and food animals), target organisms (*Salmonella, Campylobacter* spp.), sampling design, and laboratory testing methodologies have been proposed. Laboratory testing methodologies includes bacterial culture identification and characterization, standardized antimicrobial susceptibility methods, testing of recommended antimicrobials for particular bacteria, and the use of uniform susceptibility standards such as those from CLSI and EUCAST for result interpretation. A final part of these guidelines is data management, validation, analysis, and reporting. Establishing a common set of comparable programs for integrated surveillance is difficult due to the differences in public health infrastructure, agricultural production systems, food supply systems, and veterinary practices. Hence, to achieve this objective, a minimum set of requirements has been proposed in the WHO guidelines; these include (i) the presence of appropriate medical and veterinary infrastructure allowing routine collection and microbiological analysis of human and animal clinical samples; (ii) the availability of established human, veterinary, and food safety laboratory facilities and workforce; (iii) a quality management system for laboratories; and (iv) the capacity to analyze and report AMR surveillance data.

The second component of the WHO guidelines for integrated surveillance of AMR involves monitoring the use of antimicrobials in humans and food animals at national and international levels. Surveillance of antimicrobial use and consumption in humans and animals can be divided into three main activities, viz., measuring the quantity of different antimicrobials sold, obtaining information on the prescription practices for antimicrobials in human and veterinary medicine, and collecting information on the actual intake of antimicrobials by humans and animals. Various approaches are available to monitor the use of antimicrobials in humans and animals, which vary in methodology and output reporting. Strategies recommended for surveillance of antimicrobial usage in humans include monitoring of national antimicrobial sales data, point prevalence surveys on antimicrobial use in hospitals, and longitudinal studies of antimicrobial use in hospitals and the community. However, collecting data on antimicrobial usage in animals is more difficult than in humans due to variation in antimicrobial consumption across different species and production types (e.g., for meat or milk). Nevertheless, similar approaches, such as surveillance of national antimicrobial sales data and longitudinal studies at the farm level, have been recommended to monitor antimicrobial usage in animals. Additionally, collection of species-wide consumption data of antimicrobials is recommended.

Inappropriate prescription of antimicrobials in humans and animals is another important issue related to antimicrobial surveillance (5). According to the authors of a recent study, the American Medical Association has diagnosed the prevalence of prescriptions and estimates for the use of antibiotics that may be inappropriate in adults and children in the United States (5). The outcome of this study suggested that in 2010 to 2011 there was an estimated antibiotic prescription rate of 506 per 1,000 population; however, only 353 antibiotic prescriptions were likely appropriate during this time. This is an indication that antibiotic misuse is not only a problem associated with the treatment of animals. The authors suggest that a 15% reduction in overall oral antibiotic prescriptions would be necessary to meet the goal of the White House National Action Plan for Combating Antibiotic Resistant Bacteria of reducing inappropriate antibiotic use (5). The situation is far more threatening in middle- and low-income countries such as India, where no control on the sale and use of antimicrobials exists. In countries like India, many antimicrobials can be purchased without a medical prescription. A recent study estimates that in India during 2007 to 2012, increases of 73% and 174% were recorded in the sale of watch group and reserve group antibiotics, respectively, compared to a 20% increase in key access antibiotics (6). The steep rise in sales of watch group and reserve group antibiotics was mainly attributed to the rampant use of fixed-dose combinations that contain watch group antibiotics. Apart from government inactivity in implementing strict antibiotic control regulations, the unauthorized and unscientific use of antibiotics by quacks and the prescription of reserve antibiotics by medical doctors for financial benefits are the major causes of the emergence of AMR in India. The prescription of higher-end or high-generation antibiotics is known to carry higher financial margins.

ANTIMICROBIAL USE IN HUMANS

According to the CDC, at least 2 million people become infected and at least 23,000 die annually due to antimicrobial-resistant bacteria in the United States alone (https://www.cdc.gov/drugresistance/index.html). A report recently concluded that if unchecked, the number of annual deaths due to drug-resistant pathogens will swell to 10 million by 2050 and could cost $100 trillion in terms of lost global production (7). We are witnessing a steep increase in the global consumption of antimicrobials. For example, Van Boeckel and coworkers reported a substantial increase of 35% in global antibiotic consumption between 2000 and 2010 (8). Seventy-six percent of the overall increase in the global consumption of antimicrobials between 2000 and 2010 was attributed to BRICS nations (Brazil, Russia, India, China, and South Africa) (8). Interestingly, during these years BRICS countries had only a 31% share in the global increase in the human population. Among all nations, India ranked first in the consumption of antibiotics in 2010, with 12.9×10^9 units (10.7 units/person) followed by China ($10{\cdot}0 \times 10^9$ units and 7·5 units/person) and the United States ($6{\cdot}8 \times 10^9$ units and 22·0 units/person). The increase in the global consumption of antimicrobials has resulted in the emergence of a high frequency of resistance in bacterial pathogens/strains that were previously considered susceptible (9).

USE OF ANTIMICROBIALS IN FOOD ANIMALS

The emergence of AMR in foodborne pathogens is one of the most important aspects of preharvest food safety, which is mainly attributed to the irrational use of antimicrobials in food animals. Prolonged use of subtherapeutic doses of antimicrobials in food animals for growth promotion and disease prevention results in the development of multidrug resistance in bacterial pathogens. Subtherapeutic doses of antimicrobials are used on these farms for rapid growth of animals and disease prevention. Antimicrobials in low doses kill the majority of the gut microbiota, but some resilient bacteria are able to survive and become resistant over time. Extensive usage of important antibiotics in production animals speeds up this process and creates selective pressure on bacteria that eventually develop resistance to survive. These farms become a reservoir for the selection of new AMR bacteria that can be exchanged between animals and humans (10). The resistant bacteria originating from production systems can travel to surrounding human dwellings and slaughter and processing units and transfer antimicrobial-resistant determinants to the native microbial population. Domestic sales and distribution of antimicrobials approved for use in food animals in the United States in 2015 amounted to approximately 15.6 million kg, of which tetracylines alone constituted 6.86 million kg (44%) (11). Moreover, it has been reported that 30 to

90% of the antimicrobials consumed by animals are released into manure and urine (12). Animal manure has been shown to contaminate the environment with antibiotic-resistant bacteria and genes (13).

People working and living close to animal feeding operations are at greater risk of acquiring AMR infections. Recently, a study evaluated differences in occupational risks of methicillin-resistant *Staphylococcus aureus* (MRSA) for farm workers working in industrial livestock operations (ILOs) and antibiotic-free livestock operations (AFLOs) (10). Although the prevalence of *S. aureus* and MRSA was similar among workers on both ILO and AFLO farms, *S. aureus* clonal complex (CC) 398, which is a livestock-associated MRSA clone, was predominately detected among ILO workers compared to AFLO workers. Moreover, only ILO workers carried *scn*-negative MRSA CC398 and *scn*-negative multidrug-resistant *S. aureus* CC398 strains, which confirms the presence of livestock-associated MRSA in ILO workers. Another study reported a higher risk of community-associated MRSA, skin, and soft-tissue infection in human populations living close to high-density swine production operations in Pennsylvania (14). The outcomes of these studies highlight the extent of the danger posed by the irrational use of antimicrobials in food animal farms and the environmental route of transmission of such organisms to the human population.

ENVIRONMENTAL AMR

AMR is inextricably linked to the three basic domains of One Health—that is, human, animal, and environmental health. While AMR affects human, animal, and environmental health equally, human health and to some extent animal health get the spotlight. Environmental health is the least-explored domain of One Health. While our overall knowledge of soil microbes is limited, genes encoding antibiotic resistance are not a new phenomenon in response to extensive use of antimicrobials in humans or animals. Within the 1% of all soil microbes that have been identified, there appears to be a global "resistome" (2, 15, 16). Selective pressures placed on environmental bacteria forced these bacteria to express resistance genes to survive. Resistance genes are known to persist in nature and can be transmitted that way among bacteria in sewage, soil, water, and waste. Perhaps humans can learn from the environment since antibiotic resistance genes are ancient, predating the selective pressure of modern antibiotic use. Such genes have been identified in Alaskan permafrost sediments that date back to the Late Pleistocene era. Microbes that are metabolically active readily have the potential for evolving or acquiring resistance genes (17).

In most low- and middle-income countries, the use of antibiotics for animal growth promotion and treatment is unregulated, which results in the high proportion of antimicrobial-resistant bacteria, antimicrobial resistance genes (ARGs), and antimicrobial residues in foods of animal origin and farm waste. Resistant bacteria and genes emanating from food animal farms are a public health threat. These resistance genes could enter the human food chain through water, soil, and manure. In particular, manure is known to harbor various AMR bacteria and genes. Despite the huge volumes of antimicrobials used on food animal farms, very little information is available regarding the presence of AMR genes in manure and soil around these farms. The dissemination of manure on soil increases the risk of resistance gene exposure to crops, surface water bodies, groundwater, and human populations living close to the farms (18–20). Few studies have explored the presence of AMR bacteria or ARGs in soil/manure samples and their persistence in the outside environment (21–23). In one such study, various *Salmonella* serotypes were found to persist in the soil outside swine farms for up to 21 days post-manure application (22). The authors confirmed the role of manure as a reservoir in the dissemination and persistence of antimicrobial-resistant bacteria in the environment. Recently, another study detected 149 unique resistance genes at three large-scale (10,000 animals per year) commercial swine farms in China. The 63 most prevalent ARGs were enriched 192-fold up to 28,000-fold compared to the antibiotic-free manure or soil controls. The diversity and abundance of ARGs reported in this study is alarming and clearly indicates that unmonitored use of antibiotics and metals on swine farms has expanded the diversity and abundance of the antibiotic resistance reservoir in the farm environment (23).

Another major environmental reservoir of ARGs is the wastewater effluents from drug industries and municipalities. It is known that a large proportion of the antimicrobials consumed by humans are excreted in their bioactive form in feces and urine (24). These biologically active components in wastewaters exert selective pressure and force bacteria to develop resistance. The presence of antibiotic resistance genes and antibiotics in wastewater effluent selects for novel combinations of AMR. These resistance determinants can be transferred to susceptible microorganisms by horizontal gene transfer via plasmids, integrons, and transposons (25).

ONE HEALTH, PREHARVEST FOOD SAFETY, AND *SALMONELLA*

As a foodborne pathogen, nontyphoidal *Salmonella* results in an estimated 1.2 million illnesses, 19,000 hospitalizations, and 370 deaths each year (26). There are over 2,600 recognized serovars of *Salmonella enterica* that have a broad range of hosts and can infect a wide variety of animals, as well as grow or survive in plants, protozoa, soil, and water. These serovars can be divided into three general groups based on host range: broad host range or generalist, host adapted, and host restricted (27). Selective pressures and the science behind host range evolution and specific reservoirs are not well understood. *Salmonella* has a complex environmental life cycle in multiple host species (mammals, reptiles, birds, and insects) that is affected by many factors of the external environment (sunlight, nutrients, temperature, moisture) (28).

The large number of possible *Salmonella* strains, coupled with their ability to be supported by a myriad of hosts, likely contributes to the continuing growth and potential for contamination by *Salmonella* in our food supply. The public health impact of *Salmonella* continues to grow, because we missed the mark on significant decreases for the Healthy People 2010 goal and remain a distance from the Healthy People 2020 goal at this time (https://www.cdc.gov/nchs/healthy_people/hp2020.htm). With the increasing use of whole-genome sequencing, more information is documented and shared about the presence of specific *Salmonella* strains in the environment as well as those isolated from animals and human clinical strains. The U.S. Food and Drug Administration supports the GenomeTrakr Network, where isolates from foodborne outbreak investigations and federal- and state-funded research projects are identified and maintained in public databases. Similar databases are supported by the World Health Organization, with information from the Global Salm-Surv *Salmonella* surveillance program, including AMR information. These types of data are being used to approach *Salmonella* One Health integration programs around the globe. For example, a Canadian surveillance program from 2005 to 2012 determined how integrated surveillance can identify an issue in an exposure source and link it to trends in human disease (29).

In their analysis of the impact of *Salmonella* on contaminated foods, the Centers for Disease Control and Prevention ranks *Salmonella* contamination of poultry among the top contributors to outbreaks and *Salmonella* on vine-stalk vegetables as the second-highest contributor to outbreak-associated hospitalizations; fruit and nut contamination with *Salmonella* is a leading cause of outbreak-associated deaths (30). *Salmonella* is consistently challenging the dogma of foodborne disease and food microbiology with its ability to persist and cause illness associated with a large range of foods, including peanut butter, dried spices, and other powders. *Salmonella* plays a unique role in preharvest food safety because it is a concern for all commodities. *Salmonella* can cause disease in foods due to its complicated association with the three domains of One Health.

HUMAN AND ANIMAL DOMAIN INTERFACE

Numerous case studies have explored the interactions of zoonotic *Salmonella*. Perhaps the longest history is human salmonellosis linked to live poultry, which is historic but also is an increasing public health concern. In 2012 to 2013, an investigation traced *Salmonella* Branderup infections to a mail-order hatchery in the United States (31). The authors describe 8 unrelated outbreaks of human salmonellosis linked to live poultry contact which resulted in more than 517 illnesses. In this study, a case was defined as an infection with the outbreak strain determined by pulsed-field gel electrophoresis with onset from 25 July 2012 to 27 February 2013. Epidemiological evidence supported an environmental investigation of one hatchery where sampling yielded the outbreak strain. The investigation concluded that improved sanitation and pest control were warranted. The authors state that similar to other outbreaks, industry knowledge and involvement are needed to solve outbreaks. Nakao et al. state that in this case a "One Health" approach leveraged the necessary expertise in human, animal, and environmental health (31).

The type of outbreak described above is not rare. The CDC stated in a 2016 report by the National Center for Emerging and Zoonotic Infectious Diseases that backyard flocks caused a record number of illnesses, with 900 human infections associated with keeping live chickens and ducks in backyard flocks. Of these 900, 200 people were hospitalized. In 2017, 10 separate multistate outbreaks of human *Salmonella* infection resulted in the largest number of illnesses linked to contact with live poultry ever recorded in the United States, affecting 48 states. According to the CDC, 1,120 people were infected, 249 of which were hospitalized, and 1 death was reported from North Carolina. These *Salmonella* outbreaks occurred between 4 January and 22 September 2017. Epidemiological and laboratory investigations linked the 10 outbreaks to contact with live poultry in backyard flocks. Importantly, 70% of the ill people reported contact with live poultry a week before their illness (https://www.cdc.gov/Salmonella/live-poultry-06-17/index.html). The CDC response to these events included education for flock owners and industry about staying healthy while maintaining a backyard flock. It is essential to educate individuals about the potential for the transmission of *Salmonella* as well as about the acute and chronic symptoms associated with salmonellosis.

ENVIRONMENTAL DOMAIN INTERFACE

A wide variety of bacteria have close relationships with plants and in some cases promote plant growth and nitrogen uptake. The study of human pathogens on plants has become increasingly popular and important over the past 20 years (32). Agricultural crops may become contaminated with bacterial and viral pathogens that are a threat to human health but not necessarily to plant health. Research has shown that plants can become contaminated with human pathogens in the preharvest environment through a variety of outlets, including soil, irrigation water or water used to apply pesticides and fertilizers, dust, insects, land-applied manures and biosolids, and directly from wild or domestic animals (33). In particular, manure from food animal farms is a major reservoir of antibiotic-resistant *Salmonella*. Plants may receive higher concentrations of *Salmonella* if the farm-generated manure is contaminated with *Salmonella*. Leafy greens are considered a high-risk food crop because they have been epidemiologically linked to foodborne illness (34) and they are commonly consumed in their raw state, with little or no processing taking place to reduce contaminants. Leafy greens are not all the same when it comes to potential contamination. For example, spinach forms a canopy that may serve as a trap for zoonotic pathogens if the plants become contaminated. The outer leaves of lettuce plants may provide a reservoir for pathogens, but this route of contamination may not lead to illness. Herbs may be cultivated differently, and some have smooth leaf surfaces while others have rougher structures.

Addressing safer ways to manage crop growth and harvest is essential to minimizing microbial contamination (35). It is important to determine the type of relationships that human pathogens have with plants, whether they be symbiotic, endophytic, or antagonistic (32). *Salmonella* is a classic human pathogen that has a unique environmental niche and the potential for fierce and continuous interactions with plants. Schikora et al. suggested that human pathogenic *S. enterica* serotype Typhimurium infects and intracellularly proliferates within tissues of *Arabidopsis thaliana* through both the root and shoot of the plant (36). These same researchers noted that *Salmonella* infection may result in disease symptoms in the plants, including wilting, chlorosis, and death of infected plant organs (36). *Arabidopsis* plants responded immunologically similarly to *Salmonella* as they would to the plant pathogen *Pseudomonas syringae* (DC3000), by induction of the mitogen-activated protein kinase cascades as well as by enhanced expression of pathogenesis-related genes (36). Numerous studies have focused on *A. thaliana*, a small flowering plant that is widely used as a model organism in plant biology, demonstrating that bacteria normally pathogenic to humans and other mammals can infect plants including *S. enterica*, *Pseudomonas aeruginosa*, *Burkholderia cepacia*, *Erwinia* spp., *S. aureus*, *Escherichia coli* O157:H7, and *Listeria monocytogenes* (37–41).

A wide range of plants and agricultural produce have been linked with *Salmonella* infection in the past. Notably, the 2017 multistate outbreak of *Salmonella* in the United States linked to Maradol papayas imported from Mexico highlights the importance of fresh produce contamination and the risk of *Salmonella* infection. Several *S. enterica* serotypes such as Urbana, Newport, Infantis, Anatum, Thompson, and Agona were identified from Papayas imported from Mexico. According to the CDC, this outbreak resulted in 251 illnesses and 2 deaths (1 each in New York and California) across 26 states. Four farms located in Mexico were implicated in this *Salmonella* outbreak. Recalls were issued after the *Salmonella* strains from papayas and human cases matched. In 2016, two separate *Salmonella* outbreaks were linked to the consumption of contaminated alfalfa sprouts. In the first outbreak, 26 people in 12 states were infected with the outbreak strains of *S. enterica* Muenchen (25 people) or *S. enterica* Kentucky (1 person). The outbreak resulted in the hospitalization of 8 people. The second *Salmonella* outbreak resulted in 36 illnesses and 7 hospitalizations in 9 states. Outbreak strains of *S. enterica* serotypes Reading and Abony were involved. No deaths were reported in either outbreak. Infected people in both of these outbreaks reported eating or possibly eating alfalfa sprouts in the week before the illness started. Table 1 summarizes the various food commodities and *Salmonella* serotypes involved in multiple *Salmonella* outbreaks during 2006 to 2017.

Salmonella is a resilient bacterium, surviving extremely successfully in the outside environment and finding its way into the food chain. However, it is not clear how *Salmonella* survives so efficiently in the outside environment. To understand this, Bleasdale and coworkers demonstrated that amoebae living freely in the environment are a potent

TABLE 1 Listing of select multistate outbreaks associated with fresh produce where preharvest contamination was suspected from 2006 to 2017[a]

Year	Commodity	*S. enterica* serotype
2017	Maradol papayas	*S.* Urbana, *S.* Newport, *S.* Infantis, *S.* Anatum
2016	Cucumbers	*S.* Oslo
2016	Alfalfa sprouts	*S.* Reading, *S.* Abony, *S.* Muenchen, *S.* Kentucky
2015	Cucumbers	*S.* Poona
2014	Bean sprouts	*S.* Enteritidis
2013	Cucumbers	*S.* Saintpaul
2012	Mangoes	*S.* Branderup
2012	Cantaloupe	*S.* Typhimurium, *S.* Newport
2011	Papayas	*S.* Agona
2011	Alfalfa and spicy sprouts	*S.* Enteritidis
2011	Cantaloupe	*S.* Panama
2010	Alfalfa sprouts	*S.* I 4,[5], 12:i:
2010	Alfalfa sprouts	*S.* Newport
2009	Alfalfa sprouts	*S.* Saintpaul
2008	Jalapeno peppers, serrano peppers, tomatoes	*S.* Saintpaul
2008	Cantaloupe	*S.* Litchfield
2006	Tomatoes	*S.* Typhimurium

[a]Adapted from reference 32 and original source (http://www.cdc.gov/foodsafety/outbreaks/multistate-outbreaks/outbreaks-list.html).

reservoir of *S.* Typhimurium (42). The authors highlighted the role of *Salmonella* pathogenicity island 2 in the survival of *S.* Typhimurium inside amoebae (*Acanthamoeba polyphaga*). They concluded that amoebae may be a major source of *Salmonella* within the environment and could play a significant role in the transmission of *Salmonella* to humans and animals.

CONCLUSION

The One Health approach is the best way to comprehensively address preharvest food safety issues such as AMR and foodborne salmonellosis. It is important to understand that there cannot be a single solution to the problem of AMR that is applicable to all nations. Regulating antibiotic use in humans may not be the only solution for middle- and low-income countries, where people need easy access to the cheap life-saving antibiotics and where antibiotics are often used as an alternative in the absence of basic healthcare and sanitation facilities. Putting a complete ban on the prescription of antibiotics will not work in such nations. Instead, a holistic approach embodying the principles of One Health is needed which includes judicious and unbiased antibiotic prescription in humans, regulation of antibiotic use in food animals, and prevention and monitoring of antibiotic resistance in environmental reservoirs. Similarly, foodborne salmonellosis can be transmitted by a wide variety of food and environmental sources. Foods of animal origin (meat, milk, and their products) and fresh produce (lettuce, papaya, cucumbers, and sprouts) have been implicated in several *Salmonella* outbreaks in the past. Also, efficient survival of *Salmonella* in the adverse environmental conditions in a variety of reservoirs is a well-known fact. Hence, a One Health approach addressing *Salmonella* in humans, animals, and the environment is envisaged.

CITATION

Kniel KE, Kumar D, Thakur S. 2018. Understanding the complexities of food safety using a "one health" approach. Microbiol Spectrum 6(1):PFS-0021-2017.

REFERENCES

1. **Hippocrates.** On Airs, Waters, and Places. http://classics.mit.edu/Hippocrates/airwatpl.html.
2. **Kahn LH.** 2016. *One Health and the Politics of Antimicrobial Resistance*. Johns Hopkins University Press, Baltimore, MD.
3. **Atlas RM.** 2013. One Health: its origins and future. *Curr Top Microbiol Immunol* **365:**1–13.
4. **King LJ.** 2013. Combating the triple threat: the need for a One Health approach. *Microbiol Spectr* **1:**OH-00122012.
5. **Fleming-Dutra KE, Hersh AL, Shapiro DJ, Bartoces M, Enns EA, File TM Jr, Finkelstein JA, Gerber JS, Hyun DY, Linder JA, Lynfield R, Margolis DJ, May LS, Merenstein D, Metlay JP, Newland JG, Piccirillo JF, Roberts RM, Sanchez GV, Suda KJ, Thomas A, Woo TM, Zetts RM, Hicks LA.** 2016. Prevalence of inappropriate antibiotic prescriptions among US ambulatory care visits, 2010–2011. *JAMA* **315:**1864–1873.
6. **McGettigan P, Roderick P, Kadam A, Pollock AM.** 2017. Access, watch, and reserve antibiotics in India: challenges for WHO stewardship. *Lancet Glob Health* **5:**e1075–e1076.
7. **O'Neill J.** 2016. *Tackling Drug-Resistant Infections Globally: Final Report and Recommendations*. The Review on Antimicrobial Resistance. HM Government and the Wellcome Trust, London, United Kingdom.
8. **Van Boeckel TP, Gandra S, Ashok A, Caudron Q, Grenfell BT, Levin SA, Laxminarayan R.** 2014. Global antibiotic consumption 2000 to 2010: an analysis of national pharmaceutical sales data. *Lancet Infect Dis* **14:**742–750.
9. **Bonten MJ, Willems R, Weinstein RA.** 2001. Vancomycin-resistant enterococci: why are they here, and where do they come from? *Lancet Infect Dis* **1:**314–325.
10. **Rinsky JL, Nadimpalli M, Wing S, Hall D, Baron D, Price LB, Larsen J, Stegger M, Stewart J, Heaney CD.** 2013. Livestock-associated methicillin and multidrug resistant *Staphylococcus aureus* is present among industrial, not antibiotic-free livestock operation workers in North Carolina. *PLoS One* **8:**e67641.
11. **Food and Drug Administration (FDA).** 2015. Summary report on Antimicrobials Sold or Distributed for Use in Food-Producing Animals. https://www.fda.gov/downloads/forindustry/userfees/animaldruguserfeeactadufa/ucm534243.pdf.
12. **Berendsen BJA, Wegh RS, Memelink J, Zuidema T, Stolker LAM.** 2015. The analysis of animal faeces as a tool to monitor antibiotic usage. *Talanta* **132:**258–268.
13. **Wichmann F, Udikovic-Kolic N, Andrew S, Handelsman J.** 2014. Diverse antibiotic resistance genes in dairy cow manure. *MBio* **5:**e01017.
14. **Casey JA, Curriero FC, Cosgrove SE, Nachman KE, Schwartz BS.** 2013. High-density livestock operations, crop field application of manure, and risk of community-associated methicillin-resistant *Staphylococcus aureus* infection in Pennsylvania. *JAMA Intern Med* **173:**1980–1990.
15. **Monier JM, Demanèche S, Delmont TO, Mathieu A, Vogel TM, Simonet P.** 2011. Metagenomic exploration of antibiotic resistance in soil. *Curr Opin Microbiol* **14:**229–235.
16. **Perry JA, Wright GD.** 2013. The antibiotic resistance "mobilome": searching for the link between environment and clinic. *Front Microbiol* **4:**138.
17. **D'Costa VM, King CE, Kalan L, Morar M, Sung WWL, Schwarz C, Froese D, Zazula G, Calmels F, Debruyne R, Golding GB, Poinar HN, Wright GD.** 2011. Antibiotic resistance is ancient. *Nature* **477:**457–461.
18. **Singer AC, Shaw H, Rhodes V, Hart A.** 2016. Review of antimicrobial resistance in the environment and its relevance to environmental regulators. *Front Microbiol* **7:**1728.
19. **Ferro G, Polo-López MI, Martínez-Piernas AB, Fernández-Ibáñez P, Agüera A, Rizzo L.** 2015. Cross-contamination of residual emerging contaminants and antibiotic resistant bacteria in lettuce crops and soil irrigated with wastewater treated by sunlight/H2O2. *Environ Sci Technol* **49:**11096–11104.
20. **Heuer H, Schmitt H, Smalla K.** 2011. Antibiotic resistance gene spread due to manure application on agricultural fields. *Curr Opin Microbiol* **14:**236–243.
21. **Kumar D, Pornsukarom S, Sivaraman GK, Thakur S.** 2017. Environmental dissemination of multidrug methicillin-resistant *Staphylococcus sciuri* after application of manure from commercial swine production systems. *Foodborne Pathog Dis*. [Epub ahead of print.]
22. **Pornsukarom S, Thakur S.** 2016. Assessing the impact of manure application in commercial swine farms on the transmission of antimicrobial resistant *Salmonella* in the environment. *PLoS One* **11:**e0164621.
23. **Zhu YG, Johnson TA, Su JQ, Qiao M, Guo GX, Stedtfeld RD, Hashsham SA, Tiedje JM.** 2013. Diverse and abundant antibiotic resistance genes in Chinese swine farms. *Proc Natl Acad Sci USA* **110:**3435–3440.
24. **Zhang QQ, Ying GG, Pan CG, Liu YS, Zhao JL.** 2015. Comprehensive evaluation of antibiotics emission and fate in the river basins of China: source analysis, multimedia modeling,

and linkage to bacterial resistance. *Environ Sci Technol* **49:**6772–6782.

25. **Xu J, Xu Y, Wang H, Guo C, Qiu H, He Y, Zhang Y, Li X, Meng W.** 2015. Occurrence of antibiotics and antibiotic resistance genes in a sewage treatment plant and its effluent-receiving river. *Chemosphere* **119:**1379–1385.
26. **Scallan E, Hoekstra RM, Angulo FJ, Tauxe RV, Widdowson MA, Roy SL, Jones JL, Griffin PM.** 2011. Foodborne illness acquired in the United States: major pathogens. *Emerg Infect Dis* **17:**7–15.
27. **Silva C, Calva E, Maloy S.** 2014. One Health and food-borne disease: *Salmonella* transmission between humans, animals, and plants. *Microbiol Spectr* **2:**OH-0020-2013.
28. **Winfield MD, Groisman EA.** 2003. Role of nonhost environments in the lifestyles of *Salmonella* and *Escherichia coli*. *Appl Environ Microbiol* **69:**3687–3694.
29. **Parmley EJ, Pintar K, Majowicz S, Avery B, Cook A, Jokinen C, Gannon V, Lapen DR, Topp E, Edge TA, Gilmour M, Pollari F, Reid-Smith R, Irwin R.** 2013. A Canadian application of One Health: integration of *Salmonella* data from various Canadian surveillance programs (2005–2010). *Foodborne Pathog Dis* **10:**747–756.
30. **CDC.** 2013. Surveillance for Foodborne Diseases Outbreaks: United States, 1998–2008. *MMWR* **62**(SS-02)**:**1–34.
31. **Nakao JH, Pringle J, Jones RW, Nix BE, Borders J, Heseltine G, Gomez TM, McCluskey B, Roney CS, Brinson D, Erdman M, McDaniel A, Behravesh CB.** 2015. 'One Health' investigation: outbreak of human *Salmonella* Braenderup infections traced to a mail-order hatchery: United States, 2012–2013. *Epidemiol Infect* **143:**2178–2186.
32. **Markland SM, Kniel KE.** 2015. Human pathogen-plant interactions: concerns for food safety, p 115–135. *In* Bais H, Sherrier J (ed), *Advances in Botanical Research: Plant Microbe Interactions*, **vol. 75**. Elsevier, Amsterdam, The Netherlands.
33. **Beuchat LR.** 2002. Ecological factors influencing survival and growth of human pathogens on raw fruits and vegetables. *Microbes Infect* **4:**413–423.
34. **Mercanoglu Taban B, Halkman AK.** 2011. Do leafy green vegetables and their ready-to-eat [RTE] salads carry a risk of foodborne pathogens? *Anaerobe* **17:**286–287.
35. **Sobsey MD, Dean CH, Knuckles ME, Wagner RA.** 1980. Interactions and survival of enteric viruses in soil materials. *Appl Environ Microbiol* **40:**92–101.
36. **Schikora A, Carreri A, Charpentier E, Hirt H.** 2008. The dark side of the salad: *Salmonella typhimurium* overcomes the innate immune response of *Arabidopsis thaliana* and shows an endopathogenic lifestyle. *PLoS One* **3:**e2279.
37. **Haapalainen M, van Gestel K, Pirhonen M, Taira S.** 2009. Soluble plant cell signals induce the expression of the type III secretion system of *Pseudomonas syringae* and upregulate the production of pilus protein HrpA. *Mol Plant Microbe Interact* **22:**282–290.
38. **Holden N, Pritchard L, Toth I.** 2009. Colonization outwith the colon: plants as an alternative environmental reservoir for human pathogenic enterobacteria. *FEMS Microbiol Rev* **33:**689–703.
39. **Milillo SR, Badamo JM, Boor KJ, Wiedmann M.** 2008. Growth and persistence of *Listeria monocytogenes* isolates on the plant model *Arabidopsis thaliana*. *Food Microbiol* **25:**698–704.
40. **Plotnikova JM, Rahme LG, Ausubel FM.** 2000. Pathogenesis of the human opportunistic pathogen *Pseudomonas aeruginosa* PA14 in *Arabidopsis*. *Plant Physiol* **124:**1766–1774.
41. **Prithiviraj B, Bais HP, Jha AK, Vivanco JM.** 2005. *Staphylococcus aureus* pathogenicity on *Arabidopsis thaliana* is mediated either by a direct effect of salicylic acid on the pathogen or by SA-dependent, NPR1-independent host responses. *Plant J* **42:**417–432.
42. **Bleasdale B, Lott PJ, Jagannathan A, Stevens MP, Birtles RJ, Wigley P.** 2009. The *Salmonella* pathogenicity island 2-encoded type III secretion system is essential for the survival of *Salmonella enterica* serovar *Typhimurium* in free-living amoebae. *Appl Environ Microbiol* **75:**1793–1795.

Index